This book describes current research in the field of cataclysmic variables and low mass X-ray binaries. The book is based on material presented at the 11th North American workshop on Cataclysmic Variables and Low Mass X-ray binaries held in Santa Fe, New Mexico, in October 1989.

Accretion-Powered Compact Binaries

Accretion-Powered Compact Binaries

Proceedings of the 11th North American Workshop on
Cataclysmic Variables and Low Mass X-ray Binaries,
Santa Fe, NM, October 9–13, 1989

Edited by

Christopher W. Mauche

Space Astronomy and Astrophysics Group,
Los Alamos National Laboratory

and

Laboratory for Experimental Astrophysics,
Lawrence Livermore National Laboratory

The right of the
University of Cambridge
to print and sell
all manner of books
was granted by
Henry VIII in 1534.
The University has printed
and published continuously
since 1584.

CAMBRIDGE UNIVERSITY PRESS
Cambridge
London   New York   Port Chester
Melbourne   Sydney

PUBLISHED BY THE PRESS SYNDICATE OF THE UNIVERSITY OF CAMBRIDGE
The Pitt Building, Trumpington Street, Cambridge, United Kingdom

CAMBRIDGE UNIVERSITY PRESS
The Edinburgh Building, Cambridge CB2 2RU, UK
40 West 20th Street, New York NY 10011–4211, USA
477 Williamstown Road, Port Melbourne, VIC 3207, Australia
Ruiz de Alarcón 13, 28014 Madrid, Spain
Dock House, The Waterfront, Cape Town 8001, South Africa

http://www.cambridge.org

First published 1990
First paperback edition 2003

A catalogue record for this book is available from the British Library

ISBN 0 521 40212 3 hardback
ISBN 0 521 54575 7 paperback

TABLE OF CONTENTS

1.2 Nonmagnetic Cataclysmic Variables

1.3 Magnetic Cataclysmic Variables

DQ Her Stars

AM Her Stars

2 ACCRETION THEORY

2.1 Nonmagnetic

2.2 Magnetic

3 NOVAE

4 EVOLUTION

Contents

PREFACE

On October 9–13, 1989, the Space Astronomy and Astrophysics Group of Los Alamos National Laboratory hosted the 11th in a series of workshops held in North America on the subject of cataclysmic variables and low mass X-ray binaries. This Workshop was itself the sixth in a series of workshops on Space Physics and Astrophysics hosted by Los Alamos National Laboratory (LANL). Recent LANL workshops have been on Gamma-Ray Stars (1986), Multiwavelength Astrophysics (1987), and Quasi-Periodic Oscillations (1988). On this occasion, the workshop took place at the Santa Fe Hilton in Santa Fe, New Mexico during the week of Albuquerque, New Mexico's annual International Hot Air Balloon Fiesta. The Workshop was sponsored by the Laboratory's Earth and Space Science Division and Institute of Geophysics and Planetary Physics and by the National Aeronautics and Space Administration. The local organizing committee consisted of myself, France Córdova, and Bill Priedhorsky. The scientific program was organized by the local organizing committee with the assistance of Keith Horne, Jim Imamura, Warren Sparks, Sumner Starrfield, and Paula Szkody.

The 11th North American Workshop on CVs and LMXRBs differed in a significant way from its predecessors in the shear number of participants. In all, the Workshop was attended by 127 scientists from institutions in 16 countries. With fully 60% of the participants from institutions in the US, the Workshops continue to be predominantly "North American." However, the 11th also had a strong international component. Significant numbers attended from institutions in the UK (17) and Germany (8); three participants attended from each of India, Italy, Japan, and the Netherlands; two attended from each of the Canary Islands, France, Israel, Spain, and the USSR. Participants from Chile, New Zealand, Norway, and South Africa rounded out the total. Unfortunately, one individual from Czechoslovakia who wanted to attend the Workshop could not be present because he was denied a visa by his government. Hopefully, given the recent events in his country and in eastern Europe in general, this sort of nonsense is a thing of the past.

The size of the Workshop can be ascribed both to the charm and uniqueness of Santa Fe and to the support we were able to provide many participants. However, it also speaks for the vitality of the field of study of accretion-powered white dwarf (CV) and neutron star (LMXRB) binaries. Given the size of the Workshop and the increasing reputability of our field of research, one participant (R. Wade) wondered aloud whether it was about time we stopped beginning all or papers with: "Cataclysmic variables (or, "Low Mass X-ray binaries) are semi-detached, mass-exchanging binaries consisting of an accreting white dwarf (or, neutron star) primary and a low mass secondary ..."

The form of the Workshop and these Proceedings are the result of the local organizing committee's philosophy that everyone should and would be given an equal opportunity to present their results. The difference between talks and posters then became largely a matter of individual preference. We even attempted to offset the obvious advantage of oral papers over poster papers by leaving the posters up for the entire Workshop and by scheduling 2 minutes for the author of each poster to present the highlights of his or her results. Given the length of the Workshop and the number of participants, it quickly became apparent that the Workshop would be a failure to the degree that insufficient time was being allotted for participants to get together in an informal manner to actually *work* on interesting problems; in this regard, an extra day would have been greatly useful. Still, there was time for a hike to Lake Peak or an excursion to Bandelier National Monument or Taos, and a banquet at the La Fonda Hotel (complete with strolling Spanish guitarist and a performance by the enthusiastic Baila! Baila! dance troupe). Nevertheless, the next meeting probably should be smaller and more focused than the 11th, since, as France Córdova points out in her contribution to these Proceedings, it is important to have a mix of both large and small meetings: "small meetings on specific topics to preserve the flavor of a 'workshop,' and large meetings to introduce more young people to the field and to examine connections between subtopics."

The individual contributions to these proceedings were typeset with TeX and LaTeX using a macro package put together by myself from a template supplied by Cambridge University Press. This macro package was "e-mailed" to the individual authors, who used it to write and typeset their own contributions. The "final" drafts of these contributions were then passed back and forth between the various authors and myself via e-mail. It is is powerful statement that a significant number ($\gtrsim$90%) of the authors of these Proceedings either use or have access to TeX or LaTeX *and* electronic mail. These standard tools have made possible the quick and relatively painless production of this "camera-ready" volume.

One of the more exciting group of talks during the Workshop concerned observations of GS2023+338, the X-ray transient discovered by the *Ginga* team in May of 1989. Since no one from the *Ginga* team could be present at the Workshop, an informal report on the published X-ray observations was given by Diane Roussel-Dupré. At Diane's request, Shunji Kitamoto prepared a formal contribution for these Proceedings. Both individuals are kindly thanked for their efforts.

Thanks are due to NASA, W. Doyle Evans (then Division Leader of the Laboratory's Earth and Space Science Division), and Chick Keller (Director of the Laboratory's Institute of Geophysics and Planetary Physics) for their support of this Workshop; Elaine Ruhe, for her excellent and unfailing efforts as conference coordinator; Margaret Flaugh, Deanna Davis, and Trisha Allen, the best and brightest group of secretaries I have ever worked with, for their help with all the many details; John Tubb and Ruth Robichaud for their assistance with the Workshop logo; Jean Johnson, Pita Valencia, and Lynne Chamberlin for their assistance with the (blessedly few) manuscripts which needed to be retyped and/or reformatted; and Florian Geyer for kindly supplying the beautiful illustration on the cover of these Proceedings.

Christopher Mauche

April, 1990

11th North American Workshop on CVs and LMXRBs

Santa Fe, NM Oct. 9-13 1989

THE WORKSHOP LOGO

The Workshop logo combines two design elements from the Southwest United States:

(1) A stylization of the Zia Sun symbol. This symbol is a decorative element on historic Zia pottery and has been adopted as the symbol of the state of New Mexico.

and

(2) The spiral petroglyph at the "Sun Dagger" site in Chaco Canyon in northwest New Mexico. This petroglyph was used by Anasazi Indians a millennium ago as a calendric device to mark the summer solstice, the winter solstice, the spring and fall equinoxes, and the northern minor and major standstills in the 18.6-year north-south cycle of the Moon.[1] Sadly, perhaps for the first time in three times the interval since Galileo first trained his telescope on the heavens, the summer solstice of 1989 passed without a dagger of sunlight slicing down the center of the petroglyph.[2] This discovery was made only a few months after the local organizing committee began using this symbol as the Workshop logo. For our purposes, it is representative of a history of astronomy in New Mexico. However, it also happens to look a bit like an accretion disk.

[1] Sofaer, A., Zinser, V., and Sinclair, R. M. 1979, *Science*, **206**, 283.
[2] *Albuquerque Journal*, June 22, 1989.

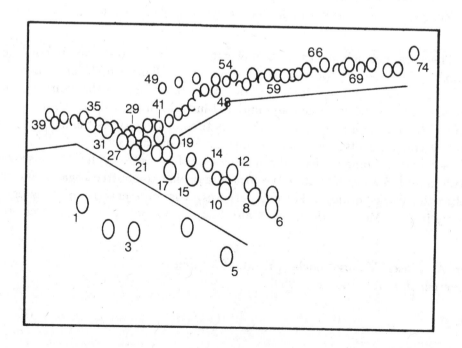

KEY TO PHOTO

1 K. Beuermann
2 H. Johnston
3 F. Hessman
4 J. Hayes
5 A. Shafter
6 S. Starrfield
7 E. Schlegel
8 R. Polidan
9 J. Raymond
10 C. Mauche
11 P. Schmidtke
12 S. Larsson
13 G. Williams
14 R. Hjellming
15 A. King
16 N. Hawkins
17 J. Frank
18 N. Vogt
19 J. Cannizzo
20 M. Hjellming
21 F. Córdova
22 J. Brainerd
23 B. Laubscher
24 J. Singh
25 C. Haswell

26 P. Higbie
27 D. Roussel-Dupré
28 J. Wood
29 H. Ritter
30 T. Marar
31 W. Priedhorsky
32 S. Litchfield
33 J.-M. Hameury
34 S. Howell
35 W. Sparks
36 P. Hertz
37 A. Parmar
38 A. Norton
39 J. Krautter
40 R. M. Wagner
41 R. Whitehurst
42 J. Bailey
43 J. Osborne
44 G. Schmidt
45 A. Tomaney
46 J. White
47 L. Molnar
48 C. Moss
49 A. Silber

50 C. Hellier
51 D. Buckley
52 I. Iben, Jr.
53 G. Chanmugam
54 M. Shara
55 R. Wade
56 M. Mouchet
57 M. Kaisig
58 R. Webbink
59 A. Bianchini
60 R. Popham
61 K. Mason
62 W. Kley
63 G. Miller
64 F. Bateson
65 M. Cropper
66 M. Watson
67 R. Smith
68 G. Machin
69 P. Callanan
70 E. Harlaftis
71 C. Bailyn
72 J. Woods
73 K. Horne
74 E. Robinson
75 J. Patterson
(behind camera)

~ 50 participants are not shown

LIST OF PARTICIPANTS

Timothy Abbott .... University of Texas at Austin (US)
Lorella Angelini .... EXOSAT Observatory (Netherlands)
Jeremy Bailey .... Joint Astronomy Centre (US)
Charles Bailyn .... Harvard-Smithsonian Center for Astrophys. (US)
Frank Bateson............ Astronomical Research, LTD (New Zealand)
George Berzins .... Los Alamos National Laboratory (US)
Klaus Beuermann .... Technische Universität Berlin (FRG)
Antonio Bianchini........ Osservatorio Astron. di Padova (Italy)
Jeffrey Bloch .... Los Alamos National Laboratory (US)
James Brainerd .... University of Chicago (US)
David Buckley .... University of Cape Town (South Africa)
Wolfram Bunk .... Max-Planck-Inst. für Extraterr. Phys. (FRG)
Paul Callanan .... University of Oxford (UK)
John Cannizzo .... Kenyon College (US)
Jorge Casares .... Inst. de Astrofisica de Canarias (Canary Islands)
Ganesar Chanmugam ..... Louisiana State University (US)
France Córdova .... Pennsylvania State University (US)
Anne Cowley............. Arizona State University (US)
Mark Cropper............ Mullard Space Science Laboratory (UK)
Vikram Dhillon .... University of Sussex (UK)
Richard Epstein.......... Los Alamos National Laboratory (US)
W. Doyle Evans .... Los Alamos National Laboratory (US)
Juhan Frank .... Max-Planck-Inst. für Phys. und Astrophys. (FRG)
Florian Geyer .... Universität Tübingen (FRG)
Franco Giovannelli .... Inst. de Astrofisica Spaziale, CNR (Italy)
Jean-Marie Hameury ...... Observatoire de Paris (France)
XiaoHong Han .... National Radio Astronomy Observatory (US)
Tomoyuki Hanawa .... Nagoya University (Japan)
Emilios Harlaftis .... University of Oxford (UK)
Carole Haswell .... University of Texas at Austin (US)
Nigel Hawkins............ University of Sussex (UK)
John Hayes............... University of Illinois (US)
Coel Hellier .... Mullard Space Science Laboratory (UK)
Paul Hertz .... Naval Research Laboratory (US)
Frederic Hessman .... Max-Planck-Inst. für Astron. (FRG)
Paul Higbie .... Los Alamos National Laboratory (US)
Michael Hjellming .... Northwestern University (US)
Robert Hjellming.......... National Radio Astronomy Observatory (US)
Keith Horne.............. Space Telescope Science Institute (US)
Steve Howell .... Planetary Science Institute (US)

Icko Iben, Jr. .... Pennsylvania State University (US)
James Imamura .... University of Oregon (US)
Jordi Isern .... Centre d'Estudis Avancats de Blanes (Spain)
Helen Johnston .... California Institute of Technology (US)
Derek Jones .... Royal Greenwich Observatory (Canary Islands)
Michael Kaisig .... University of Texas at Austin (US)
Ronald Kaitchuck .... Ohio State University (US)
Timothy Kallman .... NASA/Goddard Space Flight Center (US)
Andrew King .... University of Leicester (UK)
Ray Klebesadel .... Los Alamos National Laboratory (US)
Wilhelm Kley .... Universität München (FRG)
Wlodzimierz Kluzniak ..... Columbia University (US)
Joachim Krautter .... Landesstern. Heidelberg-Königstuhl (FRG)
Olga Kuhn .... Harvard-Smithsonian Center for Astrophys. (US)
Ping-Wai Kwok .... Whipple Observatory (US)
Javier Labay .... Universidad de Barcelona (Spain)
Stefan Larsson .... University of Tromsø (Norway)
Bryan Laubscher .... University of New Mexico (US)
Amir Levinson .... Ben-Gurion University (Israel)
James Liebert .... Steward Observatory (US)
Vladimir Lipunov .... Sternberg State Astron. Inst. (USSR)
Simon Litchfield .... University of Leicester (UK)
Graham Machin .... University of Oxford (UK)
Cathy Mansperger .... Ohio State University (US)
T. M. K. Marar .... ISRO Satellite Centre (India)
Thomas Marsh .... Space Telescope Science Institute (US)
Phillip Martell .... Ohio State University (US)
Keith Mason .... Mullard Space Science Laboratory (UK)
Christopher Mauche .... Los Alamos National Laboratory (US)
John Middleditch .... Los Alamos National Laboratory (US)
Guy Miller .... Los Alamos National Laboratory (US)
Shin Mineshige .... University of Texas at Austin (US)
Shigeki Miyaji .... Chiba University (Japan)
Lawrence Molnar .... University of Iowa (US)
Martine Mouchet .... Observatoire de Meudon (France)
Koji Mukai .... Mullard Space Science Laboratory (UK)
S. Naranan .... Tata Institute of Fund. Research (India)
Ramesh Narayan .... Steward Observatory (US)
Andrew Norton .... University of Southampton (UK)
Julian Osborne .... EXOSAT Observatory (Netherlands)
Arvind Parmar .... EXOSAT Observatory (Netherlands)
Joseph Patterson .... Columbia University (US)

Ronald Polidan........... Lunar and Planetary Lab. – West (US)
Michael Politano .... Arizona State University (US)
Robert Popham .... Steward Observatory (US)
Konstantin Postnov .... Sternberg State Astron. Inst. (USSR)
William Priedhorsky ..... Los Alamos National Laboratory (US)
John Raymond .... Harvard-Smithsonian Center for Astrophys. (US)
Hans Ritter .... Max-Planck-Inst. für Phys. und Astrophys. (FRG)
Edward Robinson .... University of Texas at Austin (US)
Diane Roussel-Dupré ..... Los Alamos National Laboratory (US)
Jonathan Schachter .... University of California, Berkeley (US)
Brad Schaefer .... NASA/Goddard Space Flight Center (US)
Eric Schlegel .... Harvard-Smithsonian Center for Astrophys. (US)
Gary Schmidt .... Steward Observatory (US)
Paul Schmidtke .... Arizona State University (US)
Pierluigi Selvelli .... Osservatorio Astron. di Trieste (Italy)
Allen Shafter .... University of Texas at Austin (US)
Anurag Shankar .... University of Illinois (US)
Michael Shara .... Space Telescope Science Institute (US)
Giora Shaviv .... Israel Institute of Technology (Israel)
Andrew Silber............ Massachusetts Institute of Technology (US)
Jyoti Singh .... Tata Institute of Fund. Research (India)
Edward Sion .... Villanova University (US)
Robert Smith .... University of Sussex (UK)
Warren Sparks .... Los Alamos National Laboratory (US)
Sumner Starrfield .... Arizona State University (US)
Paula Szkody............. University of Washington (US)
Ronald Taam.............. Northwestern University (US)
Austin Tomaney .... University of Texas at Austin (US)
Masami Uchida .... Ecole Secondaire, Osaka (Japan)
John Vallerga .... University of California, Berkeley (US)
Nikolaus Vogt .... Universidad Católica de Chile (Chile)
Richard Wade............. Steward Observatory (US)
R. Mark Wagner .... Ohio State University (US)
Michael Watson .... University of Leicester (UK)
Ronald Webbink .... University of Illinois (US)
J. Craig Wheeler .... University of Texas at Austin (US)
James White II............ Indiana University (US)
Robert Whitehurst .... University of Leicester (UK)
Glen Williams .... Central Michigan University (US)
Michael Wolff .... Naval Research Laboratory (US)
Janet Wood .... University of Cambridge (UK)
John Woods............... University of Oxford (UK)

A Short History of the CV Workshops

France A. Córdova

Department of Astronomy,
The Pennsylvania State University

The North American Workshops on Cataclysmic Variable Stars ("CV" Workshops) are topical meetings held nearly every year. While the cataclysmic binaries comprise the largest focus of these Workshops, other, similarly active close binaries, especially symbiotic stars and low mass X-ray binaries, are frequently included. The Workshops are sited wherever there is interest in organizing them. This, the Santa Fe Workshop, is the eleventh in the series.

Perhaps surprisingly, no astronomer has been present at all of the CV Workshops. The prize for highest attendance goes to Sumner Starrfield (Arizona State Univ.), who, because of an *IUE* run, missed only the Seattle Workshop. Over half of the ∼130 Santa Fe Workshop participants were new to the Workshops. This means that the history of the Workshops is in danger of becoming extinct. To preserve this history, we wanted to include a brief account of the past CV Workshops. This record is also meant to be a tribute to Workshop organizers who have provided invaluable forums for dissemination of knowledge in the field and notably enhanced the climate for interaction and new research directions. The following table is a synopsis of the CV Workshop particulars.

No.	Year	Place	Organizers
1	1976	Urbana, IL	S. Starrfield, J. Truran
2	1977	Boulder, CO	C. Hansen, B. Warner
3	1978	Tempe, AZ	S. Starrfield
4	1979	Rochester, NY	M. Savedoff, H. Van Horn
5	1980	Austin, TX	E. Robinson, E. Nather
6	1981	Santa Cruz, CA	J. Faulkner
7	1983	Cambridge, MA	J. Patterson, D. Lamb
8	1984	Baton Rouge, LA	H. Bond, G. Chanmugam, and J. Tohline
9	1985	Seattle, WA	P. Szkody
10	1987	Aspen, CO	J. Patterson, D. Lamb
11	1989	Santa Fe, NM	F. Córdova, C. Mauche, and W. Priedhorsky

The idea for the first CV Workshop is attributed to Sumner Starrfield and Jim

Truran (Univ. of Illinois) who got together about 15 people in 1976 to discuss, informally, their research on cataclysmic variables. Jim Pringle was visiting Illinois that year from the Institute of Astronomy at Cambridge, England, so even the first Workshop had an international flavor. There were no formal talks. The "hot" topic was the quasi-periodic oscillations detected in the optical by Rob Robinson and Ed Nather (Univ. of Texas, Austin) and Brian Warner (Univ. of Capetown). This was the year that Robinson and Warner independently published their comprehensive reviews of the field.

The first meeting was inspiring enough that Carl Hansen (Univ. of Colorado) and Brian Warner, who was spending a sabbatical at Boulder, organized the second CV Workshop a year later at the University of Colorado. Again, there were no formal talks.

The first formal talks on CVs were given the following year at the third CV Workshop, this time in Tempe. The meeting was organized by Sumner Starrfield. It was at this meeting that new wavelengths were introduced into the discussions: the IR was represented by Paula Szkody (Univ. of Washington, Seattle) and Ed Ney (Univ. of Minnesota); the X-ray was represented by France Córdova (then at CalTech); and the radio by Bob Hjellming (NRAO).

In 1979, Malcolm Savedoff organized an IAU Colloquium on "White Dwarfs and Variable Degenerate Stars" in Rochester. Hugh Van Horn (Univ. of Rochester) was a co-organizer for this meeting and appended a CV Workshop. The Proceedings of the Colloquium, edited by Van Horn and V. Weidemann, included the papers presented at the Workshop. It was dedicated to Jesse Greenstein (CalTech) on the occasion of his seventieth birthday, and represented the first published CV Workshop proceedings.

A cataclysmic event associated with the Rochester meeting occurred during Howard Bond's (then at Louisiana State Univ.) presentation. Howard was saying something suspicious like, "all CVs are descended from wide binaries that have gone through a common-envelope interaction," when a loud thunderclap sounded, lending the speaker an unusual degree of authority. The hot, pulsating, C-O white dwarf PG 1159−035 was first announced at this Workshop. It was also anticipated at the meeting that the 1980s would see the launch of XUV instruments that would survey the XUV sky—an expectation that still has not been realized a decade later. The excitation of CV QPOs eluded us then, as it continues to do.

One remembrance of the Rochester meeting that continues to inspire us is that of John Whelan (Univ. of Cambridge), who was the popular choice to give the Workshop summary. It was a job only a great diplomat like John could do because some of the debate had been fractious and a few issues were unresolved. A young man when he died soon thereafter of cancer, John was simply the nicest person many of us have ever met. He reflected a deep joy and serenity that were captivating.

The next year Rob Robinson and Ed Nather (Univ. of Texas, Austin) organized the fifth CV meeting. A summary of the meeting appears in *Nature*, 1980, Vol. 286. Joe Patterson (then at Univ. of Texas, Austin) gave three talks (on 2A 0311-227,

HT Cas, and rapid oscillations), establishing a pattern of conspicuous presence at CV meetings, and a prescription for outrageous artwork and patter in scientific presentations. Joe's aphorisms have taken on a life of their own, resurrecting their paraphrased selves at subsequent CV Workshops. The Austin meeting was the first one in which we confronted the problem that the interpretation of the amplitudes of the radial velocity curves of CVs was more difficult than had been assumed. We are beginning to understand the absorption-line velocities, but, even now, only limited progress has been made in understanding the emission-line velocities. At this meeting, there was a lot of interest in the new *IUE* data as well as the problem of interpreting the SU UMa phenomena. Ron Webbink (Univ. of Illinois) led a complex path through common-envelope evolution with graphically explicit viewgraphs in living color.

The longest CV Workshop was organized by John Faulkner (UC, Santa Cruz) as part of Santa Cruz's series of summer workshops. It lasted three weeks, included a partial lunar eclipse (which went unrecognized by some X-ray astronomers who apparently didn't know which way the Earth rotated), and provided plenty of time for roller coaster rides on the boardwalk and hiking in the surrounding forested countryside. This meeting is famous for its genesis (in the person of Jim Pringle) of the idea for the effective temperature – surface density diagram ("S"-curve) for accretion disk instabilities. It is also memorable for Bob Kraft's (UC, Santa Cruz) introductory, historical talk, which was the story of the founding of the subject by the founder himself. The audience especially enjoyed his tale of holding up the single-trailed spectrogram of WZ Sge, still dripping wet, and discovering the emission-line "S-wave." Kraft's talk was followed by an extraordinary review of the current state of the subject by Brian Warner, who shares responsibility for shaping the field. Ron Webbink presented a scholarly paper on very old novae, casually mentioning having consulted the original sources (written, of course, in Latin and French).

One of the traditional social events at the Santa Cruz Conference is the wine-tasting organized and led by Joe Miller, a noted Santa Cruz researcher in both AGNs and oenology. The Santa Cruz CV Workshop started, like so many Santa Cruz Workshops before it, with Miller's presentation of young red and white wines chosen from a number of local vineyards. CV people, however, have sometimes been accused of being a bit boorish and their behavior that evening was consistent with this reputation: not long into an "unassuming little Red," the group went unstable. The Italian CV astronomers had brought an ocarina and, under the influence of an exuberant Antonio Bianchini, led the group in a number of dances. A highlight was Jim Pringle's exhibition of his own instability during the "bunny hop." The evening was brought to a finish when Bob Williams, who runs circles around CTIO, yelled to Miller "I'll drink no wine that I can't unscrew." This evening was also noteworthy for the introduction of Paula Szkody's daughter, Allison, born one week earlier.

Joe Patterson and Don Lamb (both then at the Harvard-Smithsonian CfA) organized the seventh, and coldest, CV Workshop, which followed the 1983 Boston

AAS meeting. The organizers ambitiously welcomed the obstreperous low mass X-ray binary crowd to share ideas about accretion onto compact stars. They noted that in spite of the similarities between the parameters of white dwarf and neutron star binaries, there was little overlap in the research programs on these active binaries. The desire was to stimulate interaction by bringing the two groups together. In the preface to their proceedings, the organizers asked, "Did it work? Well, we had a well-attended and highly successful conference. We heard a number of excellent review talks on various aspects of CV's and LMXB's ... But we are inclined to think that the basic *goal* of the conference, a synthesis of the two fields, didn't really quite happen ... we identified few exchanges in which the CV's met the LMXB's to their mutual enlightenment. On at least two occasions, questioners who proposed such a comparison were actually scolded for comparing 'apples and oranges.' So it seems pretty clear that the time for such a synthesis has not yet come ... " Since that meeting, LMXRB talks have figured in the CV Workshops.

The loudest, most colorful meeting to date was the Baton Rouge CV Workshop, organized to coincide with the 1984 Mardi Gras season by Howard Bond (now STScI), Ganesh Chanmugam and Joel Tohline (both Louisiana State Univ.). Many of the participants arrived in a classic thunderstorm and were diverted to other airports, having to fend for themselves to get to the meeting on time. The other problem that plagued the meeting was the nature of the soft X-ray emission from the AM Herculis variables. Observational data presented by John Heise (Utrecht) revealed that the hard and soft X-rays might have their origins in different poles of the magnetic, accreting white dwarf star. Keith Horne (STScI) introduced the CV world to Doppler tomography of accretion disks. During an expedition to New Orleans, Bill Priedhorsky (Los Alamos) distinguished himself by collecting the most Mardi Gras necklaces, thrown by revelers from parade floats—not a mean feat, considering the competition from other CV festival goers. The following day, back in Baton Rouge, Mike Shara (STScI) maintained the festive mood when he opened his CV talk by throwing trinkets from a cigar box to an audience of scrambling astronomers. John Faulkner got his comeuppance for years of bad jokes from an anonymous poster paper debunking "The Theories that Jack Built."

Paula Szkody provided the most rural setting for a CV Workshop: a 330-acre army fort on the Olympic Peninsula which had been converted to a State Park. The logistics of hiring rental cars and ferries to reach this remote area did not preclude a strong showing by the European contingent. The Victorian houses which used to be the Officers Quarters provided some different housing accommodations (there was a "Janet House" in which Janet Drew, Janet Wood, and Janet Mattei stayed). The meeting took place in the army Chapel. The mess hall brought back bad memories of school cafeterias. The Parade Grounds was the site of a CV softball game, which followed in the tradition of cricket games in the U.K. at close binary star meetings. The Olympic game ended in an unfortunate "bouncer ball" striking a player, establishing a tradition of sports injury which was continued at the later Aspen Workshop.

A cold wind made the salmon barbecue on the beach the shortest conference dinner in Workshop history as everyone scurried for warm shelter.

The science of this ninth CV Workshop involved continuing discussions of the disk instability and superhump problems in dwarf novae and the oscillations and evolution of magnetic systems. For the 65 participants, there was a large concentration of the primary people involved modeling spectral line formation: Allen Shafter, Keith Horne, Rick Hessman, John Clarke, Don Ferguson, John Raymond, and Janet Drew provided descriptions of the complexities of the optical and UV emission within a single session. This was also the first Workshop in which several members of the AAVSO participated, since the AAVSO semi-annual meeting had taken place the previous weekend in Seattle. Paula edited a proceedings for this, the ninth CV Workshop.

There was not another North American CV Workshop for two years. In the intervening year, 1986, a CV Colloquium was held in Bamberg, West Germany.

Some organizers seem to enjoy the job: Joe Patterson and Don Lamb organized their second CV Workshop in 1987, this one in Aspen, Colorado. Afternoons were reserved for hiking in the beautiful Aspen mountains. CV evolution and the recent Supernova 1987A were important focuses of this, the tenth Workshop.

The Los Alamos, New Mexico, contingent of CV astronomers had been promising for a long time to host a Workshop and finally made good on this in October of 1989. These Proceedings contain much of the work presented at this meeting, held in Santa Fe. The meeting was organized by France Córdova, Christopher Mauche, and Bill Priedhorsky (Los Alamos). It was so large, with ~130 astronomers from 16 countries, that it gave some participants pause to wonder where the field was headed. It was generally concluded that meetings both large and small, interspersed, would be best for the discipline: small meetings on specific topics to preserve the flavor of a "workshop," and larger meetings to introduce more young people to the field and to examine connections between subtopics. The meeting was held during Albuquerque's famous International Hot Air Balloon Fiesta. The day after the meeting found some amazed participants lost in a great tide of colorful, expanding, and rising balloons. Jets carrying other participants home carefully negotiated their way past the precisely drifting balloons, which formed a giant box pattern in a perfect blue sky.

In 13 years and 11 North American CV Workshops, the field of cataclysmic variables has seen dramatic change.

X-ray space research has enhanced significantly a whole new field of CV inquiry: the magnetic, accreting white dwarf binary. Today, observational and theoretical research on AM Herculis and DQ Herculis variables in every wavelength band forms a significant contribution to the literature on CVs. Another important result on CVs during this interval was the discovery, using *HEAO 1*, of X-ray oscillations from dwarf novae, an affirmation that the optical QPOs reflect the presence of a high-energy source. The surprise was that the X-ray oscillations were also quasi-periodic. The *Einstein* satellite observations confirmed that X-ray emission is ubiquitous among all classes of CVs. *EXOSAT* satellite results more recently revealed that low mass X-ray binaries also exhibit quasi-periodic X-ray oscillations, although with much smaller amplitudes and higher frequencies than the CV pulsations.

The technique of Doppler imaging has given us remarkable new pictures of the CV accretion disk and sources of the optical emission. New theoretical inquiries into accretion disk instabilities, evolution, and the production of winds from disk geometries, have greatly enhanced our naïve understanding of a decade ago. The *IUE* satellite has given us a new picture of the nature of the dwarf nova outburst, the contribution of the white dwarf, the relative importance of X-ray heating and disk reprocessing, as well as the genesis of the high-velocity winds seen in luminous CVs.

This volume testifies to the enormous growth in the field. It looks to the 1990s for new understanding of CVs that should result from space- and ground-based ventures of unprecedented resolution, sensitivity, or temporal coverage in virtually every wavelength band.

Topical meetings like the CV Workshops have more purposes than sharing information. As one CV astronomer put it, "The nicest feature of the five Workshops that I have attended was the opportunity to meet again with old friends, and to make new ones. There has always been a wonderful feeling of community and camaraderie among those astronomers fascinated by the marvelous zoo of cataclysmic variables."

The organizers of the CV Workshops gratefully acknowledge their host institutions, the NSF, NASA, and the DOE for financial support for the meetings. The entire CV community acknowledges a debt to Janet Mattei and Frank Bateson who lead the largest amateur astronomer organizations in this country and New Zealand, respectively. The amateur astronomers have provided essential data for many kinds of research; their support for satellite UV and X-ray observations of CVs has been invaluable.

Nothing would please that indefatigable punster Howard Bond more than my acknowledgment of his contribution of remembrances to this history. Remembrances were also kindly shared by Ganesh Chanmugam, Rob Robinson, Sumner Starrfield, and Paula Szkody.

An Eclipsing Black Hole (?) in the LMC

A. P. Cowley[1], P. C. Schmidtke[1], D. Crampton[2], and
J. B. Hutchings[2]

[1]Department of Physics and Astronomy, Arizona State University
[2]Dominion Astrophysical Observatory

1 INTRODUCTION

One of the great difficulties in studying X-ray sources in the Galaxy is that often their distances and reddening are very poorly determined, and therefore their true luminosities and other parameters are also not known. However, sources in the Magellanic Clouds can be studied without these difficulties, since the distance of the LMC is well established and there is little reddening in that direction. Thus, it is possible to study the X-ray sources with some certainty that the sample is complete to a known X-ray luminosity.

A survey of the Large Magellanic Cloud carried out at the Columbia Astrophysics Lab (Long, Helfand, and Grabelsky 1981) was complete to approximately $L_X > 3 \times 10^{35}$ erg s^{-1}. Making optical identifications for these sources has been difficult, and only a small number of X-ray binaries in the LMC has been found (see Cowley et al. 1984). These include four bright, massive binaries: LMC X-1, LMC X-3, LMC X-4, and 0538-66. Although in the Milky Way the low mass X-ray binaries (LMXRBs) are an order of magnitude more numerous than the massive binaries, to date only three LMXRBs are known in the LMC (LMC X-2, CAL 83, and CAL 87, with only limited data available for each). The small number of LMXRBs may be partially due to the difficulty of making identifications in very crowded LMC fields, but additionally there may also be fewer such sources, since all X-ray sources with $L_X > 10^{37}$ erg s^{-1} have already been identified.

2 PHOTOMETRY

CAL 87 is a relatively weak X-ray source ($L_X \sim 10^{36}$ erg s^{-1}) with an extremely soft X-ray spectrum ($kT \sim 0.1$ keV). The source was identified with a 19th magnitude blue star by Pakull et al. (1987). Its spectrum shows He II (4686 Å) and weak Hα emission superimposed on a blue continuum. At minimum light, the absorption spectrum of an F–G star is weakly visible. Photometry by Pakull et al. (1988) suggested that the system might be eclipsing, and, independently, Naylor et al. (1989) and Cowley et al. (1989) discovered the period to be 10.6 hours — considerably shorter than had been expected. Combining photometry from 1985, 1987, and 1988, we derive an ephemeris

$$T_{\min} = \text{HJD}\,2447506.8021(\pm0.0002) + N \times 0.442683(\pm0.000005)\ \text{days}.$$

The light curve, shown in Figure 1, has a range of slightly over 1 magnitude in V. At maximum light, the light curve is fairly flat, which implies that we are observing an eclipse, rather than the variation of a heated secondary star (as is seen in Her X-1, for example). The primary eclipse is very wide (wings extend to $\pm0.25P$), indicating that the object being eclipsed is extremely large and has strong limb darkening. It is also consistent with a large mass ratio in favor of the eclipsed object. The eclipse does not appear to be total. However, its large depth suggests the orbital inclination must be quite high ($> 70°$). The primary minimum is deeper in B than in R, indicating that the central part of the eclipsed source is hotter than the average temperature for the system. All of the data from the primary eclipse are consistent with the interpretation that a large, thick accretion disk is being eclipsed by the secondary F–G star. Both our light curves and those of Callanan *et al.* (1989) show the ingress to be broader and more variable than the egress. Such pre-eclipse variations are seen in other LMXRBs (e.g., in 1822-37, see Mason *et al.* 1980) and are thought to be due to variations in the thickness of the accretion disk (Parmar *et al.* 1986). The secondary eclipse is considerably narrower than the primary. Both its depth and position are variable, again suggesting that it is caused by changing structure in the accretion disk, which lies between the observer and secondary star at this phase.

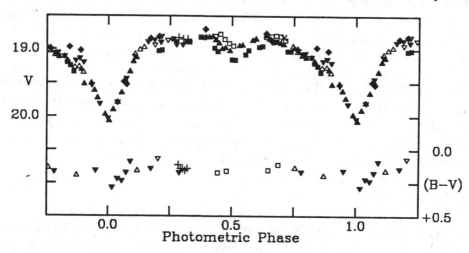

Fig. 1—V light curve and $(B - V)$ color curve of CAL 87 plotted on its 10.6-hour orbital period, using the ephemeris given in the text. Different symbols show behavior on different nights: crosses, 1985 Nov. 12/13; pluses, 1987 May 3/4; filled diamonds, 1987 Nov. 23/24–26/27; open upwards triangles, 1988 Dec. 7/8; open downward triangles, 1988 Dec. 8/9; filled squares, 1988 Dec. 9/10; filled upward triangles, 1988 Dec. 10/11; filled downward triangles, 1988 Dec. 11/12. Estimated errors are comparable to the size of the symbols.

Thus, the light curve can be best modeled by a luminous, thick accretion disk which is eclipsed by the secondary F–G star at primary minimum. The secondary star appears to be well shielded from the X-ray source, since it does not show any X-ray heating. This thick accretion disk may also prevent us from viewing the central source directly and explain why the observed L_X is so low.

3 SPECTROSCOPY

Spectra taken simultaneously with the photometry show the He II emission line varies in velocity with a semi-amplitude $K = 40$ km s^{-1}. The maximum positive velocity occurs at photometric phase 0.75, exactly when expected if the emission arises in the accretion disk and moves with the compact star. Two other lines of evidence suggest that the He II emission lines are formed near and move with the degenerate object. First, the line profiles are > 2000 km s^{-1} wide, suggesting formation in the inner regions of an accretion disk. Second, although the equivalent width of He II is somewhat variable, it shows no systematic variation with orbital phase. This means that the line-forming region is eclipsed in the same way as the continuum, again indicating that He II is formed in the accretion disk (rather than in a stream or isolated hot spot).

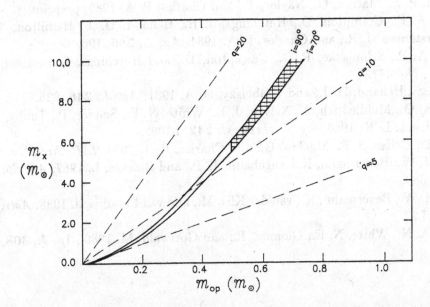

Fig. 2—Mass of the X-ray source versus mass of the optical secondary star, for extreme values of the orbital inclination i. This figure is based on the mass function computed using a semi-amplitude of the He II emission line, $K = 40$ km s^{-1}. Dashed lines show various values of the mass ratio, q. Most likely masses lie in the cross-hatched region where the mass ratio is above 10.

If the emission-line velocity curve represents the motion of the compact star, we can derive masses for the component stars using the mass function. Figure 2 shows how the mass of the X-ray source varies for different values of the optical-star mass when the orbital inclination lies between 70° and 90°. Analysis of the light curve, using Roche geometry, shows that the inclination cannot be lower than $\sim 70°$. The absolute magnitude of the secondary star is estimated to be about +2.4, if it contributes about half of the brightness at minimum light. An F star of this magnitude has a mass of about 1.0 to 1.5 solar masses. However, in many X-ray binaries the mass-losing star is undermassive for its luminosity. Even allowing for this, by assuming the secondary star is only $\sim 0.5\,M_\odot$, implies $M_X \sim 5\,M_\odot$! Thus, the data suggest that CAL 87 may be an eclipsing black-hole binary. If this is the case, we have a unique opportunity to study the structure of an accretion disk near a black hole, by observing the eclipse at many wavelengths.

The data presented here were taken at CTIO, and we thank the staff for their assistance. A. P. C. and P. C. S. acknowledge support from the National Science Foundation.

REFERENCES

Callanan, P. J., Machin, G., Naylor, T., and Charles, P. A. 1989, preprint.

Cowley, A. P., Crampton, D., Hutchings, J. B., Helfand, D. J., Hamilton, T. T., Thorstensen, J. R., and Charles, P. A. 1984, *Ap. J.*, **286**, 196.

Cowley, A. P., Schmidtke, P. C., Crampton, D., and Hutchings, J. B. 1989, *IAU Circ.*, No. 4772.

Long, K. S., Helfand, D. J., and Grabelsky, D. A. 1981, *Ap. J.*, **248**, 925.

Mason, K. O., Middleditch, J., Nelson, J. E., White, N. E., Seitzer, P., Tuohy, I. R., and Hunt, L. K. 1980, *Ap. J. (Letters)*, **242**, L109.

Naylor, T., Callanan, P., Machin, G., and Charles, P. A. 1989, *IAU Circ.*, No. 4747.

Pakull, M. W., Beuermann, K., Angebault, L. P., and Bianchi, L. 1987, *Ap. Sp. Sci.*, **131**, 689.

Pakull, M. W., Beuermann, K., van der Klis, M., and van Paradijs, J. 1988, *Astr. Ap.*, **203**, L27.

Parmar, A. N., White, N. E., Giommi, P., and Gottwald, M. 1986, *Ap. J.*, **308**, 199.

The Light Curve of A0620-00, the Black Hole Binary

Carole A. Haswell[1], Edward L. Robinson[1], and Keith D. Horne[2]

[1]The University of Texas at Austin and McDonald Observatory
[2]Space Telescope Science Institute

1 INTRODUCTION

In 1975 A0620-00 went into outburst, reaching $V \sim 12$ (Boley *et al.* 1976), becoming the brightest X-ray source in the sky. Fifteen months later the object returned to its pre-outburst brightness of $V \sim 18.3$ (Ciatti and Vittone 1977) and the X-ray brightness declined beyond the limits for detection (Whelan *et al.* 1977).

The quiescent spectrum of A0620-00 shows features of a K4 V – K7 V star. McClintock and Remillard (1986) obtained a radial velocity curve for the K dwarf: the velocity semi-amplitude was 457 ± 8 km s^{-1}, leading to a mass function for the primary of $3.18 \pm 0.16 \, M_\odot$. The stiffest possible nuclear equation of state produces a maximum neutron star mass of $2.7 \, M_\odot$; as the mass function for A0620-00 is above this limit, the compact object in this system is a prime black hole candidate.

To deduce the mass of the primary in A0620-00 it is necessary to determine the inclination of the system. A standard way to measure the inclination of interacting binary stars is to model the "ellipsoidal" variations in the light curve, which arise as a result of the non-spherical shape of the secondary star: orbital modulation of the apparent brightness of the secondary occurs because its projected area varies with orbital phase. The effect is strongest when the inclination is high and has zero amplitude if the system is viewed pole-on.

McClintock and Remillard (1986) obtained photometry of A0620-00 that clearly showed ellipsoidal variations, but the signal-to-noise ratio of their data was insufficient to allow a detailed analysis. We have obtained higher-quality multicolor photometry of the object. We present the light curves here and give a preliminary discussion of their properties.

2 OBSERVATIONS

2.1 Procedure

We observed A0620-00 using the 2.7-m telescope at McDonald observatory and the Stiening multicolor photometer over New Year 1987. This instrument is a 5-channel aperture photometer that accumulates data simultaneously in the U, B, V, and R bands on the program star and monitors a comparison star in a fifth channel. In all observations we used an integration time of 1 second.

As the seeing was poor, we used a large aperture, so the flux due to A0620-00 in the R band was as little as 10% of the observed sky brightness; in the other colors the object was even fainter. To facilitate accurate sky subtraction, we nodded the telescope between "star + sky" and "sky" continuously throughout the night, spending 30 seconds in each position. A nearby comparison star was monitored every twenty to thirty minutes; we also obtained flux standard observations.

The data were reduced using DRAGON, an interactive analysis program for reduction of simultaneous multicolor photometry. Extraction of the U, B, and V data was made possible only by the simultaneous R band data, in which the difference between "star" and "star + sky" was large enough to allow discrimination. Sky subtraction was performed by fitting a cubic spline function. A least squares fit to the comparison star observations was used to determine the extinction coefficients for each run, and the absolute fluxes were obtained by using the flux standard star observations.

Using the photometric ephemeris of McClintock and Remillard (1986), the flux-calibrated data were folded in orbital phase space to produce a mean orbital light curve in each color. The quality of the data varied significantly from night to night, so we used a weighting scheme to average the data. We assigned weights to 100 second chunks of data by calculating the standard deviation of the data points from the mean value for that chunk; orbital variations are insignificant compared to the noise over such short times. For ease of interpretation we then rebinned the data into a small number of orbital phase bins.

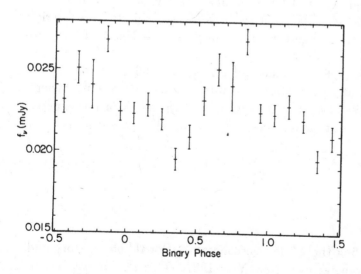

Fig. 1—U band light curve. The data have been divided into 10 phase bins and the light curve has been plotted twice.

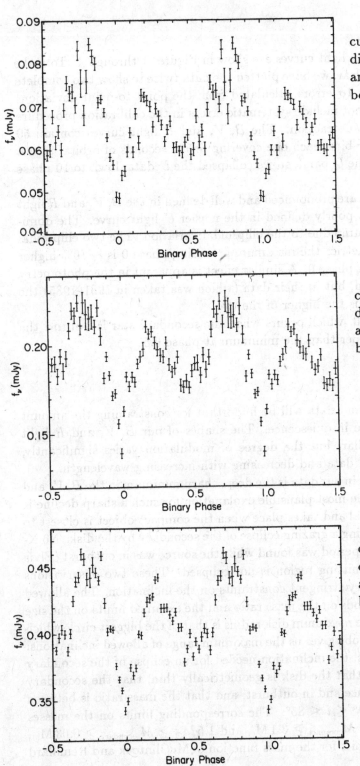

Fig. 2—B band light curve. The data have been divided into 50 phase bins and the light curve has been plotted twice.

Fig. 3—V band light curve. The data have been divided into 50 phase bins and the light curve has been plotted twice.

Fig. 4—R band light curve. The data have been divided into 50 phase bins and the light curve has been plotted twice.

2.2 Results

The flux-calibrated mean light curves are show in Figures 1 through 4. To make the orbital variations more clear, we have plotted the data twice to show two complete orbits. The error bars show 1σ errors calculated from the point-to-point deviations in the light curves: they do not include systematic errors in the calibration procedure or uncertainties in the sky subtraction. The B, V, and R light curves contain 50 equally-spaced orbital phase bins, each one covering ~ 550 seconds of orbital phase. The object is very faint in the U band, so we collapsed the U data down to 10 phase bins.

The orbital modulations are pronounced and well defined in the B, V, and R light curves; they are present but poorly defined in the noisier U light curve. The dominant feature is the double-hump due to the ellipsoidal variation. The two ellipsoidal maxima are strongly asymmetric: the maximum preceding phase 0 is $\sim 10\%$ higher than the maximum following phase 0. A similar effect is apparent in the photometry of McClintock and Remillard, but in their data (which was taken in 1981–1985) the maximum following phase 0 is the higher of the two.

The minimum at phase 0, which occurs when the secondary star is **behind** the compact object, is much deeper than the minimum at phase 0.5.

3 DISCUSSION

Although noisy, the U band data will be important for constraining the amount of mass transfer taking place in quiescence. The shapes of our B, V, and R light curves are qualitatively similar, but the degree of modulation varies significantly, being highest for the B band data and decreasing with increasing wavelength.

The most striking feature in our data is the deep, sharp minimum in the B, V, and R light curves at phase 0. The most plausible explanation for such a sharp decline is an eclipse; the eclipse is partial and takes place when the compact object is closest to us, suggesting that we are seeing a grazing eclipse of the secondary by the disk. No X-ray modulation at the orbital period was found while the source was in outburst (Elvis *et al.* 1975), so the X-ray-emitting region is not eclipsed. These two observations when taken together place very stringent constraints on the inclination. The allowed range for the inclination depends on the mass ratio and the assumed limits on the size of the disk. Assuming that the maximum disk radius is that of the biggest circle which fits inside the primary Roche lobe gives us the maximum range of allowed inclinations: if the disk is smaller than this, the inclination needed for an eclipse of the secondary increases. If we also assume that the disk is geometrically thin, that the secondary fills its Roche lobe in quiescence and in outburst, and that the mass ratio is between 5 and 100, then we obtain $59° \leq i \leq 85°$. The corresponding limits on the masses of the two stars are $7.6\,M_\odot \geq M_{compact} \geq 3.1\,M_\odot$ and $1.5\,M_\odot \geq M_{secondary} \geq 0.03\,M_\odot$, where we have used the 1σ limits for the mass function of McClintock and Remillard

(1986). For smaller mass ratios, smaller inclinations would be permitted, but a mass ratio of 5 already implies an implausibly large secondary star mass for the lower limit on the inclination.

The secular change which is apparent when our curves are compared with those of McClintock and Remillard (1986) is intriguing. There are two possible causes for the reversal in the asymmetry of the two maxima: (*a*) migrating RS CVn-type spots (analogous to sunspots) on the surface of the secondary star, or (*b*) a substantial change in azimuth of a hot spot at the disk/stream intersection.

The RS CVn hypothesis implies that the K star is magnetically active, and that magnetic braking of orbital angular momentum may be playing an important role in the evolution of the system. If this hypothesis is correct, the decrease in the amplitude of the modulation with increasing wavelength must be due to a steady source of long-wavelength flux which dilutes the variations at longer wavelengths: such flux could emanate from cool regions of the accretion disk.

The hot spot hypothesis requires the azimuth where the stream impacts the disk to have changed by $\sim 180°$ over a few years. For this to occur, the disk radius would have to change radically, and we would expect to see a corresponding change in the depth of the eclipse. Such a change would be unprecedented, but in view of the dramatic outburst A0620-00 underwent in 1975, the possibility needs to be explored. In this model, some of the *B* band modulation is due to changes of the visible projected area of the hot spot or possibly by an eclipse of the hot spot by the secondary.

REFERENCES

Boley, F., Wolfson, R., Bradt, H., Doxsey, R., Jernigan, G., and Hiltner, W. A. 1976, *Ap. J. (Letters)*, **203**, L13.

Ciatti, F, and Vittone, A. 1977, *IBVS*, No. 1261.

Elvis, M., Page, C. G., Pounds, K. A., Ricketts, M. J., Turner, M. J. L. 1975, *Nature*, **257**, 656.

McClintock, J. E., and Remillard, R. A. 1986, *Ap. J.*, **308**, 110.

Whelan, J. A. J., Ward, M. J., Allen, D. A., Danziger, I. J., Fosbury, R. A. E., Murdin, P. G., Penston, M., V., Peterson, B. A., Wampler, E. J., and Webster, B. L. 1977, *M. N. R. A. S.*, **180**, 657.

Optical Spectroscopy of A0620–00 and GS2023+338

Helen M. Johnston and Shrinivas R. Kulkarni

California Institute of Technology

1 INTRODUCTION

We present spectroscopic observations of two soft X-ray novae — A0620–00 (Nova Monocerotis 1975) and GS 2023+338 (V404 Cygni).

2 THE X-RAY NOVA A0620–00

A0620–00 was discovered on 3 August 1975 as an X-ray nova, and was the first X-ray nova to be identified with an optical object, a counterpart which reached twelfth magnitude in brightness. Since then, A0620–00 has been X-ray quiet. The optical spectrum was found to consist of a K4 V–K7 V stellar spectrum plus an emission line component from the accretion disk. McClintock and Remillard (1986, henceforth MR), in a photometric and spectroscopic analysis, determined an orbital period of 7.75 hr and measured the mass function to be $3.18\,M_\odot$. This limit exceeds the maximum mass of a neutron star under any present model, and hence they concluded that the primary star in A0620–00 is a black hole.

2.1 Low-Resolution Data and Analysis

In order to confirm this claim, we observed A0620–00 for a total of three nights in 1986 using the Double Spectrograph on the Hale 5-m telescope. Our results are presented elsewhere (Johnston, Kulkarni, and Oke 1989), but confirm the results of McClintock and Remillard — we measured a mass function of $3.30 \pm 0.95\,M_\odot$ (1σ), consistent with MR's value. However, by analyzing the emission lines, which show the double-peaked Smak profiles characteristic of disk emission, we can measure the projected velocity of the outer edge of the disk. We measured this velocity to be 540 ± 15 km s^{-1}, compared to $K = 460 \pm 40$ km s^{-1} for the dwarf K companion. By comparing these two velocities, and arguing that the velocity of the outer edge of the disk cannot be less than the velocity of a particle in the largest non-overlapping, stable, simple periodic orbit, we can put an upper limit on the mass ratio of the system $\mu = M_2/(M_1 + M_2)$: $\mu \leq 0.07$ (3σ). We required an S-wave in the hot spot, with amplitude β and constant phase with respect to the K star $\psi_{\rm hs}$. $\psi_{\rm hs}$ was found

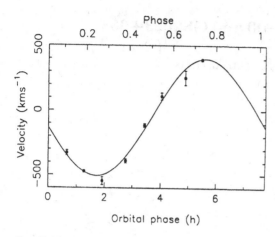

Fig. 1—Radial velocity of the K dwarf of A0620–00 derived from a correlation with a comparison K5 V star. The error bars are measurement errors derived from the width of the cross-correlation peak.

to be 0.02, so the hot spot is very close to the line joining the stars, thus indicating the disk must be very large.

We decided to obtain new, higher velocity resolution data of the system, the motivation being both to improve the errors on the velocity of the companion and hence improve our measurement of the mass ratio, and possibly to measure the variations in the disk velocity and even the orbital motion of the black hole itself, thus unambiguously determining its mass.

2.2 High-Resolution Data and Results

We observed A0620–00 on the night of 13 February 1989 UT, using the Double Spectrograph on the Hale 5-m telescope. We used a 1200 lines mm^{-1} grating in the red camera, yielding a resolution of 1.8 Å. Conditions were less than ideal, with $\sim 3''$ seeing, so it should prove possible to obtain better data. The data from the blue camera will not be discussed as the signal-to-noise ratio was too poor; the wavelength coverage of the red data was 5985–6640 Å. Nine 1800 s exposures were taken, together with spectra of four K dwarf radial velocity standards. Data reduction followed exactly the procedure described in Johnston, Kulkarni, and Oke (1989).

Figure 1 shows a substantially improved radial velocity curve for the secondary. The sine curve was fit assuming the photometric period from MR of 7.75234 hr. The velocity semi-amplitude is $K = 453 \pm 11$ km s^{-1}, giving a value for the mass function of $f(M) = 3.1 \pm 0.2$ M$_\odot$ (1σ), the best determination of the mass function so far. Figure 2 shows the Hα lines as a function of orbital phase. As before, we found we needed to include an azimuthal term in the standard Smak profile in order to fit our data. However, the amplitude of this term was substantially reduced, from 21% to 9%, and was consistent with zero at the 1σ level. This would seem to indicate the disk is still varying on several year timescales. The full details of the fit are presented in Table 1, with the values obtained from the 1986 low-resolution data for comparison. The velocity at the outer edge of the disk was measured to be $v_{\mathrm{disk}} = 580 \pm 20$ km s^{-1} (1σ), giving $v_{\mathrm{disk}}/K = 1.28 \pm 0.08$. This puts a 1σ upper limit on μ of $\mu \le 0.01$, or a 3σ upper limit of $\mu \le 0.07$, as before. The quality of the present data prevented us

Fig. 2—Variation in the Hα emission of A0620−00 over the orbit. The histogram shows the data, the solid line the model fit. The data were normalized by a low-order polynomial fit to the local continuum and then a constant 1.0 was subtracted. The horizontal scale is ±2500 km s^{-1}.

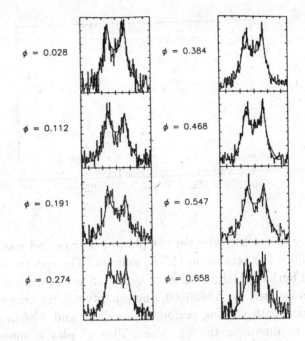

from measuring the disk orbital motion.

Thus, our original analysis would appear to be borne out. Our observations yield the best determination of the mass function, showing that the central object of the A0620−00 system is almost certainly a black hole, possibly even a very massive black hole. It must be noted, of course, that scenarios can be imagined where the above analysis, assuming a Keplerian disk in the orbital plane of the system, might break down.

TABLE 1
PARAMETERS DERIVED FROM FIT

	Low resolution	High resolution
K	460 ± 40 km s^{-1}	453 ± 11 km s^{-1}
$f(M)$	$3.30 \pm 0.95\,M_\odot$	$3.1 \pm 0.2\ M_\odot$
v_{disk}	540 ± 40 km s^{-1}	580 ± 20 km s^{-1}
β	0.21 ± 0.05	0.09 ± 0.08
ψ_{hs}	0.02 ± 0.18	0.02 ± 0.15

3 GS 2023+338 = V404 CYGNI

The X-ray nova GS 2023+338 was discovered by the *GINGA* satellite when it underwent a nova explosion on 22 May 1989 (Makino 1989). This was soon identified with V404 Cygni, a nova of 1938 (Wagner, Starrfield, and Cassatella 1989). Wagner *et al.* (1989) reported a possible 10 min-photometric period. Spurred by this, we obtained time-resolved spectroscopy with the Hale telescope on 29 August 1989 UT to

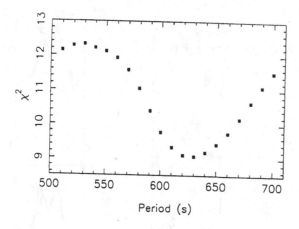

Fig. 3—The χ^2 value for a sine wave of varying period fit to our velocity data of GS 2023+338, as a function of period. There is a minimum near the observed photometric period of 10 min.

try to confirm whether the photometric period was an orbital period, taking thirteen 150 s, low-resolution (15 Å) spectra. The spectra we obtained showed emission lines of both the Balmer series of hydrogen and ionized helium. Applying the same analysis as we used for A0620−00, our velocity results were somewhat equivocal. We fit a sine wave with varying period to our data, and measured the goodness of fit, looking for a minimum in the χ^2 value. This χ^2 plot is shown in Figure 3; there is indeed a minimum near 10 min (at 630 s) but when this best-fit sine curve is plotted with the data and the errors on the velocity measurement, the result is less than convincing (Fig. 4).

Fig. 4—The sine curve with 630 s period fit to our velocity data of GS 2023+338, with error bars obtained from the width of the cross-correlation function. The semi-amplitude of the velocity is $80 \pm 30\ \mathrm{km\,s^{-1}}$.

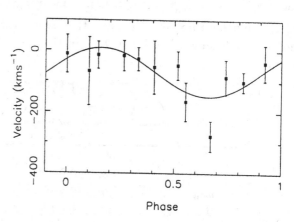

REFERENCES

Johnston, H. M., Kulkarni, S. R., and Oke, J. B. 1989, *Ap. J.*, **345**, 492.
Makino, F. 1989, *I.A.U. Circ.*, No. 4782.
McClintock, J. E., and Remillard, R. A. 1986, *Ap. J.*, **308**, 110.
Wagner, R. M., Starrfield, S. G., and Cassatella, A. 1989, *I.A.U. Circ.*, No. 4783.
Wagner, R. M., Kreidl, T. J., Howell, S. B., Collins, G. W., and Starrfield, S. G. 1989, *I.A.U. Circ.*, No. 4797.

The X-ray Light Curve of GS2023+338 (=V404 Cyg)

Shunji Kitamoto

Department of Physics, Faculty of Science, Osaka University, 1-1,
Machikaneyama-cho, Toyonaka Osaka, 560, Japan

1 INTRODUCTION

We present the X-ray light curve obtained with the Large Area Counter (LAC) and All Sky Monitor (ASM) on board the *GINGA* satellite (Makino and the Astro-C Team 1987) of the bright X-ray nova GS2023+338 discovered with the *GINGA* ASM on 21 May 1989 (Makino and the Ginga Team 1989). Ten observations each lasting from 1 day to 5 days were made with the LAC from 23 May to 1 November and it was monitored with the ASM until it reached the detection limit in September. In this paper, we present the light curve obtained with the LAC and ASM experiments. The spectral evolution and the temporal behavior with up to 1 msec resolution are in progress and will be published elsewhere.

2 THE X-RAY LIGHT CURVE

On 21 May 1989 (JD 2447668) a transient X-ray source in the constellation of Cygnus was discovered during routine monitoring by the ASM experiment. This source was designated GS2023+338 (Makino and the Ginga Team 1989). Later, this source was identified as V404 Cyg by Wagner, Starrfield, and Cassatella (1989).

The All Sky Monitor (Tsunemi *et al.* 1989*a*) consists of a pair of proportional counters filled with a Xe-CO$_2$ gas mixture. Each proportional counter is divided into three independently operating chambers with an effective area of 70 cm^2. Each chamber has a fan-beam collimator ($1° \times 45°$) slanted at different angles with respect to the Z-axis of the satellite. The entrance window is 50 mm-thick beryllium. For a normal ASM scanning observation, the satellite performs a 20-min 360° rotation, once per day. The ASM usually observes 1–20 keV X-rays in 16 energy channels with 0.5-s time resolution. The Large Area Counter (Turner *et al.* 1989) is comprised of eight identical proportional counters with 62 mm-thick beryllium windows. These detectors are filled with an Ar-Xe gas mixture and have a total effective area of 4000 cm^2. The maximum time resolution is 1 msec. The energy range is 1–37 keV.

The X-ray light curve of GS2023+338 obtained during the 1989 outburst is shown in Figure 1, where we have plotted both the 64 s averaged X-ray fluxes in the energy band 1.2–18.6 keV obtained with the LAC during pointed observations and

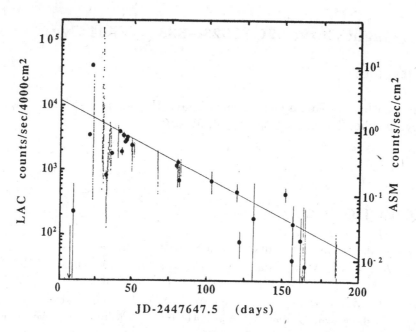

Fig. 1—X-ray light curve of GS2023+338 as a function of JD−2447647.5. The LAC light curve (*small points*) is in units of LAC counts s^{-1} 4000 cm^{-2} in the energy range 1.2–18.6 keV averaged 64-s observation. The ASM light curves (*filled circles*) is in units of ASM counts s^{-1} cm^{-2} in the energy range 1.0–20 keV for each scanning observation. Only the ASM data observed near the center region of the field of view (source elevation in satellite coordinates less than 26°) are plotted. The scale for the ASM data is adjusted to be 4000 times of that of the LAC data. The solid line shows an exponential decay with a time constant of 36 days.

the 1–20 keV ASM fluxes obtained during scanning observations as a function of JD−2447647.5. The ASM data observed at the edge of the field of view (where the source elevation in satellite coordinates exceeds 26°) are not shown in this figure. The ASM data is adjusted to be 4000 times that of the LAC to account for the differences in effective area. An expanded light curve during the early observations which include all the ASM data is shown in Figure 2. During a normal scanning observation, each ASM detector observes the target for only 3.3 s; the total time on target with six detectors ranges from several seconds to several tens of seconds. Therefore, the ASM data shown in Figures 1 and 2 represent 'snap-shots' of this source. On the other hand, the LAC observation usually extends for 10–30 min. Unfortunately, between 11 May and 21 May the Cygnus region was not observe by ASM because Earth occulted the Cygnus region during satellite rotations. On 11 May the upper limit of the flux in the energy range 1–6 keV was 50 mCrab (1.0×10^{-9} erg s^{-1} cm^{-2}).

The light curves in Figure 1 and 2 suggest that there are two phases to this outburst. The first phase, before 4 June (JD 2447682), shows large intensity variations, with variations ranging from 21 Crab (2–10 keV) on May 30 and a local minimum of

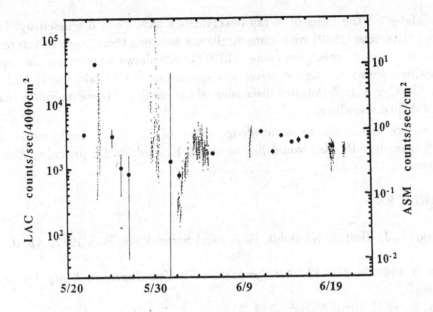

Fig. 2—An expanded light curves from 20 May to 24 July. The LAC data (*small points*) are the same as in Figure 1. For ASM data (*filled circles*), both low efficiency data and the data shown in Figure 1 are also plotted.

about several tens mCrab around 1 June. Before 1 June, the data coverage is sparse, but our data suggests that there were several fluctuations of more than one order of magnitude in brightness. The second phase shows a much more gradual decline after recovery on 4 June from the minimum on 1 June and continuing to the most recent observations on 1 November. The decline with time in brightness during this phase can be represented by an exponential decay with a time constant of about 36 days as shown in Figure 1.

3 DISCUSSION

We report here the X-ray light curve obtained with the LAC and the ASM experiments on *GINGA*. The violent intensity variation in the first phase suggests unstable mass accretion or inhomogeneities in the circumstellar matter. After 4 June, the large-scale variations cease and the light curve is roughly represented by an exponential decay with a time constant of about 36 days, which is similar to GS2000+25 (Tsunemi *et al.* 1989*b*) and A0620-00 (Kaluzienski *et al.* 1977). A0620-00 was identified as a binary system composed of a black hole and a late-type star (McClintock and Remillard 1986). The similarity of the decay time suggests that GS2023+338 might be similar to A0620-00 and GS2000+25.

The uniqueness of this source as a transient X-ray nova has been pointed out by Kitamoto *et al.* (1989). In addition, a comparison between this source and black hole candidates on the basis of X-ray observations have been discussed by Tanaka (1989) and Kitamoto (1989) with some similarity between them, such as the roughly power-law type energy spectrum (note: GS2023+338 shows a more complex spectra and sometimes shows a roughly power-law spectrum; see Tanaka 1989) and rapid flickering like Cyg X-1. A detailed discussion of the results of these on-going analyses will be published elsewhere.

The author would like to acknowledge many of the Ginga Team members for valuable comments. He also would like to thank K. Yoshida for providing the LAC light curves.

REFERENCES

Kaluzienski, L. J., Holt, S. S., Boldt, E. A., and Serlemitsos, P. J. 1977, *Ap. J.*, **212**, 203.

Kitamoto, S. 1989, in *23rd ESLAB Symposium On Two Topics in X-ray Astronomy*, in press.

Kitamoto, S., *et al.* 1989, *Nature*, **342**, 518.

McClintock, J. E., and Remillard, R. A. 1986, *Ap. J.*, **308**, 110.

Makino, F., and the Astro-C Team. 1987, *Ap. Letters Comm.*, **25**, 223.

Makino, F., and the Ginga Team. 1989, *IAU Circ.*, No. 4782.

Tanaka, Y. 1989, in *23rd ESLAB Symposium On Two Topics in X-ray Astronomy*, in press.

Tsunemi, H., *et al.* 1989a, *Pub. Astr. Soc. Japan*, **41**, 391.

Tsunemi, H., Kitamoto, S., Okamura, S., and Roussell-Dupré, D. 1989b, *Ap. J. (Letters)*, **337**, L81.

Turner, M. J. L., *et al.* 1989, *Pub. Astr. Soc. Japan*, **41**, 345.

Wagner, R. M., Starrfield, S. G., and Cassatella, A. 1989, *IAU Circ.*, No. 4783.

Radio Observations of V404 Cyg (=GS2023+338)

Xiaohong Han[1,2] and R. M. Hjellming[1]

[1]National Radio Astronomy Observatory, Socorro, NM 87801
[2]Beijing Observatory, Beijing, P. R. C.

1 INTRODUCTION

We report VLA radio observations of V404 Cyg. Initially discovered as the bright X-ray "nova" GS2023+338 by the *Ginga* Team on May 21, the radio source (Hjellming and Han 1989) is coincident with the X-ray source (Makino 1989) and is at the optical position of nova V404 Cygni 1938 (Wagner *et al.* 1989).

We made VLA observations of V404 Cyg during 22 epochs between 1989 May 30 and September 1. A strong synchrotron-radiation source was initially observed in optically thin decay, followed by a decay of $\approx (t - t_0)^{-0.83}$, going from 0.15 to 0.01 Jy in a hundred days. Evolution between flat and inverted spectral states occurs on timescales of hours. Most observations showed significant variations on timescales of minutes. One of the few continuous observations, made at 14.9 GHz, shows sinusoidal variations with a period of ~ 20 minutes. A neutral hydrogen absorption experiment performed on 1989 June 3, when the source was 0.072 Jy, determined an H I column density of $\approx 5 \times 10^{21}$ atoms cm^{-2} and a distance ≥ 3 kpc.

2 OBSERVATIONAL RESULTS

The radio light curves for V404 Cyg, obtained from daily averages of 1.49, 4.9, 8.4, and 14.9 GHz data, are shown in Figure 1 in the form of flux density plotted as a function of JD − 2440000.5. These VLA data cover the roughly 100-day period from 1989 May 30 to September 1. The radio light curve shows a "dip" and secondary peak at all frequencies. The initial two days of observations on May 30 and June 1 sample the end of a transient, very short timescale synchrotron-radiation "bubble" event, of the type seen in the X-ray transients A0620-00, Cen X-4, and GS2000+25 (Hjellming *et al.* 1988), with the characteristic spectrum and time decay given by $S_\nu \approx 5900 \cdot \nu_{GHz}^{-0.45} \cdot (JD - 2447654.5)^{-2\gamma}$, with the relation between the spectral index $\alpha = -0.45$ and $\gamma = -2\alpha + 1 \ (= 1.9)$, as expected for such an event. The rest of the light curve is dominated by a slowly decaying non-thermal component, which, at 4.9 GHz, can be fitted by $(t - t_0)^{-0.83}$, as shown in the log-log plot in Figure 2.

Radio spectra from the daily averages at 1.49, 4.9, 8.4, and 14.9 GHz are plotted for the indicated days of observations in Figure 3. The May 30 and June 1 spectra are

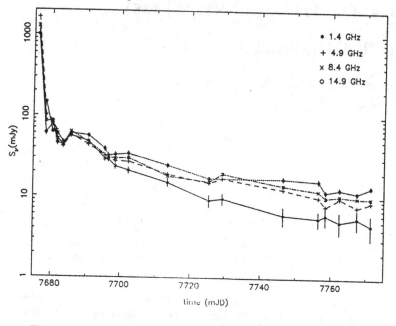

Fig. 1—Radio light curves of V404 Cyg at four frequencies for the period between 1989 May 30 and September 1.

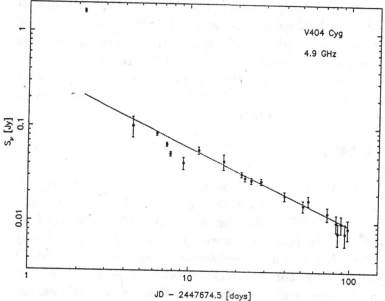

Fig. 2—Radio light curve of V404 Cyg at 4.9 GHz plotted in log-log form, together with a power-law decay curve.

characteristic of an optically thin synchrotron radiation source, mainly reflecting late sampling of a strong, transient synchrotron bubble event as seen in X-ray transients and commonly occurring in the strong outbursts of Cyg X-3. The spectra then vary from flat to "inverted," with a strong overall tendency to become more and more inverted during the later part of the 100 days of observations. However, it is difficult to separate an overall trend of evolution from a sparse and uneven sampling of the flat-inverted-flat evolution seen over a few hours on 1989 June 8.

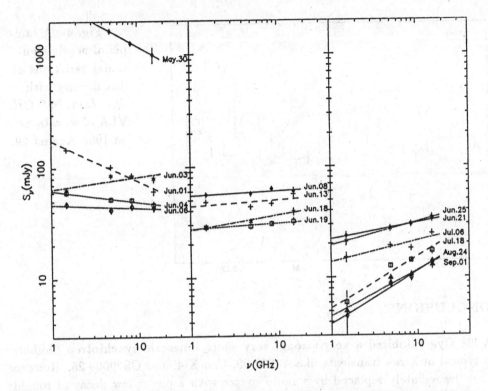

Fig. 3—Radio spectral evolution of V404 Cyg from 1989 May 30 to Sept. 1.

On 1989 June 3, when the V404 Cyg radio source was 0.072 Jy at 1.42 GHz, the 21-cm hydrogen absorption line spectrum of V404 Cyg was measured with the VLA. Two other sources in the field were an unidentified background source 4′ north of V404 Cyg, and the BL Lac-like radio source 2023+336. For all three spectra, a feature at +40 km s^{-1} is consistent with complete absorption, so one can only say that $N_H \geq (5.1 \pm 0.8) \times 10^{21} (T_s/100)$ atoms cm^{-2} for V404 Cyg, where T_s is the H I spin temperature. Since the background sources are almost certainly extra-galactic, one can say that there is no significant H I absorption beyond V404 Cyg, except for a very weak feature at +10 km s^{-1} which is present in both background sources but seems to be absent in V404 Cyg. Conventional interpretation of the N_H/T_s for V404 Cyg would imply a distance ≥ 3.2 kpc.

When it was realized that the V404 Cyg radio source was varying significantly on timescales of minutes in addition to the obvious timescales of hours, days, weeks, etc., continuous sampling at one frequency over time intervals of 60–90 minutes was carried out wherever possible. The flux density at 14.9 GHz on Aug. 29 is plotted as a function of time in Figure 4. The variations appear nearly periodic with a period ~ 20 minutes.

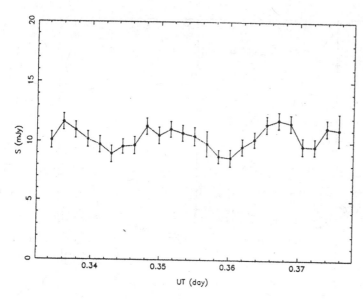

Fig. 4—A sample of nearly sinusoidal variations of flux density with time from 14.9 GHz VLA observations on 1989 August 29.

3 CONCLUSIONS

V404 Cyg exhibited a very strong, very short timescale synchrotron "bubble" event typical of X-ray transients like A0620-00, Cen X-4, and GS2000+25. However, this was immediately replaced by a radio source with a power-law decay of roughly $\sim (t-t_0)^{-0.83}$, during which variations on timescales of hours and minutes were almost always observed. Continuous observations over periods of 60–90 minutes in July and August almost always showed variations on timescales of the sampling window, but, in addition, some cases show quasi-periodic or periodic variations with timescales between several minutes and twenty minutes — near the 10 minute timescale reported for a periodicity in the optical light variations (Wagner et al. 1989).

V404 Cyg behaves somewhat like the radio emission in the X-ray binary Cyg X-3 in that it has continuous variations on timescales of hours and changes of radio spectra. However, the initial outburst event in V404 Cyg was very brief and was followed by a long, slow decay, during which it exhibited significant short timescale variations on timescales of minutes, and occasional sinusoidal variations; nothing like this has yet been found in the radio emission from any other X-ray binary.

The National Radio Astronomy observatory is operated by Associated Universities, Inc. under a cooperative agreement with the National Science Foundation.

REFERENCES

Hjellming, R. M. and Han, X. 1989, *IAU Circ.*, No. 4790.
Hjellming, R. M., Calovini, T. A., and Han, X. 1988, *Ap. J. (Letters)*, **335**, L75.
Makino, F. 1989, *IAU Circ.*, No. 4782.
Wagner, R. M., Starrfield, S. G., and Cassatella, A. 1989, *IAU Circ.*, No. 4783.
Wagner, R. M, *et al.* 1989, *IAU Circ.*, No. 4797.

Optical Observations of the X-ray Nova V404 Cygni

R. M. Wagner[1], S. G. Starrfield[2], T. J. Kreidl[3], S. B. Howell[4], A. Cassatella[5], G. W. Collins[1], R. H. Kaitchuck[1], R. Bertram[1], and R. Fried[6]

[1]Department of Astronomy, Ohio State University
[2]Department of Physics and Astronomy, Arizona State University
[3]Lowell Observatory
[4]Planetary Science Institute
[5]ESA *IUE* Observatory, VILSPA
[6]Braeside Observatory

1 INTRODUCTION

On May 22, 1989 the Japanese *Ginga* Team discovered a new bright transient X-ray source that was cataloged as GS 2023+23 (Makino *et al.* 1989). Upon their announcement, there was an immediate attempt to identify the optical counterpart of the X-ray source. Unfortunately, as was discovered later, the position of the object actually lay outside the initial X-ray error box. However, B. Marsden noticed that there was a previously-known nova listed in Duerbeck (1987) and cataloged as V404 Cygni that lay close to the boundary of the error box. Its last recorded outburst was in 1938 when it rose to about 12th magnitude. Subsequent photometric observations with the FES on the *IUE* satellite at VILSPA and spectroscopic observations with the Perkins 1.8-m telescope and Ohio State CCD spectrograph at Lowell Observatory by Wagner, Starrfield, and Cassatella (1989) confirmed that this was indeed the object.

After the original X-ray discovery by *GINGA* and the optical spectroscopic and photometric confirmation that V404 Cyg was the object in outburst, it was monitored for the next few months from γ-ray to radio wavelengths. In this contribution, we summarize some of our results of optical photometric and spectroscopic observations of V404 Cyg obtained from late May through early November 1989.

2 THE LIGHT CURVE

The optical light curve of V404 Cygni in the *V*-band obtained during the 1989 outburst is shown in Figure 1, where we have plotted the optical flux in mJy as a function of Julian Day. These data were obtained from the following sources: visual and photographic magnitudes reported on the *IAU Circulars* and to the AAVSO (open symbols), photoelectric measurements obtained at Braeside Observatory (before JD2447694), and CCD measurements reported on the *IAU Circulars* and those obtained at the Perkins Telescope and at Kitt Peak (filled symbols).

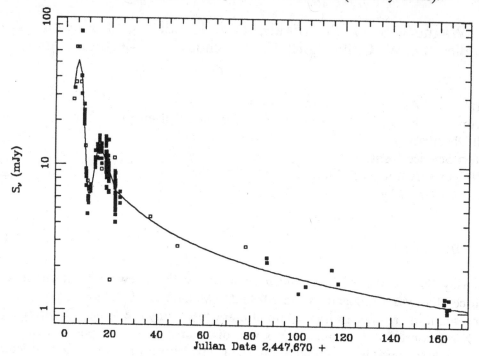

Fig. 1–Optical light curve of V404 Cygni during the outburst of 1989 from late May through early November. Photoelectric and CCD measurements are plotted as filled squares, whereas visual and photographic measurements are plotted as open squares. Note the strong flickering on short timescales and the power-law decline after JD47695.

The light curve in Figure 1 suggests that we can identify two phases of this outburst. The first phase consisted of a rise to a maximum brightness of 11.6 mag on May 30 (JD47677), the steep decline to a local minimum at 14.5 mag on June 2 (JD47680), a subsequent brightening and plateau on June 7 (JD47685), and a moderately-steep decline through mid-June (JD47694). The minimum on June 2 was first noticed by Jones and Carter (1989). The second phase refers to the much more gradual decline after mid-June and continuing to the most recent observations in early November. The decline with time in brightness during this phase can be represented by a power law proportional to $t^{-0.89 \pm 0.08}$. The slow decline observed in the radio (Han and Hjellming 1989) is also consistent with this power law within the uncertainties. Also, as can be seen in Figure 1, strong flickering on short timescales is superposed on these longer-term variations.

3 SPECTROSCOPY

The first spectroscopic data obtained on May 27.4 (Wagner, Starrfield, and Cassatella 1989) and subsequent spectra on the following nights (Wagner and Starrfield 1989; Charles *et al.* 1989) showed a reddened continuum superposed with strong emission lines and interstellar absorption features (Fig. 2). Emission lines present in the spectrum included both the Balmer and Paschen series of hydrogen and the neutral helium lines at 4471 Å, 5876 Å, 6678 Å, and 7065 Å. In addition, He II 4686 Å was comparable in strength to Hβ. Charles *et al.* also reported that the He I 3888 Å, 4471 Å, and 5876 Å showed weak P Cygni absorption, suggesting the presence of an expanding shell surrounding the source. Weak emission lines of Fe II and O I were also present. All of these emission lines have remained visible throughout the optical decline, although the Fe II and O I lines seemed to have decreased considerably in strength since early in the outburst.

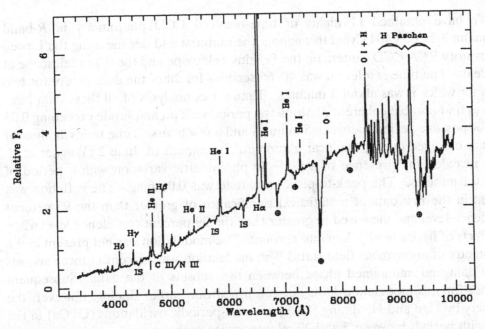

Fig. 2–Optical spectrum of V404 Cyg obtained on May 31.4. The spectrum is dominated by strong lines of H, He I, and He II early in the outburst.

Line profiles of Hγ, Hδ, and He I 4471 Å obtained on May 28.4 at 2 Å resolution revealed that these emission lines were single-peaked with a FWHM of about 800 km s^{-1}. Charles *et al.* reported that on June 1.2 Hα had a FWHM of about 550 km s^{-1} and that He II 4686 Å sometimes showed a double-peaked structure. On June 11, our moderate-resolution spectra revealed that both He II and Hβ were double-peaked. Thus there was considerable evolution of the line profiles early in the outburst. Time-resolved moderate-resolution spectrophotometry on May 31.4 indicated that changes in the line profiles of He II and Hβ were visible on timescales of about 5 minutes. This places an upper limit of about 9×10^{12} cm on the dimension of the region responsible of the line emission. The blackbody limit applied to the peak Hα flux places a lower limit of about 2×10^{11} cm for the size of the emission region.

We have obtained spectra on 11 nights suitable for radial velocity analyses. Unfortunately, no coherent periodic radial velocity variation is apparent in the data with an amplitude exceeding 30–40 km s^{-1} for possible periods between 30 minutes and 8 hours. This analysis was however very preliminary and a more detailed one is in progress.

4 TIME-RESOLVED CCD PHOTOMETRY

We have obtained 11 nights of time-resolved CCD photometry in R-band continuum light of V404 Cygni throughout the outburst and decline using the Lowell Observatory RCA CCD camera on the Perkins Telescope and the 0.9-m telescope at Kitt Peak. The time resolution was 40–66 seconds for all of the data except for two nights for which it was about 3 minutes. Time-series analyses of all these data have failed to find a stable coherent photometric period with an amplitude exceeding 0.05 mag for possible periods between 5 minutes and a few hours. Time series analysis of the data obtained during the local photometric minimum on June 2 (Wagner *et al.* 1989) revealed the presence of a periodic photometric variation with a period of 10.0 ± 0.1 minutes. The peak-to-peak amplitude was 0.06 mag. The variation was present in the raw data at a statistical significance of greater than the 95-percent confidence level and increased to greater than the 99-percent confidence level when the effects of flaring were taken into account. The modulation was not present in the photometry of anonymous field stars. The modulation was present in three subsets of the data and maintained phase between two subsets of the data. Subsequent searches for this modulation have not confirmed our earlier finding; however, the discovery by Han and Hjellming (1989) of quasi-periodic oscillations (QPOs) in the radio with periods between 5 and 20 minutes suggests that a similar optical periodic modulation could have been present for a short time but may have been difficult to detect. The optical minimum on June 2 may also have increased the probability of finding such a periodic modulation.

5 DISCUSSION

The 1989 outburst of V404 Cyg has turned out to be the most unique of all of the well-studied X-ray novae. Neither its optical spectral development nor its radio development resemble any other such outburst. For example, the presence of strong emission lines at maximum was unlike that of two other X-ray novae: A0620-00 and GS 2000+25. However, the spectra of V404 Cyg resembled those of A0620-00 obtained some two to three weeks after maximum (Whelan *et al.* 1977) and those of the recurrent nova U Sco near maximum (Barlow *et al.* 1981), although that object was not seen as an X-ray source at either its 1979 or 1987 outbursts. The behavior of V404 Cyg at X-ray and radio wavelengths is similar in some respects to Cyg X-1, Cyg X-3, and SS433.

V404 Cyg is almost certainly a low mass X-ray binary system consisting of a secondary filling its Roche lobe and a compact object, either a neutron star or black hole. The X-ray nova outburst must be associated with a period of increased mass transfer onto the compact object or the result of an accretion disk instability. The detection of QPOs in the radio and the hard X-ray emission argues for a neutron star, although the observed rapid X-ray variability is more characteristic of the likely black hole in Cyg X-1. There are spectroscopic indications that the secondary star may be evolved. Detailed comparisons of the wealth of multiwavelength data obtained during the 1989 outburst of V404 Cyg will be required to distinguish between these cases.

REFERENCES

Barlow, M. J., *et al.* 1981, *M. N. R. A. S.*, **195**, 61.

Charles, P., *et al.* 1989, *IAU Circ.*, No. 4794.

Duerbeck, H. W. 1987, *A Reference Catalogue and Atlas of Galactic Novae* (Berlin: Springer).

Han, X. H., and Hjellming, R. M. 1989, *IAU Circ.*, No. 4879.

Jones, D., and Carter, D. 1989, *IAU Circ.*, No. 4794.

Makino, F., *et al.* 1989, *IAU Circ.*, No. 4782.

Wagner, R. M., and Starrfield, S. G. 1989, *IAU Circ.*, No. 4786.

Wagner, R. M., Starrfield, S. G., and Cassatella, A. 1989, *IAU Circ.*, No. 4783.

Wagner, R. M., Kreidl, T. J., Howell, S. B., Collins, G. W., and Starrfield, S. G. 1989, *IAU Circ.*, No. 4797.

Whelan, J. A. J., *et al.* 1977, *M. N. R. A. S.*, **180**, 657.

Light and Color Variations of 1556-605

Paul C. Schmidtke

Department of Physics and Astronomy, Arizona State University

1 INTRODUCTION

The optical counterpart of the low mass X-ray binary (LMXRB) 1556-605 was identified by Charles *et al.* (1979) with a UV-excess star, showing He II emission on a flat continuum. Compared to other LMXRBs, however, this source has been little studied due to its optical faintness ($V \approx 19$) and lack of distinguishing X-ray phenomena such as bursts or dips. Motch *et al.* (1985) presented an initial report on CCD photometry from May 1984 that revealed nightly variations of 0.2–0.5 mag in V. A complete synopsis of these observations, as well as additional photometry from April 1985, was published in conjunction with an analysis of *EXOSAT* monitoring (Motch *et al.* 1989, hereafter MOT89). Although both energy regimes showed intensity changes on a timescale of hours, neither the optical nor X-ray data were modulated on a recognizable orbital period. To better define the optical variations, we initiated an independent program of multicolor photometry. Details regarding these new data were given by Schmidtke (1990, hereafter SCH90).

2 OBSERVATIONS

Photometric observations of 1556-605 in $BVRI$ bandpasses were obtained on five nights in June 1988 at Cerro Tololo Inter-American Observatory. Due to poor weather conditions, only a few observations were taken on the first two nights, while the estimated errors on the fourth night were larger than those normally found in differential CCD photometry. The data were reduced using DAOPHOT (Stetson 1987), following the guidelines of Schmidtke (1988). The reduced photometry is tabulated in SCH90 and is shown here in Figure 1. The V filter light curve ranges between 18.75 and 19.19 mag and shows nightly excursions of 0.3 mag, similar to MOT89. Small variations may be present in $(V-R)$, but the $(B-V)$ and $(V-I)$ data appear constant within the estimated errors. Therefore, the variations seen in $(V-R)$ are most likely only statistical. The weighted-mean values of the colors for 1556-605 are $(B-V) = +0.513 \pm 0.046$, $(V-R) = +0.356 \pm 0.033$, and $(V-I) = +0.742 \pm 0.026$.

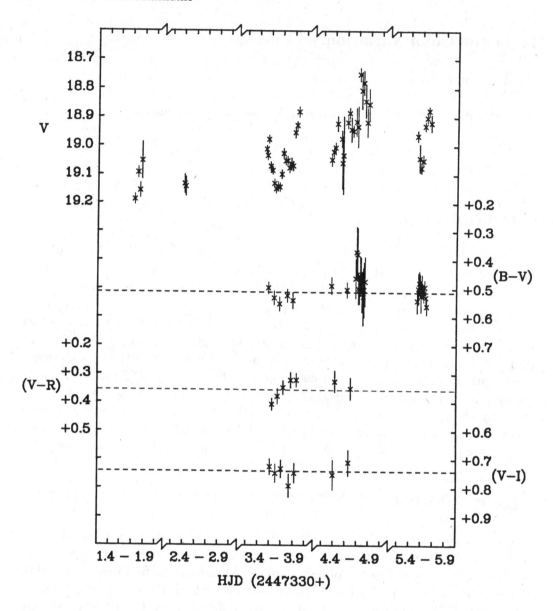

Fig. 1—Light and color curves of the LMXRB 1556-605 in June 1988, using data from Schmidtke (1990). The dashed lines indicate the weighted-mean values for the system's colors: $(B - V) = +0.513$, $(V - R) = +0.356$, and $(V - I) = +0.742$.

3 PERIOD SEARCHES

The V filter photometry was searched for periodicities using both the periodogram technique (Horne and Baliunas 1986) and the least-scatter method (Morbey 1978). The initial search failed to identify a meaningful short period (i.e., in the range $P = 4$–24 hours) for 1556-605. Subsequently, the data were examined after subtraction of nightly mean magnitudes to remove gross changes in the system's brightness that might conceal low-level orbital modulation. Again, no definitive period was identified. Prominent peaks in the power spectrum of the 1988 data are not found in data sets from earlier epochs. Thus, the system's light is dominated by random flickering, and a coherent short-term modulation, if it exists, remains unidentified.

SCH90 suggested that 1556-605 may have a very long period, making this system similar to the LMXRBs LMC X-2 ($P = 8.96$ days; Crampton *et al.* 1989) and 0921-630 ($P = 8.99$ days; Chevalier and Ilovaisky 1982). The combined photometry of MOT89 and SCH90 are consistent with values near $P = 20$ days for 1556-605, but the data set is severely undersampled for these periods to be conclusive. Additional observations are needed to identify the underlying orbital period.

We gratefully thank the CTIO staff for their assistance and acknowledge the support of the National Science Foundation.

REFERENCES

Charles, P. A., Thorstensen, J. R., Bowyer, S., Griffiths, R. E., Grindlay, J. E., and Schwartz, D. A. 1979, *Bull. AAS*, **11**, 720.

Chevalier, C., and Ilovaisky, S. A. 1982, *Astr. Ap.*, **112**, 68.

Crampton, D., Cowley, A. P., Hutchings, J. B., Schmidtke, P. C., and Thompson, I. B. 1989, *Ap. J.*, submitted.

Horne, J. H., and Baliunas, S. L. 1986, *Ap. J.*, **302**, 757.

Morbey, C. L. 1978, *Pub. Dom. Ap. Obs.*, **15**, 105.

Motch, C., Chevalier, C., Ilovaisky, S. A., and Pakull, M. W. 1985, *Space Sci. Rev.*, **40**, 239.

Motch, C., Pakull, M. W., Mouchet, M., and Beuermann, K. 1989, *Astr. Ap.*, **219**, 148, MOT89.

Schmidtke, P. C. 1990, *Pub. Astr. Soc. Pac.*, **102**, in press, SCH90.

Stetson, P. B. 1987, *Pub. Astr. Soc. Pac.*, **99**, 191.

MS 1603.6+2600: An Unusual X-ray Selected Binary System at High Galactic Latitude

Simon L. Morris[1], James Liebert[1], Isabella Gioia[2], Tommaso .
Maccacaro[2], Rudy Schild[2], and John Stocke[3]

[1]Carnegie Observatories
[2]Harvard-Smithsonian Center for Astrophysics
[3]CASA, University of Colorado

1 INTRODUCTION

MS1603.6+2600 is an X-ray source in the *Einstein* Observatory Extended Medium Sensitivity Survey (EMSS, Gioia *et al.* 1989). This survey contains the 835 serendipitous sources with statistical significance $\geq 4\sigma$ found in *Einstein* IPC images with $b(II) > 20°$. The EMSS is thus a flux-limited and homogeneous sample of high Galactic latitude X-ray sources. An optical identification program is approximately 90% complete, with $\sim 74\%$ of the identifications being extragalactic. Of the Galactic objects, the majority are late-type main-sequence stars with coronal emission (Fleming, Gioia, and Maccacaro 1989). Excluding MS1603.6+2600, there are currently 5 cataclysmic variables in the EMSS, including 2 AM Herculis systems (Morris *et al.* 1987).

The error circle for MS1603.6+2600 contained three candidate objects visible on the sky survey. Two of these were found to be normal stars. The third was initially classified as a possible BL Lacertae-type active galaxy, based on its optical spectrum and X-ray-to-optical flux ratio. However, VLA observations placed a very low upper limit on the radio flux from this object, giving a radio-to-optical flux ratio well below that of EMSS BL Lac objects (Stocke *et al.* 1990). Further optical spectroscopy and photometry show that this object is some kind of short period Galactic binary system with very weak emission lines, either an unusual type of cataclysmic variable (CV) or (more likely) a low mass X-ray binary (LMXRB).

2 OBSERVATIONS

2.1 X-ray Observations

MS1603.6+2600 was discovered as a serendipitous source during a 2112 second IPC exposure. 51 counts were recorded, a number insufficient to analyse either for variability or spectral shape. This corresponds to a flux of 1.14×10^{-12} ergs cm^{-2} s^{-1} in the 0.3–3.5 keV energy band.

2.2 Optical Photometry

MS1603.6+2600 was observed at the Palomar 60-inch telescope in white light (no filter) for 4 nights (May 23–26 1989). Exposure times were 5 minutes, with gaps between exposures of 1 minute for CCD readout and preparation. Long-term monitoring with the Mt. Hopkins 24-inch telescope shows MS1603.6+2600 has a median V magnitude of 19.5.

2.3 Optical Spectroscopy

Spectroscopic observations were obtained at the Palomar 200-inch on August 10 1988 at 4.78 UT. Two half hour exposures were obtained, which were enough to see the HeII and Hα emission lines near rest, and hence classify the object as a Galactic binary system. One month after the Palomar 60-inch photometry run described above, two more nights of spectroscopic data were obtained on the Palomar 200-inch (June 27–28 1989). A mean spectrum was constructed from these observations. Weak emission lines of HeII, Hα, Hβ, a blend of CIII and NIII at 4600Å, and also weak HeI and CaII absorption lines can be seen.

2.4 Radio Data

Observations of MS1603.6+2600 were obtained on Oct 24 1987 at 6 cm in 'snapshot mode' with the VLA in A/B configuration. 5 minutes of data were obtained. From this data, a 5σ upper limit of < 0.3 mJy can be placed on any source at the optical position.

3 ANALYSIS AND DISCUSSION

During the four nights of monitoring, MS1603.6+2600 gradually brightened. On the first night, clear sharp eclipses occurred. As the object brightened, the eclipses weakened, until by the fourth night they more or less disappeared. In fact, local minima can be identified for 10 eclipses during the four nights. A least squares fit to these eclipses with an assumed error of 2 minutes for each eclipse gives an ephemeris (with phase zero defined as the bottom of the eclipse) of:

$$JD = 2447670.72709(\pm 0.00075) + 0.077108(\pm 0.000029)E,$$

i.e., a period of 111.04 minutes is found. With the above ephemeris, the photometry can be phased. As the object brightens, the eclipse (initially covering phase 0.8–0.1) is seen to weaken, and a broad maximum from phase 0.1 to 0.4 appears. A sharp spike also appears near phase 0.9 which can be seen in 4 orbits.

The observations presented here require that MS1603.6+2600 be modelled as an accreting, close binary system whose primary is a compact object. The derived orbital period of just less than two hours implies that the mass donor is $\lesssim 0.2\ R_\odot$, the Roche lobe radius for a low-mass companion at this period. This size corresponds to a cool main-sequence star of very low mass, i.e., $\lesssim 0.2\ M_\odot$. The short-timescale variability suggest that the optical continuum and emission lines are emitted by an accretion disk around the compact primary star and that the erratic eclipse feature in our photometry is due to the passage of the darker companion in front of it. The photometry and spectrophotometry indicate that this eclipse is partial, since there is evidence for neither a flat bottom in the light curve nor for a change in the color of the energy distribution during minimum light. The emission lines also may not disappear completely. For these reasons we envision that the disk is most likely of the same scale as the secondary star.

Models of this system involving a white dwarf primary have a number of difficulties. First, the F_X/F_{opt} ratio for this system is at the extreme high end for CV systems (Córdova and Mason 1983), in a region inhabited almost exclusively by AM Her systems. These magnetic systems almost invariably display strong emission lines, and show polarization in their optical spectra. In MS1603.6+2600, no evidence for Zeeman splitting in the HeI or CaII absorption lines is seen. Also, in order to explain the weak Balmer lines and the HeI absorption lines, an unusual abundance must be assumed for the accretion disk, which implies that the secondary must be helium enriched. The period, and the evidence for mass transfer, require that the secondary fills its Roche lobe. Thus a non-degenerate helium-rich low-mass secondary is required.

The optical spectrum and rather large ratio of X-ray-to-optical flux both are more easily understood if the accreting primary is a neutron star rather than a white dwarf, indicating that the appropriate model for MS1603.6+2600 is that of a low mass X-ray binary (LMXRB) with a low-mass M-type companion. MS1603.6+2600 does have a ratio F_X/F_{opt} at the extreme low end of the distribution of LMXRBs from van Paradijs (1983), but, as discussed in White and Mason (1985), for example, eclipsing LMXRBs are expected to have low values of F_X/F_{opt} relative to other LMXRBs as the X-ray flux from these systems is largely the result of scattering from hot ionized gas above the plane of the accretion disk (an accretion disk corona). The direct X-rays from the primary star are absorbed by the thick disk rim. LMXRBs do show quite similar line spectra to MS1603.6+2600, generally characterized by HeII, a broad blend of features due to highly-ionized CNO species at wavelengths around 4640–4660 Å, and weak hydrogen lines. All the lines are relatively weak against the strong continuum (van Paradijs and Verbunt 1984), as observed here.

There are several similarities between this system and the globular cluster LMXRB AC 211 (Naylor *et al.* 1988). Of particular interest is the fact that the HeI absorption lines in MS1603.6+2600 are found to be systematically blueshifted relative to the HeII and Hα emission lines by ~ 300 km s^{-1} during phase 0.5–0.8 (the only phases when it could be reliably measured). This confirms models which placed the absorbing material in an outflowing wind or as part of an outflow from the L_2 point (Fabian, Guilbert, and Callanan 1988; Bailyn, Garcia, and Grindlay 1989).

If the LMXRB model is correct, the distance to MS1603.6+2600 is very uncertain but must be large. Assuming the system to be a typical LMXRB with $M_V = 1.2 \pm 1.0$ (van Paradijs 1983) puts the system at a distance of ~ 50 kpc, 45° out of the Galactic plane. Galactic reddening in this direction is small ($E_{B-V} \leq 0.06$, Burstein and Heiles 1982), and will not significantly reduce this distance estimate. Alternatively, MS1603.6+2600 could be another example of an underluminous LMXRB of the type described by Chevalier and Ilovaisky (1987). Also, if confirmed as a LMXRB, MS1603.6+2600 will be the first in the period range 1–2 hours. The absence of LMXRB systems with periods matching the CV SU UMa and AM Her systems has been commented on by several authors (e.g., White and Mason 1985), although no selection effect against such systems is known.

We would like to thank Rex Saffer for help with radial velocity calibrations. SLM is funded through NASA contract NAS5-30101. JL acknowledges with pleasure the support and hospitality of Carnegie Observatories for his sabbatical stay. JS acknowledges grants AST 8715983 and NAG 8-658. The EMSS survey at CfA is supported though NAS 8-30751 and SS 88-03-87.

REFERENCES

Bailyn, C. D., Garcia, R., and Grindlay, J. E., 1989, *Ap. J.*, **344**, 786.

Burstein, D., and Heiles, C., 1982, *A. J.*, **87**, 1165.

Chevalier, C., and Ilovaisky, S. A., 1987, *Astr. Ap.*, **172**, 167.

Córdova, F. A., and Mason, K. O., 1983, in *Accretion-Driven Stellar X-ray Sources*, ed. W. H. G. Lewin and E. P. J. van den Heuvel (Cambridge: Cambridge University Press), p. 147.

Fabian, A. C., Guilbert, P. W., and Callanan, P. J., 1987, *M. N. R. A. S.*, **225**, 29p.

Fleming, T. A., Gioia, I. M., and Maccacaro, T., 1989, *Ap. J.*, **340**, 1011.

Gioia, I. M., Maccacaro, T., Morris, S. L., Schild, R. E., Stocke, J. T., and Wolter, A. 1989, *Ap. J.*, in press.

Morris, S. L., Schmidt, G. D., Liebert, J., Stocke, J., Gioia, I. M., and Maccacaro, T. 1987, *Ap. J.*, **314**, 641.

Naylor, T., Charles, P. A., Drew, J. E., and Hassall, B. J. M., 1988, *M. N. R. A. S.*, **233**, 285.

Stocke, J. T., Morris, S. L., Gioia, I., Maccacaro, T., Schild, R. E., and Wolter, A., 1990, *Ap. J.*, in press.

van Paradijs, J., 1983, in *Accretion-Driven Stellar X-ray Sources*, ed. W. H. G. Lewin and E. P. J. van den Heuvel (Cambridge: Cambridge University Press), p. 189.

van Paradijs, J., and Verbunt, F., 1984, in *High Energy Transients in Astrophysics*, ed. S. E. Woosley (New York: Amer. Inst. of Phys.), p. 49.

White, N. E., and Mason, K. O., 1985, *Sp. Sci. Rev.*, **40**, 167.

X-ray Observations of Scorpius X-1 During a Multiwavelength Campaign

P. Hertz[1], K. S. Wood[1], J. P. Norris[1], B. A. Vaughan[2],
P. F. Michelson[2], K. Mitsuda[3], and T. Dotani[3]

[1]E. O. Hulburt Center for Space Research, Naval Research Laboratory
[2]Department of Physics, Stanford University
[3]Institute of Space and Astronautical Sciences

1 THE MULTIWAVELENGTH CAMPAIGN

On 9–11 March 1989, a multiwavelength campaign to observe Sco X-1 was conducted. An international collaboration obtained coverage in the X-ray (*GINGA*, MIR), ultraviolet (*IUE*), optical (Wise, McDonald, ESO, Dodaira), and radio (VLA, Westerbork, Australia) bands. We summarize here some early results from the *GINGA* X-ray data (see Wood *et al.* 1990).

The goals of the X-ray observation included (*i*) the study of quasi-periodic oscillations (QPOs) with high photon count rates, including short timescale fluctuations, time lags, and correlations with gross spectral properties, (*ii*) a search for millisecond X-ray pulsations using coherence recovery techniques to account for binary motion, and (*iii*) a search for signatures of non-axisymmetric, gravitational radiation induced, neutron star instability.

Ground-based and space-based observations covering the electromagnetic spectrum from radio to hard X-ray were scheduled in order to search for correlations of non-X-ray properties with the X-ray spectral mode of Sco X-1 (Fig. 1). The multiwavelength data base includes radio flux and spectral index, optical flux and color, and ultraviolet continuum and line strengths.

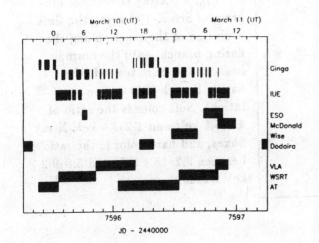

Fig. 1—Timeline for the Sco X-1 Multiwavelength Campaign showing observing windows by instrument.

2 X-RAY OBSERVATIONS

In Figure 2 we show the X-ray light curve of Sco X-1 during the *GINGA* observations. The source is on the flaring branch (FB) before approximately 03:00 March 11 UT, and on the normal branch (NB) afterwards. X-ray, radio, and ultraviolet observations show that Sco X-1 returned to the flaring branch after 07:30 March 11 UT.

In Figure 3 we show Sco X-1's temporal evolution on the color-color diagram. Before 03:00 March 11 UT, the source wanders up and down the flaring branch (circles and triangles). Afterwards, it moves out the normal branch and then returns to the FB/NB junction (stars). The last data indicate that the source is returning to the flaring branch.

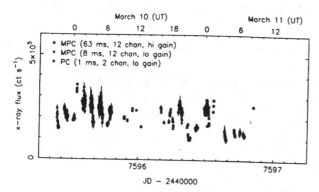

Fig. 2—X-ray light curve of Sco X-1 from *GINGA* data (1–18 keV flux) shown with 30 s time resolution. Differences in detector gain and energy channels cause slight systematic differences between observations taken with different detector modes.

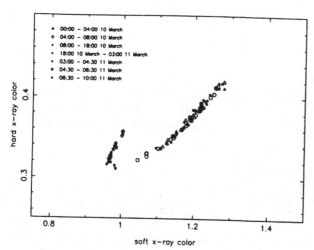

Fig. 3—X-ray color-color diagram of Sco X-1 showing the time evolution of the source along the flaring branch, onto the normal branch, and back towards the flaring branch (240 s time resolution). Soft color is the ratio of 3.5–5.8 keV and 1.2–3.5 keV X-ray fluxes, and hard color is the ratio between 9.2–18.4 keV and 5.8–9.2 keV bands.

3 QUASI-PERIODIC OSCILLATION ACTIVITY AND LOW FREQUENCY NOISE

High time resolution data (1 ms) from the observation have been searched for QPO activity over frequencies up to 128 Hz. QPO activity is weak or undetectable during the first, long interval on the flaring branch (23:00 March 9 to 03:00 March 11 UT). This interval covers all of the *GINGA* PC mode data with time resolution better than 8 ms.

Data were summed on 40 and 120 s intervals for the low (1–6 keV), high (6–16 kev), and total (1–16 keV) energy channels. No QPO were seen above 0.5% over 1–128 Hz. Data segments were then sorted by hardness ratio (6–16 keV / 1–6 keV) and summed in integrations with total accumulated times of 1056 s. No strong QPO activity was seen. A possible 5 Hz QPO was seen in this analysis mode, but it remains a marginal detection.

Low frequency noise (LFN) is seen throughout the high time resolution observation, varying from sample to sample. The LFN appears to vary systematically as the source moves along the flaring branch (Fig. 4). Near the NB/FB junction, the LFN is more pronounced. As the source moves further out the flaring branch, the LFN gets steeper and begins to resemble very low frequency noise (VLFN).

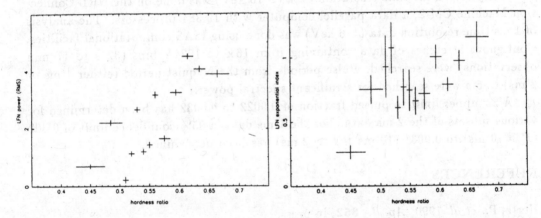

Fig. 4—The correlation of low frequency noise with hardness ratio. (*a*) The rms power in the LFN is seen to increase as Sco X-1 moves out the flaring branch and the hardness ratio increases. (*b*) The steepness of the LFN, as determined from the best fitted exponential index, also increases as Sco X-1 moves out the flaring branch.

Near the end of the observation (08:30–08:33 March 11 UT), when the source had returned to the flaring branch after time on the normal branch, a 15 ± 3 Hz QPO was detected. This QPO was not present in flaring branch data from the earlier part of the observation and was seen only after the interval spent in the normal branch (approximately 03:00–07:30 March 11 UT).

Unfiltered optical photometry at 100 Hz was obtained with the McDonald Observatory 36-inch telescope on the last day of the campaign. The 3.4 hours of optical data include data simultaneous with the 4 minutes of 8 ms X-ray photometry in which the 15 Hz QPO was seen. The 4 min X-ray observation shows that Sco X-1 had moved from the NB to the FB at this time. We have computed summed power spectra of the optical data (courtesy K. Horne and E. Robinson) during the observation. No optical QPO is present.

4 SEARCH FOR MILLISECOND X-RAY PULSATIONS

Models of QPOs and evolutionary schemes for millisecond radio pulsars require low mass X-ray binaries to have millisecond neutron star spin periods. Low magnetic fields, gravitational lensing, and magnetospheric smearing imply the pulse fraction is small. Binary motion will introduce unknown frequency modulation of pulsations. An optimal one-dimensional coherence recovery (CR) technique is used to search the phase space of possible orbits, correct data in the time domain, FFT the data, and search for a significant peak in the power spectrum (Hertz *et al.* 1990). The CR technique is equivalent to approximating the orbit of the pulsar with a quadratic curve.

Analysis of 2 ms time resolution data (1–16 keV) was done on the NRL Connection Machine CM-2, a data parallel computer with 16,384 processors. The analysis of 1 ms time resolution data (1–6 keV) was done using ISAS computational facilities. Contiguous stretches of data containing from 16 k to 1024 k bins (32 s to 17 min observations) were searched. Pulse periods from the Nyquist period (either 4 ms or 2 ms) to 5 s were searched for significant spectral power.

A 3σ upper limit to pulsed fraction of 0.0022 to 0.0032 has been determined for various subsets of the 2 ms data. For the 1 ms data, a 95% confidence limit of 0.002 ($P > 10$ ms) to 0.0031 (10 ms $> P > 2$ ms) has been determined.

REFERENCES

Hertz, P., *et al.* 1990, *Ap. J.*, **352**, in press.
Wood, K. S., *et al.* 1990, in *Proc. 23rd ESLAB Symposium on X-ray Binaries*, ed. N. E. White, in press.

EXOSAT Observations of Five Luminous Globular Cluster X-ray Sources

Arvind N. Parmar[1], Luigi Stella[1,2], and Paolo Giommi[1]

[1]EXOSAT Observatory, Astrophysics Division,
 Space Science Department of ESA, ESTEC, The Netherlands
[2]Osservatorio Astronomico di Brera, Italy

1 INTRODUCTION

Although the luminous ($\gtrsim 10^{36}$ ergs s^{-1}) X-ray sources located in globular clusters have similar properties to some of the low-mass X-ray binaries (LMXRBs) in the galactic disk (e.g., Lewin and Joss 1983) their binary nature remained unconfirmed until the discovery of an 11.4 minute modulation in the X-ray flux of X 1820-303 located in NGC 6624 (Stella, Priedhorsky, and White 1987). The stability of this quasi-sinusoidal $\leq 3\%$ (peak-to-peak) modulation confirms that it is an orbital period, the shortest of any binary system. A second globular cluster X-ray source was shown to be in a binary system following the identification of X 2127+12 (M 15) with the variable blue object AC 211 by Aurière, Le Fèvre, and Terzan (1984) and the subsequent discovery of a 9 hr optical modulation by Naylor et $al.$ (1986) and Ilovaisky et $al.$ (1986).

Models for the formation of LMXRBs indicate that systems containing main-sequence companions should also be present in globular cluster cores (Verbunt 1988; Verbunt and Meylan 1988; Bailyn and Grindlay 1987). The distribution of orbital periods of galactic LMXRBs, with a broad maximum between 5–10 hrs, indicates that this is the case in the galactic disk (Parmar and White 1988).

2 OBSERVATIONS

We report on $EXOSAT$ observations of the five luminous X-ray sources located in the globular clusters NGC 1851, NGC 6441, NGC 6712, Terzan 1, and Terzan 2. Figure 1 shows the medium energy instrument (ME; Turner, Smith, and Zimmermann 1981) 1–10 keV light curves of the five X-ray sources discussed here. The mean ME count rates range from ~ 30 counts s^{-1} (X 1746-371 in NGC 6441) to ~ 5 counts s^{-1} (X 0513-401 in NGC 1851). During the 1985 September 9 observation of X 1746-371 the background counting rate in the ME was unusually high between 18:05 and 18:15 UT and this interval is excluded from subsequent analysis. Figure 1 shows that five of the six light curves are remarkably featureless, with no evidence for any intensity dips, eclipses, or other variations. The exception is the light curve of X 1746-371 which shows significant variability as well as the two X-ray bursts

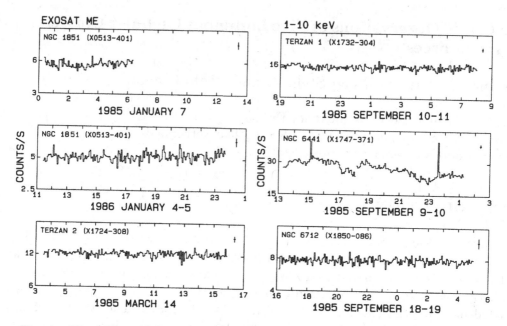

Fig. 1—The ME 1–10 keV light curves of each of the X-ray sources discussed in the text plotted with a time resolution of 240 s. Each plot has been scaled so that the count rate extrema correspond to approximately 0.5 and 1.5 times the mean value.

discussed in Stzajno *et al.* (1987).

The X 1746-371 light curve shows a gradual decrease in count rate from ~ 35 counts s^{-1} close to the start of the observation to ~ 25 counts s^{-1} at the end. Superposed on this decline are two dips in X-ray intensity centered on 1985 September 9 17:45 and 22:45 UT. A third dip may have been in progress when the observation started. Given the small number of observed dips and their irregular profiles, it is difficult to reliably estimate the dip separation. However, by estimating the earliest and latest possible ingress and egress times, we obtain a mean separation of 5.0 hrs with an uncertainty of ± 0.5 hrs. The maximum reduction in flux during the dips is $\sim 15\%$ compared to the level adjacent to the dips. The mean dip duration is ~ 90 minutes. Figure 2 shows the 1–10 keV light curve of X 1746-371 in more detail and Figure 3 the expanded light and hardness ratio curves of the dip centered on 1985 September 9 at 17:45 UT. There is no significant change in hardness ratio during the dips.

Although it is not possible to produce a reliable ME light curve of the entire 1985 April 24 observation of X 1746-371 due to the highly variable background counting rate, we have examined the light curve during intervals when the background appeared stable to see if dips are present. We find evidence for two narrow dips separated by 4.3 hr. It is possible that these narrow dips represent the same phenomenon as the broader dips seen in the 1985 September 9 observation. However, because of the unpredictable background close to these dips, we cannot be sure that the dips themselves are not the result of undetected background variations.

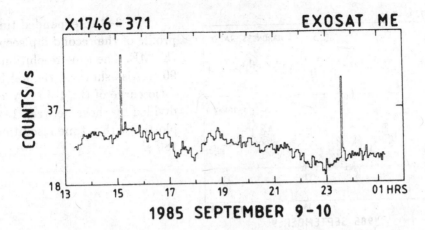

Fig. 2—The ME 1–10 keV light curve of the 1985 September 9 *EXOSAT* observation of X 1746-371 plotted with a time resolution of 180 s.

3 DISCUSSION

While we cannot rule out the possibility that the X 1746-371 dips result from changes in the intrinsic luminosity of the source, it is useful to compare their properties with the orbital variations seen from other LMXRBs (e.g., Mason 1986; Parmar and White 1988). In the case of the dip sources, the variations result from obscuration of a point-like central X-ray source by material that is not significantly ionized. The energy independence of the X 1746-371 dips rules out such a possibility here since the overall abundance of NGC 6441 is only a factor of 4–10 less than solar material (Pilachowski 1984). Even allowing for a possible factor ~ 3 abundance variations within the cluster (e.g., Cohen 1978; Sandage and Katem 1977), this is inconsistent with the abundance derived here of at least a factor 150 below that of solar material. In the case of the accretion disk corona (ADC) sources, the central X-ray source is hidden from direct view and only X-rays scattered into the line of sight in the ADC are observed. However, the X-ray bursts observed from X 1746-371 have peak luminosities consistent with the Eddington luminosity for a $1.4\,M_\odot$ object at the likely distance to NGC 6441 of 6.8–11.1 kpc (Sztajno *et al.* 1987) ruling out this possibility.

If the dips are periodic then there are at least two mechanisms which could produce energy-independent variations. The first is that the elements responsible for photoabsorption in the X-ray waveband may be completely ionized. This possibility was examined by Mason, Parmar, and White (1985) to account for the shallow energy-independent dips seen from X 1755-338. For a luminosity of 9×10^{36} ergs s^{-1}, material closer than 6×10^9 cm will be completely photoionization. However, for solar mass stars in a 5 hr binary, the minimum size of the accretion disk is 2×10^{10} cm (Lubow and Shu 1975). Thus, this possibility requires that the accretion disk be much smaller than expected, or that the obscuring material be located well inside the outer edge of the disk. Alternatively, the emission from X 1746-371 may come from two (or more) regions. If one of these regions is point-like and responsible for most of the observed

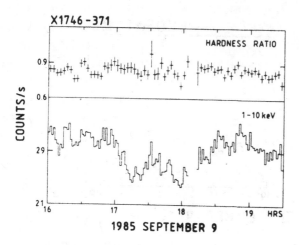

Fig. 3—An expanded time profile of the second dip seen in the ME. The time resolution is 90 s. Also shown is the hardness ratio curve of the 4–10 keV counts divided by those between 1–4 keV plotted with a time resolution of 180 s.

luminosity and the second is an extended homogeneous ADC that contributes $\gtrsim 15\%$ of the total luminosity, then periodic obscuration of part of the ADC by material in an azimuthally structured accretion disk could produce energy-independent dips.

REFERENCES

Aurière, M., Le Fèvre, O., and Terzan, A. 1984, *Astr. Ap.*, **138**, 415.

Bailyn, C., and Grindlay, J. E. 1987, *Ap. J. (Letters)*, **316**, L25.

Cohen, J. G. 1978, *Ap. J.*, **223**, 487.

Ilovaisky, *et al.* 1986, *IAU Circ. No.* 4263.

Lewin, W. H. G., and Joss, P. C. 1983, in *Accretion Driven X-ray Sources*, ed. W. H. G. Lewin, E. P. J. van den Heuvel (Cambridge: Cambridge University Press), p. 41.

Lubow, S. H. G., and Shu, F. H. 1975, *Ap. J.*, **198**, 383.

Mason, K. O. 1986, in *Lecture Notes in Physics*, Vol. **266**, *The Physics of Accretion onto Compact Objects*, ed. K. O. Mason, M. G. Watson, and N. E. White (Berlin: Springer-Verlag), p. 29.

Mason, K. O., Parmar, A. N., and White, N. E. 1985, *M. N. R. A. S.*, **216**, 1033.

Naylor, T., *et al.* 1986, *IAU Circ. No.* 4263.

Parmar, A. N., and White, N. E. 1989, in *X-ray Astronomy with EXOSAT, Mem. Soc. Astron. Italiana*, **59**, No. 1-2, 147.

Pilachowski, C. A. 1984, *Ap. J.*, **281**, 614.

Sandage, A., and Katem, B. 1977, *Ap. J.*, **512**, 62.

Stella, L., Priedhorsky, W., and White, N. E. 1987, *Ap. J. (Letters)*, **312**, L17.

Sztajno, M., *et al.* 1987, *M. N. R. A. S.*, **226**, 39.

Turner, M. J. L., Smith, A., and Zimmermann, H. U. 1981, *Space Sci. Rev.*, **30**, 513.

Verbunt, F. 1988, *Adv. Sp. Res.*, Vol. **8**, No. 2, p. 529.

Verbunt, F., and Meylan, G. 1988, *Astr. Ap.*, **203**, 297.

The Low Mass X-ray Binary in M15

P. J. Callanan[1], T. Naylor[2], and P. A. Charles[3]

[1]Department of Astrophysics, University of Oxford,
 Keble Road, Oxford OX1 3RH, U. K.
[2]Institute of Astronomy, University of Cambridge,
 Madingley Road CB3 OHA, U. K.
[3]Royal Greenwich Observatory, Apartado 321,
 38780 Santa Cruz de La Palma, Tenerife, Canary Islands

1 INTRODUCTION

Theory predicts that considerable numbers of binary systems should be formed in the cores of globular clusters, and that they are capable of significantly influencing the subsequent evolution of their host cluster (Elson, Hut, and Inagaki 1987). However, at present we have only circumstantial observational evidence for their existence. The increasing number of millisecond radio pulsars found in globular clusters, and the presence of X-ray sources apparently akin to galactic LMXRBs (Hertz and Grindlay 1983), indicates the presence of neutron star binaries, formed by tidal interactions of stars in the core (Fabian, Pringle, and Rees 1975).

The only optically identified globular cluster LMXRB is that in M15 (Aurière *et al.* 1984). Initial observations revealed low amplitude periodic radial velocity variations ($\sim$40 km s^{-1}) superimposed on a $\sim$150 km s^{-1} systemic velocity shift with respect to the cluster, initially interpreted as the true space velocity of AC211 (Naylor *et al.* 1988). Photometric variability consists of deep eclipse-like events and permitted the derivation of an accurate orbital period of 8.54 hours (Ilovaisky *et al.* 1987). A modulation was also present in the *HEAO-1* X-ray data, although at a level which was only significant in light of the known optical period (Hertz 1987). However, such a period was not observed during $\sim$ 1.5 days of subsequent *EXOSAT* observations (Callanan *et al.* 1987).

We present here a preliminary analysis of the first simultaneous spectroscopy and photometry of this system, as well as present the *GINGA* X-ray light curve of the source taken 2 months after our optical observations. More complete results, as well as optical observations taken simultaneously with the X-ray observations, will be presented elsewhere.

2 OBSERVATIONS

We performed spectroscopic observations of AC211 with the INT and IPCS simultaneously with WHT CCD photometry (with a blue-sensitive GEC chip) from

Fig. 1—The relative U magnitude of AC211 observed with the WHT, and the radial velocity of the He I line measured simultaneously. The CCD photometry has an integration time of 300 s, and the error of the photometry is $\sim \pm 0.1$ magnitudes, dominated by variable contamination from nearby stars. The spectroscopic time resolution is ~ 1 hour, with a typical error in line centroid determination of ~ 0.3 Å.

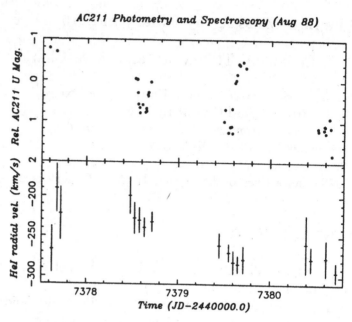

AC211 Photometry and Spectroscopy (Aug 88)

La Palma on the nights of August 4–7, 1988. The seeing was always 1 arcsec or better. The CCD photometry was performed using the stellar profile-fitting package STARMAN (Penny 1989). M15 was also observed by *GINGA* during October 20–24, 1988. The long time base of this X-ray observation was optimal for the detection of the orbital period.

3 DISCUSSION

The INT spectra indicate a gradual change in the γ velocity of AC211 as measured from the He I absorption line from night-to-night. A weak radial velocity variation ($\pm \sim 20$ km s^{-1}) on the orbital timescale is also visible in our data. These observations are in agreement with the results obtained by Illovaisky (1989), who further points out that the apparent night-to-night change in the mean γ velocity indicates that it does not represent the true space velocity of AC211: certainly such changes cannot be due to binary motion. The exact origin of the site of the He I absorption is debatable: formation in a wind leaving the disk (Fabian, Guilbert and Callanan 1987), or the disk edge (Naylor *et al.* 1988) has been proposed. A third possibility is formation in material thrown out of the orbital plane at the point of impact of the accretion stream from the secondary and the disk, in a manner similar to that observed in FO Aqr (Hellier, Mason and Cropper 1989). Why such behaviour has not been observed in any other LMXRB is unclear, but we draw attention to the spectroscopy of GX339-4 discussed by Callanan *et al.* (1990).

Fig. 2—The *GINGA*
X-ray light curve of M15,
taken 2 months after our
optical observations. The
time resolution is 100 s, and
the selected energy range is
2–12 keV. The gaps in the
data are caused by Earth
occultation due to the low
Earth orbit and passage
through the South Atlantic
Anomaly. ∼ 50% variability
can clearly be seen. The
data as plotted have not
been aspect-corrected.

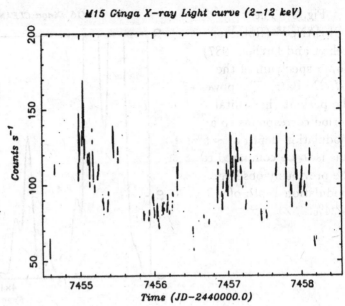

The lack of a discontinuity in the radial velocity curve at phase 0.0 (e.g., during the deep eclipse event at JD 2447379.5) does not agree with the behaviour predicted by the L_2 mass-loss model of Bailyn, Garcia, and Grindlay (1989), and thus casts serious doubts on the validity of their model.

The *GINGA* X-ray behaviour indicates that the modulation is reduced considerably in comparison to that previously observed with *HEAO-1*. If the X-ray modulation is due to the obscuration of X-rays scattered in a large accretion disk corona (ADC) then the dramatic variation of the X-ray orbital modulation depth may simply imply a considerable change in the size of the ADC. The hardness ratio plotted as a function of orbital phase does not show any obvious modulation, supporting the ADC model. If the size of the ADC is related to the overall X-ray luminosity, then the depth of modulation may be anticorrelated with the latter, and indeed we note that the data presented by Hertz (1987) are consistent with a much lower X-ray luminosity (by a factor of ∼ 10) than that observed with *EXOSAT*.

Clearly, we are just beginning to untangle the complexities of this system; however, the success of our task depends critically on future even-more prolonged campaigns at optical, UV, and X-ray energies.

The Isaac Newton Group of telescopes is operated on the Island of La Palma by the Royal Greenwich Observatory in the Spanish Observatorio del Roque de los Muchachos of the Instituto de Astrofísica de Canarias. This work is being carried out in collaboration with Tadayasu Dotani and the ISAS *Ginga* team.

54   P. J. Callanan, *et al.*

Fig. 3—The "CLEANed" (Roberts, Lehar, and Dreher 1987) power spectrum of the *GINGA* data. The power of the peak at the orbital period corresponds to a modulation depth of ~ 6%: this is to be compared to the previously-observed modulation depth of ~ 24%.

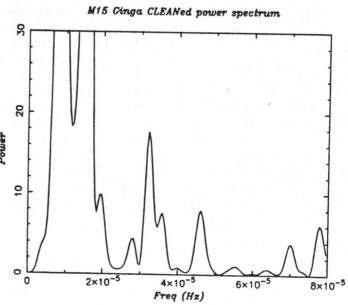

M15 Ginga CLEANed power spectrum

REFERENCES

Aurière, M., le Fèvre, O., and Terzan, A. 1984, *Astr. Ap.*, **138**, 415.

Bailyn, C. D., Garcia, M. R., and Grindlay, J. E. 1989, *Ap. J.*, **344**, 786.

Callanan, P. J., Fabian, A. C., Tennant, A. F., Redfern, R. M., and Shafer, R. A. 1987, *M. N. R. A. S.*, **224**, 781.

Callanan, P. J., *et al.* 1990, this volume.

Elson, R., Hut, P., and Inagaki, S. 1987, *Ann. Rev. Astr. Ap.*, **25**, 565.

Fabian, A. C., Pringle, J. E., and Rees, M. J. 1975, *M. N. R. A. S.*, **172**, 15p.

Fabian, A. C., Guilbert, P. W., and Callanan, P. J. 1987, *M. N. R. A. S.*, **225**, 29p.

Hellier, C., Mason K. O., and Cropper, M. 1989, *M. N. R. A. S.*, in press.

Hertz, P., and Grindlay, J. E. 1983, *Ap. J.*, **275**, 105.

Hertz, P. 1987, *Ap. J. (Letters)*, **315**, L119.

Ilovaisky, S. A., *et al.* 1987, *Astr. Ap.*, **179**, L1.

Ilovaisky, S. 1989, in *Proc. of the 23rd ESLAB Symposium (Part I)*, in press.

Margon, B., and Cannon, R. 1989, *Observatory*, **109**, 82.

Naylor, T., Charles, P. A., Drew, J. E., and Hassall, B. J. M. 1988, *M. N. R. A. S.*, **233**, 285.

Penny, A. J. 1989, private communication.

Roberts, D. H., Lehar, J., and Dreher, J. W. 1987, *A. J.*, **93**, 968.

Optical Counterparts for Globular Cluster X-ray Sources

Charles D. Bailyn

Harvard-Smithsonian Center for Astrophysics
Visiting Astronomer, Cerro Tololo Inter-American Observatory

1 INTRODUCTION

Since the confirmation of AC211 as the counterpart of the X–ray source in M15 (Aurière *et al.* 1984; Ilovaisky *et al.* 1987; Hertz 1987; Callanan, Naylor, and Charles 1990), three other optical counterparts for globular cluster X-ray sources have been suggested. Cudworth (1988) and Bailyn *et al.* (1988) independently discovered a faint UV-bright star in the X-ray error circle of NGC 6712, Grindlay (1986) reported a similar object in the error circle of NGC 5824, and Aurière *et al.* (1989) suggested a blue object near the X-ray source in 47 Tuc as the optical counterpart. Here I present a progress report on optical observations of these stars from data obtained (except where noted) July 28–31 1989 at CTIO with the 4-m PF/CCD and TI chip.

2 NGC 6712

Bailyn *et al.* (1988) and Cudworth (1988) have suggested that an apparently UV-bright star of approximately $V = 20$ (hereafter referred to as star x) may be the optical counterpart of the X-ray source, but due to the crowding at the center of the cluster they were unable to obtain good magnitudes. We have obtained U, B, and V exposures of the center of NGC 6712 in excellent seeing (1.0 arcseconds FWHM for the V and B frames and 1.2 arcseconds for the U frames), and have performed a photometric analysis using DAOPHOT (Stetson 1987) of the region near the X-ray source. The results from the DAOPHOT group including star x are given in Table 1. Due to the crowding, the magnitudes are only accurate to 0.2, except for the horizontal branch star C28, which is accurate to ≈ 0.04. Star x is clearly unusually UV bright. However, the relative centroid of this star appears to move from filter to filter (see Fig. 1), suggesting that this star is in fact a blend of two faint stars of differing colors.

We monitored star x over a two-hour period for luminosity variations. To do this we used 600 s U exposures to maximize the relative brightness of star x. However, in the several frames with seeing $> 1.3''$, DAOPHOT would not converge on star x. We therefore subtracted stars with the relative positions and magnitudes given in Table 1 for stars other than star x, and then fit a star to the position of star x. Using

TABLE 1
PHOTOMETRY OF NGC 6712

star id[a]	X_v[b]	Y_v[b]	V[c]	$B - V$[c]	$U - B$[c]
C28	99.9	178.1	16.00	1.00	+0.54
KC62	105.0	168.5	17.82	1.05	+0.35
x	104.1	172.5	19.28	1.15	−0.36
1	93.7	180.1	18.55	0.77	+0.21
2	98.4	182.7	19.26	0.73	+0.13
3	105.0	165.2	19.26	1.04	+0.03
4	103.4	161.6	19.37	1.03	−0.14
5	110.7	169.6	19.61	1.03	−0.09

[a]Star id's are from Sandage and Smith (1966) for C28 and Cudworth (1988) for KC62. Star x is the proposed X-ray counterpart.

[b]X_v and Y_v are the x and y pixel coordinates of the star centroid in the V frame. Pixel scale is 0.29″ pixel^{-1}.

[c]Magnitudes have errors of ≈ 0.2 except for C28, which has errors of ≈ 0.05.

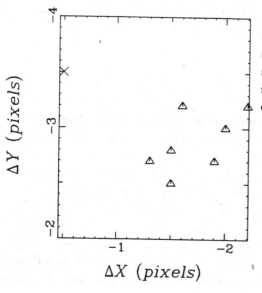

Fig. 1—Pixel offsets between the V and U frames of stars near the X-ray source. Star x (marked x) has an offset significantly different from the other stars, suggesting that it may be a blend of two stars of different colors.

this method, no significant luminosity variations are detected down to $\Delta U \approx 0.2$. H. Cohn and P. Lugger have obtained B and R frames of this object in excellent seeing (0.5″) at the CFHT which are currently being analyzed for possible duplicity and variability of star x.

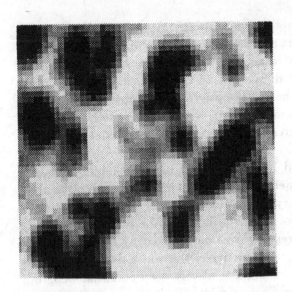

Fig. 2—U frame of the region near the X-ray source in 47 Tuc. Shown here is a 15″ square region centered on the X-ray position from the best seeing frame from our run, with FWHM = 1.15″. Comparison between this image and Fig. 7 of Aurière *et al.* (1989) clearly shows the absence of star 9 in our data. North is up and east to the left in this picture; the pixel size is 0.29″ pixel^{-1}.

3 47 TUC

As reported by Bailyn and Cool (1989), we observed 47 Tuc and found that the blue object (star 9) suggested by Aurière *et al.* (1989) as the optical counterpart has disappeared. We summed the 6 best-seeing U frames (exposed for 60 s each) to form a deep exposure with stellar images of FWHM= 1.26″. We then performed a DAOPHOT analysis of this frame, successfully recovering all the stars noted by Aurière *et al.* except star 9.

The rms variation between our U magnitudes and those of Aurière *et al.* (1989) was 0.17, substantially greater than the errors given in the DAOPHOT output. We attribute this to systematic errors in making and fitting a point spread function in this crowded field, as well as to differences due to the different filters and detectors used. We have not completed reductions of standards for our data, so we cannot yet say if the mean magnitudes for our observations agree with those of Aurière *et al.* — the rms difference quoted above was computed after forcing the mean magnitude for the 32 stars involved to agree. We note that star 31, directly east of star 9, is actually a blend of two stars. After subtracting all of these stars from the frame, we examined the residuals for traces of star 9 and found none. Experiments with adding artificial stars to the frame indicate that any star brighter than $U = 18$ (using the magnitudes determined by Aurière *et al.* as standards) at the position of star 9 would have easily been detected. Thus, star 9 has dimmed at least 1.7 magnitudes between 1986 and 1989.

4 NGC 5824

The star suggested by Grindlay (1986) as the counterpart to the X-ray source has $V = 20.6$, $B - V = 0.2$, and $U - B = -0.6$ with errors of about 0.15. The ratio of X-ray to optical flux is thus 40, almost exactly midway between the values expected for an accreting neutron star and an accreting white dwarf. We monitored this star for luminosity variations over several hours and found none. *UBV* data taken in 1987 agree with the colors and magnitudes quoted above, while data from 1985 taken with an RCA chip show marginally significant differences, in the sense that the star was fainter by 0.6 ±0.3 magnitudes then. We have attempted spectroscopy of this object; a spectrum of 2.3 hours from the CTIO 4-m 2D/Frutti shows a blue continuum with no obvious features. Weak emission features like those found by Morris *et al.* (1990) in MS1603.6+26 are not excluded.

Empirical correlations (Patterson and Raymond 1985) suggest that high mass-accretion rates in CVs will give rise to spectra with weak emission lines due to the increased optical thickness of the boundary layer. However, this would also predict a relatively low X-ray-to-optical flux ratio. If this object is to be interpreted as an accreting white dwarf, perhaps this discrepancy can be resolved by a strong magnetic field which disrupts the inner disk. Alternatively, the white dwarf might be particularly massive, and hence small, providing a relatively deep potential well. In this case this white dwarf might exceed its Chandresekhar limit at some point in its future. If the accretor is a neutron star, the low mass-accretion rate required by its low X-ray luminosity would suggest the presence of relatively strong emission lines and large X-ray-to-optical flux ratios (for LMXRBs) which are not observed. The absence of short-term luminosity variations would seem to exclude the possibility that the system is viewed side-on and the X-ray flux is obscured by an accretion disk.

I thank A. Cool, who participated in obtaining the data reported here, and my other collaborators on these projects: J. Grindlay, H. Cohn, and P. Lugger. Cerro Tololo Inter-American Observatory is operated by the Association of Universities for Research in Astronomy Inc. under contract with the National Science Foundation.

REFERENCES

Aurière, M., Le Fevre, O., and Terzan, A. 1984, *Astr. Ap.*, **128**, 415.

Aurière, M., Koch-Miramond, L., and Ortolani, S. 1989, *Astr. Ap.*, **214**, 113.

Bailyn, C. D., Grindlay, J. E., Cohn, H., and Lugger, P. M. 1988, *Ap. J.*, **331**, 301.

Bailyn, C. D., and Cool, A. 1989, *IAU Circ.*, No. 4835.

Callanan, P. J., Naylor, T., and Charles, P. A. 1990, this volume.

Cudworth, K. 1988, *A. J.*, **96**, 105.

Grindlay, J. E. 1986, in *Evolution of Galactic X-ray Binaries*, NATO ASI Series, **167**, 25.

Hertz, P. 1987, *Ap. J. (Letters)*, **315**, L119.

Ilovaisky, S. A., *et al.* 1987, *Astr. Ap.*, **179**, L1.

Morris, S. L., *et al.* 1990, this volume.

Patterson, J., and Raymond, J. C. 1985, *Ap. J.*, **292**, 535.

Sandage, A., and Smith, L. L. 1965, *Ap. J.*, **144**, 886.

Stetson, P. B. 1987, *Pub. Astr. Soc. Pac.*, **99**, 191.

The 14.8-Hour Orbital Period of GX339-4

P. J. Callanan[1], W. B. Honey[1], P. A. Charles[2], R. H. D. Corbet[3],
B. J. M. Hassall[4], K. Mukai[5], A. P. Smale[6], and J. Thorstensen[7]

[1]Department of Astrophysics, University of Oxford,
 Keble Road, Oxford OX1 3RH, U. K.
[2]Royal Greenwich Observatory, Apartado 321,
 38780 Santa Cruz de La Palma, Tenerife, Canary Islands.
[3]Institute of Space and Astronautical Science,
 1-1 Yoshinodai 3-chrome, Sagamihara-shi, Kanagawa 229, Japan
[4]Royal Greenwich Observatory, Cambridge.
[5]Mullard Space Science Laboratory, University College London,
 Holmbury St. Mary, Dorking, Surrey RH5 6NT, U. K.
[6]Laboratory for High Energy Astrophysics, NASA/Goddard Space Flight Center,
 Greenbelt, Maryland 20771, U. S. A.
[7]Department of Physics and Astronomy, Dartmouth College,
 Hanover, NH 03755, U. S. A.

1 INTRODUCTION

The X-ray transient GX339-4 has long been considered a strong black hole candidate, primarily on the basis of its short-term X-ray variability ($\tau \sim 10\text{--}100$ ms) and soft ($kT \sim 1\text{--}2$ keV) high-state X-ray spectrum (Markert *et al.* 1973; Samimi *et al.* 1979). The source has been observed in high, low and off X-ray states with corresponding X-ray intensities of 80–350, 40–200 and <2 μJy; the corresponding optical V band magnitudes are $\sim$17, 15.4, and 18.7–20 respectively (Corbet *et al.* 1987). The slope of the X-ray spectrum varies as the source moves between the low and high state, pivoting about ~ 6 keV (Ricketts 1983); the X-ray spectrum also exhibits a hard X-ray tail which extends up to 100 keV.

Although we now know that extreme variability is not in itself a unique signature of a black hole (e.g., Stella *et al.* 1985), the similarities of GX339-4 with systems like Cyg X-1 clearly make it a system warranting close study. Thus, we embarked on a campaign of spectroscopy and CCD photometry while the system was in an off state, in the hope of determining the orbital period by detecting the ellipsoidal variations of the secondary.

2 OFF-STATE OBSERVATIONS

These observations consisted of near-contiguous runs in Chile and South Africa during 1987 July 23–29. The mean R magnitude of GX339-4 during the observations was $\sim$ 19.4; the relative photometry was carried out using stellar profile-fitting so as

Fig. 1—The power spectrum of the R band data. These spectra have been generated by using a one-dimensional version of a CLEAN deconvolution algorithm (Roberts, Lehar, and Dreher 1987). We see a strong peak at a period of 0.618 days; a similar peak is present in I band data taken during the same run.

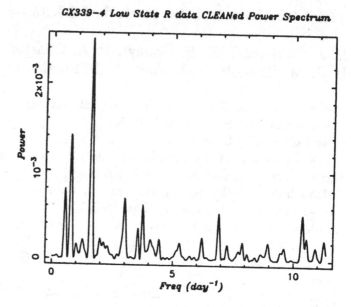

to properly deconvolve the counterpart from a nearby (~ 1.1 arcsec) contaminator. Spectroscopy was also carried out with the AAT during 1987 May 10–11 and June 18–19 in poorer (> 2 arc second) seeing.

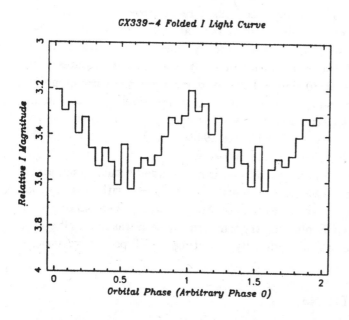

Fig. 2—The I band data folded on 0.618 days and rebinned in phase. The error in the estimate of relative magnitude is ~ 0.025 mags. The folded optical light curve does not exhibit any obvious flattening at maximum, suggesting that there may be a significant contribution from either an X-ray heated bulge or secondary at maximum light.

The size of the optical modulation can be produced by a comparatively small change in effective temperature; such a small change is consistent with the fact that we do not see any significant $R - I$ color variation. Honey *et al.* (1990) provide evidence

Fig. 3—The Hα pro-
file obtained with the AAT
during the off-state spectroscopy we performed during 1987 May. We used the
270R and 600V gratings,
giving a wavelength cov-
erage of 6100–8200 Å in
total. Only ∼ 3 spectra per
night could be obtained,
due to the faintness of the
source. The lower right-
hand panel is the sum of
the first three profiles.

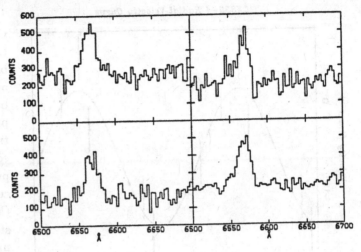

for the 0.618 day period in high-state optical photometry — further confirmation of
0.618 days as the true orbital period of GX339-4.

3 DISCUSSION

The size of the system as determined from the orbital period, combined with
the upper limit which we can place on the luminosity of the secondary from the low
state magnitude, implies that the secondary is likely to be an evolved subgiant of
mass ∼ $1M_\odot$ (Honey *et al.* 1990). Knowledge of the period also allows us to fold our
spectroscopy and that previously reported by other authors on the orbital period.
Unfortunately, orbital phase coverage obtained by us does not extend to an entire
orbital cycle, but the two high-state data sets presented by Cowley *et al.* (1987) do.

Curiously, we find that one of the Cowley *et al.* data sets has an average γ (systemic) velocity of 96 km s^{-1}, the other −49 km s^{-1}. The average velocity estimated
from each *limited* data set obtained from the AAT is 238 km s^{-1} and 92 km s^{-1}. Thus
the data do not appear consistent with a constant γ velocity for GX339-4. However,
as shown in Fig 3, the Hα profile is extremely variable, and systematic errors may well
be introduced by fitting, for example, symmetric model line profiles to asymmetric
spectral lines.

Alternatively if these variations are real, their origin may lie for example in the
formation of the lines in a variable-velocity wind being driven from the system, in a
manner analogous to the model for AC211 in M15 (Fabian, Guilbert, and Callanan
1987). This wind is associated with an accretion disk corona (ADC). We note that
evidence has recently come to light indicating that the γ velocity shift of AC211 also
changes (see, for example, Callanan, Naylor, and Charles 1990). An ADC could also

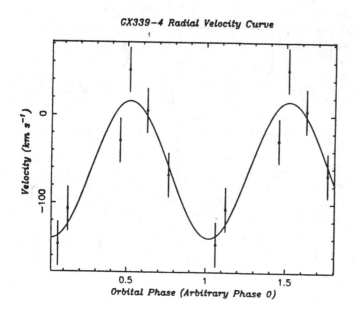

GX339-4 Radial Velocity Curve

Fig. 4—A subset of GX339-4 high-state spectroscopic data published by Cowley *et al.* folded on our orbital period with the best-fit sine-wave superimposed. Although this spectroscopy does not tightly constrain the mass of the primary, the 6 Hz QPO recently detected by *GINGA* (Makino *et al.* 1988) is strong evidence for the primary being a neutron star.

account for the hard X-ray tail seen in the X-ray spectrum.

4 CONCLUSIONS

We have determined the orbital period of GX339-4 to be 14.8 hours. We have evidence for variation in the apparent γ velocity of the system, but note that care must be taken in fitting such complex line profiles. The presence of an ADC explains many of the characteristics of this system, but a more detailed study must await a more sustained campaign of spectroscopy and photometry over several orbital cycles, particularly in the high state.

REFERENCES

Callanan P. J., Naylor, T., and Charles P. A. 1990, this volume.
Corbet, R. H. D., *et al.* 1987, *M. N. R. A. S.*, **227**, 1055.
Cowley, A. P., Crampton, D., and Hutchings, J. B. 1987, *A. J.*, **93**, 195.
Fabian, A. C., Guilbert, P. W., and Callanan, P. J. 1987, *M. N. R. A. S.*, **225**, 29p.
Honey, W. B., *et al.* 1990, *M. N. R. A. S.*, in preparation.
Makino, F., *et al.* 1988, *I. A. U. Circular*, No. 4653.
Markert, T. H., *et al.* 1973, *Ap. J. (Letters)*, **184**, L67.
Ricketts, M. J. 1983, *Astr. Ap.*, **118**, L3.
Roberts, D. H., Lehar, J., and Dreher, J. W. 1987, *A. J.*, **93**, 968.
Samimi, J., *et al.* 1979, *Nature*, **278**, 434.
Stella, L., *et al.* 1985, *Ap. J. (Letters)*, **288**, L45.

The 8-Second Quasi-Periodic Oscillations in GX 339-4

J. N. Imamura[1], T. Steiman-Cameron[2], J. Kristian[3], and
J. Middleditch[4]

[1]Dept. of Physics and Institute of Theoretical Science, University of Oregon
[2]Theoretical Studies Branch, NASA/Ames Research Center
[3]Observatories of the Carnegie Institution of Washington
[4]Computer Applications, Los Alamos National Laboratory

1 INTRODUCTION

The black hole candidate GX 339-4 exhibits a rich spectrum of temporal activity. Aperiodic, quasi-periodic, and periodic variability on timescales ranging from milliseconds to months have been reported by a variety of observers over the last several years. Such a panoply of phenomena suggests that GX 339-4 would be one of the better understood of the compact X-ray sources. This is not the case, however, because much of the reported phenomena have only been seen once, making their interpretations ambiguous and their value as probes limited. As a consequence, even some of the fundamental properties of GX 339-4 are not known. For example, whether the compact object is a black hole or a neutron star remains uncertain (Dolan *et al.* 1987). Further, it has only recently become apparent that GX 339-4 is, in fact, a binary star system (Honey *et al.* 1988; Callanan 1990). Some phenomenon which appears more than once would be very instructive. Here, we report the detection of 8-second optical quasi-periodic oscillations (QPOs). An optical feature with a similar timescale has previously been reported by Motch, Ilovaisky, and Chevalier (1985). X-ray QPOs are known to be a common feature of the neutron star low mass X-ray binary systems (van der Klis *et al.* 1985), and so this feature has the potential of settling the issue of the nature of the compact object in GX 339-4.

2 OBSERVATIONS AND ANALYSIS

High-speed white light photometry of GX 339-4 was obtained on 1989 August 1 (UT) using the 1.5-m telescope of the Cerro Tololo Inter-American Observatory with the Automatic Single Channel Aperture Photometer (ASCAP) and a cooled GaAs phototube. A 7-arcsec aperture was used and the data were recorded at a rate of 5 kHz using the "Li'l Wizard Pulsarator" data acquisition system. The visual magnitude of GX 339-4 at the beginning of our observation was $m_V = 17.7$.

Coherent and incoherent features were searched for using Fourier techniques. In particular, we searched for the 885 Hz feature proposed by Imamura, Steiman-Cameron, and Middleditch (1987) and the 5.25 mHz feature found by Steiman-Cameron *et al.* (1989). Neither were detected at rms upper limits of $\sim 1\%$. The

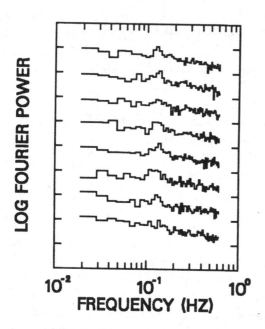

Fig. 1—Arbitrarily-scaled power spectra of successive 30-minute segments of the time-resolved photometric data for GX 339-4. Time increases from top to bottom.

only coherent features detected were at 120 Hz and aliases and harmonics of 120 Hz. The 120 Hz feature could have been caused by stray light entering the photometer or by a faulty coldbox heater. Power spectra calculated for consecutive 30-min segments of the time series data are presented in Figure 1. The time increases from top to bottom in this figure and the power spectra are arbitrarily scaled to aid in their presentation.

The parameters of the QPOs are determined by fitting functions to the power spectra for two-hour segments of the time series data. We use a fitting function composed of a constant, an exponential, and a Lorentzian profile (van der Klis *et al.* 1985). The free parameters in the fitting function are the constant power level, the amplitude and decay constant of the exponential, and the amplitude, frequency centroid, and width of the Lorentzian profile. The fits to the power spectra have reduced χ^2 on the order of 1.1 to 1.4. The results of our analysis are tabulated in Table 1. The amplitudes are root-mean-square (rms) pulsed amplitudes found by integrating the Lorentzian profile over the frequency range $f = 0$ Hz to f_{Nyquist}. Using a Gaussian profile also leads to acceptable fits. Gaussian fits yield somewhat lower rms pulsed amplitudes than do the Lorentzian fits, 3% vs. 5%, but lead to the same centroid frequencies.

3 DISCUSSION

GX 339-4 exists in distinct on and off X-ray states; each state lasting for times on the order of months. In the X-ray on-state, the X-ray luminosity is roughly 10^{37} to 10^{38} ergs s^{-1}. In the X-ray off-state, the X-ray luminosity can drop to as low as

TABLE 1
RESULTS OF ANALYSIS

Time (UT)	Total Counts	Amplitude (%)	Centroid (Hz)	FWHM (Hz)
01:35:00–03:40:50	3,615,131	4.71(.85)	0.127(.001)	0.0249(.0023)
03:50:00–05:59:00	3,460,784	5.46(1.0)	0.125(.001)	0.0405(.0037)

5×10^{34} ergs s^{-1} (Ilovaisky *et al.* 1986). In addition, GX 339-4 also shows two distinct modes of X-ray on-state behavior. It shows a soft X-ray state where its spectrum is dominated by a component which resembles a 1 keV Comptonized blackbody, and it shows a hard X-ray state where its energy spectrum is adequately fit by a power law of index 0.6. The optical luminosity of GX 339-4 varies depending on the X-ray mode. The optical luminosity ranges from $m_V = 17.7$ to $m_B > 21$ during the X-ray off-state, from $m_V = 15.4$ to 17.5 during the X-ray hard state, and from $m_V = 16.5$ to 18 during the X-ray soft state. The optical luminosity is thus not a single-valued function of the X-ray state and so the X-ray state cannot be unambiguously determined without X-ray observations. No such observations were made during our run. GX 339-4 was $m_V = 17.7$ during our observations and could have been in any of the X-ray states. We suggest that GX 339-4 was in the X-ray off-state based on similarities to the observations of Motch, Ilovaisky, and Chevalier (1985).

The QPO power spectra averaged over two-hour timescales have roughly the same properties. In the second half of the data, when GX 339-4 had a 5% lower count rate, the QPO centroid frequency was slightly lower, 0.125 Hz vs. 0.127 Hz, and the QPO bump was wider, 0.041 Hz vs. 0.025 Hz. The difference in the power spectrum of the second two hours of data is due primarily to the fact that the QPO centroid frequency moves around and its width varies, not that it is intrinsically wider (see Fig. 1). In the final 30-minute interval, the fitting routine was unable to find the feature.

A 7-second optical QPO was detected in data taken during 1982 May when GX 339-4 was $m_V = 17.7$ and in an X-ray off-state (Motch, Ilovaisky, and Chevalier 1985). Our measured visual magnitude of 17.7 and the 8-s timescale for the QPO suggests that we also observed GX 339-4 when it was in an X-ray off-state. The QPO mechanism in GX 339-4 is thus capable of generating QPOs even when the accretion ceases or is at a very low level. Further, the magnitude estimate of 17.7 corresponds to an optical brightness which is comparable to the off-state X-ray luminosity. This is unusual for a low mass X-ray binary where the optical-to-X-ray ratio is usually much less than 0.001.

GX 339-4 has also exhibited 20-s optical and X-ray QPOs (Motch, Ilovaisky, and Chevalier 1982; Motch *et al.* 1983). The optical QPO had a 30–40% full amplitude and appeared when GX 339-4 was in the hard X-ray state and was optically much brighter ($m_V = 15.4$). The optical QPO thus decreases in amplitude and increases in centroid frequency as the source luminosity decreases.

4 SUMMARY

We detected an 8-second QPO in the optical emission from GX 339-4. The feature had an rms pulsed amplitude of roughly 5% and a width of 0.02–0.04 Hz. GX 339-4 had a visual magnitude of 17.7 during our observations and we suggest that it was in its X-ray off-state based on similarities to the previously-reported 7-s optical QPOs (Motch, Ilovaisky, and Chevalier 1985). Because there were no simultaneous X-ray observations, the exact X-ray state during our observations is uncertain, however.

This research was partially supported through the AAS Small Research Grants Program by a grant to one of the authors (TSC) from the Margaret Cullinan Wray Charitable Lead Annuity Trust, and by grants from the Research Corporation and the Dudley Observatory. The authors also wish to thank the staff of CTIO for its usual outstanding assistance in obtaining these observations. Cerro Tololo Inter-American Observatory is operated by AURA, Inc. under contract with the National Science Foundation.

REFERENCES

Callanan, P. 1990, this volume.
Dolan, J. F., Crannell, C. J., Dennis, B. R., and Orwig, L. E. 1987, *Ap. J.*, **322**, 324.
Honey W. B., Charles, P. A., Thorstensen, J. R., and Corbet, R. H. D. 1988, *I.A.U. Circ.*, No. 4532.
Ilovaisky, S. A., Chevalier, C., Motch, C., and Chiapetti, L. 1986, *Astr. Ap.*, **164**, 67.
Imamura, J. N., Steiman-Cameron, T. Y., and Middleditch, J. 1987, *Ap. J. (Letters)*, **314**, L11.
Motch C., Ilovaisky, S. A., and Chevalier, C. 1982, *Astr. Ap.*, **109**, L1.
Motch C., Ricketts, M. J., Page, C. G., Ilovaisky, S. A., and Chevalier, C. 1983, *Astr. Ap.*, **119**, 171.
Motch C., Ilovaisky, S. A., and Chevalier, C. 1985, *Sp. Sci. Rev.*, **40**, 219.
Steiman-Cameron, T. Y., Imamura, J. N., Middleditch, J., and Kristian, J. 1989, *Ap. J.*, submitted.
van der Klis, M., *et al.* 1985, *Nature*, **316**, 225.

QPOs and Preferred Variability Timescales in X-ray Pulsars

Lorella Angelini[1], Luigi Stella[1,2], and Arvind N. Parmar[1]

[1]EXOSAT Observatory, Astrophysics Division, Space Science Dept. of ESA
[2]Osservatorio Astronomico di Brera, Milano, Italy

1 INTRODUCTION

In binary systems containing an X-ray pulsar, the X-ray flux shows aperiodic noise components in addition to the neutron star rotation signal. In the power spectra of their light curves, the noise variability appears in the form of continuum components underlying the discrete peaks arising from the coherent periodic signals. The study of these continuum components can add to our knowledge of accretion processes onto magnetized neutron stars; models for quasi-periodic oscillations (QPOs) which require the presence of a rotating magnetosphere interacting with an accretion disk can be tested (see, e.g., Lamb 1988). We report here on the timing analysis of the *EXOSAT* data of two X-ray pulsars, EXO 2030+375 and SMC X-1.

2 EXO 2030+275

The 42 s transient X-ray pulsar EXO 2030+375 was discovered with *EXOSAT* in 1985 and the evolution of two outbursts was followed through 14 observations. During the first outburst, the luminosity in the 1–20 keV energy band decreased by a factor ≥ 2500 from its initial value of $\sim 10^{38}$ ergs s^{-1}. Because of the large range of luminosity, it has been possible to study in unprecedented detail the dependence of the observed parameters on luminosity (e.g., Parmar *et al.* 1989; Parmar, White, and Stella 1989). Figure 1 shows the average power spectra of EXO 2030+375 from the first four observations when the source was close to its maximum observed intensity. The coherent signal associated with the neutron star rotation has been subtracted using a new technique to model in detail the corresponding power spectrum peaks (see Angelini, Stella, and Parmar 1989).

A feature around 0.2 Hz with a FWHM $\simeq 0.05$ Hz, far broader than that expected from a coherent modulation, is clearly seen in the *cleaned* power spectra obtained in this way. The rms fractional strength associated with it is $\epsilon_{rms} \simeq 3.5\%$. The peak was only detected in the first four observations, when the source luminosity was within $\sim 25\%$ of its observed maximum. Correspondingly, the centroid frequency decreased by 12%. No evidence for a dependence of the peak strength on energy was found. We interpret the 0.2 Hz peak as QPOs. During these four observations, the source was

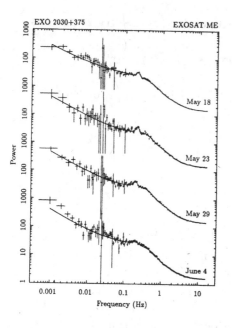

Fig. 1—*Cleaned* power spectra from the 1–15 keV light curve of the four observations of EXO 2030+375 during which the source was brightest and ~ 0.2 Hz QPOs were detected. The solid lines represent the best-fits obtained by using a Gaussian to model the QPO peak in addition to the other continuum components. The residuals of the coherent peaks have large errors bars because the power associated with the peaks and the corresponding errors were very high.

rapidly spinning up, indicating the presence of an accretion disk. In the Keplerian frequency model (KFM), QPOs are due to the orbital motion of matter orbiting close to the magnetospheric radius, r_m, between the accretion disk and the neutron star magnetosphere and $\nu_{\mathrm{KFM}}(r_m) = 1.05 \mu_{30}^{-6/7} L_{37}^{3/7}$ Hz. By using the best-fit value for the luminosity and the magnetic dipole moment derived by Parmar *et al.* (1989), QPO frequencies in the range 0.19–0.22 Hz are predicted for the four observations in which QPOs were detected. In the beat frequency model (BFM), QPOs arise from the modulation of the accretion onto the neutron star at the beat frequency between $\nu_k(r_m)$ and the spin frequency ν_s of the neutron star, i.e., $\nu_{\mathrm{BFM}} = \nu_k(r_m) - \nu_s$. Close to the maximum of the outburst of EXO 2030+375, $\nu_s \ll \nu_k(r_m)$ and the QPO frequencies predicted by the BFM are similar to those calculated for the KFM. For both models the observed QPO frequency and frequency-luminosity dependence are in remarkable agreement with the measured values, especially in view of the fact that there are no free parameters to adjust. However, the KFM has difficulties in accounting for the QPO rms strength of ~ 3.5% because the gravitational energy released up to r_m is at most ~ 0.2%. In the BFM, instead, the interaction between the magnetosphere and the inner disk modulates the accretion rate at the QPO frequency and a large fraction of the gravitational energy released at the neutron star surface is available to generate the QPO signal.

3 SMC X-1

The 0.72 s X-ray pulsar SMC X-1 is in a 3.9-day eclipsing binary system with a

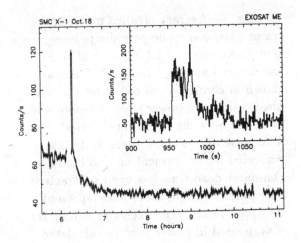

Fig. 2—SMC X-1's background-subtracted light curve (1–14 keV) in 80 s bins of the first observation when the burst-like event was detected (18 Oct 1984). The insert shows the burst-like event with a time resolution of 0.711 s, corresponding to the pulsar period. Seconds are counted from 6:00 UT.

B0 I supergiant companion. The pulse period shows a monotonic spin-up, indicating that an accretion disk mediates the flow. The X-ray luminosity varies from $\leq 10^{37}$ up to 10^{39} ergs s^{-1}, making SMC X-1 one of the brightest X-ray pulsars known. SMC X-1 was observed five times in a period ~ 20 days with *EXOSAT*, always at an orbital phase $\phi \simeq 0.5$ (Bonnet-Bidaud and van der Klis 1981). Figure 2 shows the light curve of the first observation. A burst-like event was detected that lasted ~ 80 s, with a rise time less than 0.7 s (i.e., less than a pulsar period), and a peak count rate ~ 10 times the persistent level. Following the burst, the persistent emission decreased by $\sim 35\%$, corresponding to a luminosity change from 1.8×10^{38} to 1.2×10^{38} ergs s^{-1} (1–14 keV). No evidence for cooling, as observed in type I bursts, was found during the decay. The shape of the pulsar signal and its relative amplitude remained unchanged before, during, and after the burst-like event, excluding the possibility that a thermonuclear flash involving the whole neutron star surface occurred. In the four following observations no bursts or other major changes in the intensity level were seen. The luminosity ranged between 1.5×10^{38} and 1.9×10^{38} ergs s^{-1}.

The power spectra of SMC X-1 calculated for each observation are plotted in chronological order in Figure 3. A highly-significant broad turnover was seen during the last three observations for frequencies < 0.01 Hz. The turnover appeared during the observation when the source flux was highest. In the two following observations a 20% decrease in luminosity occurred and this broad feature became correspondingly more marked, with a very broad peak extending over two decades around a frequency of ~ 0.01 Hz. An excess in the same frequency range was also seen during the first observation before the burst-like event occurred (Fig. 3). The variability associated with the broad feature in the power spectrum at ~ 0.01 Hz is clearly seen in the light curves which are characterized by rather symmetric, sharply-peaked, flare-like events on a timescale of ~ 100 s. The burst-like event detected during the first observation seems to be an extreme manifestation of this activity.

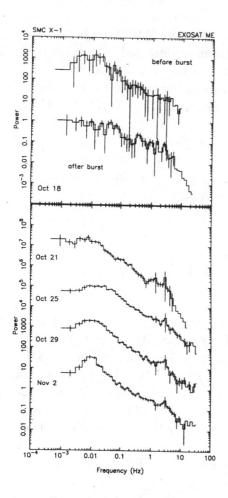

Fig. 3—Average *cleaned* power spectra of SMC X-1. The periodic pulsar signal at frequency 1.4 Hz with all the harmonics have been removed. They are shown in chronological order; for the first observation two power spectra before and after the burst-like event are plotted. The noise due to the Poisson statistics corrected for instrument and on board computer dead time has been subtracted. The points without error bars represent 2σ upper limits. The excess at ~ 0.01 Hz disappeared in power spectra calculated after the burst-like event when the intensity declined by $\sim 35\%$.

4 CONCLUSIONS

The analysis of the continuum power spectrum of X-ray pulsars reveals features with different characteristics. The power spectrum peaks discovered in EXO 2030+375 and Cir X-1 (Tennant 1988) have width/frequency $\sim 20\%$, closely resemble those in the power spectra of LMXRBs and are best interpreted in terms of QPOs. The QPO properties of EXO 2030+275 are in good agreement with the predictions of the BFM. The very broad peaks in the power spectra of SMC X-1 have no clear analogs in the power spectra of LMXRBs (Lewin, van Paradijs, and van der Klis 1988). These peaks are likely to originate from the 100 s flaring activity in the light curve of SMC X-1; the burst-like event seems to be an extreme manifestation of this phenomenon.

REFERENCES

Angelini, L., Stella, L., and Parmar, A. N. 1989, *Ap. J.*, in press.

Bonnet-Bidaud, J. M., and van der Klis, M. 1981, *Astr. Ap.*, **97** 134.

Lamb, F. K. 1988, in *Physics of Neutron Stars and Black Holes*, ed. Y. Tanaka (Tokyo: Universal Academy), p. 21.

Lewin, W. H. G., van Paradijs, J., and van der Klis, M. 1988, *Space Sci. Rev.*, **46**, 273.

Parmar, A. N., White, N. E., and Stella L., Izzo, C., and Ferri, P. 1989, *Ap. J.*, **338**, 359.

Parmar, A. N., White, N. E., and Stella L. 1989, *Ap. J.*, **338**, 373.

Tennant, A. F. 1988, paper presented at the Workshop on QPO, La Cienega, New Mexico, October 2–7.

Optimal Filter Techniques for Quasi-Periodic Oscillations

J. P. Norris[1], P. Hertz[1], K. S. Wood[1], B. A. Vaughan[2],
P. F. Michelson[2], K. Mitsuda[3], and T. Dotani[3]

[1]E. O. Hulburt Center for Space Research, Naval Research Laboratory
[2]Department of Physics, Stanford University
[3]Institute of Space and Astronautical Sciences

ABSTRACT

Optimal filter analysis techniques are employed in order to set constraints on the nature of possible relationships between low frequency noise (LFN) and quasi-periodic oscillations (QPOs) in GX5-1 on timescales near the QPO coherence length. Models are explored in which LFN shots modulate sinusoidal QPOs for shot rates up to 400 Hz and shot clustering fractions up to ~50%. Such models are found to be constrained by comparison with the data.

1 INTRODUCTION

Three low mass X-ray binaries (GX5-1, Sco X-1, Cyg X-2) have been observed in states where spectral hardness is approximately independent of intensity ("horizontal branch") and strong correlations between QPO centroid frequency and source flux are evident. Observations of QPOs and theories advanced to explain them are reviewed by Lewin *et al.* (1988). A beat-frequency modulated-accretion (BFMA) model has been developed to explain the strong correlations observed in GX5-1 and in other horizontal branch QPO sources (Alpar and Shaham 1985; Lamb *et al.* 1985). The beat (QPO) frequency is postulated to arise from the difference between the instantaneous Keplerian orbital frequency of clumped accreting matter and the neutron star rotation frequency. The X-ray temporal signature of the clumped matter envelope is the LFN.

The BFMA model predicts a correlation between LFN amplitude and QPO amplitude. Optimal filters — in which a moving filter with shape matching that of the expected waveform is convolved with the input data — probe the LFN/QPO relationship to short timescales, affording an improved discriminant for correlation tests over the split power test (Norris *et al.* 1990*b*). With presently-available observations, the individual QPO waveforms are still overwhelmed by low counting statistics. However, *GINGA*'s Large Area Counter (LAC) instrument (Turner *et al.* 1989) affords sufficiently high count rates to justify an optimal filter search, on a timescale near the QPO coherence length, for correlations between QPO and LFN fluctuations. We have applied optimal filter techniques to a *GINGA* LAC observation of GX5-1 and the results constrain the range of BFMA model parameters which may be viable.

2 THE OPTIMAL FILTER METHOD

An optimal filter is one whose pulse shape matches that of the expected waveform in the input data. By convolving the filter with the input, one can obtain pulse amplitudes and times of occurrence. QPO coherence lengths are inferred from the signal width ($\Delta\nu/\nu \simeq 0.125$ [FWHM] in spectral power for the present observation) to be short, typically a few cycles. In order to preserve phase and amplitude information within the QPO wavetrain, the optimal filter should be shorter than the coherence length. A good first trial for an optimal QPO filter is a 1 cycle sine-and-cosine kernel with frequency equal to the QPO centroid frequency. This is essentially a moving Fourier filter with one frequency component. Evaluated for the input data, $I(t)$, the function yields the power at the centroid frequency averaged over 1 cycle,

$$P(\tau) = \int \left[\{I(t)\sin(\nu t)\}^2 + \{I(t)\cos(\nu t)\}^2 \right] dt ,$$

where τ is the coordinate of the smoothed data. The 1 cycle averaged power is then converted to amplitude. Filters for portions of the LFN band are also constructed. Their outputs are correlated against the filtered QPO data by sorting QPO amplitudes according to LFN amplitude, or vice versa. Comparison is then made between the resulting distributions for optimally filtered real data and models (see Norris *et al.* 1990*a* for details).

In analyses where the signal-to-noise ratio is relatively low, a large noise bias is present which tends to level any correlation among parameters. In order to calibrate the noise bias and constrain theory, we construct shot models whose averaged 1 s power spectra are made to match that of GX5-1 by adjusting parameter values such as LFN amplitude and lifetime distributions, QPO coherence length, QPO/LFN amplitude distribution, correlation of QPO lifetime with LFN lifetime, etc. The simulations also include the important option of creating a fraction of the shots with QPOs which are coherent across (incoherent) overlapping shots, thus realizing shot clustering (Shibazaki and Lamb 1987). Higher QPO coherence across shots reduces the LFN/QPO correlation.

3 OPTIMAL FILTER RESULTS

In our first test of shot models using optimal filters, we searched for correlations between QPO and LFN fluctuations. The amplitude distributions which result from application of 24 Hz (QPO) and 6 Hz (LFN) filters to GX5-1 and models are binned according to LFN amplitude, i.e., all QPO amplitudes occurring when the LFN amplitude is $A_{\mathrm{LFN}}(i) \pm \Delta A_{\mathrm{LFN}}$ are averaged to yield $A_{\mathrm{QPO}}(i)$.

The resulting correlation plots are illustrated in Figure 1. Part (a) shows a calibration run for a model in which QPO and LFN amplitudes and occurrences are

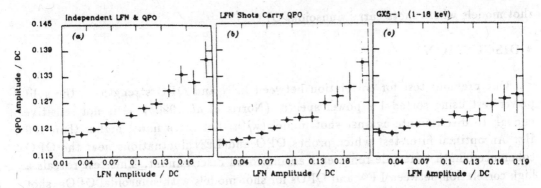

Fig. 1—Optimal filter search for correlation between QPO and LFN amplitudes. (a) Null hypothesis model in which QPO and LFN occurrences and amplitudes are independent. Correlation is nevertheless present because QPO amplitude variation introduces LFN. Filter algorithm output for (b) 400 Hz shot model with 50% shot clustering fraction (LFN shots carry QPO), and (c) GX5-1 data.

independent, an "acausal" model. In this simulation, the timescale and strength of the QPO amplitude variability are purposely set comparable to those of the "causal" model illustrated in part (b). Thus, the model QPO itself manifests LFN associated with its amplitude fluctuations, but not associated with the shape of shot envelopes. Part (b) shows the filter output for a 400 Hz shot model in which LFN shots carry the QPO waveforms, a "causal" model. Indeed, as constructed, the correlation slopes for the causal and acausal models are comparable — shot models which carry QPO and acausal models which do not can be constructed to yield similar signatures. Part (c) shows the filter algorithm output for GX5-1. The correlation slope is significantly less than those of the causal and acausal models, indicating that much lower short timescale amplitude variability is contained in the GX5-1 QPO.

Several adjustments can be made to parameter values of the acausal model so that a match with the GX5-1 optimal filter output can be had, while still retaining a match to the average 1 s power spectrum of GX5-1. However, not as much freedom is available in the case of the causal model since QPO and LFN amplitudes are necessarily related through the LFN envelope shape.

Note that the averaged power spectra for both models fit that of GX5-1 and thus all three have nearly the same variance as a function of frequency on a 1 s timescale. It is in phasing information on shorter timescales, basically, that the models differ from each other. Phase optimal filters and/or skewness tests should be capable of distinguishing the two model signatures. Figure 1 demonstrates that they also differ from GX5-1 in amplitude information on a short timescale.

It may be possible to fit models to GX5-1 such that the LFN/QPO correlations on short timescales are similar. However, we already see that short timescale QPO amplitude variability in GX5-1 is significantly less than that of 400 Hz rectangular shot models where shots carry sinusoidal QPO.

4 DISCUSSION

Our previous test for correlation between LFN and QPO strength — the split power test using sorted 1 s power spectra (Norris *et al.* 1990*b*) — is not sensitive enough to discriminate against shot models with shot rates much higher than 100 Hz. An optimal filter test which probes QPO and LFN fluctuations near the QPO coherence length timescale represents an improved discriminant. This test reveals a high correlation between LFN and QPOs for shot models with sinusoidal QPOs, shot rates of ~ 400 Hz, and shot clustering fractions of ~ 50%. Such a model is similar to some proposed BFMA models (Lamb *et al.* 1985). The GX5-1 data also exhibit a correlation, but the correlation slope is significantly less than that of the 400 Hz shot model, and acausal models are easily constructed to fit GX5-1.

Tests for correlation between local mean and LFN power and for skewness on LFN timescales also show that models with shot rates up to 400 Hz are inconsistent with the GX5-1 data (Norris *et al.* 1990*b*). Because the LFN/QPO correlation originally explained by the BFMA model develops on a timescale longer than ~ 8 s (Mitsuda *et al.* 1990), it may be that LFN and QPOs are only phenomenologically connected by a third parameter, perhaps accretion rate. Since some degree of QPO coherence across overlapping shots is required to fit the QPO peak, some short timescale QPO attributes which depend on factors other than LFN fluctuations may provide more sensitive tests of the BFMA model. For instance, QPO frequency modulation (FM) may be manifested as shots spiral through the magnetospheric transition region (Shibazaki and Lamb 1987). A test for the presence of FM in GX5-1 on a cycle-to-cycle timescale (~ 40 ms) using phase optimal filters and frequency return maps yields a null result; simulations are needed to calibrate its significance.

REFERENCES

Alpar, M. A., and Shaham, J. 1985, *Nature*, **316**, 239.
Lamb, F. K., *et al.* 1985, *Nature*, **317**, 681.
Lewin, W. H. G, *et al.* 1988, *Space Sci. Rev.*, **46**, 273.
Mitsuda, K., *et al.* 1990, *Pub. Astr. Soc. Japan*, submitted.
Norris, J. P., *et al.* 1990*a*, in *Proc. 23rd ESLAB Symposium on X-ray Binaries*, ed. N. E. White, in press.
Norris, J. P., *et al.* 1990*b*, *Ap. J.*, submitted.
Shibazaki, N., and Lamb, F. K. 1987, *Ap. J.*, **318**, 767.
Turner, M. J. L., *et al.* 1989, *Pub. Astr. Soc. Japan*, in press.

A Study of EUV Emission from Cataclysmic Variables

Ronald S. Polidan[1], Christopher W. Mauche[2], and Richard
A. Wade[3]

[1]Lunar and Planetary Laboratory – West, University of Arizona and
 Lab. for Astronomy and Solar Physics, NASA/Goddard Space Flight Center
[2]Space Astronomy and Astrophysics Group, Los Alamos National Laboratory
[3]Steward Observatory, University of Arizona

1 INTRODUCTION

The far ultraviolet (FUV, 912–1200 Å) band and the extreme ultraviolet (EUV, 200–912 Å) band are important regions in which to study cataclysmic variables (CVs). A major portion of the flux, λF_λ, produced by the accretion disk is predicted to be emitted shortward of 1200 Å. The EUV spectral region also contains the peak of the spectral energy distribution expected from the boundary layer, and the accreting white dwarfs in some of these systems are hot enough to produce significant amounts of flux shortward of 1200 Å.

In this paper we summarize the results of a study by Polidan, Mauche, and Wade (1990, hereafter PMW) of EUV (500–912 Å) emission from cataclysmic variables. Five CVs, the dwarf novae SS Cyg and VW Hyi and the novalike variables V3885 Sgr, RW Sex, and IX Vel, were chosen for study. *Voyager* far- and extreme-ultraviolet (500–1200 Å) spectrophotometry and estimated neutral hydrogen column densities derived from high-resolution *IUE* spectra were used to place upper limits on the emitted flux in the 600–700 Å EUV band. The reader should refer to PMW for complete details of the study.

2 OBSERVATIONS

The *Voyager* data were obtained between 1983 and 1986 using both *Voyager 1* and *2*. The ultraviolet spectrometer (UVS) on each spacecraft is sensitive in the 500–1700 Å region with an effective resolution of ~ 18 Å for a well-observed point source. Pertinent instrument characteristics and performance data are discussed in Polidan and Carone (1987) and Polidan and Holberg (1987).

The *Voyager* spectra used in this study are time-averaged spectra obtained over periods of from less than one day to over 30 days. The IX Vel and V3885 Sgr spectra result from single observations obtained in 1985. The RW Sex spectrum is an average of two observations obtained in 1985 and 1986. The VW Hyi observations were obtained during one normal outburst and one superoutburst. For both outbursts, which were treated independently, the *Voyager* spectrum used is an average of the data taken near the peak of the optical light curve. For SS Cyg, *Voyager* data from multiple outbursts are available, and the spectrum chosen was that with the best signal-to-noise ratio, obtained during the early decline from maximum of a short outburst that began on 1984 June 26.

The FUV spectra of the novalike stars IX Vel and RW Sex in the 912–1200 Å band differ from those observed in the third novalike star, V3885 Sgr, and the dwarf novae SS Cyg and VW Hyi. The spectral slopes in this region of the former CVs are positive, whereas the slopes of the spectra of the latter are flat. The origin of this difference is unknown.

TABLE 1
N_{HI} Values, Far-UV Flux Densities, and EUV Upper Limits

Star	N_{HI} (cm^{-2})	$F_\lambda(1000 \text{ Å})^a$	$F_\lambda(650 \text{ Å})^a$	$R_\lambda{}^b$
SS Cyg	3.5×10^{19}	2.4×10^{-11}	$< 1 \times 10^{-13}$	< 0.4
VW Hyi { outburst	6.0×10^{17}	3.8×10^{-12}	$< 3 \times 10^{-13}$	< 8
{ superoutburst		2.4×10^{-11}	$< 1 \times 10^{-13}$	< 0.4
V3885 Sgr	5.6×10^{19}	7.0×10^{-12}	$< 5 \times 10^{-14}$	< 0.7
RW Sex.................	8.9×10^{19}	4.1×10^{-12}	$< 2 \times 10^{-13}$	< 5
IX Vel	2.0×10^{19}	1.2×10^{-11}	$< 3 \times 10^{-13}$	< 3

^{a}The units of F_λ are (erg cm^{-2} s^{-1} Å^{-1}).
$^b R_\lambda \equiv 100 \times F_\lambda(650 \text{ Å})/F_\lambda(1000 \text{ Å})$, the ratio of observed flux densities.

No EUV flux was detected in any of the stars. Table 1 lists the mean flux density observed in the 925–1075 Å band, the observed 90% confidence upper limit to the 600–700 Å flux density, and the 90% confidence upper limit to the ratio of the flux density in these bands for each cataclysmic variable. Upper limits to the observed EUV flux density in quiescence from the two dwarf novae here are 5 to 10 times higher than those listed in Table 1.

Neutral hydrogen column densities for the five targets were derived from curve-of-growth analyses of interstellar absorption lines in high-resolution *IUE* spectra. Results for the four objects other than VW Hyi have been discussed by Mauche, Raymond, and Córdova (1988). For VW Hyi a curve of growth consistent for both N I and Si II can be constructed, indicating $\log N_{HI} \approx 17.78 \pm 0.2$. Details of the column density estimates for VW Hyi can be found in PMW; the hydrogen column density value for each star appears in Table 1.

3 MODELING THE SPECTRAL ENERGY DISTRIBUTION

The modeling of the accretion disk spectrum followed a procedure described by Wade (1984, 1988 and references therein). We adopted the standard prescription of the run of effective temperature and gravity with radius for the emitting surfaces of the disk. We modeled the emission from each surface element of the disk using either Planck spectra or specific intensities from model stellar atmosphere computations by Kurucz (1979 and private communication). Edge effects, such as a thick outer rim or a disrupted inner disk, were ignored. The integrated spectrum of the disk was computed by summing over annuli.

Steady-state models were calculated for a system viewed with the disk face-on at a distance of $d = 100$ pc, using white dwarf masses of $M_{wd} = 0.65$ and 1.30 $M_\odot$. Scaling by $1/d^2$ and $\cos i$, where i is the inclination of the disk, is straightforward (ignoring any complications due to limb-darkening or a flared disk surface). A mass-transfer rate onto the white dwarf of $\dot{M} = 10^{-8}$ $M_\odot$ yr^{-1} was adopted, appropriate for luminous CVs. These choices of parameters are adequate for a general discussion of the *Voyager* results since the quantity of interest, the flux density ratio $F_\lambda(650\,\text{Å})/F_\lambda(1000\,\text{Å})$, varies relatively slowly with changes in the parameters. For the specific modelling of VW Hyi below, values of d, M_{wd}, i, *etc.* appropriate to that system were used.

The boundary layer was modelled as a single-temperature region emitting a black-body spectrum. Its outward-directed luminosity is parameterized by ζ and its emitting area by f:

$$L_{BL} = \zeta \frac{GM_{wd}\dot{M}}{2R_{wd}}$$

$$= 7.0 \times 10^{34} \zeta \left(\frac{M_{wd}}{1M_\odot}\right) \left(\frac{\dot{M}}{10^{-8}\ M_\odot\ \text{yr}^{-1}}\right) \left(\frac{R_{wd}}{6 \times 10^8\ \text{cm}}\right)^{-1}\ \text{erg s}^{-1}$$

$$= 4\pi R_{wd}^2 f \times \sigma T_{BL}^4.$$

The factor ζ can be used to account for boundary layer radiation that is directed inward to the white dwarf or to account for non-radiative losses from the boundary layer, such as the kinetic energy carried away by a wind. The factors f and ζ together determine the temperature of the boundary layer, which is assumed to radiate isotropically. The boundary layer and disk spectra were added without regard to any obscuration or occultation effects. A somewhat surprising result is that in the range 400–912 Å it is often the disk, and not the boundary layer, that dominates in these models. This result is true for cases involving high M_{wd} or high $\dot{M}$ and would be true even more often for models in which the boundary layer has a lower luminosity or a complex temperature structure. Thus, even in the absence of interstellar extinction, the spectral region accessible to *Voyager* generally conveys little information about the boundary layer. For low M_{wd} and low $\dot{M}$, however, the disk can be sufficiently cool that the dominance is reversed. The boundary layer will also dominate if the inclination of the disk is sufficiently high.

The theoretical models predicted values for the ratio of intrinsic flux densities $R_{\lambda,o} \equiv 100 \times F_\lambda(650\,\text{Å})/F_\lambda(1000\,\text{Å})$ that vary between 92 and 144 for models with $M_{wd} = 1.30$ $M_\odot$, and between 18 and 84 for models with $M_{wd} = 0.65$ $M_\odot$.

The effects of interstellar extinction were evaluated using the models of Seaton (1979), Morrison and McCammon (1983), and Cruddace *et al.* (1974). Representative neutral hydrogen column densities of $N_{HI} = 0$, 6×10^{17}, 5×10^{18}, and 3.5×10^{19} cm^{-2} were considered. The lowest non-zero value corresponds to the dwarf nova VW Hyi and the highest to SS Cyg; the middle value would correspond to a favorably-placed nearby CV. Dust is the primary agent of extinction at 1000 Å and neutral atomic or molecular hydrogen accounts for almost all of the extinction at 650 Å. In particular, photoelectric absorption by helium and heavier atoms does not operate at 650 Å. For the three non-zero values of N_{HI} given above, the effect of extinction is to make the ratio of *observed* flux densities, R_λ, smaller than the ratio of *intrinsic* flux densities $R_{\lambda,o}$ by factors of about 4.5, 3×10^5, and more than 10^{38}, respectively. Thus, due simply to the strength of extinction at these wavelengths by neutral hydrogen, the typical CV will not be visible between 912 Å and 504 Å, regardless of the specifics of the model.

4 RESULTS

For four of the CVs under study, SS Cyg, V3885 Sgr, RW Sex, and IX Vel, the absence of observable flux near 650 Å as established by *Voyager* can be easily understood in terms of the relatively high neutral hydrogen column estimated for these stars. The limits on their *intrinsic* 650 Å flux density are too high to provide a scientifically interesting test of existing accretion disk models. Specifically, for these four CVs, the upper limits placed on $R_{\lambda,o}$ are more than 20 orders of magnitude larger than the expected values. It is impossible to discuss *any* aspect of the EUV spectrum of these four stars without observations at shorter wavelengths. For the remaining object, VW Hyi, the upper limit to the 650 Å flux density is significantly below the predictions of the theoretical models, and this system must therefore be considered further.

5 VW HYI

The *Voyager* data for VW Hyi during superoutburst show no detected EUV emission; the upper limit on the 650 Å flux density can be expressed as $R_\lambda < 0.4$. With $N_{HI} = 6 \times 10^{17}$ cm^{-2}, the models overpredict R_λ by more than an order of magnitude, and an explanation is needed.

Detailed results for VW Hyi can be found in PMW. In brief, models were constructed specifically for VW Hyi using the best available system parameters and using system parameters that were modified to *minimize* the EUV flux with respect to the FUV flux. In no case was it possible to get the predicted R_λ closer than ~ 6 times the observed upper limit. It therefore seems rather certain that the deficit of EUV flux from VW Hyi cannot be accounted for simply by "fine-tuning" of the model parameters.

Possible causes for the failure of the models are (1) that the opacity in the 600–700 Å band is higher than indicated by the curve-of-growth analysis of interstellar absorption lines, or (2) that the model stellar atmosphere spectra used in the discussion are inappropriate for the modelling of disks.

The observations require that in VW Hyi the outward-directed radiative flux from a (distinct) boundary layer must be small or else the problem of reconciling models and observations is aggravated. Whether the "missing" flux is directed inward to heat the white dwarf, whether it emerges mechanically as a wind, or whether the flux does not exist in the first place (e.g., because the white dwarf is rapidly rotating) cannot be answered until better observations and better theories are available.

6 CONCLUDING REMARKS

We emphasize that the EUV "deficiency" found for VW Hyi could be a common feature of cataclysmic variables. The remaining four CVs in our study have *no* information available on the nature of their EUV spectra. They could, in principle, exhibit the same EUV deficiency that we observe in VW Hyi, or this deficiency could be unique to VW Hyi.

The results of this investigation, especially when combined with the conclusions reached by Wade (1988), strongly argue that existing calculations of models for cataclysmic variable disks appear incapable of producing a self-consistent picture of an accretion disk that is compatible with modern multispectral observations. In addition, if VW Hyi can be considered typical of the non-magnetic cataclysmic variables, then CVs are *not* a major a source of EUV photons for the interstellar medium.

This work has been supported by NASA contract 956729 to the Jet Propulsion Laboratory, NASA grants NAGW-587 and NAG5-8565 and National Science Foundation grants AST-8514778 and AST-8818069 to the University of Arizona, and the US Department of Energy.

REFERENCES

Cruddace, R., Paresce, F., Bowyer, S., and Lampton, M. 1974, *Ap. J.*, **187**, 497.

Kurucz, R. L. 1979, *Ap. J. Suppl.*, **40**, 1.

Morrison, R., and McCammon, D. 1983, *Ap. J.*, **270**, 119.

Mauche, C. W., Raymond, J. C., and Córdova, F. A. 1988, *Ap. J.*, **335**, 829.

Polidan, R. S., and Carone, T. E. 1987, *Ap. Sp. Sci.*, **130**, 235.

Polidan, R. S., and Holberg, J. B. 1987, *M.N.R.A.S.*, **225**, 131.

Polidan, R. S., Mauche, C. W., and Wade, R. A. 1990, *Ap. J.*, in press.

Seaton, M. J. 1979, *M.N.R.A.S.*, **187**, 73p.

Wade, R. A. 1984, *M.N.R.A.S.*, **208**, 381.

———. 1988, *Ap.J.*, **335**, 394.

Observed Properties of S-Waves

R. H. Kaitchuck[1], E. M. Schlegel[2,3], and P. A. Hantzios[1]

[1]The Ohio State University
[2]Harvard-Smithsonian Center for Astrophysics
[3]NASA-GSFC/Universities Space Research Association

ABSTRACT

We examine some commonly-held ideas about S-waves. We find that some of these properties may not apply to all S-waves. In particular, S-waves may not be solely responsible for distorted radial velocity curves typical of cataclysmic variables.

1 INTRODUCTION

S-waves have been part of the literature of cataclysmic variables since their appearance in trailed spectra of WZ Sge, and were first discussed by Kraft (1961) and Kraft, Matthews, and Greenstein (1962). Those images showed a separate emission line component oscillating between the two Doppler disk components at the orbital period. The origin of S-waves has usually been associated with the gas stream or a location on the outer disk edge near the stream impact point. Some commonly-held ideas found in the literature include the following:

1. S-waves are sinusoidal in their radial velocity variations.
2. S-waves are responsible for the distortions in the radial velocity curves of the disk lines.
3. The strength of the S-wave is probably correlated with the strength of the continuum hot spot because both phenomena are related to the impact of the stream on the disk.
4. The S-wave velocity vector is the same as Keplerian motion at a point on the outer edge of the disk.

We will now consider these in turn.

2 S-WAVES HAVE SINUSOIDAL VELOCITY VARIATIONS

That this is not true in many cases becomes apparent from a study of modern, digital-trailed spectra. For example, an examination of the images of U Gem and WZ Sge in the grey-scale atlas by Honeycutt, Kaitchuck, and Schlegel (1987) clearly shows non-sinusoidal S-waves. This indicates that the emitting region is something more complex than a localized, radiating spot.

3 S-WAVES AND RADIAL VELOCITY CURVE DISTORTIONS

The image of U Gem mentioned above shows an S-wave whose strength is comparable to that of the Doppler disk components. This makes it easy to believe that the S-wave, with its distinct amplitude and phasing, will contaminate the disk components in such a way as to make the determination of the white dwarf radial velocity unreliable.

U Gem 1983 H–Beta

U Gem 1986 H–Beta

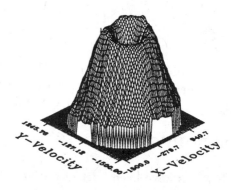

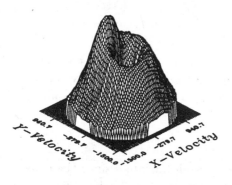

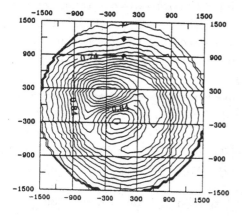

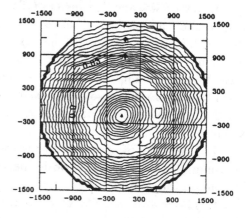

Fig. 1—Doppler map of the 1983 Hβ emission in U Gem.

Fig. 2—Doppler map of the 1986 Hβ emission in U Gem.

In a data set obtained in 1986 (Kaitchuck, unpublished), the S-wave in U Gem was very weak and only barely detectable. The expectation is that the 1986 data set should yield a nearly undistorted white dwarf radial velocity curve. Figures 1 and 2 show the Doppler map (Marsh and Horne 1988, without maximum entropy treatment) for the Hβ emission line for the 1983 and 1986 data sets, respectively. In Figure 1 the line flux shows the expected "volcano-like" structure of a disk plus the added emission from the S-wave. In the bottom portion of Figure 1 the S-wave emission can be seen to come from an extended region in velocity space which produces the non-sinusoidal radial velocity variations. The S-wave in this data set has recently been analyzed by Marsh *et al.* (1990). Figure 2 shows that the S-wave is almost absent from the 1986 data set, leaving almost pure disk emission. Even so, problems are still apparent in the lower frame of Figure 2. The inner-most contours (low-velocity gas) have a different centroid than the outer contours (high-velocity gas). That is to say, the line wings and core indicate different radial velocity amplitudes.

A radial velocity study confirms this problem. The radial velocity amplitude was found by the double-Gaussian technique of Schneider and Young (1980). The basic idea is to slide two Gaussians of fixed width and separations over the profile in order to find a position that equalizes the flux in the two Gaussians. This gives different weights to the line core and wings depending on the Gaussian separations. Figure 3 shows the radial velocity amplitude for various displacements of the two Gaussians (5 Å FWHM) from line center. The 1983 data show a convex curve with the radial velocity amplitude peaking at a half separation of 17 Å. The expected behavior for an undistorted data set would be that of a flat curve at the value of the true radial velocity amplitude. Figure 3 shows the results for the 1986 data set. The radial velocity amplitude continues to increase to the maximum Gaussian displacements. It is difficult to look at these two curves and conclude that the 1986 data set is significantly improved over the 1983 data. It is still impossible to specify the true radial velocity amplitude with any great certainty. This suggests that the S-wave itself may not be fully responsible for the radial velocity curve distortion and disk asymmetries that are still present.

4 S-WAVE/HOT SPOT CORRELATION

U Gem is an ideal system to test this correlation because it has both a strong signature of a hot spot in its light curve (the pre-eclipse bump) and a prominent S-wave. Certainly, if the strengths of the hot spot and the S-wave were correlated, the hot spot should be much weaker in the 1986 data. Figure 4 compares the light curves (obtained from the absolute spectral photometry) of 1983 and 1986. To within the observational uncertainties, the hot spot strength is unchanged.

5 S-WAVE VELOCITY VECTOR IS THE SAME AS THE DISK ROTATION VECTOR

Hantzios (1988*a, b*) has shown that an S-wave model based on localized emission at the outer disk edge, with a velocity vector matching that of the disk, fails to account for non-sinusoidal velocity variations, and for the amplitude and phasing of the S-wave in some systems. Smak (1985) suggested that in some cases the S-

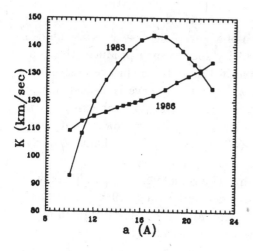

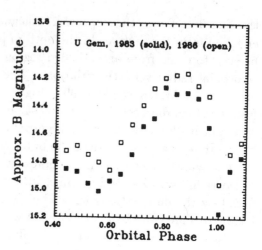

Fig. 3—The radial velocity amplitude as a function of the double-Gaussian separation for 1983 and 1986. Neither epoch suggests a "true" amplitude.

Fig. 4—The light curve near the bright spot phases. There is very little difference apparent between the two epochs.

wave originates in the stream gas which overflows the top of the disk (the stream is thicker than the disk). This overflow has been recently modeled by Lubow (1989). Hantzios (1988a,b) has constructed S-wave models of an emitting region confined to the stream overflow plus a transition zone on the disk surface where the velocity vectors gradually change from those of the stream to those of the disk. The model produces a reasonably good match to the radial velocity and flux variations with orbital phase for the S-waves in U Gem, WZ Sge, and LX Ser. Lubow (1989) has also explained the S-wave in Z Cha as emission from the stream overflow. More recent work by Marsh *et al.* (1990) on U Gem suggests that the post-shock gas in the stream overflowing the disk has velocities between those of the stream and the disk rotation.

6 SUMMARY

S-waves do not have all the properties commonly associated with them. They may not be correlated with continuum hot spots and they may not be solely responsible for the distorted radial velocity curves of cataclysmic binaries. We need to be more careful when assuming a particular location or velocity vector of the S-wave emitting region, especially when the S-wave phasing or velocity is used to obtain some physical parameters such as the disk radius.

REFERENCES

Hantzios, P. A. 1988a, Ph.D. thesis, The Ohio State University.

Hantzios, P. A. 1988b, *Pub. Astr. Soc. Pac.*, **100**, 1188.

Honeycutt, R. K., Kaitchuck, R. H., and Schlegel, E. M. 1987, *Ap. J. Suppl.*, **65**, 451.

Kraft, R. P. 1961, *Science*, **134**, 1433.

Kraft, R. P., Matthews, J., and Greenstein, J. L., 1962, *Ap. J.*, **136**, 312.

Lubow, S. H. 1989, *Ap. J.*, **340**, 1064.

Marsh, T. R., and Horne, K. 1988, *M. N. R. A. S.*, **235**, 269.

Marsh, T. R., Horne, K., Honeycutt, R. K., Kaitchuck, R. H., and Schlegel, E. M. 1990, *Ap. J.*, in press.

Schneider, D. P., and Young, P. 1980, *Ap. J.*, **238**, 946.

Smak, J. 1985, *Acta Astr.*, **35**, 351.

The O I λ7773 Å Line as a Diagnostic of Disk Properties

Robert Connon Smith

Astronomy Centre, Division of Physics and Astronomy
University of Sussex, Falmer, Brighton BN1 9QH, U. K.

1 INTRODUCTION

Relatively little work has been done on the absorption line spectrum arising from the cool secondary in cataclysmic variables. Since 1985 an RGO/Sussex collaboration has been exploiting the red sensitivity of CCDs to look for evidence of the red star in the optical infrared, where the disk emission is falling off and the red star continuum is increasing. Our aim is to measure radial velocities to determine masses with the minimum of assumptions. An ideal choice for this purpose is the Na I doublet at λ8183, 8195 Å which is well-defined over a wide range of spectral types from mid-K to late-M. Our initial survey (Friend *et al.* 1988) used a resolution of 2 to 3 Å over a wavelength range of ≈ 7000 to 8500 Å. As well as the Na I doublet characteristic of red dwarfs, various other interesting features emerged. The commonest feature, which appeared (unresolved) in 28 out of the 57 systems with good signal-to-noise ratio, was the O I triplet at λλ7771.9, 7774.2, and 7775.4 Å. In 18 cases the feature was in emission and in 10 in absorption. There is an interesting anti-correlation with the Na I feature: O I emission lines were seen in 11 of the 16 systems that showed Na I absorption; O I absorption lines were seen only in systems with no Na I absorption.

2 INTERPRETATION OF THE O I TRIPLET OBSERVATIONS

What is the origin of the O I features and why do they appear in both emission and absorption? Some of the emission lines are double-peaked (e.g., in EX Hya, cf. Fig. 7 of Friend *et al.* 1988), suggesting an origin in the disk. Friend *et al.* argued that strong O I absorption corresponded to an optically thick disk; in that case, the disk continuum would be strong and completely veil the red star spectrum, explaining the non-detection of the Na I doublet. If the O I is in emission, the disk is optically thin, the disk continuum is weak and the red star is easier to detect. This interpretation is supported by our observations of AM Her and IP Peg. AM Her has no disk, but shows strong O I emission, presumably from the optically thin accretion stream or column; the red star is readily detectable (Friend *et al.* 1988, Fig. 1). In IP Peg, O I is visible as a weak emission feature, probably originating in the bright spot (Martin *et al.* 1987), when the system is in quiescence. During decline from outburst (Martin *et al.*

1989), O I is visible as a strong absorption feature, while the Na I from the red star is barely detectable. We therefore conclude that the O I feature gives an immediate qualitative impression of the strength of the accretion disk in cataclysmic variables; we plan to look for a more quantitative relationship between absorption line strength and mass-accretion rate when we have more accurate spectrophotometry.

There is another way in which O I observations can be used if the line at 8446 Å is also observed. The energy spacing between the $2\,^3$P level and the $3\,^3$D level in O I is almost exactly equal to the energy of a Lyβ photon (Keenan and Hynek 1950). The $3\,^3$D level decays to the upper level ($3\,^3$P) of the 8446 Å transition with the emission of a 1.1287 μm photon. Thus, if Lyβ is in emission it can cause overpopulation of the upper level of λ8446 Å and lead to anomalous ratios between the strengths of the 8446 and 7773 Å features. Our existing spectra did not extend far enough to the red to include the 8446 Å feature, but future observations should look at the line strength ratios as a test of whether Lyβ is in emission.

3 SOME APPLICATIONS

3.1 IP Pegasi

Martin *et al.* (1987) showed that in quiescence the O I feature was visible in emission, but only near phase 0.75, consistent with an origin in the hot spot. In outburst, O I is seen in absorption at all phases (Martin *et al.* 1989). A trailed spectrum (Martin 1988, private communication) shows that the O I absorption has a maximum blueshift near phase 0.25, consistent with the motion of the white dwarf. This suggests that the absorption arises uniformly throughout the disk, so that the average motion mimics that of the white dwarf. The trailed spectrum also shows that the feature is not eclipsed near phase 0, which rules out confinement to the bright spot. We conclude that in quiescence the emission is from the bright spot but that in outburst the disk becomes thicker and absorption occurs throughout the disk. There are no surprises here.

3.2 AM Herculis

In AM Her, O I is visible as a strong emission feature all round the orbit. The radial velocity curve (Hawkins 1988) is noisy and the amplitude depends on what mean template is used in the cross-correlation. However, it is certainly small (60–85 km s^{-1}) and the blueshift has a maximum near phase 0.5. These properties are inconsistent with an origin in either stellar component, but are consistent with the expected origin in the accretion stream or column. Again, there are no surprises.

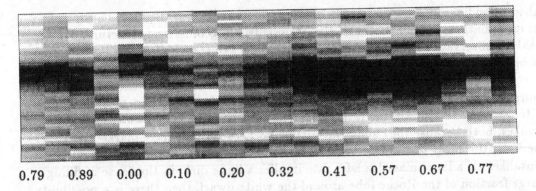

0.79   0.89   0.00   0.10   0.20   0.32   0.41   0.57   0.67   0.77

Fig. 1—A trailed spectrum of V1315 Aql, showing about 10 Å around the O I triplet (blue at the top). Phase is plotted horizontally (the scale is not quite linear). The white-light eclipse occurs at phase zero. The O I feature may be eclipsed slightly earlier, and is very weak between phases 0.1 and 0.3.

3.3 V1315 Aquilae

V1315 Aql is an eclipsing novalike variable (Downes *et al.* 1986). We obtained 40 intermediate-dispersion red spectra in 1985 August/September (Fiddik 1987; Hawkins 1988; Smith *et al.* 1990). These spectra show a very strong absorption feature of O I λ7773 Å, the strongest in our survey (cf. Fig. 5 of Friend *et al.* 1988). This feature is visible around most of the orbit, but varies considerably in strength with phase. A trailed spectrum (Fig. 1) shows a hint of an eclipse at phase ≈ 0.94 and very weak absorption between phases 0.1 and 0.3; unfortunately, we do not have spectrophotometry and cannot quantify the variation in strength. Interestingly, the absorption is a maximum near phase 0.45, which is about the phase where an absorption feature appears in the center of the Balmer emission lines in this and in other novalikes (Dhillon, Marsh, and Jones 1990; Honeycutt, Kaitchuck, and Schlegel 1987).

We attempted to measure the radial velocity of the O I feature by cross-correlating it with a mean template formed by adding suitably-shifted spectra and smoothing the continuum outside the O I lines (Smith *et al.* 1990). The resultant radial velocity curve has a large scatter, and the amplitude is not reliable, but, as is also apparent in the trailed spectrum in Figure 1, the maximum blueshift is near phase 0, when the disk is eclipsed. Unlike IP Peg, this is inconsistent with the mean motion of the disk. Naively, the absorption feature appears to be situated on the leading edge of the disk. The difficulty with this interpretation is that the absorption is weakest exactly at the phases when the leading edge of the disk is most easily visible. We do not have a definite proposal to explain these observations. However, one thing seems fairly clear from the large variations in absorption line strength: the absorption is not uniform over the surface of the disk. Alternatively, there is emission filling in the absorption lines at some phases. In either case, it seems that the disk is asymmetric in its radiation, perhaps because it is asymmetric in geometrical structure. We are making observations of a number of other novalikes to see whether this is a common property.

Asymmetric disks have previously been suggested in the context of superoutbursts in SU UMa systems, where the evidence takes the form of superhumps (Honey et al. 1988). Although these systems appear to be rather different from novalikes, both because they show outbursts and because they have markedly shorter periods, the mass-transfer rates in superoutburst may be comparable to those in novalikes. If a large mass-transfer rate implies a large disk, in the sense of filling a large fraction of the Roche lobe around the white dwarf, then there is a possibility of tidal effects similar to those suggested for SU UMas; certainly, the great strength of the O I absorption in V1315 Aql suggests a large disk.

4 CONCLUSIONS

- The O I triplet near $\lambda 7773$ Å is a useful qualitative diagnostic of the optical depth of an accretion disk in cataclysmic variables.
- The ratio of the O I features at $\lambda 8446$ and $\lambda 7773$ Å could reveal whether Lyβ is in emission.
- The $\lambda 7773$ Å triplet in dwarf novae and polars reveals no unexpected properties.
- The novalike variable V1315 Aql shows variations in the O I triplet that suggest an asymmetric disk. This may be related to the variations in the Balmer emission lines which were found by Dhillon, Marsh, and Jones (1990).

It is a pleasure to thank my colleague Derek Jones and our students Bob Fiddik, Malcolm Friend, Nigel Hawkins, and Jackie Martin for their (large) contributions.

REFERENCES

Dhillon, V. S., Marsh, T. R., and Jones, D. H. P. 1990, this volume.
Fiddik, R. J. 1987, M.Sc. thesis, University of Sussex.
Friend, M. T., Martin, J. S., Smith, R. C. and Jones, D. H. P. 1988, *M. N. R. A. S.*, **233**, 451.
Hawkins, N. A. 1988, M.Sc. thesis, University of Sussex.
Honey, W. B., Charles, P. A., Whitehurst, R., Barrett, P. E., and Smale, A. P. 1988, *M. N. R. A. S.*, **231**, 1.
Honeycutt, R. K., Kaitchuck, R. H., and Schlegel, E. M. 1987, *Ap. J. Suppl.*, **65**, 451.
Keenan, P. C., and Hynek, J. A. 1950, *Ap. J.*, **111**, 1.
Martin, J. S., Friend, M. T., Smith, R. C., and Jones, D. H. P. 1989, *M. N. R. A. S.*, **240**, 519.
Martin, J. S., Jones, D. H. P., and Smith, R. C. 1987, *M. N. R. A. S.*, **224**, 1031.
Smith, R. C., Hawkins, N. A., and Fiddik, R. J. 1990, in preparation.

The Bright Spot in U Geminorum

Janet H. Wood[1], F. Hessman[2], and A. Fiedler[3]

[1]Institute of Astronomy, University of Cambridge
[2]Max-Planck-Institut für Astronomie, Heidelberg
[3]Universitäts-Sternwarte München

1 INTRODUCTION

Eclipses of accretion disks in cataclysmic variables (CVs) have been studied extensively by light curve fitting and surface brightness reconstruction methods (e.g., Zhang and Robinson 1987; Horne and Stiening 1985). However, there have been no attempts to reconstruct the surface brightness distribution in the bright spots. There is usually more information about the bright spot than the disk in the light curves of quiescent systems, since it is constrained by both the eclipse and the shape of the orbital hump. Here we describe a method of mapping the bright spots on the disk rims. We chose to use our method first on U Gem because it has a strong orbital hump and bright spot eclipse but no sign of a disk eclipse. We have multicolor photometry of U Gem, on the decline from one outburst into quiescence, and in another quiescent interval. The observations are described in Fiedler (1989). Here we will concentrate on the light curves in the B-band, and leave a complete analysis of the $UBVRI$ data to a later paper.

2 THE METHOD AND APPLICATION TO U GEM

Assuming that the bright spots in CVs lie on the disk rims, we can reconstruct the surface brightness distribution in the bright spot. This is done in a similar way to the maximum entropy eclipse-mapping method for the disk (Horne 1985). Instead of using a grid covering the disk in the orbital plane, we define a grid on a cylinder around the rim of the disk perpendicular to the orbital plane. At a particular phase k the flux F_k is given by

$$F_k = \Sigma f_i R_{ik},$$

where f_i is the flux in the i'th pixel on the grid and R_{ik} is the visibility factor of the i'th pixel at phase k. The visibility factor is the area of the pixel not eclipsed by the secondary star, projected onto the plane perpendicular to the line of sight. We then find the bright spot map with the maximum entropy, measured relative to a specified default map, such that the light curve fits the data with a specified value of χ^2. We use default maps which are symmetric about the orbital plane with the flux decreasing with height in the rim.

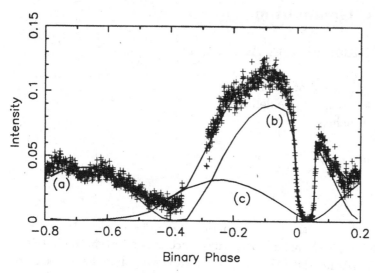

Fig. 1—Fit to a quiescent light curve. The data are shown with 10 s time resolution. Also shown are the model light curves resulting from each individual spot in the fitted map (labelled a, b and c).

We first concentrate on our most complete quiescent light curve. Here we take phase zero to be 0.0026 earlier than mid bright spot eclipse (Wade 1981) and use a mass ratio of 0.475 and inclination of 69.5° (Zhang and Robinson 1987). The contact phases for this eclipse show that the bright spot does not penetrate far into the disk, and hence our assumption that it lies on the rim is reasonable. The contact phases determine the distance of the bright spot from the center of the white dwarf and we use this distance as the radius of the cylinder for our grid. The flat-bottomed shape of the bright spot eclipse has led to the conclusion that it is totally eclipsed (e.g., Smak 1984) and we therefore subtract a constant flux from the light curve, before fitting, so that the flux at mid bright spot eclipse is zero.

Initially, we use a disk radius, $R_D = 0.46R_L$ and a rim of height $0.01R_L$, where R_L is the distance of the inner Lagrangian point to the center of the white dwarf. We assumed that grid elements on the far side of the disk do not contribute to the light curve. For the default map we used a map in which the rim is uniformly bright in azimuth, θ, but where the light falls off exponentially with height, Z, from the orbital plane ($Z = 0$). The resulting map of the bright spot will be as close to the default map as the data will allow.

Figure 1 shows the light curve with the fit we obtained. Figure 2 shows a contour map of the brightness distribution which produced this fit and Figure 3 shows the total flux in each vertical strip of our grid. The bright spot itself is clearly visible at $R_D\theta \sim 0.3R_L$ but, surprisingly, so are two other peaks in the surface brightness distribution, approximately 180° apart. The individual contribution of each of these spots to the light curve is shown in Figure 1. The extra two peaks were necessary because the bright spot, as determined by the shape of eclipse, is incompatible with the shape of the orbital hump.

To test whether the presence of the extra two peaks is spurious and due to inaccurate assumptions, we recalculated the surface brightness distribution on the rim of the disk for many different cases. We varied phase zero, the mass ratio, and

Fig. 2—Contour
map of the surface
brightness distri-
bution on the rim
of the disk. The
contours are on a
log scale with 0.1
decades between
each one.

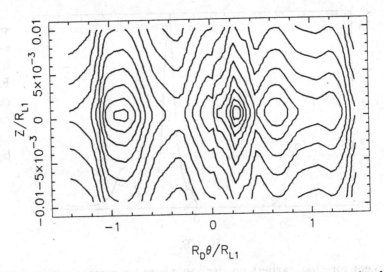

the inclination, but for each change all three peaks in the distribution remained
(though different in detail). For each change in geometry we changed the value of
R_D. We found that it was very difficult to fit the ingress and egress of the eclipse if
a disk radius different from that required by the observed contact phases was used.

We used many different default maps but the three peaks were always present.
We only have vertical information about the part of the rim which is eclipsed (near
the bright spot). In the other parts we know the flux in each vertical strip but its
distribution depends on the default map. Thus, the two extra peaks could be due
either to a change in intensity or a change in the height of the rim. This might tie in
with the X-ray dips seen in U Gem (Mason *et al.* 1989).

We also tried allowing certain fractions of the light to shine through the back
of the disk. This did alter the relative brightnesses of the two extra peaks but they
were still needed. The maximum at phase ~ 0.25 is not half a cycle from the orbital
hump maximum and thus cannot simply be due to the bright spot shining through
the disk. We tried fitting to only part of the light curve, between phases -0.35 and
0.15, and found that three peaks were still obtained. This shows that it is the shape
of the orbital hump itself which requires the extra two peaks to be present and not
only the secondary maximum at phase ~ 0.25.

Finally, we varied the flux at mid bright spot eclipse, F_0. Since the bright spot
is not the only contributor to the light from the rim in our maps, there is no reason
why F_0 should be zero during bright spot eclipse, as we previously assumed. We
therefore recalculated the maps for a variety of values of F_0. The bright spot is no
longer constrained to be as small as originally, and the other peaks are no longer
forced to have zero contribution during bright spot eclipse. All the spots spread out,
with the two extra peaks each becoming wide and double-peaked structures which
are extremely similar in shape and remain $180°$ apart.

We now return to our original assumptions and calculate the distributions on the
rim for our other B light curves. We found very similar results for all the light curves,

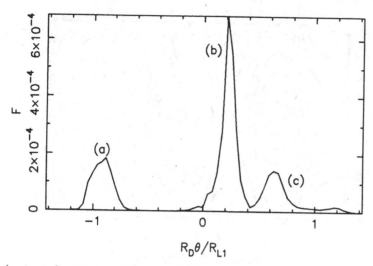

Fig. 3—The summed fluxes in each vertical strip of the map shown in Figure 2. The peaks corresponding to the model light curves in Figure 1 are labelled a, b and c.

(except for that earliest on decline where an extra peak in the map was required), with the two extra peaks remaining in the same positions.

3 CONCLUSIONS

The bright spot in U Gem is not simply a spot painted onto the edge of a circular rim. Some other effect is coming into play. Possibly there are two extra bright spots or regions where the rim bulges $\sim 180°$ apart on the rim, or perhaps a more sophisticated model of the bright spot is necessary. Another alternative would be the existence of a double-peaked variation in the light curve, similar to ellipsoidal variations but strong in U, B, and V as well as redder bandpasses. This could well be unrelated to the bright spot. Whatever the effect, it remains stable even during outbursts.

In the future we intend to investigate other possible causes of this effect, e.g., to look for a geometrical shape for the rim of the disk which will allow the eclipse and hump to be compatible, or to vary the assumed radiation pattern of the bright spot. However, in addition to the orbital hump and eclipse, the models must be able to account for the secondary maximum.

REFERENCES

Fiedler, A. 1989, Diplom. thesis, Universität München.

Horne, K. 1985, *M. N. R. A. S.*, **213**, 129.

Horne, K., Stiening, R. F. 1985, *M. N. R. A. S.*, **216**, 933.

Mason, K. O., Córdova, F. A., Watson, M. G., and King, A. R. 1989, *M. N. R. A. S.*, **232**, 779.

Smak, J. 1984, *Acta Astr.*, **34**, 93.

Wade, R. A. 1981, *Ap. J.*, **246**, 215.

Zhang, E. H., Robinson, E. L. 1987, *Ap. J.*, **321**, 813.

A Sixty Night Campaign on Dwarf Novae – A Progress Report

D. H. P. Jones[1], P. A. Charles[1], M.-J. Arevalo[2],
J. E. F. Baruch[3], R. Biernicowicz[12], P. J. Callanan[4], J. Casares[2],
J. Cepa[2], V. S. Dhillon[1], A. Gimenez[5], I. Gonzalez[2], R. Gonzalez[2],
E. H. Harlaftis[4], B. J. M. Hassall[6], C. Hellier[7], P. Johnson[3],
M. R. Kidger[2], D. L. King[1], C. Lazaro[2], K. O. Mason[7], K. Mukai[7],
T. Naylor[8], V. Reglero[9], R. G. M. Rutten[10], and J. van Paradijs[11]

[1] Royal Greenwich Observatory, Apartado 321,
  38780 Santa Cruz de la Palma, Tenerife, Canary Islands
[2] Instituto de Astrofisica de las Canarias
[3] Department of Physics, University of Leeds
[4] Department of Astrophysics, University of Oxford
[5] Instituto de Astrofisica de Andalucia
[6] Royal Greenwich Observatory, Cambridge
[7] Mullard Space Sciences Laboratory, University College, London
[8] Institute of Astronomy, Cambridge
[9] Dept de Matematica Aplicada y Astronomia, Universidad de Valencia
[10] Kapteyn Observatory, Roden
[11] Sterrenkundig Institut "Anton Pannekoek," Universiteit te Amsterdam
[12] Dublin Institute of Advanced Studies

1 OBSERVATORIO DEL ROQUE DE LOS MUCHACHOS

The Spanish Observatorio del Roque de los Muchachos has several international adherents, among them the United Kingdom, the Netherlands, and Ireland as well as Spain. Under the international agreement, five per cent of the observing time on every telescope each year is reserved for a large project or projects on which all the collaborating countries can cooperate and which cannot be done under the normal time-assignment procedures. In 1988/9 this time was awarded to P. A. Charles as Principal Investigator for his project 'Accretion Disc Evolution in Cataclysmic Variables.' The multinational scope of this venture is clear from the thundering herd of authors of this and other papers arising from the project.

The telescopes used in this program were:

Name of Telescope	Place	Aperture (m)
Isaac Newton	La Palma	2.5
Jacobus Kapteyn	La Palma	1.0
Carlos Sanchez	Tenerife	1.5

Unfortunately, the 4.2-meter William Herschel Telescope had not yet been commissioned at the time of this campaign and could not take part. However, we were awarded twenty half-shifts of *IUE* time to make ultraviolet observations in support of this ground-based campaign.

2 SCIENTIFIC OBJECTIVE

Our objective was to carry out continuous observations of one or more dwarf novae covering a complete outburst cycle. This would allow us to study the evolution of the properties of the accretion disk on a timescale of about a month and to attempt to distinguish between different trigger mechanisms for the outburst. The two rival mechanisms for dwarf novae outbursts are: first, a mass-transfer instability (MTI) on the secondary star which suddenly increases the mass-loss rate, and, second a disk instability (DI) in which the disk catastophically alters between different viscosity states thereby changing the mass-transfer rate through the disk (see, e.g., the review by Bath 1985). In the DI case the mass transfer from the secondary remains constant and hence the disk should show a gradual brightness increase between outbursts; an increase which has not been observed (Pringle *et al.* 1987). However, this could be counterbalanced by a cooling white dwarf. We therefore decided to investigate this further by a major systematic campaign of photometric and spectroscopic observations of a dwarf nova during all phases of an outburst-quiescence cycle.

A secondary objective was to study how normal outbursts evolve into superoutbursts in the SU UMa variables.

3 STRATEGY

These objectives could be achieved with a variety of different strategies. To fit within the confines of the International Time Conditions we had to choose a system with, first, short enough period to be able to easily remove orbital effects, and second, a short enough inter-outburst time to be able to guarantee at least one, and preferably two, outbursts during the campaign. After much discussion we adopted the following approach:

(*a*) That we would not mix different outburst cycles because they might be intrinsically different.

(*b*) That the best division of time was to take 60 consecutive nights at three hours each night.

(*c*) On any one night to cover one orbital period of one variable if at all possible.

(*d*) That our primary target should be SU UMa. We succeeded in obtaining about 100 hours of coverage of this star. This was the only object observed in our allocated 80 hours of *IUE* time although a substantial fraction of this time was used in slewing the satellite and reading out the cameras. Our secondary target was YZ Cnc on which we accrued 30 hours of ground-based observations. In addition to these two targets we secured smaller numbers of observations on IP Peg, U Gem, TY Psc, RX And, AF Cam, and IR Gem.

4 IUE OBSERVATIONS

The campaign lasted from 1988 Nov 29 to 1989 Jan 25 (JD 2447495 to 2447553). During this period there were three major outbursts of SU UMa at about JD-2440000 7516, 7532, and 7543 with a much smaller outburst at 7526.

Koji Mukai has searched for orbital modulation in SU UMa by dividing the ultraviolet fluxes $\lambda\lambda$ 1750–1850 Å and $\lambda\lambda$ 1420–520 Å by the mean for each day after discarding those days on which the star was varying too fast. He has folded these ratios on each of the 12 possible periods proposed by Thorstensen, Wade, and Oke (1986) and finds that the light curves with the least scatter correspond to their ninth period: 0.07628 days. The amplitude is 38 per cent at $\lambda\lambda$ 1420–1520 Å, which is surprisingly large for what is believed to be a low-inclination system.

On one occasion during a period of five hours we made four consecutive SWP observations interleaved with LWR observations. During these observations, SU UMa rose by a factor of two from quiescence at λ 1400 Å but by smaller factors at λ 2950 Å and in the visual as measured by the FES on *IUE*. The disk appeared to be becoming hotter as it brightened.

Because of our large number of spectra at quiescence we can derive a mean spectrum at high signal-to-noise. It shows the expected lines of N V, Si III, C II, Si IV, C IV, He II, Al III, and Mg II. During outburst C IV λ1550 Å shows a pronounced P Cygni profile which, if the system has a low inclination, would imply that we are seeing an expanding wind or corona above a bright disk (cf. Drew 1987). Similar line profiles have been observed in TW Vir (Córdova and Mason 1982) and EK TrA (Hassall 1985).

5 ISAAC NEWTON TELESCOPE

We devoted the 2.5-meter INT to optical spectroscopy with the Intermediate Dispersion Spectrograph. We opted to use the grism cross disperser so that the data appear in orders 3 to 7 of an echelette format covering the whole optical spectrum from λ9500 to λ3300 Å (Terlevich, Terlevich, and Charles 1989). We usually used the Image Photon Counting System (IPCS) but some of the observations were made with a conventional CCD. The IPCS has an S20 cathode with an effective long-wavelength limit of λ7000 Å. On each night we would follow one of our target variables continuously, taking a spectrum every two minutes. The sum of this data represents a challenging reduction problem which we are tackling with Koji Mukai's DILEMMA (**D**irect **I**mage **L**inearization and **E**xtraction **M**ethod for **M**aximum **A**ccuracy) optimal extraction routine.

6 JACOBUS KAPTEYN TELESCOPE

During the campaign the 1-meter JKT was fitted with a CCD camera with *UBVRI* filters. We used a GEC P8603 chip and binned the pixels 2×2 so that there were 200×300 elements, each of which were 0.6 arcsec square. On each night the 1-meter telescope observed the same variable as the nearby 2.5-meter. Each variable was placed on the chip so that there were two convenient comparison stars as well. On most nights we took a single set of colors followed by a continuous run in V. The primary function of these observations is to reduce the spectroscopic observations to absolute fluxes.

7 CARLOS SANCHEZ TELESCOPE

The 1.5-meter CST was used with an infrared JHK photometer. This could not adequately follow SU UMa and YZ Cnc through minimum so the star most frequently observed was U Gem.

8 CURRENT STATUS

There is still a substantial amount of work to do on reducing the data which must be completed before we can commence on the discussion. We hope to have significant results to announce in six months to a years time.

REFERENCES

Bath, G. T. 1985, *Rep. Prog. Phys.*, **48**, 483.
Córdova, F. A., and Mason, K. O. 1982, *Ap. J.*, **260**, 716.
Drew, J. E. 1987, *M. N. R. A. S.*, **224**, 595.
Hassall, B. J. M. 1985, *M. N., R. A. S.*, **216**, 335.
Pringle, J. E., *et al.* 1987, *M. N. R. A. S.*, **225**, 73.
Terlevich, E., Terlevich, R. J., and Charles, P. A. 1989, *Spectroscopy with the CCD on the INT*, I. N. G. User Manual *VII*.
Thorstensen, J. R., Wade, R. A., and Oke, J. B. 1986, *Ap. J.*, **309**, 721.

Spectroscopic Monitoring of SU UMa during Quiescence and Outburst

J. Casares[1], P. A. Charles[2], and J. van Paradijs[3]

[1]Instituto de Astrofisica de Canarias, Universidad de La Laguna
[2]Royal Greenwich Observatory, La Palma
[3]Sterrenkundig Institut "Anton Pannekoek," Universiteit te Amsterdam

1 INTRODUCTION

SU UMa is the prototype of those CVs which exhibit superoutbursts (more extended and brighter than normal outbursts) and superhumps, which have periods related to the orbital period (see Warner 1985). SU UMa also has frequent normal outbursts with typical recurrence times in the range 5 to 33 days, in which it undergoes V mag variations from 14.8 (quiescence) to 12.2 (outburst). With an inclination of $\sim 44°$, we would expect no significant orbital modulation and single-line profiles (Thorstensen, Wade, and Oke 1986). Two basic models (see, e.g., review by Bath 1985) have been invoked to explain the outburst behavior of dwarf novae:

1. Mass Transfer Instability (MTI) Model
The outburst is the result of a sudden increase in mass loss from the companion.

2. Disk Instability (DI) Model
The mass-transfer rate from the secondary is essentially constant. Nevertheless, the accretion disk follows a limit cycle, skipping between two α-states of high and low mass-transfer rates through the disk.

The two models make different predictions for the observed behaviour of the disk during the period between outbursts. Because the surface density increases in the DI model, we expect a secular brightness increase between outbursts. Furthermore, the relative size of the optically thin region decreases simultaneously and hence the equivalent width (EW) decrease monotonically. In the MTI model we do not expect any such systematic variation in brightness and EW between outburst.

There are relatively few observations that substantially constrain these models. One is the well-known UV delay, relative to the optical, in the onset to outburst for many dwarf novae (Verbunt 1987). In particular, VW Hyi undergoes a delay of 0.5 to 1 days, which is too long for the DI model predictions. Meyer and Meyer-Hofmeister (1987), however, could explain such behavior in the DI framework by including a modified limit cycle.

Furthermore, from the results of a large observing campaign (comprising *EXOSAT*, *Voyager*, *IUE*, and optical telescopes) on VW Hyi (Pringle *et al.* 1987), no evidence was found in this system for a secular brightness increase during quiescence. It is however possible to argue that the expected increase (in the DI model) in the quiescent disk brightness is counterbalanced by the decreasing brightness of the cooling white dwarf (heated during the previous outburst).

We therefore decided to undertake spectroscopic monitoring of a dwarf nova during the quiescent phase between outbursts in order to use EW measurements to test the model predictions. SU UMa was an ideal target because of its brightness and frequent normal outbursts, making the observing program tractable over a period of a few weeks.

2 OBSERVATIONS

SU UMA was observed with the 2.5-m Isaac Newton Telescope on La Palma equipped with the IDS and IPCS detector (for details, see Unger *et al.* 1988). The observations took place in the last two weeks of 1987, from December 14 to 28. We obtained a few spectra each night with typical integration times of 5 min and succeeded in following an entire quiescent cycle between outbursts, except for a gap in the nights 19 and 20 caused by a failure in the data acquisition system.

The wavelength range used was $\lambda\lambda$ 3500–5000, which covered the principal Balmer and He I lines with a typical resolution of 2 Å.

Fig. 1—Spectral evolution during quiescence. Note the double-peaked profiles on Dec 21 (*lower*) and how features change between the two spectra. Notice the increase of the peak separation with excitation level.

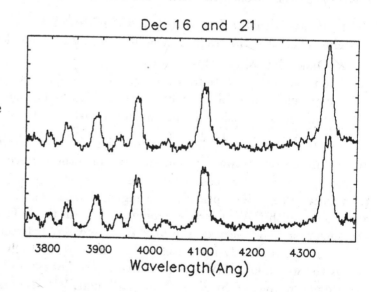

3 REDUCTION

Spectra were summed for each night, giving us a sample of 14 bins to cover the complete inter-outburst and outburst evolution. For Dec 15 we co-added the spectra in orbital phase bins in order to compare them and search for any orbital modulation in the line profiles which might otherwise confuse the interpretation of the secular variations.

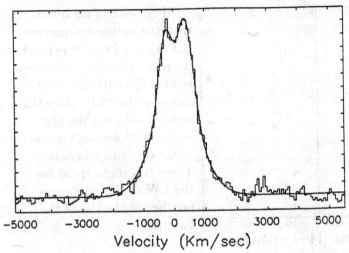

Fig. 2—This figure shows a three-component fit to Hγ: a broad component has been used for the wings and two narrow ones for the peaks. The velocity separation between peaks is $2V_d \sin i = 667 \pm 143$ km s^{-1}.

Looking at the general features, we note the presence of double-peaked profiles in Ca II K and high members of the Balmer lines, contrary to the results reported by Thorstensen, Wade, and Oke (1986). The double peak appears to be enhanced on Dec 21, and becomes especially clear in Hγ (Fig. 1). Multi-component fitting to Hε and Hγ gives a peak separation of $\sim$ 670 km s^{-1} (Fig. 2), of the same order as the value found by Persson (1988) in the similar system YZ Cnc.

Spectra were normalized relative to the continuum level and we then derived EWs for the main Balmer and He I lines using two different methods:

a) Multi-component fitting to the core and wings.

b) Total flux integration.

The results are shown in Figure 3.

4 DISCUSSION

We describe the inter-outburst behavior of the line intensities in SU UMa by dividing the outburst cycle into three different phases:

1. A steep rise during the first 2-3 days of quiescence after the previous outburst, the EWs increasing typically by 75%; the same behavior has been observed by Shafter and Hessman (1988) in YZ Cnc.

2. Stationary or "standstill" evolution with fluctuations < 20%; also observed by Thorstensen, Wade, and Oke (1986) in their orbitally-resolved data.

3. For Hε and the narrow component of Hβ we see a trend for the EW to increase $\sim$ 20-30 % just before the onset of the outburst.

Since these are preliminary results, we are not able to reach significant conclusions at this time until further analysis has been done and a theoretical framework established. However, it does seem that these results do not agree with the DI model

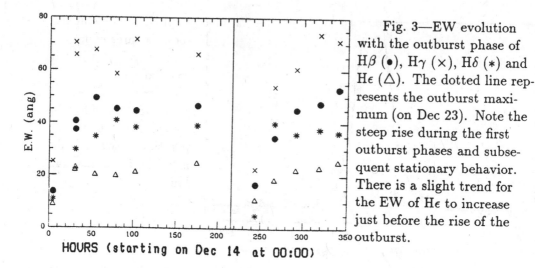

Fig. 3—EW evolution with the outburst phase of Hβ ($\bullet$), Hγ ($\times$), Hδ ($*$) and Hϵ ($\triangle$). The dotted line represents the outburst maximum (on Dec 23). Note the steep rise during the first outburst phases and subsequent stationary behavior. There is a slight trend for the EW of Hϵ to increase just before the rise of the outburst.

predictions. Nevertheless, we consider the detection of a clear double-peaked profile in Hγ to be highly relevant and indicative of short-term mass motions within the system at particular phases of outburst. We have insufficient data at present to determine whether this effect is an intrinsic feature of the outburst onset or an orbital phenomenon. If the velocity separation of the peaks were associated with Keplerian motion in the disk then the orbital parameters of Thorstensen, Wade, and Oke (1986) would imply that the disk radius was comparable to the size of the Roche lobe. However, as noted by Thorstensen, Wade, and Oke, it is notoriously difficult to determine orbital parameters on the basis of emission line motions.

We expect to confirm and strengthen these results with the substantial amount of data obtained in following up this work in the 1988 International Time Project. (see Jones *et al.* 1990).

REFERENCES

Bath, G. T. 1985, *Rep. Prog. Phys.*, **48**, 483.

Drew, J. E. 1987, *M. N. R. A. S.*, **224**, 595.

Jones, D. H. P., *et al.* 1990, this volume.

Meyer, M., and Meyer-Hofmeister, E. 1987, *Astr. Ap.*, **175**, 113.

Persson, S. E. 1988, *Pub. Astr. Soc. Pac.*, **100**, 710.

Pringle, J. E., *et al.* 1987, *M. N. R. A. S.*, **225**, 73.

Shafter, A. W., and Hessman, F. V. 1988, *A. J.*, **95**, 178.

Thorstensen, J. R., Wade, R. A., and Oke, J. B. 1986. *Ap. J.*, **309**, 721.

Unger, S. W., Brinks, E., Laing, R. A., Tritton, K. P., and Gray, P. M. 1988, ING Observer's Guide, La Palma. RGO.

Warner, B. 1985, In *Interacting Binaries*, ed. J. E. Pringle and P.P. Eggleton (Cambridge: Cambridge University Press), p. 367.

Z Cha : Superoutburst - Normal Outburst Comparison[†]

E. Harlaftis[1], T. Naylor[2], G. Sonneborn[3], B. J. M. Hassall[4], and
P. A. Charles[5]

[1]Department of Astrophysics, Oxford University, Keble Road, Oxford OX1 3RH
[2]Institute of Astronomy, Madingley Road, Cambridge CB3 0HA, U. K.
[3]Laboratory for Astronomy and Solar Physics, NASA Goddard Space Flight
  Center, Code 681, Greenbelt, MD 20771, U. S. A.
[4]Royal Greenwich Observatory, Madingley Road, Cambridge, CB3 0HA, U. K.
[5]Royal Greenwich Observatory, Apartado 321, 38780 Santa Cruz de La Palma,
  Tenerife, Canary Islands

1 INTRODUCTION

It was previously thought that normal outbursts and superoutbursts in SU UMa
systems were spectroscopically indistinguishable in the UV; VW Hyi, the prototype of
low-inclination SU UMa systems, shows such behaviour (Hassall *et al.* 1983; Verbunt
et al. 1987). We wished to investigate if this idea is true for eclipsing systems, where
the line-of-sight passes very close to the disk surface. There were indications from
limited observations extracted from the *IUE* archive that this is not the case for
eclipsing SU UMa systems. To resolve this problem, we obtained *IUE* observations
of the 1988 January normal outburst and 1987 superoutburst of Z Cha.

2 RESULTS

2.1 Flux Distribution

Figure 1 shows representative SWP spectra of the two outburst states. The normal outburst spectra have strong emission lines and a strong blue continuum. The
continuum has a similar flux distribution to a main-sequence B3–B4 star (T_{eff} =
18,000–20,500 K) (Heck *et al.* 1985). The superoutburst spectra also have strong
emission lines but continuum distributions which are much cooler and weaker, characteristic of a B8–B9 spectral type flux distribution (T_{eff} = 11,000–12,200 K).

2.2 Variation with Time and Orbital Phase

The 1987 *IUE* superoutburst observations gave very good orbital coverage on

---

† Based on observations by the *International Ultraviolet Explorer* collected at the
Villafranca Tracking Station of the European Space Agency and at the Goddard
Space Flight Center of the National Aeronautics and Space Administration.

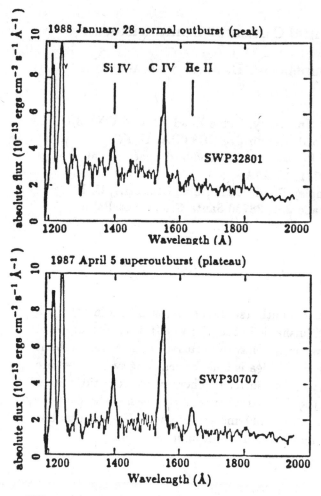

Fig. 1—Normal and super-outburst *IUE* spectra of Z Cha. The normal outburst spectrum is the upper spectrum taken at the peak of the outburst on 28 January with an FES magnitude of 12.78 at mean orbital phase 0.47. The superoutburst spectrum (lower panel) was taken on 5 April 1987 (the plateau stage) with an FES magnitude of 12.80 at mean orbital phase 0.73. The superoutburst spectrum at similar optical magnitude has a fainter and flatter continuum flux distribution with stronger emission lines (especially He II 1640 Å and N V 1240 Å).

different stages of the superoutburst. In previous work (Harlaftis *et al.* 1988), we reported a 50% drop of the mean out-of-eclipse continuum flux at orbital phase 0.75 at the peak of the superoutburst (1–2 April 1987). Figure 2 shows how this orbital modulation evolves through the superoutburst. The dip at orbital phase 0.75 is weak in the first light curve (31 March–1 April), very strong in the second (1–2 April), and fades away through the other two curves (5–7 April). Additional variability at around orbital phase 0.2 on 5 and 7 April might also be present.

The normal outburst observations in January 1988 also gave us good orbital coverage on 29 and 30 January. These observations show no obvious modulation of the continuum flux with orbital phase except the one caused by the disk eclipse (see Fig. 3). The continuum flux on 29 January shows considerable scatter compared to the 30 January data.

3 DISCUSSION

The superoutburst spectra of Z Cha are similar to those observed in the 1985 OY Car superoutburst (Naylor *et al.* 1987). The "cool" spectra observed there were

Fig. 2—The
1987 superoutburst
data of Z Cha plot-
ted against orbital
phase show a broad
dip at orbital phase
0.8 similar to that
seen in OY Car and
in WZ Sge (Nay-
lor *et al.* 1988; Nay-
lor 1989) This dip
is weak in the first
light curve, very
strong in the sec-
ond, and fades away
through the other
two curves. In all
the plots, points
near phase 0.1–
0.3 are lower than
points at around
phase 0.5. The con-
tinuum band used is
1470–1515 Å.

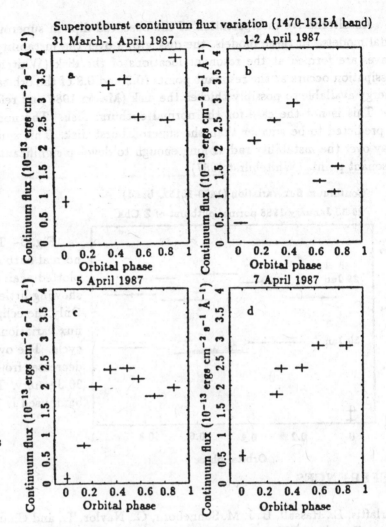

Superoutburst continuum flux variation (1470-1515Å band)

interpreted as being caused by extended azimuthal structure at the disk edge occulting
the hot disk center. The UV dips may then be associated with an azimuthal variation
of the disk's height. We interpret the 0.8 orbital phase dip in our data as being caused
by the impact region of the gas stream with the disk. A parallel to this can be found
in LMXRB systems (Parmar and White 1987) and in other CVs in the X-ray and
ultraviolet band (Mason 1986; Naylor 1988). Since the dip becomes shallower with
time as the superoutburst progresses, the mass transfer from the secondary may be
decreasing with time. The interpretation of the 0.2 orbital phase variability is not
clear, although there may be a parallel with the anomalous LMXRB dips.

In normal outburst, the spectra are "hotter" because the hot inner-disk regions
are not obscured. This implies that there is no significant opening angle of the disk
(i.e., it is a thin disk). The observed orbital phase variation implies no extended
azimuthal structure in the disk (i.e., it is flat).

An alternative interpretation of the observed dips in superoutburst is given by tidal models. In these models, the disk grows over the instability radius and spiral waves are formed at the resonant locations of the disk (Whitehurst 1988). Tidal dissipation occurs at the resonant points (0.2 and 0.8 of the orbital phase), releasing energy available to possibly thicken the disk (Mason 1986 and references therein).

This is not the case for the normal outburst disk. The normal outburst disk is predicted to be smaller than the superoutburst disk. This smaller disk does not stay over the instability radius long enough to develop significant dissipation at the resonant points (Whitehurst 1987).

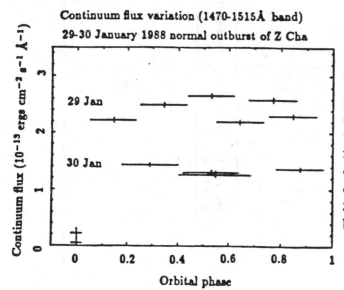

Fig. 3—The 1988 January normal outburst data of Z Cha plotted against orbital phase showing little orbital variation. Only the eclipse points cause flux variation in every orbital cycle. The overall flux level decreases from 29 January to 30 January. The continuum band used is 1470–1515 Å.

REFERENCES

Harlaftis, E., Hassall, B. J. M, Sonneborn, G., Naylor, T., and Charles, P. A. 1988, in *A Decade of UV Astronomy with the IUE Satellite*, ESA SP-281, Vol. 1, p. 187.

Hassall, B. J. M., *et al.* 1983, *M. N. R. A. S.*, **203**, 865.

Heck, A., *et al.* 1984, *IUE Low-Dispersion Spectra Reference Atlas – Part 1. Normal Stars*, ESA Spec. Publ. No. 1052.

Mason, K. O. 1986, in *Physics of Accretion on Compact Objects*, ed. K. O. Mason, M. G. Watson, and N. E. White, (Berlin: Springer-Verlag), p. 29.

Naylor, T., *et al.* 1988, *M. N. R. A. S.*, **231**, 237.

Naylor, T. 1989, *M. N. R. A. S.*, **238**, 587.

Parmar, A. N., and White, N. E. 1988, *Mem. Soc. Astron. Ital.*, **59**, No. 1–2, 147.

Verbunt, F., Hassall, B. J. M., Pringle, J. E., Warner, B., and Marang, F. 1987, *M. N. R. A. S.*, **225**, 113.

Whitehurst, R. 1987, Ph.D. thesis, University of Oxford.

Whitehurst, R. 1988, *M. N. R. A. S.*, **232**, 35.

SS Cygni on the Rise to an Anomalous Outburst

Keith Horne[1], Constanze A. la Dous[2], and Allen W. Shafter[3]

[1]Space Telescope Science Institute
[2]NSSDC, NASA Goddard Space Flight Center and
 Institute of Astronomy, University of Cambridge
[3]Astronomy Department, San Diego State University and
 Astronomy Department, University of Texas at Austin

1 INTRODUCTION

SS Cyg is the brightest known dwarf nova, with a visual magnitude of about 12 during quiescence and 8.2 during outburst. The mean interval between outbursts is roughly 50 days. SS Cyg has three types of outbursts — *short* outbursts lasting for several days and *long* outbursts lasting for over a week have fast rises shorter than a day, while *anomalous* outbursts (which in spite of their name are a rather frequent phenomenon in SS Cyg and other dwarf novae) have rises taking several days followed by an immediate decline.

We present optical spectrophotometry of SS Cyg on 18 nights, 12 covering the last half of a quiescent period, and 6 covering the entire rise into an anomalous outburst. An 8-hour period during the first magnitude of the rise is covered in detail.

2 THE OUTBURST LIGHT CURVE

The outburst light curve of SS Cyg spanning the time of our observations was very thoroughly covered and kindly provided to us by J. Mattei, Director of the AAVSO. In Figure 1 we plot the AAVSO light curve along with synthetic Johnson V magnitudes computed from each of our observed spectra.

Our first observation occurred almost in the middle of an 18-day quiescent phase following a long normal outburst. During most of this quiescent phase the system's light appears to have been stable around 11.9 mag. There is, however, some evidence for variations less than 0.5 mag, and a gradual rise in the general brightness level for 3–4 days or more before onset of the actual outburst.

The following outburst was of the anomalous type, i.e., it had a slow rise, which our spectra cover, and an immediately following decline. The rise to outburst develops in three distinct stages, first a fast rise of 1 mag in 1 day, then a slower rise by 1 mag in 3 days, and finally a fast rise of 1.5 mag in 1.5 day.

We find generally good agreement between our spectrophotometry and the AAVSO observations. There is a minor systematic difference, in the sense that our data are 0.1 mag fainter in quiescence and 0.1 to 0.2 mag brighter in outburst, which may arise from spectral changes in SS Cyg combined with somewhat different response curves

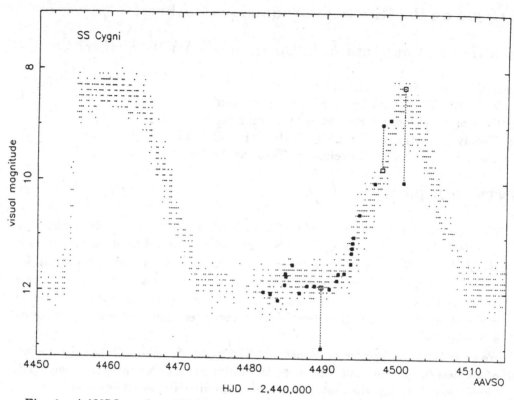

Fig. 1—AAVSO outburst light curve of SS Cyg during the time of our observations. The times when individual spectra were taken are indicated by black squares. The fluxes of three non-photometric observations were corrected by applying grey shifts as indicated.

between the Johnson V passband and the eyes of AAVSO observers. Three of our spectra were taken on non-photometric nights, and we have applied grey magnitude shifts, as indicated in Figure 1, to bring these spectra into better accord with the AAVSO data.

3 SPECTRAL EVOLUTION ON THE RISE TO OUTBURST

Figure 2 shows 13 spectra of SS Cyg on the rise to outburst. The spectra fainter than 11th magnitude, covering quiescence and the first night of the rise, were taken at 21 Å resolution with the SIT spectrograph on the Palomar 60-inch telescope. Those brighter than 11th magnitude, covering the remainder of the rise to outburst, have 10 Å resolution and were taken with the IDS spectrograph on the 40-inch Nickel reflector at Lick Observatory.

The lowest three spectra in Figure 2 represent averages of spectra taken near the beginning, middle, and end of our coverage of SS Cyg in quiescence. These quiescent spectra have a red continuum slope due to the contribution of the late-type

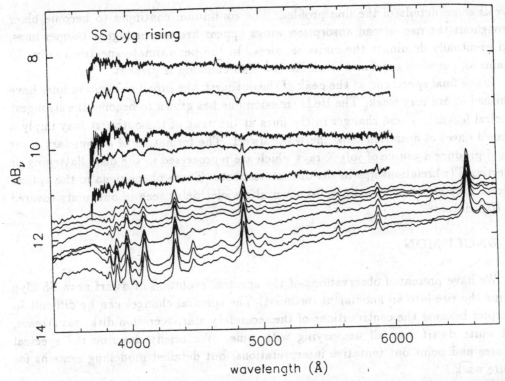

Fig. 2—Spectra of SS Cyg rising into an anomalous outburst. The fainter spectra have 21 Å resolution and were taken at Palomar. The lowest three are averages of spectra taken during the quiescence; from bottom to top are Q1 (= average of spectra 1, 2, 3), Q3 (= average of 8, 9, 10), and Q4 (= average of 12, 13, 14, 15). The next 5 spectra (16 through 20) were taken in an 8-hour period on the first night of the rise. The highest five spectra (21 through 25) have 10 Å resolution and were taken from Lick observatory. Spectrum 25 was taken at maximum optical brightness.

secondary star. H I and He I emission lines and Balmer continuum emission are from the accretion disk. The continuum becomes somewhat bluer, and the emission line fluxes and Balmer jump increase somewhat through this quiescent interval.

More striking spectral changes are evident in the next 5 spectra, which were taken in an 8-hour period on the first night of the rise. In this first magnitude of the rise, the continuum becomes progressively bluer while the emission line and Balmer continuum fluxes decrease. These changes can be interpreted as the appearance of new source of light which has a strong blue continuum, Balmer jump, and lines in absorption. Alternatively, the changes could represent a rising featureless blue continuum coupled with a decrease in the optically thin region that produces the emission lines and Balmer continuum emission.

The top 5 spectra in Figure 2 cover the remainder of the rise. The Balmer jump and Hα are not covered in these Lick spectra, but the increased resolution reveals some details of the line profiles. The continuum continues to become bluer throughout the rise. Broad absorption wings appear first in the higher Balmer lines, and eventually dominate the emission cores. In the penultimate spectrum only Hβ retains an emission core, and weak He II λ4686 emission is present.

In the final spectrum, at the peak of the outburst, the broad absorption lines have vanished or are very weak. The He II emission line has grown to become the strongest spectral feature. These changes in the lines at the peak of the outburst may imply a delayed onset of accretion onto the white dwarf. The formation of a boundary layer would produce a source of soft X-rays which are reprocessed in the disk, flattening or inverting $T(\tau)$ relationships in the disk atmosphere. These phenomena in the optical spectra may be related to the well-known ultraviolet delay seen in outbursts covered by *IUE* data.

4 CONCLUSION

We have presented observations of the spectral evolution of dwarf nova SS Cyg during the rise into an anomalous outburst. The spectral changes can be difficult to interpret because the contributions of the secondary star, accretion disk, gas stream, and white dwarf may all be varying with time. We briefly describe the spectral changes and point out tentative interpretations, but detailed modelling remains for future work.

We expect these observations to be useful for testing quantitative predictions of dwarf nova outburst models, in particular the disk instability models which attribute the outburst to a modulation in the viscosity and hence the mass-transfer rate of the accretion disk. More complete details of the observations are in preparation for publication elsewhere.

The Secondary in WW Ceti

N. A. Hawkins[1], R. C. Smith[1], and D. H. P. Jones[2]

[1]Astronomy Centre, University of Sussex
[2]Royal Greenwich Observatory, Herstmonceux

1 INTRODUCTION

We present results from a series of high resolution ($\sim$1.45Å/pixel) near infra-red (7700 Å–8500 Å) spectra of WW Ceti taken over 3 nights on the AAT during November 1988. The CCD spectra were optimally extracted (Horne 1986) using PAMELA and FIGARO on the Royal Greenwich Observatory STARLINK node.

The spectra were cross-correlated with a series of field red dwarf spectra to obtain radial velocity measurements of the secondary star from the strong features in the region (principally the NaI doublet at $\lambda\lambda$ 8183.3, 8194.8 Å). The red dwarf spectrum which gave the least noisy results was Gliese 176 which gives us a rough estimate of the spectral type of the secondary (M2.5).

The table below lists the (heliocentric) observation times and measured (heliocentric) radial velocities. The radial velocity of Gliese 176 was taken to be +25.97 km s^{-1} (Marcy *et al.* 1987), and all exposures were 1024 seconds long.

UT mid-exposure	Radial Velocity (km s^{-1})
22nd Nov. 1988 12:41:05	27.8
22nd Nov. 1988 13:11:16	180.0
22nd Nov. 1988 13:29:17	220.3
22nd Nov. 1988 13:47:17	209.0
23rd Nov. 1988 12:46:42	−194.5
23rd Nov. 1988 13:05:07	−167.6
24th Nov. 1988 12:28:09	193.8
24th Nov. 1988 12:47:47	149.7
24th Nov. 1988 13:05:45	41.7
24th Nov. 1988 13:23:53	−154.7
24th Nov. 1988 13:42:16	−189.7
24th Nov. 1988 14:00:24	−221.5

2 RESULTS: PERIOD DETERMINATION

Thorstensen and Freed (1985, hereafter TF) give two possible periods for WW

Fig. 1—Average
spectrum of
WW Cet. Individual
spectra have been
shifted to correct for
orbital RV variations
before adding.

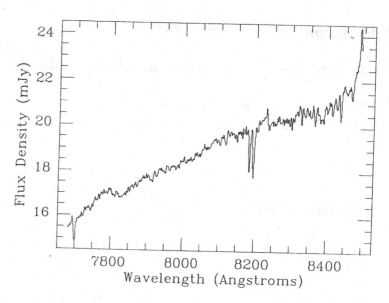

Cet due to cycle-count alias problems. These periods are 0.17578 days (±0.00013 days) and 0.14945 days (±0.00011 days), with a slight preference being given to the longer of the two.

In an attempt to choose between these 2 periods, a string length method was used to find possible periods. A Monte Carlo technique similar to that of TF was then used to check the probability of each period being the correct one.

Fig. 2—String length
diagram for WW Cet. All
minima below the dotted line
are significant, and the
dashed line shows the
theoretical minimum string
length.

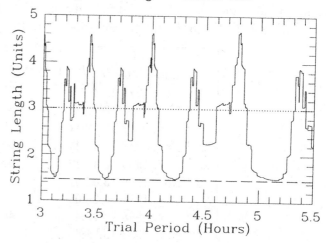

Figure 2 shows the string length diagram over the range 3–5.5 hours (0.125–0.208 days). All of the minima except the 4 deepest were rejected simply by examining the orbital fits at these periods. These 4 minima correspond to periods of 3.123, 3.586, 4.236, and 5.168 hours, respectively. The second and third of these periods match the two periods of TF to within error limits.

To test the significance of the 4 periods, a Monte Carlo method was used. A sinusoid at our 'best' period (4.236 hours—the one with the smallest string length) was sampled at the same phases as our original observations with noise added at a median value for the 4 trial periods (28 km s^{-1}). The string length at each of the trial periods was then calculated and the *contrast ratio* for each then found. The contrast ratio is defined here as the ratio of the string lengths at the 'best' and alias periods. This was repeated 1000 times, with a record being kept of the contrast ratio found and of which period gave the shortest string length. The percentage of results close to the observed contrast ratio which also select the 4.236 hour period as best is called the *correctness likelihood* of the data set and is an estimate of the probability of our chosen best period being the true period.

Using this technique on our data gave correctness likelihoods for the 4.236 hour period of 75%, 62%, and 87% when compared with the 3.12, 3.58, and 5.16 hour periods, respectively.

This result does not give any real indication as to which of the 2 periods is correct, but does show a slight preference for the longer period. The poor quality of this result is due to the low number of observations which were obtainable.

Fig. 3—Circular orbit fit for WW Cet at the shorter (3.586 hour) period. Zero phase is defined as the average observation time, and the bar shows the size of the RMS residuals.

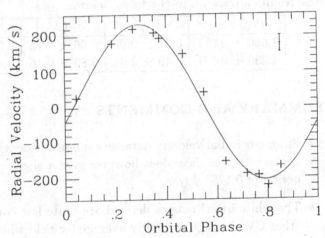

3 RESULTS: SYSTEM PARAMETERS

Circular and elliptical fits were made to the data at the two possible periods, though the elliptical fits were not significantly better than the circular ones. The values of K_2 found from the fits, the RMS deviation, and the estimates of the systemic velocity are shown in the table below together with the relevant values of K_1 from TF. Figures 3 and 4 show the circular orbital fits at the two periods.

P (hours)	K_2 (km s^{-1})	V_0 (km s^{-1})	σ (km s^{-1})	K_1 (km s^{-1})
3.586 ± 0.017	220 ± 9	14 ± 7	23	111 ± 9
4.236 ± 0.021	222 ± 13	-4 ± 15	33	108 ± 7

From the above data, we can calculate the mass ratio and then, by assuming the secondary is a main-sequence star (Patterson 1984) which exactly fills its Roche lobe

Fig. 4—Circular orbit fit
for WW Cet at the longer
(4.236 hour) period. Zero
phase is defined as the
average observation time,
and the bar shows the size of
the RMS residuals.

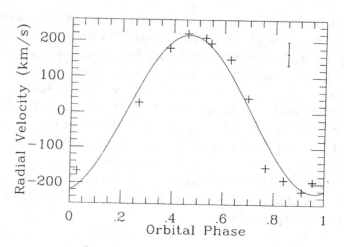

(Eggleton 1983), we can find estimates of M_2, M_1, and the inclination of the system. These results are shown in the table below.

P (hours)	q	M_1 $(M_\odot)$	M_2 $(M_\odot)$	i
3.586 ± 0.017	0.51 ± 0.05	0.66 ± 0.13	0.33 ± 0.06	$56° \pm 9°$
4.236 ± 0.021	0.49 ± 0.04	0.83 ± 0.16	0.41 ± 0.07	$54° \pm 9°$

4 SUMMARY AND COMMENTS

- From our radial velocity data we cannot decide which of the two periods of TF is correct. Our data does however give a slight preference for their preferred period of 0.1758 days.

- The white dwarf masses derived are quite low compared with those found for other CVs, though are fairly average for field white dwarfs. This also tends to suggest that the longer period is correct, as it gives the larger mass.

- Allen (1973) gives the mass of an M2 dwarf as 0.39 $M_\odot$, so our derived masses are consistent with a main-sequence star.

NAH wishes to thank the S.E.R.C. for his grant. Thanks also to staff at the AAO and RGO and to J. Martin and M. Friend for use of their programs.

REFERENCES

Allen, C. W. 1973, *Astrophysical Quantities (3rd edition)* (London: Athlone Press).
Eggleton, P. P. 1983, *Ap. J.*, **268**, 368.
Horne, K. 1986, *Pub. Astr. Soc. Pac.*, **98**, 609.
Marcy, G. W., Lindsay, V., and Wilson, K. 1987, *Pub. Astr. Soc. Pac.*, **99**, 490.
Patterson, J. 1984, *Ap. J. Suppl.*, **54**, 444.
Thorstensen, J. R., and Freed, I. W. 1985, *A. J.*, **90**, 2082.

LB1800: An Eclipsing CV Counterpart of a Transient X-ray Source

David A. H. Buckley[1] and Denis J. Sullivan[2]

[1]Department of Astronomy, University of Cape Town, South Africa
[2]Physics Department, Victoria University of Wellington, New Zealand

1 INTRODUCTION

The $V \sim 13$ star LB1800 (Luyten and Anderson 1958) has been found to be a deeply *eclipsing* ($\Delta V \sim 2.4$) novalike cataclysmic variable. An orbital period of 5.566 hr has been determined from the photometric and radial velocity variations. LB1800 has also been identified as the optical counterpart of the transient hard X-ray source 4U0608-49.

2 IDENTIFICATION AND X-RAY OBSERVATIONS

LB1800 was identified as a cataclysmic variable following spectroscopy carried out with the Mount Stromlo 1.9-m telescope in 1986. The discovery spectra showed typical CV characteristics: Balmer, He I, and He II emission lines on a blue continuum. The equivalent widths are typically: Hδ = 16 Å, Hγ = 12 Å, He I λ4471 = 4 Å, He II λ4686 = 17 Å, Hβ = 17 Å, and Hα = 36 Å.

LB1800 was found to be inside the error box of the previously unidentified *Uhuru* X-ray source, 4U0608-49. Subsequent analysis of *HEAO-1* data confirmed the existence of LB1800 as a transient, hard X-ray source. The 2–10 keV flux varies from below $\sim 1 \times 10^{-11}$ erg cm^{-2} s^{-1} to 7.4 $\times 10^{-11}$ erg cm^{-2} s^{-1}. With an estimated 5000–6000 Å optical flux of $\sim 1.2 \times 10^{-11}$ erg cm^{-2} s^{-1}, the ratio $F_x/F_{opt} \sim 4$ (though the observations were *not* simultaneous). This ratio, typical of magnetic CVs (Córdova and Mason 1984), led us to investigate whether LB1800 was significantly polarized. We obtain an upper limit of 0.5% over the region 3400 Å–1.6 μm.

3 RADIAL VELOCITY AND SPECTRAL LINE VARIATIONS

Time-resolved spectroscopic observations were undertaken on 1987 January 27 and 1988 January 24 using the Mt. Stromlo 1.9-m telescope. Several independent methods were used to derive radial velocities of the emission lines: (1) cross correlations, (2) Gaussian profile fitting, and (3) the Gaussian convolution scheme (e.g., Schneider and Young 1980; Shafter 1985). Figure 1 shows the 1987 radial velocity curve and sine fit for the Balmer cross correlations. The 1988 data are similar, but show evidence of a rotational distortion at superior conjunction of the white dwarf. This corresponds to within phase 0.03 ± 0.02 of the predicted photometric minimum (Section 4). Gaussian convolution velocities indicate that the upper limits of K-velocity and zero phase variations are ~ 20 km s^{-1} and 0.05, respectively, for line wing positions from ~ 400 to 1200 km s^{-1} from line center.

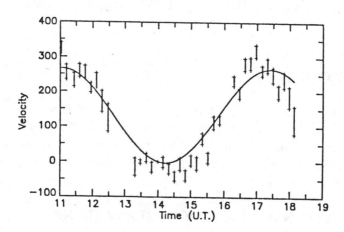

Fig. 1—Radial velocity variations of the Balmer lines derived using the cross correlation method for the 1987 Jan 27 data.

The line parameter variations (intensity, equivalent width, and sigma) for the better quality 1988 data show that the lines become both weaker and narrower during eclipse. Also, the eclipse widths are the same for both the high- and low-excitation lines (e.g., He II and Hβ). This behavior is in marked contrast to other high-excitation eclipsing novalikes, for example V1315 Aql (KPD1911+1212; Downes *et al.* 1986) and SW Sex (PG1012+029; Honeycutt, Schlegel, and Kaitchuck 1986). In the former case, the radial extent of the high-excitation region is less than for the low, while in the case of SW Sex, there must be a vertical extension of the line-emitting regions.

In an effort to understand the emission profile variations, the spectra were phase binned over the 5.566 hr photometric (i.e., orbital) period. While the lines do exhibit temporal variations in their structure, the *persistent symmetric* double-peaked profiles, typical of high-inclination CVs, are not always observed. A double peaked, or 'splitting,' of the profile is observed at certain orbital phases (e.g., 0.8–1.0). This is coincident with an increase in the strength of Ca II absorption. The fact that the absorption reversal in the Balmer lines increases in strength towards shorter wavelengths is perhaps evidence of an opacity effect rather than a dynamical line splitting. The occurrence of this around phase 0.8–1.0 occurs when the accretion stream is seen projected against the bright disk.

   Also noticeable is the variation in the relative strength of the red and blue peaks of
the Balmer lines. Spectra in the phase interval 0.6–0.8 have blue peaks considerably
stronger than red, and possibly *vice versa* for the phase interval 0.3–0.4. Such a
variation could be explained by the superimposition of a component produced in the
disk bright spot.

4 PHOTOMETRY AND LIGHT CURVES

   Photometric observations have been conducted on LB1800 since Feb 1987. These
data have primarily been obtained with the Mt. John University Observatory (MJUO)
0.6- and 1.0-m telescopes, employing a two-star photometer operating in white light.
A single *UBVRI* light curve was obtained with the ANU 2.3-m telescope. The typical
(out of eclipse) magnitude and colors of LB1800 were: $V \sim 13.4$, $U - B = -0.6$,
$B - V = 0.16$, $V - R = 0.25$, and $V - I = 0.60$. An eclipse of LB1800, in white light,
is shown in Figure 2.

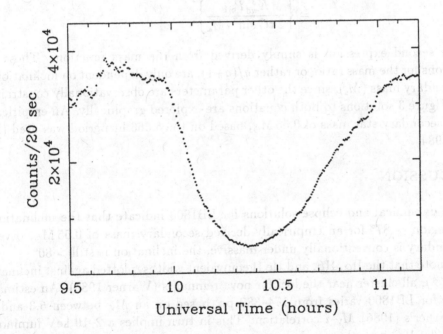

Fig. 2—White light eclipse curve of LB1800 obtained on 1987 Feb 20 using the
0.6-m telescope at MJUO.

   Onset of the eclipse is quite sudden, and relatively easily measured, unlike many
other systems whose eclipse wings merge into the out-of-eclipse flickering. The phase
of first contact, and half-width at half eclipse depth were estimated as $\phi_f = 0.105 \pm$
0.005 and $\phi_{1/2} = 0.052 \pm 0.002$, respectively.

The orbital ephemeris, based on the 1987 photometry, is given as:

$$T_{min}(H.J.D.) = 2446836.9620\,(\pm 0.0010) + 0.231928\,(\pm 0.000020)\,E\,.$$

5 MODEL AND DYNAMICAL SOLUTION

The K-velocity of 134 ± 9 km s^{-1}, obtained from the relatively uncomplicated 1987 radial velocity curve, and orbital period of 5.566 hr were used to obtain a secondary mass function of $5.8 \pm 0.3 \times 10^{-2}\,M_{\odot}$. The two formulae for the secondary radius, based on the filled Roche lobe (Paczynski 1971) and thin disk geometries (Penning et al. 1984), respectively, were used to derive the formulae

$$\frac{q}{(q+1)} = \left\{ \frac{\cos^2 i + \sin^2 i \sin^2(2\pi\phi_{1/2})}{(0.462)^2} \right\}^{3/2} \tag{1}$$

$$= \left\{ \frac{K_{wd}^3 P_{orb}}{2\pi G \sin^3 i M_s} \right\}^{1/2}\,. \tag{2}$$

The second expression is simply derived from the mass function. These two expressions for the mass ratio, or rather $q/(q+1)$, are only dependent on inclination (i) and secondary mass (M_s), since the other parameters are observationally constrained.

In Figure 3 solutions to both equations are explored graphically. An empirically-derived secondary star mass of $0.55\,M_{\odot}$, based on the 5.566 hr period was used (Patterson 1984).

6 DISCUSSION

The dynamical and eclipse solutions for LB1800 indicate that the inclination is in the region $\sim 87°$ for an empirically-derived secondary mass of $0.55\,M_{\odot}$. Even if the secondary is conventionally under-massive, the inclination is still $> 80°$.

We note that the Hα, Hβ, and He II equivalent widths, plotted against inclination (~ 80–$87°$), all cluster near the loci for nova remnants (Warner 1986). An estimated distance for LB1800 varies from 174–380 pc, based on an M_V between 5.3 and 7.0 from Warner's (1986) M_V-i correlation. This in turn implies a 2–10 keV luminosity of ~ 3–13×10^{33} erg s^{-1}.

LB1800 exhibits strong emission lines of the high-excitation species He II $\lambda4686$ and C III/N III $\lambda\lambda4640$–4650, with strengths characteristic of magnetic variables, although spectropolarimetry rules out an AM Her classification. In many respects, LB1800 resembles the other high-excitation novalike systems, although the radial velocity variations, and phasing, are better behaved. A detailed paper on this system will appear shortly (Buckley et al. 1990).

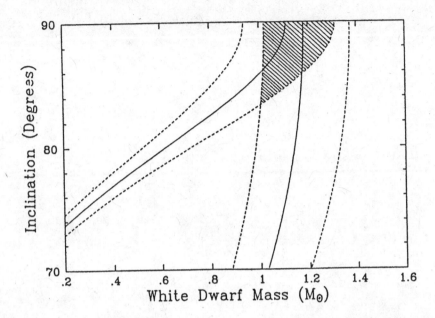

Fig. 3—Diagnostic diagram for the eclipse and dynamical solutions for LB1800 assuming an empirical secondary star mass of $0.55\,M_\odot$. The shaded region is the solution 'phase space,' defined by the uncertainties on the K-velocity and orbital period.

REFERENCES

Buckley, D. A. H., *et al.* 1990, *Ap. J.*, submitted.

Córdova, F. A., and Mason, K. O. 1984, *M. N. R. A. S.*, **206**, 879.

Eggleton, P. P. 1983, *Ap. J.*, **268**, 368.

Honeycutt, R. K., Schlegel, E. M., and Kaitchuck, R. H. 1986, *Ap. J.* **302**, 388.

Luyten, W. J., and Anderson, J. H. 1958, *A Search for Faint Blue Stars XII* (Minneapolis: University of Minnesota).

Paczynski, B. 1971, *Ann. Rev. Astr. Ap.*, **9**, 183.

Patterson, J. 1984, *Ap. J. Suppl.*, **54**, 443.

Penning, W. R., *et al.* 1984, *Ap. J.*, **276**, 233.

Schneider, D. P. and Young, P. 1980, *Ap. J.*, **238**, 946.

Shafter, A. W. 1985, in *Cataclysmic Variables and Low-Mass X-ray Binaries*, ed. D. Q. Lamb and J. Patterson (Dordrecht: Reidel), p. 355.

Warner, B. 1986, *M. N. R. A. S.*, **222**, 11.

High-Resolution Spectroscopy of IX Velorum

Frederic V. Hessman

Max-Planck-Institut für Astronomie, Heidelberg
Guest Observer at the European Southern Observatory, La Silla, Chile

1 INTRODUCTION

The novalike variable IX Vel (= CPD -48 1577) is one of the brightest cataclysmic variables and shows all of the classic symptoms of a UX UMa system: a strong blue continuum and broad but shallow Balmer and He I absorption lines with weak emission cores. Beuermann and Thomas (1989; hereafter BT89) have studied the orbital variations of the lines in considerable detail and have even seen a narrow component due to chromospheric emission from the secondary star (Beuermann and Thomas 1987). The broader central emission components are of unknown origin. In the hope of determining the structure and ultimately the origin of such components, I have observed IX Vel at much higher spectral resolution.

2 THE OBSERVATIONS

The high-resolution spectra were taken with the ESO 3.6-m Cassegrain echelle spectrograph (CASPEC) using an RCA CCD. Two spectra covering the wavelength region of 3900 to 4800 Å with a resolution of ~15 km s^{-1} were obtained on 1987 February 10. Two nights later, I took a short series of spectra centered around Hβ using exposure times of 5 minutes (~2% of the orbital period). Because these spectra were taken primarily to determine the feasibility of a detailed study, only 20 spectra covering the phases −0.23 to 0.22 were obtained.

The average blue spectrum is shown in Figure 1. An unexpected wealth of absorption lines is evident. In addition to the Balmer lines, a whole series of He I and other absorption lines can be seen: $\lambda\lambda3889, 4026, 4388, 4471, 4713$, and the Ca II $\lambda3934$ line (though $\lambda4471$ shows a rather complicated structure). Unfortunately, the whole region around the Hϵ line has been severely damaged by a bad CCD column, making it difficult to disentangle the blend of Hϵ and Ca II $\lambda3968$. Although the absorption lines are fairly weak (e.g., EW($\lambda4026$)=1.1 Å), and so difficult to see in low-dispersion spectra, they have great potential as probes of the accretion disk: not suffering from as much (obvious) central emission and having higher excitation potentials than the Balmer lines, these features may be accurate tracers of the disk's orbital motion.

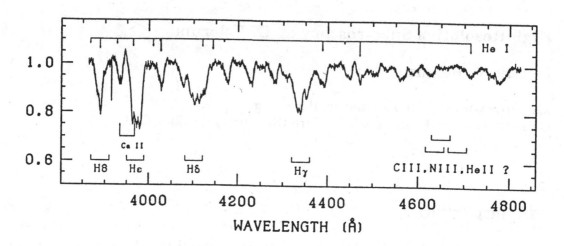

Fig. 1—The mean blue spectrum of IX Vel. Note the large number of absorption lines and the absence of any appreciable He II λ4686 or C III and N III emission. The narrow feature at λ3914 is unfortunately a particularly insidious flat-field effect.

Strangely absent in the blue spectra are the emission lines of He II λ4686 and C III λ4690 and N III λ4640. Low-resolution spectra were taken on the following night and show normal high-excitation emission levels. The echelle Hβ spectra taken 2 nights later show very weak He II emission. Wargau *et al.* (1983) also observed variations of factors of 2 in the strength of the He II line, so these observations may simply be more extreme cases of similar behavior. There is little sign of interstellar Ca II λλ3934, 3968 absorption: the limit on the equivalent width of the λ3934 line is 22 mÅ.

The mean Hβ profile — in the rest frame defined by BT89's ephemeris for the central emission component — is shown in Figure 2. The absorption wings are very broad (reaching out to around ±2500 km s^{-1}) and the main emission component has a parabolic shape. Narrower emission and absorption components have been averaged into the mean spectrum, yielding the residual structure of the central emission line. Also visible is the He I absorption/emission line at λ4922. Unlike the profiles seen by BT89, there is little sign of a central emission component in the latter line. BT89 measured a mean K-velocity of 140 km s^{-1} for both the broad absorption and narrower emission components. The radial velocities for these features in our echelle data are shown in Figure 3. Given the small portion of orbital phases covered, it is not possible to determine independent amplitudes, but the absorption line velocities are certainly consistent with those of BT89. Note, however, that the amplitude of the broad emission components is systematically smaller than seen previously by BT89 — much smaller than can easily be explained by resolution or measurement effects. This result casts doubt upon the efficacy of the central emission line as a radial velocity probe of the inner disk. There are too few spectra for the radial velocity variations

Fig. 2—The mean Hβ profile of IX Vel in the frame defined by the broad emission component in Beuermann and Thomas' (1989) spectra.

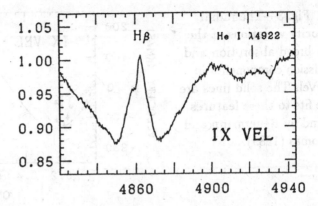

of the chromospheric emission component to be accurately measured, but the radial velocities of the peak around phase 0.1 seem to be ~90° out of phase with those seen by BT89.

3 THEORETICAL DISK SPECTRA

Given the detailed line profiles, we can now compare the observations with the theoretical atmosphere calculations described by Adam *et al.* (1989). IX Vel is a particularly good choice for this comparison, since it appears to be accreting steadily, making it possible to make detailed comparisons of the observations and the complex theoretical models. By choosing the right set of parameters, it is possible to fit the observed continuum of IX Vel quite accurately from the ultraviolet to the infrared using these models (Shaviv and Wehrse 1989).

The lower layers of the disk atmospheres are quite dense in IX Vel, creating the pressure-broadened Balmer absorption lines. Due to the assumption of a constant α viscosity parameter throughout the disk ("microphysics"), the outer layers of the disk atmospheres develop coronae where the optically thin cooling has a difficult time keeping up with the viscous heating. The "chromospheric" transition layer between the optically thick and thin parts of the atmosphere provides a natural mechanism for the production of the emission line cores. The *form* of the emission cores is determined by several processes in addition to the form of the source function, most importantly the Doppler-broadening due to the Keplerian motion of the gas, and the local effects of thermal, turbulent, and/or pressure broadening.

The width of a Doppler-broadened central emission line is set mainly by the outer radius of the emission region unless the flux distribution is highly peaked towards the center. The addition of thermal and/or turbulent broadening smooths the sharp peaks but should not fundamentally change the character of the line profiles at high spectral resolutions. Since the Keplerian velocities of the outer chromospheric regions are very large compared to the dispersion of the echelle data, we might expect to see the double-peaks from Doppler-broadened emission regions. Slight doubling of the central emission is occasionally seen, but the mean profile has more of a parabolic

126   F. V. Hessman

Fig. 3—The radial
velocity variations of the
Hβ broad absorption and
emission components in
IX Vel. The solid lines are
the fits to these features
found by Beuermann and
Thomas (1989).

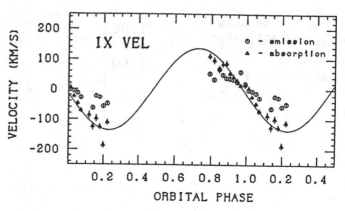

shape than that expected from simple Keplerian motion.

It is very difficult to predict the amount of smearing expected in the spectra due to pressure broadening. While Stark broadening is not unexpected in the denser, deeper layers responsible for the absorption lines, one would not immediately expect to see too much pressure broadening in upper regions of the disk which are diffuse enough to be optically thin. Indeed, the standard Wehrse-Shaviv models do not produce pressure-broadened emission cores. Given the expected temperatures of $1-2 \times 10^4$ K in the chromosphere, electron densities of over $10^{13}\,\mathrm{cm^{-3}}$ are required to totally overwhelm the Keplerian broadening. If one attributes the lack of simple double-peakedness in the central emission cores to Stark broadening, the observations may be used to probe the densities in the chromospheric regions. Since these densities depend upon the amount of dissipation in the upper atmospheres of the disks, and are not yet being produced in the current generation of disk models, we may be able to learn something about the viscous mechanisms operating in the outer, optically thin parts of otherwise optically thick disks.

REFERENCES

Adam, J., Innes, D., Shaviv, G., Störzer, and Wehrse, R. 1989, in *Theory of Accretion Disks*, NATO ASI Series C, Vol. **290** (Dordrecht: Kluwer), p. 403.

Beuermann, K., and Thomas, H.-C. 1987, *Mitt. Astr. Ges.*, **70**, 369.

Beuermann, K., and Thomas, H.-C. 1989, *Astr. Ap.*, submitted.

Shaviv, G., and Wehrse, R. 1989, in *Theory of Accretion Disks*, NATO ASI Series C, Vol. **290** (Dordrecht: Kluwer), p. 419.

Wargau, W., Drechsel, H., Rahe, J., and Bruch, A. 1983, *M. N. R. A. S.*, **204**, 35p.

Wehrse, R., Störzer, G., and Duschl, W. 1989, preprint.

V1315 Aquilae — Why are the Emission Lines Single-Peaked?

V. S. Dhillon[1,2], T. R. Marsh[3], and D. H. P. Jones[2]

[1]University of Sussex, U. K.
[2]Royal Greenwich Observatory, U. K.
[3]Space Telescope Science Institute, U. S. A.

1 INTRODUCTION

Novalike variables form a class of cataclysmic binaries which have a high mass-transfer rate and therefore particularly bright accretion disks. These disks are characterized by the strong Balmer emission lines in their spectra. In high-inclination systems, we expect to see the double-peaked line profiles that result from the rotation of an edge-on accretion disk. However, several such systems, for example V1315 Aql, PG 1012–029, and PG 1030+590, have symmetric, single-peaked profiles which show a double-peaked structure only near the inferior conjunction of the emission line source, i.e., when the disk is between the secondary and the observer. We have investigated this anomaly by means of orbitally-resolved spectrophotometry of V1315 Aql.

2 EMISSION LINE VARIATIONS

On the night of 1988 April 27/28, 139 spectra of V1315 Aql covering 1.0 orbits were taken using the 2.5-m Isaac Newton Telescope on La Palma. We cast the data into 10 binary phase bins by averaging all the spectra falling into each bin. The result is shown in Figure 1.

A comparison with the spectra of other high-excitation eclipsing novalikes (e.g., PG 1030+590 and PG 1012–029) shows the great similarity of their spectra. These objects show a number of features very different from those exhibited by high-inclination dwarf novae (e.g., IP Peg) — the Balmer and He I lines are narrower and show symmetric, single-peaked profiles, and the high-excitation lines of He II $\lambda4686$ and C III/N III $\lambda\lambda4640$–4650 are significantly stronger.

The Balmer lines are single-peaked except during phases 0.35–0.55, when they show a double-peaked structure accompanied by a significant decrease in line flux. The He I lines at $\lambda4921$ and $\lambda4471$ also exhibit the same behaviour. However, He II $\lambda4686$ and the C III/N III blend at $\lambda\lambda4640$–4650 appear to be single-peaked for the whole orbit. The primary eclipse has very little effect upon the Balmer and He I line profiles. Although there is a general decrease in line flux and width, it is a very small effect when compared to dwarf novae eclipses (e.g., IP Peg). This suggests that there is a source of low-velocity emission other than the disk, since the disk is

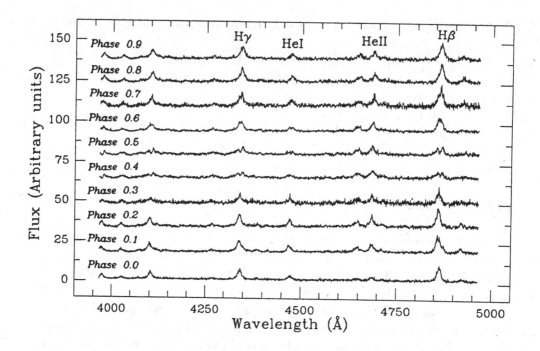

Fig. 1—Orbital emission line variations in V1315 Aql. The data has been averaged into 10 binary phase bins with a multiple of 15 added to each spectrum in order to displace the data in the y-direction.

seen to be deeply eclipsed in the continuum. The lines of He II $\lambda4686$, C III/N III $\lambda\lambda4640$–4650, and C II $\lambda4267$ are almost totally eclipsed at phase 0.0. This implies that the high-excitation lines come from close to the white dwarf, although they also remain single-peaked at all phases.

3 LIGHT CURVES

We subtracted the continuum from each spectrum so that we could derive light curves for the emission lines. Figure 2 shows the continuum light curve and those of $H\beta$, $H\gamma$, and C III/N III $\lambda\lambda4640$–4650 and He II $\lambda4686$.

The continuum eclipse is deep and symmetric. The Balmer lines are barely affected by primary eclipse, but appear to undergo a significant decrease in line flux during phases 0.35–0.55. There is no evidence for a similar feature in the continuum light curve. This suggests that the continuum and the majority of the Balmer line emissions are produced in different regions. The C III/N III $\lambda\lambda4640$–4650 and He II $\lambda4686$ lines are totally eclipsed at phase 0.0, implying an origin close to the white dwarf. The eclipse of these lines is also quite narrow, further supporting an inner-disk origin.

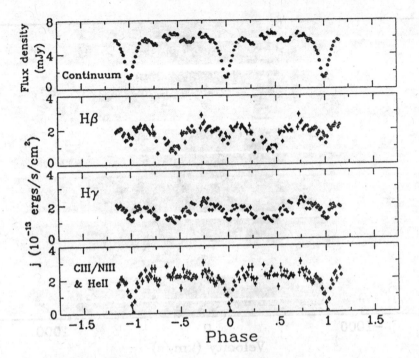

Fig. 2—Continuum and line light curves of V1315 Aql. The open circles represent points where the real data (closed circles) has been folded over.

4 TRAILED SPECTRUM AND RADIAL VELOCITIES

We binned the continuum-subtracted spectra into 70 phase bins covering one orbit. We then rebinned the Balmer lines $H\beta$, $H\gamma$, and $H\delta$ onto a uniform velocity scale of ±1500 $km\,s^{-1}$. These three contributions were then added together with weights in order to optimize the signal-to-noise ratio. The resulting trailed spectrum of the sum of the Balmer lines is shown in Figure 3.

The trailed spectrum is very 'messy,' with no obvious disk-type orbital modulation present in the line. In particular, the line appears distinctly redshifted at phase 0.0 and crosses over to zero velocity at approximately phases 0.25 and 0.75. There is no evidence for an 'S-wave' component from the hot spot

The trailed spectrum clearly shows the lack of primary eclipse at phase 1.0. The double-peaked profile at around phase 0.5 can be seen to extend from phase 0.35–0.55. The red peak of the line is stronger than the blue peak during this phase and the core of the double-peaked line appears to shift from red to blue.

We then binned the data into 20 phase bins and measured radial velocities using the standard double Gaussian method. The Gaussians were of width 300 $km\,s^{-1}$ and we varied their separation from 1400 to 2200 $km\,s^{-1}$. The Balmer and He II $\lambda4686$ radial velocity curves are shown in Figure 4.

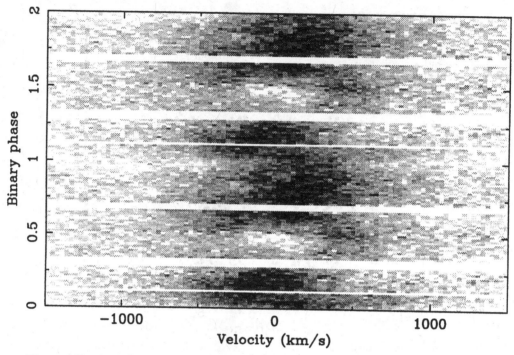

Fig. 3—Trailed spectrum of the sum of the Balmer lines in V1315 Aql. The data has been repeated twice for clarity. Gaps in the figure are due to calibration spectra.

It is surprising that the fit to the wings of He II $\lambda4686$ produces such a distorted radial velocity curve, especially since the previous results indicate a line origin close to the white dwarf. The Balmer line radial velocity curves appear very similar to the He II $\lambda4686$ curves in both cross-over phase and amplitude. However, the large scatter and uncertainty of line origin make these radial velocity curves very difficult to interpret.

5 DISCUSSION

V1315 Aql presents a puzzling set of constraints for any model to satisfy. The eclipse behaviour of the Balmer lines shows that the majority of the Balmer emission does not come from the disk. The lines display a double-peaked structure at phase 0.5, often taken as the signature of an edge-on accretion disk. The eclipse behaviour of the high-excitation lines suggests they originate from close to the white dwarf. Nevertheless, despite an inner-disk origin, the high-excitation lines appear single-peaked and exhibit a very distorted radial velocity curve. We list a number of possible models below and make a comment on whether they satisfy the constraints placed upon V1315 Aql by our observations.

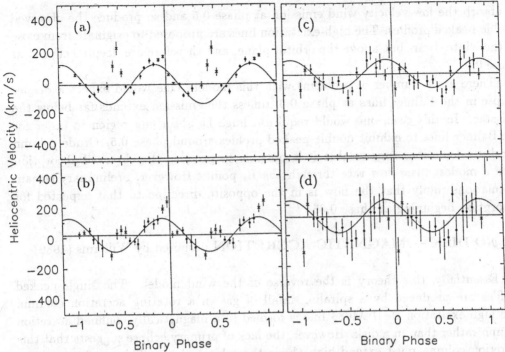

Fig. 4—Radial velocity curves: (*a*) Sum of Balmer lines with a Gaussian separation of 1400 km s^{-1}. (*b*) Sum of Balmer lines with a Gaussian separation of 2200 km s^{-1}. (*c*) He II λ4686 with a Gaussian separation of 1400 km s^{-1}. (*d*) He II λ4686 with a Gaussian separation of 2000 km s^{-1}. Points marked with a cross were not included in the radial velocity fits (due to measurement uncertainties during phases 0.9–0.1 and 0.35–0.55).

5.1 STARK BROADENING proposed by Lin, Williams, and Stover (1988)

Stark broadening is invoked to increase the intrinsic disk line-broadening to such an extent that it is comparable with the orbital Doppler broadening. This has the effect of convolving the double-peaked profile into a single maximum. This model is unlikely since the eclipse behaviour of the Balmer lines suggests a non-disk origin and Stark broadening requires rather finely-tuned disk conditions in order for the effect to be appreciable. The Balmer lines are also narrower than the double-peaked lines often seen in dwarf novae, where Stark broadening is not thought to be important.

5.2 ACCRETION DISK WIND proposed by Honeycutt, Schlegel, and Kaitchuck (1986)

A nonrotating accretion disk wind is proposed which dominates the Balmer emission — hence the single-peaked lines. The model also invokes material at the L_3 point to absorb the low-velocity wind emission at phase 0.5 and so produce the observed double-peaked profiles. The high-excitation lines are proposed to originate from close to the white dwarf but above the orbital plane, and therefore are deeply eclipsed at phase 0.0

There are a number of problems with this model. One would expect a deeper eclipse in the Balmer lines at phase 0.0, unless the emission extends far below the red star. In this case, one would require a huge L_3 absorbing region in order for the Balmer lines to exhibit double-peaked profiles around phase 0.5. Crude column density calculations suggest that it is possible to produce the observed absorption with a modest mass-flow rate through the L_3 point. However, preliminary stream calculations imply that the flow is in the opposite direction to that expected for absorption beginning at phase 0.35.

5.3 NO DISK — MAGNETIC ACCRETION proposed by Williams (1989)

Essentially, this theory is the reverse of the wind model. The single-peaked profiles are produced by a spiraling infall of gas in a rotating accretion column. The high-excitation He II $\lambda4686$ line is formed in a magnetically-confined accretion column rather than in a disk. However, the lack of primary eclipse suggests that this accretion column must extend high above the orbital plane. This would require a strong magnetic field, which would most probably disrupt the disk, so it is unclear why we still observe such a disk-like continuum eclipse. One would also expect a strong magnetic field to produce the emission line features often observed in AM Her systems — high γ velocities and very broad line wings. It is possible that the absorption at phase 0.5 is due to self-absorption of the accretion stream. However, this is very difficult to quantify without a knowledge of the accretion column geometry.

6 CONCLUSIONS

None of the proposed models can satisfactorily explain the emission line behaviour observed in V1315 Aql. Although we can rule out Stark broadening as the cause of the single-peaked lines, it is much more difficult to differentiate between the accretion disk wind and magnetic accretion models. The abundant observations of strong winds in low-inclination novalike variables as deduced from the P Cygni profile of C IV $\lambda1549$ lends credence to the wind models, but improved theoretical calculations are required in order to determine the contribution of wind emission to optical line fluxes.

REFERENCES

Honeycutt, R. K., Schlegel, E. M., and Kaitchuck R. H. 1986, *Ap. J.*, **302**, 388.
Lin, D. N. C., Williams, R. E., and Stover R. J. 1988, *Ap. J.*, **327**, 234.
Williams, R. E. 1989, *A. J.*, **97**, 1752.

Photometric Observations of the Novalike Variable KR Aurigae

Jyoti Singh[1], P. Vivekananda Rao[2], P. C. Agrawal[1], K. M. V. Apparao[1], B. B. Sanwal[3], R. K. Manchanda[1], and M. B. K. Sarma[2]

[1]Tata Inst. of Fund. Research, Homi Bhabha Road, Bombay 400 005, India
[2]Astronomy Department, Osmania University, Hyderabad 500 007, India
[3]Uttar Pradesh State Observatory, Manora Peak, Nainital, Uttar Pradesh, India

1 INTRODUCTION

KR Aur is known to be a novalike cataclysmic variable (CV) which normally varies between visual magnitudes 12 and 14 (Liller 1980). It does not show any eruptions but occasionally shows dips down to 18th magnitude. It is therefore classified as an anti-dwarf nova. It was detected by the *Einstein Observatory* as a weak X-ray source (Mufson, Wisniewski, and McMillian 1980). The X-ray-to-visual luminosity ratio is 0.2 (Shafter 1983), a value typical for optically-selected CVs. The UV observations with the *IUE* satellite (Verbunt 1987) show a flat power-law continuum. At optical wavelengths its spectrum shows broad Balmer lines and weak He I lines in emission and a blue continuum. From the time-dependent variation of the radial velocity of the H_α emission line, the orbital period was found to be 3.90714 hr (Shafter 1983). Dorochenko, Lyutyi, and Terebizh (1977) observed KR Aur photometrically with 5- to 10-min time resolution and found that it shows variability on timescales of 20–30 min in the U, B, V colors. These properties of KR Aur are similar to those of the Intermediate Polars (IPs). The white dwarf in an IP is thought to possess a magnetic field $\sim 10^6$ G. IPs are asynchronous rotators and the white dwarf rotation period is revealed as optical and X-ray pulsation. The rotation period is typically found to be 10–20 min. We undertook fast photometry of KR Aur, since 1984, to search for coherent oscillations and also to make detailed studies of its light curve.

2 OBSERVATIONS AND RESULTS

2.1 Optical Photometry

Optical photometric observations of the cataclysmic variable KR Aur were carried out with the 1.2-m telescope at the Japal-Rangapur Observatory near Hyderabad, and with the 1.0-m telescope at the Uttar Pradesh State Observatory, Nainital in India. A single-channel photometer with a RCA 8575 phototube was used. The star was observed on 11 nights during 1984–88 with integration time of 10 s. The observational log is given in Table 1 and Figure 1 shows the extinction-corrected light curves obtained in February 1987.

TABLE 1
Log of Photometric Observations

Obs. No.	Date of Obs.	Filter	Start Time (HJD−2445000)	Duration (hrs)	Ave. count rate (s⁻¹)
1	18/19 Dec 84	B	1053.340	2.30	1588.6
2	19/20 Dec 84	"	1054.350	1.93	2509.9
3	23/24 Jan 85	"	1089.127	3.73	2063.2
4	18/19 Feb 85	"	1115.187	2.10	1171.7
5	19/20 Feb 85	"	1116.132	3.05	533.9
6[a]	07/08 Feb 86	"	1469.119	4.60	3289.9
7	02/03 Feb 87	"	1829.202	2.67	3235.3
8	03/04 Feb 87	"	1830.108	4.79	2873.7
9	15/16 Mar 88	"	2236.175	0.85	716.7
10	24/25 Jan 85	U	1090.111	4.58	969.7
11	20/21 Feb 85	"	1117.140	3.22	339.0

[a] 7/8 Feb 86 observation was made at Nainital Observatory using 1-m telescope. All other observations were made using the 1.2-m telescope at the Japal-Rangapur Observatory.

The spectroscopic orbital period of KR Aur is $P = 3.90714 \pm 0.00072$ hr (Shafter 1983), but the photometric period has not been determined. The 2–3 hr long light curves of January and February 1985 show no orbital modulation. The 1986–87 light curves, however, do show a significant modulation. The light curve of 3/4 February 87 lasts for about 4 hours with a maximum at HJD 2446830.19 ± 0.01. Identifying minima at HJD 2446830.10 and 2446830.26 in the light curve leads to a period of ∼ 3.9 hr. Assuming the photometric and the spectroscopic orbital period to be the same, the phases of the light curves of 2/3 and 3/4 February 87 are compared using Shafter's (1983) ephemeris. It is found that the maxima or minima on the two days differ by 0.4 in phase (see Fig. 1). This suggests that the photometric period may not be identical to the spectroscopic orbital period. More observations of longer duration and closely spaced in time are required to determine the photometric period accurately.

Pulsations with 8–12 minute periods and with large fractional amplitude (∼ 0.15–0.37) are seen in the data. The periodograms of individual runs, using pre-whitened data, were calculated and are shown in Figure 2. Two or three peaks with significant power are present in all the periodograms. The peak with period between 500 and 800 s is due to the well-defined pulsation clearly seen in the light curve. The peaks at the longer periods can be explained as due to the broader features present in the data. Since the pulsation peaks do not occur at the same frequency even on consecutive nights, it is concluded that these represent quasi-periodic oscillations. The periodogram analysis was also performed on the auto-correlation function (ACF).

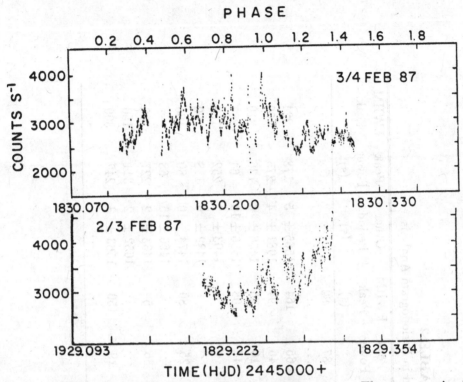

Fig. 1—B-band light curves with 10-s integration time. The top x-axis shows orbital phase according to Shafter's (1983) ephemeris.

The initial exponential decrease was removed from an ACF before calculating the periodogram. All the significant peaks agree with those seen in Figure 2. In Table 2 we have listed all the periods found in the data.

2.2 X-ray Observations

The detection of KR Aur by the *Einstein Observatory* was reported by Mufson, Wisniewski, and McMillian (1980), but a detailed analysis of this data has not been reported so far. The X-ray observations of KR Aur were made with the Imaging Proportional Counter on 20–21 September 1979, which resulted in a total exposure time of 2961 s and a net detected source counts of 295±17. For a thermal or blackbody spectrum, this corresponds to an incident source flux of 2×10^{-12} erg cm^{-2} s^{-1} in the 0.2–4 keV band. The X-ray light curve constructed by binning the counts in 100-s bins suggests marginal intensity variations, as indicated by a reduced χ^2 value of 1.85 for 32 degrees of freedom for a constant intensity hypothesis. A thermal spectrum with gaunt factor as well as a blackbody model provide acceptable fits to the data with the temperature and interstellar column density being $kT = 0.64\pm0.34$ keV and $N_{\rm H} = (2.2\pm1.0)\times10^{22}$ cm^{-2} and $kT = 0.43\pm0.31$ keV and $N_{\rm H} = (5.7\pm4.2)\times10^{21}$ cm^{-2}, respectively. The soft nature of the spectrum is also confirmed from the Monitor

TABLE 2

QPO Periods Found in Periodogram Analysis

Obs. No.	QPO Period (s)	Peak Power	FWHM Peak (s)	Other Period (s)	Peak Power	FWHM Peak (s)	Other Period (s)	Peak Power	FWHM Peak (s)
1	492 ± 3	13	30	1034 ± 4	87	82	...	...	...
	726 ± 4	27	77	...	...	...	1565 ± 5	178	224
2	662 ± 7	16	60	1135 ± 8	65	164	1981 ± 4	275	400
3	527 ± 1	47	25	999 ± 4	46	94	1456 ± 6	136	310
4	589 ± 3	40	60	909 ± 11	16	85	1556 ± 10	64	287
5	627 ± 1	119	50	996 ± 5	47	108	1403 ± 2	252	152
6	785 ± 3	25	50	...	...	...	1432 ± 6	118	200
7	758 ± 5	21	100	...	...	...	1474 ± 10	80	130
8	694 ± 9	113	39	1046 ± 5	25	80	1460 ± 15	68	...
9	...	...	...	...	...	...	1433 ± 2	227	100
10	500 ± 1	25	14	1176 ± 5	35	58	1652 ± 2	311	150
	795 ± 4	20	81	...	...	...	1563 ± 3	248	300
11	543 ± 3	10	23	1031 ± 8	19	150	...	...	...
	893 ± 6	20	68	...	...	...	...	...	...

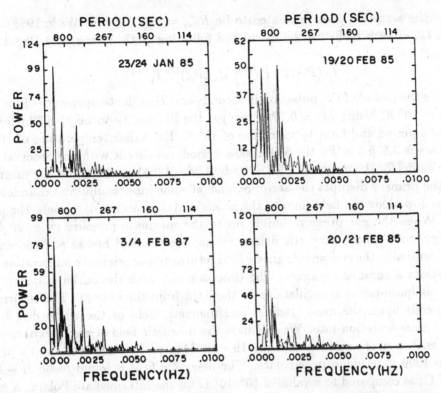

Fig. 2—Periodograms of few representative data runs. Power is in arbitrary units.

Proportional Counter observations which yielded a very low count rate of 0.19 ± 0.06 counts s^{-1} in the 1.2–4.9 keV band. Using the measured flux and a distance of 180 pc (Berriman 1987), the X-ray luminosity is derived to be 8×10^{30} erg s^{-1}. These parameters are in the same range as observed for several other CVs in the quiescent state (Córdova, Mason, and Nelson 1981).

3 DISCUSSION

Rapid quasi-periodic oscillations have been found in many dwarf novae (during eruptions), novalike variables, and the polars in the range of 1–1200 seconds (Córdova and Mason 1983). Recently, the Intermediate Polar H2215-086 has also been found to show QPOs of 100–400-s period superposed on the 20.9-min coherent rotational modulation of the white dwarf (Steiman-Cameron et al. 1989).

Since QPOs have been found in systems belonging to various classes of CVs, their explanation must be sought in some common feature of CVs like the accretion flow or the magnetic field of the primary or the late-type secondary star. We examine the possibility that the QPOs of 500 to 1000 s found in our data arise due to occultation of the blobs in the accretion disk by the white dwarf primary. If the blobs are rotating at the Keplerian velocity around the primary with 500–1000-s periods, the distance from the center is estimated to be $R = 8.3\text{–}13 \times 10^9$ cm. This is within the outer radius of the accretion disk, estimated to be $R_{disk} = 29 \times 10^9$ cm (Wade 1988). The lifetime, t_P, of such blobs can be calculated following Bath, Evans, and Papaloizou (1974) as,

$$t_P/P < 22\, P^{-1/3}(M_1/M_\odot)^{1/3}\, T_7^{-1/2}\,,$$

where P is the period of the pulsation in seconds and T_7 is the temperature of the disk in units of 10^7 K. Using $M_1 = 0.7\,M_\odot$, we get the lifetime t_P to be 11–17 hr for the 500–1000-s period and for a temperature of 10^4 K. For a disk temperature of 10^5 K, the lifetime is 3.5–5.5 hr for the 500–1000-s period, consistent with our observations.

The model for the magnetic CVs given by Livio (1984) assumes that the magnetic field of the primary disrupts the accretion disk at some inner radius and channels the accretion flow along the field lines to the pole(s). Matter accumulates onto the polar cap(s). When the gas pressure builds up to the magnetic pressure ($p \sim B^2/8\pi$), the energy stored in the magnetic field is released as the field breaks and reconnects again. Eventually, the stressed magnetic field returns to the original configuration and matter starts accumulating again. The timescale on which the outburst occurs will show up as quasi-periodic oscillations in the light from the system. The recurrence time depends upon the mass, radius, and magnetic field of the white dwarf and upon the mass-accretion rate. We calculate the magnetic field strength expected for KR Aur with a mass-accretion rate of $0.19 \times 10^{-9}\,M_\odot$ yr^{-1} (Wade 1988) and a period of 500 to 1000 s, as found in our data. The magnetic field is found to be $B \sim 0.5$–0.8×10^6 G, as compared to a value of 10^6–10^7 G for the Intermediate Polars. A more detailed discussion of the results can be found in Singh *et al.* (1989).

REFERENCES

Bath, G. T., Evans, W. D., and Papaloizou, F. 1974, *M. N. R. A. S.*, **167**, 7p.

Berriman, G. 1987, *Astr. Ap. Suppl.*, **68**, 41.

Córdova, F. A., and Mason, K. O. 1983, in *Accretion-Driven Stellar X-ray Sources*, ed. W. H. G. Lewin and E. P. J. van den Heuvel (Cambridge: Cambridge University Press), p. 147.

Córdova, F. A., Mason, K. O., and Nelson, J. E. 1981, *Ap. J.*, **245**, 609.

Dorochenko, V. T., Lyutyi, V. M., and Terebizh, V. Yu. 1977, *Sov. Astr. Lett.*, **3**, 279.

Liller, M. H. 1980, *A. J.*, **85**, 1092.

Livio, M. 1984, *Astr. Ap.*, **141**, L4.

Mufson, S. L., Wisniewski, W. Z., and McMillian, R. S. 1980, *IAU Circ.*, No. 3471.

Shafter, A. W. 1983, *Ap. J.*, **267**, 222.

Singh, J., *et al.* 1989, in preparation.

Steiman-Cameron, T. Y., Imamura, J. N., and Steiman-Cameron, D. V. 1989, *Ap. J.*, **339**, 434.

Verbunt, F. 1987, *Astr. Ap. Suppl.*, **71**, 339.

Wade, R. A. 1988, *Ap. J.*, **335**, 394.

Further Photometric Observations of AM CVn

S. Seetha[1], B. N. Ashoka[1], T. M. K. Marar[1], V. N. Padmini[1],
K. Kasturirangan[1], U. R. Rao[1], J. C. Bhattacharyya[2], B. C. Bhatt[3],
D. C. Paliwal[3], and A. K. Pandey[3]

[1]ISRO Satellite Centre, Bangalore 560 017, India
[2]Indian Institute of Astrophysics, Bangalore 560 034, India
[3]UP State Observatory, Nainital 263 129, India

ABSTRACT

About 18 hours of high-speed photometric data was collected during April 1989 to determine the true nature of the 1051 s period of AM CVn. Raw data reveals 0.05 mag amplitude variations. Results of our period analysis are presented. Combining published data with ours, we derive the rate of decrease as $(3.7 \pm 0.4) \times 10^{-12}$ s s^{-1}.

1 INTRODUCTION

The nature of the 1051 s period of AM CVn is still uncertain. Until 1979, it was thought to be the orbital period of the twin white dwarf novalike variable. Patterson *et al.* (1979) reported an increase of the period at a rate of 3.8×10^{-10} s s^{-1} which is about 1,000 times faster than the rate predicted by gravitational radiation. More recently, Solheim *et al.* (1984, hereafter SRNK), based on an extensive analysis of all available data, reported a rate of change more than two orders of magnitude lower than the value reported by Patterson *et al.* and, more importantly, that the period is in fact slowly decreasing. SRNK concluded that the 1051 s period appears to represent the rotation period of the accreting white dwarf. These extremely interesting but controversial results motivated us to monitor the object. We have redetermined the nature of change of the 1051 s period, the results of which are presented here.

2 OBSERVATIONS

The observations were carried out in white light during 1985–1987 (totalling 10.9 hours) with a single-channel photometer at the 1-m and 2.34-m telescopes of Vainu Bappu Observatory, Kavalur. The 1989 data (duration ∼18 hours) was collected with a two-star photometer (Venkata Rao *et al.* 1989) at the 1-m telescope of the UP State Observatory, Nainital. A diaphragm of 24 arcsec and typical integration times of 1 s or 10 s were employed. The two-star photometer data was collected with a PC-XT using the interface and Quilt-9 software kindly supplied by Prof. Nather of the University of Texas at Austin.

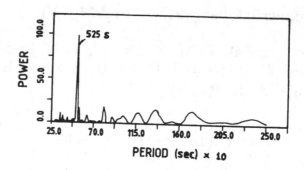

Fig. 1—Power spectrum of 2 April 1989 of AM CVn.

3 ANALYSIS AND RESULTS

The long data runs of April 1989 were subjected to a Discrete Fourier Transform (DFT) analysis (Deeming 1975). The power spectrum of the 2 April 1989 data (∼5 hours duration) is presented in Figure 1. All the runs consistently show peak power at 525 s. No significant power is observed in any of the runs at 1051 s. As our database did not permit any further refinement of the 1051 s period, we have assumed the best value of P derived by SRNK (equal to 1051.04086 s) for our O−C computations and the period change analysis.

We have used the linear ephemeris of SRNK (personal communication), viz.,

$$T_{min}(HJD) = 2437700.005036 + 0.012164824768519\, E\,.$$

Following SRNK, the minima observed in a single night of observations were noted and segregated into two sets differing by half a cycle. The average time of minima for the night was computed. The times of minima thus obtained were transformed to heliocentric times of minima and are presented in Table 1.

TABLE 1

HELIOCENTRIC TIMES OF MINIMA

46121.36812	47619.31954	46121.36119	47619.31397
46144.29815	47620.23144	46144.30440	47620.26271
46467.44355	47621.33901	46467.44995	47621.33339
46469.38245	47624.16112	46469.37557	47624.16803
46858.31126	47627.19172	46858.30663	47627.19803

These values were appended to those published in Table 2 of SRNK. At this point we wish to mention that in the analysis by SRNK and ourselves all the minima have been given equal weight irrespective of the instrumentation, data quality, and the number of cycles observed in a single night. The cycle numbers (E) rounded to the nearest integer and their corresponding O−C values in fractions of a period (cycle) were computed. The minima were then segregated into two sets: set 1 with O−C values in the range −0.5 to 0 and set 2 with O−C values in the range 0 to +0.5. A plot of the O−C and E values along with the least square fit is shown in Figure 2. It may be compared with the upper curves in Figure 9 of SRNK.

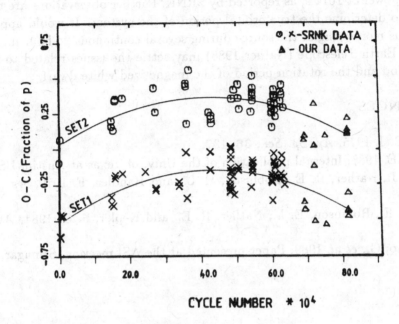

Fig. 2—O−C diagram of AM CVn.

From the above results, we derive an ephemeris for the 1051 s period as:

$$T_{\min}(\mathrm{HJD}) = T_o + 0.0121648445\,E - 2.27 \times 10^{-14}\,E^2$$
$$\phantom{T_{\min}(\mathrm{HJD}) = T_o + 0.012} \pm 19 \qquad \pm 22$$

where
$$T_o(\mathrm{HJD}) = 2437699.9987 \pm 0.0006 \text{ for set 1 and}$$
$$T_o(\mathrm{HJD}) = 2437700.0054 \pm 0.0006 \text{ for set 2 .}$$

This implies a period change $dP/dt = (-3.7 \pm 0.4) \times 10^{-12}$ s s^{-1}. Thus, our present results seem to confirm the decreasing trend of the 1051 s period with a statistical significance of 9σ.

4 DISCUSSION AND CONCLUSION

Our results confirm with better statistical accuracy the nature of the variation (decrease) of the 1051 s period of AM CVn, first reported by SRNK. We would, however, hasten to add that the above results depend on the correctness of the period (1051.04086 s) derived by SRNK. The decreasing trend of this period ($dP/dt = (-3.7 \pm 0.4) \times 10^{-12}$ s s^{-1}) supports the interpretation that the 1051 s period represents the rotation period of the magnetized white dwarf in the binary system. Detection of X-rays modulated at the rotation period by future X-ray missions will definitely help to settle the above interpretation. The orbital period of the binary has not yet been measured accurately. Our DFT analysis of a single night's data did not reveal significant power at 1011 s, as reported by SRNK. Further observations are therefore called for to determine the true orbital period of the system. It would appear that a continuous monitoring of the source during several continuous 24-hour nights with the Whole Earth Telescope (Nather 1988) may settle the issues related to the true orbital period and the rotation period of the magnetized white dwarf.

REFERENCES

Deeming, T. J. 1975, *Ap. Sp. Sci.*, **36**, 137.

Nather, R. E. 1988, Internal publication of the Univ. of Texas at Austin, USA.

Patterson, J., Nather, R. E., Robinson, E. L., and Handler, F. 1979, *Ap. J.*, **232**, 819.

Solheim, J. E., Robinson, E. L., Nather, R. E., and Kepler, S. O. 1984, *Astr. Ap.*, **135**, 1.

Venkata Rao, G., *et al.* 1989, Paper presented at the ASI meeting, Srinagar, India.

Orbitally-Modulated Optical and UV Emissions from SS Cygni in Quiescence

F. Giovannelli[1], I. González Martínez-Pais[2], S. Gaudenzi[3],
R. Lombardi[3], C. Rossi[3], and R. U. Claudi[4]

[1]Istituto di Astrofisica Spaziale, CNR, Frascati, Italy
[2]Instituto de Astrofísica de Canarias, La Laguna, Spain
[3]Istituto Astronomico, Universitá "La Sapienza," Roma, Italy
[4]Osservatorio Astronomico di Padova, Italy

1 INTRODUCTION

The dwarf nova SS Cygni is the brightest object belonging to the class of the cataclysmic variables (CVs). Its magnitude ranges from about 12 to 8 during quiescent and outburst phases, respectively. So, SS Cygni can be easily detected in the optical even with small telescopes. For this reason it is one of the most observed systems. Nevertheless, only very seldom have simultaneous (or at least coordinated) multifrequency observations been performed. The most recent review of this system can be found in the paper by Giovannelli *et al.* (1985).

In this paper we present the orbitally-modulated UV and optical emissions from SS Cygni in quiescence, essentially based on our multifrequency observations with a discussion in order to demonstrate the possibility that this system is an intermediate polar.

2 OBSERVATIONS

In this section we will briefly discuss the main behaviour of SS Cygni in quiescence at UV and optical wavelengths. This work is based on observations performed with the *IUE* (VILSPA) satellite and with the telescopes of Loiano and Roque de los Muchachos observatories.

2.1 Ultraviolet

The UV behaviour of SS Cygni in quiescence is different than in outburst. This behavior can be briefly summarized as follows:

a) *In quiescence:*

(*i*) Fitting the UV flux with a function of the form $F_\lambda \propto \lambda^{-\alpha}$, the slope α drastically changes from -4.0 to -1.2 for wavelengths shortward and longward 1450 Å, respectively (Giovannelli *et al.* 1984; Gaudenzi *et al.* 1986; Lombardi *et al.* 1987).

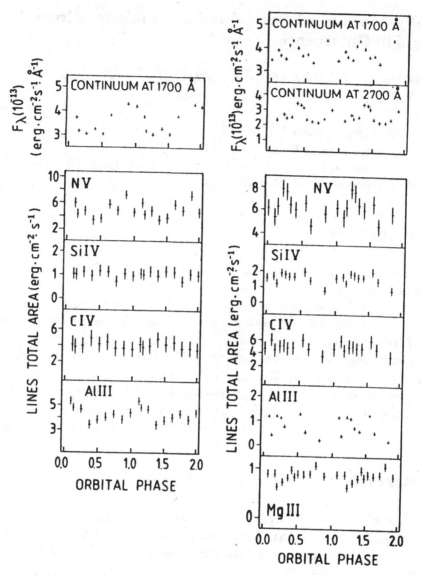

Fig. 1—Orbital modulations in the UV quiescent spectrum of SS Cygni
after long (*left*) and short (*right*) outbursts.

(*ii*) The continuum flux shows modulations at the orbital period ($P \simeq 6.6$
hr) and depends on the preceding outburst type and on the time elapsed since the
system entered quiescence. Also, the fluxes and the equivalent widths (EWs) of the
emission lines — C III, N V, O I, C II, Si IV, C IV, He II, Al III, Mg II, — show
orbital modulations with similar dependences of the continuum (Giovannelli *et al.*
1985; Gaudenzi *et al.* 1986; Lombardi *et al.* 1987). Figure 1 shows the modulations
detected in quiescence after long (*left*) and short (*right*) outbursts.

b) In outburst:

(*i*) For $\lambda < 1000$ Å, the slope α is -0.5 (Polidan and Holberg 1984). In the *IUE* range (1200–3200 Å), the slope α is -2.7 for $\lambda < 1760$ Å, -1.6 for 1760 Å $< \lambda < 2200$ Å, and -3.2 for $\lambda > 2200$ Å (Gaudenzi *et al.* 1986).

(*ii*) The spectral lines disappear or go into absorption. Some of these lines show P Cygni profiles (e.g., C IV $\lambda1550$ Å), clearly indicating the presence of a wind during this phase (Giovannelli *et al.* 1989). Typical UV spectra both in quiescence and in outburst are shown in the paper by Giovannelli et al. (1984).

2.2 Optical

Also in the optical range the spectrum of SS Cygni behaves in a different way depending on its state: quiescence or outburst.

a) In quiescence:

(*i*) The spectrum shows a strong blue continuum with wide and strong emission lines, such as the Balmer lines, He I, He II, and Ca II produced in the accretion disk. These lines are often double-peaked because of disk rotation and vary with the orbital phase (Walker and Chincarini 1968 for Ca II; Giovannelli *et al.* 1983 for the Balmer lines). From this fact and using all the multiwavelength experimental constraints, Giovannelli *et al.* (1983) derived the orbital parameters of SS Cygni: orbital inclination angle $i = 40°^{+1}_{-2}$, $M_{wd} = 0.97^{+0.14}_{-0.05} M_\odot$, $M_{sec} = 0.56^{+0.08}_{-0.03} M_\odot$, $R_{sec} = 0.68^{+0.03}_{-0.01} R_\odot$, and the inner and outer radii of the accretion disk (to order of magnitude only, since they vary with outburst phase): $R_i = 3.6 \times 10^9$ cm and $R_o = 2.9 \times 10^{10}$ cm , respectively. Recently, Honey *et al.* (1988) confirmed the values of the orbital inclination angle (40°) and the masses of the two components of the system ($M_{wd} = 1.1 M_\odot$, $M_{sec} = 0.7 M_\odot$).

From a long series of observations performed in July 1987 with the 2.5 m Isaac Newton telescope of the Roque de los Muchachos Observatory, SS Cygni shows orbital modulations of the EWs of the Balmer emission lines. Figures 2*a*, *b*, *c*, *d* show these modulations for H_α, H_β, H_γ, and H_δ, respectively. The Balmer lines often show an extra central emission feature which can mask the doubling (Giovannelli *et al.* 1984);

(*ii*) The spectrum also shows absorption lines from a late-type main-sequence star. It may be a G5 V or more likely a K5 V star (Stover *et al.* 1980; Walker 1981).

b) In outburst:

(*i*) The spectrum resembles that of an O or B star with wide absorption lines often showing a narrow central emission feature that behaves as an S-wave (Hessman *et al.* 1984).

(*ii*) There is a dependence of the shape of the central Balmer and He II emission lines with the orbital phase (Hessman 1986).

From optical photometry SS Cygni shows *UBV* modulations with amplitude of 0.2 mag in the *V* band (Voloshina and Lyutyi 1983). Modulations are also detected in *UBVRI* during quiescence and during the decline from an outburst (Bartolini *et*

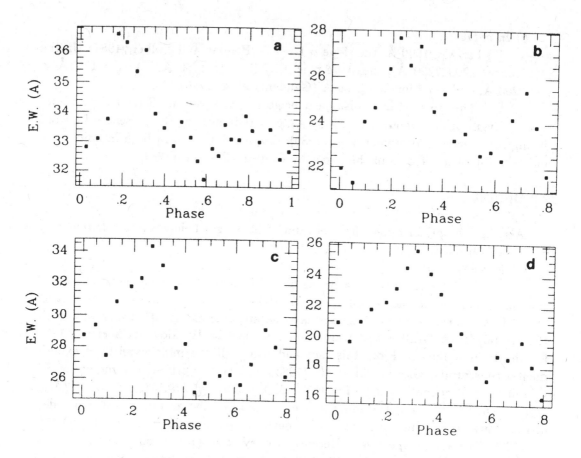

Fig. 2—Orbital modulation of H_α, H_β, H_γ, and H_δ.

al. 1985; Giovannelli *et al.* 1989). The system also shows a 12.18 minute periodicity detected in the R and I bands and less clearly in the UBV during quiescence
(Bartolini *et al.* 1985).

3 DISCUSSION AND CONCLUSIONS

We have reported the main behaviour of SS Cygni in quiescence in UV and in
optical bands. These results can be summarized as follows: (*i*) In the UV power
distributions, the continuum has a break at $\lambda \simeq 1450$ Å, with slopes -4 and -1
shortward and longward 1450 Å, respectively. Orbital modulations of the continuum
at 1700 Å and 2700 Å and of the fluxes of the emission lines have been detected. (*ii*)
In the optical, orbital modulations in $UBVRI$ and the EWs of the Balmer emission
lines have been also detected. The presence of an accretion disk around the white
dwarf is clearly demonstrated by the doubling in the Balmer lines, which is orbitally

modulated. Sometimes a third component in emission (T-component) in the doubling of the Balmer lines is present probably originated in the vicinity of the white dwarf (Smak 1985a). A periodicity at $\simeq$ 12 min during quiescence, probably associated with the rotation period of the white dwarf, has been detected.

Given these results, we conclude that the classification of SS Cygni as an intermediate polar, following the classification of Warner (1985) and Smak (1985b), is strongly supported.

In order to solve definitively and unambiguously this problem we have planned optical polarimetric and fast spectrophotometric measurements in order to detect the modulation of the emission lines with the spin period of the white dwarf which is expected to be about 12 minutes (Bartolini et al. 1985).

REFERENCES

Bartolini, C., et al. 1985, in *Multifrequency Behaviour of Galactic Accreting Sources*, ed. F. Giovannelli, Edizioni Scientifiche SIDEREA, Roma, p. 50.

Gaudenzi, S., Giovannelli, F., Lombardi, R., and Claudi, R. 1986, in *New Insights in Astrophysics: 8 Years of UV Astronomy with IUE*, ESA SP-263, p. 455.

Giovannelli, F., Gaudenzi, S., Rossi, C., and Piccioni, A. 1983, *Acta Astr.*, **33**, 319.

Giovannelli, F., et al. 1984, in *Proceedings of the 4th European IUE Conference*, ESA SP-218, p. 391.

Giovannelli, F., et al. 1985, in *Multifrequency Behaviour of Galactic Accreting Sources*, ed. F. Giovannelli, Edizioni Scientifiche SIDEREA, Roma, p. 37.

Giovannelli, F., et al. 1989, *Ap. Sp. Sci.*, in press.

Hessman, F. V., Robinson, E. L., Nather, R. E., and Zhang, E. H. 1984, *Ap. J.*, **286**, 747.

Hessman, F. V. 1986, *Ap. J.*, **300**, 794.

Honey, W. B., et al. 1989, *M. N. R. A. S.*, **236**, 727.

Lombardi, R., Giovannelli, F., and Gaudenzi, S. 1987, *Ap. Sp. Sci.*, **130**, 275.

Polidan, R .S., and Holberg, J. B. 1984, *Nature*, **309**, 528.

Smak, J. 1985a, in *Multifrequency Behaviour of Galactic Accreting Sources*, ed. F. Giovannelli, Edizioni Scientifiche SIDEREA, Roma, p. 3.

Smak, J. 1985b, in *Multifrequency Behaviour of Galactic Accreting Sources*, ed. F. Giovannelli, Edizioni Scientifiche SIDEREA, Roma, p. 17.

Stover, R. J., Robinson, E. L., Nather, R. E., and Montemayor, T. J. 1980, *Ap. J.*, **240**, 597.

Voloshina, I. B., and Lyutyi, V. M. 1983, *Soviet Astr. Letters*, **9**, 319.

Walker, M. F. 1981, *Ap. J.*, **248**, 256.

Walker, M. F., and Chincarini, G. 1968, *Ap. J.*, **154**, 157.

Warner, B. 1985, in *Cataclysmic Variables and Low-Mass X-Ray Binaries*, ed. D. Q. Lamb and J. Patterson (Dordrecht: Reidel), p. 268.

Solar-Type Cycles in Cataclysmic Variables

A. Bianchini

Osservatorio Astronomico, 35100-Padova, Italy

1 DETECTING SOLAR-TYPE CYCLES IN CVS

The long-term light curves of cataclysmic variables (CVs) are often characterized by continuous, periodic, or quasi-periodic fluctuations with timescales from tens of days to years and amplitudes of a few tenths of a magnitude.

Barring the case of steady nuclear burning on the surface of the white dwarf primary, these light variations can be attributed to cyclical variations of the luminosity of the accretion disk which, in turn, may be produced either by modulated mass transfer from the secondary (Bath and Pringle 1981) or by the disk instability mechanism (Hoshi 1979; Meyer and Meyer-Hofmeister 1984).

The mass-transfer rates of quiescent novae and novalike systems are believed to be more stable than dwarf novae. They may however show cyclical variations of their quiescent luminosity on timescales of years (Bianchini 1989a). The disk instability model does not explain this behaviour because old novae and novalike binaries possess accretion disks which are too small and viscous to justify sinusoidal light curves and recurrence times of years.

A few old novae show regular optical modulations with periods from 50 to 100 days (Shugarov 1983; Della Valle and Rosino 1987; Della Valle 1988; Della Valle and Calvani 1989; Shara, Potter, and Shara 1989), but, also in this case, the rather strictly periodic nature of these variations seem to exclude the disk instability mechanism. We conclude that, in most cases, the long-term photometric variability of old novae and novalike objects must reflect the variations of the mass-transfer rate from the secondary. So, cyclical behaviour can be easily discovered by harmonic analysis of their long-term light curves.

Detection of cyclical mass-transfer rate variations in dwarf nova systems is more complicated because the inter-outburst luminosity is not related in a simple way to the mass-transfer rate from the secondary. Bianchini (1987, 1988) suggested that the cyclical variation of the outburst intervals shown by some dwarf novae are produced by solar-type cycles of activity in the secondaries of these systems. Strong support for this hypothesis came originally from Kiplinger and Mattei (1988, private communications) who, using the digitized data of the AAVSO on the dwarf nova SS Cyg, independently found a 7.3-yr modulation of its quiescent luminosity, thus confirming the existence of modulated mass-transfer rate with almost the same period found by Bianchini (1987). Most recently, Cannizzo (1990, this conference) has demonstrated, through very long computational runs of the disk instability model, that the spread in recurrence timescale which has been observed in several objects can only be produced by secular variations of the mass-transfer rate. Furthermore, the accretion disk must have a long-term memory of these changes. This can be achieved only if the viscosity is coupled to the mass-transfer rate. Thus, we are allowed to assume that the length of the time interval between successive outbursts is a function of the mass-transfer rate from the secondary, though through a change in the viscosity of the disk. We conclude that the presence of cycles of activity in the secondaries of dwarf novae can be easily discovered by looking for cyclical modulation in the ΔT versus T diagram of the outbursts.

Another independent test for the existence of periodic mass-transfer rate variations in close binary systems is provided by observing orbital period variations. Applegate and Patterson (1987) showed that if the mean value of $B^2/8\pi$ in the convective layers of the cool component varies through an activity cycle, then the change will cause the star to expand and contract in a quadrupole-deformed Roche potential. This should lead to detectable apsidal motion and, even for circular orbits, to a change of the orbital period with the same timescale as the magnetic cycle. Spin-orbit tidal interactions have timescales which are too long compared to the observed periods, while dynamical effects such as the exchange of angular momentum due to mass transfer do not have an influence on the orbital period as large as the variation of the quadrupole moment.

Recently, Warner (1988) has demonstrated that the variation of the quadrupole moment caused by a typical magnetic cycle can actually account for the cyclical variations of the orbital periods observed in some CVs. Warner has also demonstrated that the values $\Delta R/R \sim 1 \times 10^{-4}$ derived from the other mass overflow-sensitive techniques (i.e., variations of the mean luminosity and of the outburst intervals of dwarf novae) are all in accord with the hypothesis that they are due to solar-type magnetic cycles. The only perplexing contradiction found by Warner is the response of the spin period variations of the magnetic white dwarf of the old nova DQ Her to changes of the mass-transfer rate.

Up to now, we have revealed the presence of long-term cycles in the light curves of 3 old novae, 1 pre-nova, 1 novalike system, 1 symbiotic nova (which, however, might be due to orbital motion), and in the ΔT versus T diagrams of 5 dwarf novae. Evidence for the presence of long-term periodicities in 10 more close binaries are found in the literature. Details are given by Bianchini (1989b, and references therein).

2 DISCUSSION

The main characteristics of the cycles are summarized in Table 1. Columns 1, 2, and 3 give the object name, the CV subtype, and the orbital period, respectively.

Column 4 gives the initial epoch, t_0, and Column 5 the time length, Δt, of the light curves analyzed by Bianchini (1989b).

Periods of solar-type cycles, P_{cycle}, are given in Column 6. Secondary periods are followed by a colon while equally probable values are separated by an hyphen. Full amplitudes of the observed cycles, expressed as Δm, $\Delta(\Delta T)/\langle\Delta T\rangle$ or $\Delta P_{orb}/\langle P_{orb}\rangle$ are listed in Columns 7, 8, and 9, respectively. Column 10 reports the short-term periods, P_{short}, discovered in some old novae by Della Valle (1988). References are in the last column. Rather uncertain values in Table 1 are followed by a question mark.

TABLE 1

Star	Type[1]	P_{orb} (hr)	t_0[2] (JD 2400000+)	Δt[2] (yr)	P_{cycle} (yr)	Δm (mag)	$\dfrac{\Delta(\Delta T)}{\langle\Delta T\rangle}$	$\dfrac{\Delta P_{orb}}{\langle P_{orb}\rangle}$ ($\times 10^{-7}$)	P_{short} (day)	Ref.
T Aur	N	4.91			≈23					
Q Cyg	N		22847	60.9	6.4	0.2			55–65	1,2
DQ Her	N	4.65			13.4	≈0.6		20		3
V446 Cyg	N		13840	60.9	12.2	0.8			71.5	1,2
					5.9 :	0.5 :				
V841 Oph	N	14.4[3]	22080	30.3	3.4	0.3			51.7	1,2
GK Per	N	48.0	22578	66.4[4]	7.2	0.6–0.1			400 ?	1
					3.2 :	var?				
					60?					
RR Pic	N	3.48			14.0	0.2				3
TT Ari	NL	3.30	25500	53	12.6	dips			400 ?	1,4
MV Lyr	NL	3.30			11–12	dips			300–400?	5,1
RW Tri	NL	5.57			8–14			21 or 37		3
UX UMa	NL	4.72			≈29			22		3
UU Aql	DN	1.88		≈ 48	12.9		0.18			6
SS Aur	DN	4.33	19824	22.0	2.3		0.33			1,3
					1.2 :		0.18			
SS Cyg	DN	6.63	13868	88.0	7.3		0.17			1,7
					2.2 :					
U Gem	DN	4.17	−1017	129	6.9		0.26	25		1
EX Hya	DN	1.64		≈19				4.5		3
RU Peg	DN	8.99		≈ 28	6		0.26			8
					12:					
RX And	ZC	5.08	23964	11.5	0.88		0.18			1
Z Cam	ZC	6.95	24047	11.4	≈ 8.8		0.26			1
44 i Boo	CE	6.42			3.3					9
					7 ?					
RR Tel	SN	11.4-yr?	12600	87.0	11.4	(orbital?)				1

[1] N = nova, DN = dwarf nova, NL = novalike, ZC = Z Camelopardalis, CE = common envelope, SN = symbiotic nova.
[2] All t_0's and almost all Δt's are given only for the light curves which have been analyzed by Bianchini (1989b).
[3] Orbital period given by Bianchini, Friedjung, and Sabbadin (1989).
[4] Time interval corresponds to the period following the mini hibernation phase of the old nova. The 3-yr periodicity is however more evident during the late decline portion of the outburst light curve.

References: 1) Bianchini (1989b), 2) Della Valle (1988), 3) Warner (1988, and references therein), 4) Hudec et al. (1987), 5) Andronov and Shugarov (1982), 6) Shakun (1987), 7) Kiplinger et al. (1988), 8) Saw (1983), 9) Liu et al. (1984).

The 0.2–0.8 mag oscillations observed in the light curves of old novae and novalike objects (we exclude here the peculiar cases of TT Ari and MV Lyr) and the $1.2 < \Delta T_{max}/\Delta T_{min} < 1.4$ ratios found for dwarf novae suggest mass-transfer rate variations by factors in the range 1.2–2.5. Following Osaki (1985), the radii of the secondary

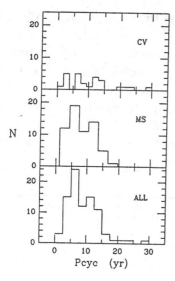

Fig. 1—Period histograms for CVs and MS stars. The two period distributions are identical. The estimated probability density function of all the periods is peaked at 6 yr.

components should then change by an amount $\Delta R/R$ in the range 0.6–3.0×10^{-4}, as expected for a typical solar-type cycle. The values of $\Delta R/R$ derived from the variations observed in the $(O-C)_{orb}$ diagrams of some eclipsing binaries are, instead, smaller by a factor 10. This has been interpreted by Warner (1988) as due to the fact that, most probably, we do not possess the correct relationship between mass-transfer rate and $\Delta R/R$.

Secondary periods have been taken only for SS Aur, V446 Her, RU Peg and GK Per. Two equally probable periods are given for MV Lyr and RW Tri. In the following, the 11-yr cycle of the symbiotic nova RR Tel has not been taken into account because it might be due to orbital motion.

The data listed in Table 1 demonstrate that the duration of the cycles of CVs are not correlated with the orbital periods and, hence, with the rotation regime of the stars. This important conclusion was already pointed out by Rodonò (1987) in his review on activity and rotation of RS CVn and BY Dra stars. It can then be argued quite generally that future theoretical models must allow for solar-type cycles (and dynamo mechanisms) for which rotation is not the key parameter.

No correlation between the length of the solar-type cycles and the shorter-term periodicities detected by Della Valle (1988) in some old novae and reported in Table 1 is observed.

The probability density distribution of the periods of CVs (top histogram of Figure 1) is found to be nonuniform. In order to compare the periods of the solar-type cycles found in CVs to those observed in single stars, we can make use of the data provided by Wilson (1978) from his survey on the long-term variations of the Ca II H-K chromospheric emissions from main-sequence stars. The distribution of the 61 periods estimated by Baliunas (1988) is peaked at 6 yr (central panel of Fig. 1). The two-sample Kolmogorov-Smirnov test indicates that the period distribution for CVs and MS stars are identical with a confidence level of 90%. This allows the simultaneous analysis of all the data. The bottom histogram of Figure 1 confirms the existence of the peak at 6 yr. A more detailed discussion on the basis of a larger number of objects, including RS CVn stars, will be given in a forthcoming paper (Maceroni *et al.* 1990).

Finally, we note that the 6-year peak shown by the histograms of Figure 1 might be related to the 2230-day (6.1-yr) period suggested for recurrent novae by Friedjung (1962). However, the statistical significance of the present result is not very large.

REFERENCES

Andronov, I. L., and Shugarov, S. Yu. 1982, *Astr. Tsirk.*, **1218**, 3.

Applegate, J. H., and Patterson, J. 1987, *Ap. J. (Letters)*, **322**, L99.

Baliunas, S. L. 1988, in *Formation and Evolution of Low Mass Stars*, ed. A. K. Dupree and M. T. V. T. Lago (Dordrecht: Reidel), p. 319.

Bath, G. T., and Pringle, J. E. 1981, *M. N. R. A. S.*, **194**, 964.

Bianchini, A. 1987, *Mem. Soc. Astr. Italiana*, **58**, No. 2–3, 245.

Bianchini, A. 1988, *Inf. Bull. Var. Stars*, No. 3136.

Bianchini, A. 1989*a*, in IAU Coll. No. 122, *The Physics of Classical Novae*, ed. A. Cassatella (Berlin: Springer), in press.

Bianchini, A. 1989*b*, *A. J.*, submitted.

Bianchini, A., Friedjung, M., Sabbadin, F. 1989, in IAU Coll. No. 122, *The Physics of Classical Novae*, ed. A. Cassatella (Berlin: Springer), in press.

Della Valle, M. 1988, Ph.D. thesis, University of Padova.

Della Valle, M., and Rosino, L. 1987, *Inf. Bull. Var. Stars*, No. 2995.

Friedjung, M. 1962, *J. Brit. Astr. Assoc.*, **72**, 276.

Hoshi, R. 1979, *Progr. Theor. Phys.*, **61**, 1307.

Hudec, R., *et al.* 1987, *Ap. Sp. Sci.*, **130**, 255.

Kiplinger, A. L., Mattei, J. A., Danskin, K. H., and Morgan, J. E. 1988, *J. AAVSO*, **17**, No. 1, p. 34.

Liu, X., and Wang, G. 1984, *Chin. Astr. Ap.*, **8**, 126.

Maceroni, C., Bianchini, A., Van't Veer, F., Rodonò, M., and Vio, R. 1989, *Astr. Ap.*, submitted.

Meyer, F., and Meyer-Hofmeister, E. 1984, *Astr. Ap.*, **140**, L35.

Osaki, Y. 1985, *Astr. Ap.*, **144**, 369.

Rodonò, M. 1987, in *Solar and Stellar Physics,*, ed. E. H. Schröder and M. Schüssler (Berlin: Springer), p. 39.

Saw, D. R. B. 1983, *J. Brit. Astr. Assoc.*, **93**, 2.

Shara, M. M., Potter, M., and Shara, D. J. 1989, *Pub. Astr. Soc. Pac.*, in press.

Shakun, L. I. 1987, *Astr. Circ.*, No. 1491, 7.

Shugarov, S. Yu. 1983, *Variable Stars*, **21**, No. 6, 807.

Warner, B. 1988, *Nature*, **336**, 129.

Wilson, O. C. 1978, *Ap. J.*, **226**, 379.

A Search for Evidence of Wind Accretion by the DA2 White Dwarf in V471 Tauri

Edward M. Sion[1], Dermott J. Mullan[2], and Harry L. Shipman[3]

[1]Department of Astronomy and Astrophysics, Villanova University
[2]Bartol Research Institute, University of Delaware
[3]Department of Physics and Astronomy, University of Delaware

ABSTRACT

We report the results of an analysis of eight high-resolution *International Ultra-violet Explorer* SWP (1200 Å–2000 Å) spectra of the Hyades eclipsing-spectroscopic pre-cataclysmic binary V471 Tauri. The technique utilized was the coaddition in velocity space of regions surrounding principal high-excitation and low-excitation ion species on a common velocity scale. Certain images were compensated for the orbital motion of the white dwarf during the exposure. Our line detections fell into three categories: (1) interstellar absorption in the line of sight to the Hyades cluster; (2) broad stellar wind features; and (3) narrow, high-velocity circumbinary absorption independent of orbital phase. In the absence of strong evidence for accreted photospheric metals and/or helium at the Einstein-redshifted velocity of the white dwarf surface layers, we favor, as a possible explanation, that the white dwarf does not accrete from the K2V stellar wind in detectable amounts due to the barrier presented by its rotating magnetosphere.

1 INTRODUCTION

A role for accretion as an important physical process in the Hyades, pre-cataclysmic, eclipsing-spectroscopic (DA2 + K2V) close binary V471 Tauri (orbital period = 12.5 hours), has been implicated recently on several fronts but its actual existence remains circumstantial. The recent discovery of a stellar wind emanating from the K2 dwarf star (Mullan *et al.* 1989) with lower limit mass-loss rate of $2 \times 10^{-11}\,M_\odot\,\mathrm{yr}^{-1}$, would suggest that through gravitational capture alone, the DA2 star may be able to accrete at a rate significantly higher than expected of interstellar values on single degenerates. The discovery of 9.25-min *EXOSAT* X-ray oscillations and optical oscillations of the same period (Jensen *et al.* 1986 and Robinson *et al.* 1988, respectively) suggest the possibility that accretion plays a role either by driving non-radial g-mode pulsations or by providing metals at the magnetic poles of a magnetic DA2 star causing X-ray *dark* magnetic accretion poles modulated by a 9.25-min white dwarf rotation period. Further indirect evidence suggesting the occurrence of accretion is the recent discovery of a cool expanding circumbinary shell at $-1200\,\mathrm{km\,s}^{-1}$

in the line of sight to V471 Tauri which may be related to an accretion-triggered thermonuclear runaway on the white dwarf (Sion *et al.* 1989). The possibility of using detached post-common envelope red dwarf-white dwarf close binaries as accretion *labs* and as an indirect means of establishing red dwarf wind/flare mass loss, was already explored by Sion and Starrfield (1984). All of the observed phenomena mentioned above, together with a critical need to acheive a more basic understanding of accretion physics, motivated an intensive search for evidence of accreted metals and/or helium at the photosphere of the DA2 star, using SWP high-resolution *International Ultraviolet Explorer* (*IUE*) spectra of the white dwarf.

2 OBSERVATIONS AND ANALYSIS

All of the observations discussed in this paper were obtained with the *IUE* satellite using the Short Wavelength Prime (SWP) camera in the high dispersion echelle mode. The telescope, instrumentation, performance characteristics, and sensitivity are described in Boggess *et al.* (1978). Three of the images were in the *IUE* archives and were originally obtained by W. Beavers. Five other high-resolution images were obtained by Sion and Bruhweiler, of which three were velocity-compensated for the orbital motion of the white dwarf. This procedure consisted of dividing the exposure into numerous segments of approximately ten minutes each. Between each ten-minute segment the exposure was interrupted so that the target could be moved to a different position in the long aperture. These positions were chosen so that the variation in velocity of the white dwarf due to its orbital motion during the exposure would be compensated in the spectrum. In Table 1, a log of the SWP spectral observations is presented with the tabulated entries as follows: SWP image number, day number and year of observation, exposure time in minutes, orbital phase at midpoint of exposure, phase range covered by the exposure, and comments.

The data were analyzed interactively at the NASA Goddard Space Flight Center Regional Data Analysis Facility (RDAF) using standard software and via a phone link between the RDAF VAX408 computer and Villanova University. In order to optimize the possibility of detecting weak features arising in the white dwarf photosphere and avoid confusion with any K2V emission, or with interstellar/circumstellar absorption features, all but two of the images were taken at points of maximum radial velocity (quadrature phases). In order to further enhance the possibility of detecting weak velocity-shifted features due to the white dwarf, we coadded ions of the same multiplet or of similar excitation in velocity space. This procedure is fully described in Sion *et al.* (1989).

3 RESULTS

Using the procedures described in Section 2, we carried out an exhaustive examination of the images in Table 1 for any evidence of absorption features at the correct

TABLE 1

High-Resolution SWP Observations of V471 Tauri

SWP Image No.	Day No. /Year	Exp. Time (min)	Phi-	ΔPhi	Comments
15898	362/1981	181	0.75	0.24	Poor S/N
15899	363/1981	114	1.0	0.15	Poor S/N
15900	363/1981	207	0.24	0.28	Fair S/N
28826	216/1986	250	0.71	0.32	Noisy
31611	234/1987	190	0.22	0.25	vel. comp.
31630	236/1987	390	0.33	0.52	circumbin.
32649	001/1988	186	0.77	0.25	vel. comp.
32659	003/1988	217	0.75	0.29	vel. comp.

(orbital plus gravitational redshift) velocity of the white dwarf, corresponding to the phased radial velocity curve of Young and Nelson (1972). An Einstein redshift of 50–60 km s^{-1} was assumed based upon their widely-accepted mass estimate for the white dwarf of 0.8 M$_\odot$. Our line detections fell into three categories: (a) interstellar absorption in the line of sight to the Hyades cluster, (b) broad stellar wind features, and (c) narrow, high-velocity circumbinary absorption independent of orbital phase.

First, it is surprising that our search for highly-ionized, high-temperature absorption at the photosphere of the hot ($T_{\rm eff} = 34,000$ K) DA2 star in the strong resonance doublets of C IV, Si IV, and N V, was entirely negative. No features in any of the four categories above were detected with any confidence. The ultraviolet metallic abundance calculations of Henry, Shipman and Wesemael (1984) reveal that solar composition material at the photosphere of a 35,000 K DA star should show detectable (300mÅ–1000mÅ) C IV and Si IV features. The complete absence of N V as well as Einstein-redshifted He II (1640) in any of the spectra in Table 1 is less puzzling because of the very high temperatures required. Despite the fact that image SWP 32659 was carefully compensated for the white dwarf orbital velocity and its exposure time was sufficiently long to have good signal to noise, there is little to support evidence of any real features except for a possible broad Si IV absorption feature which agrees with the wind absorption components detected by Mullan *et al.* (1989) in cooler ions.

There is no evidence for any real redshifted absorption. A comparison of the V471 Tauri white dwarf at $T_{\rm eff} = 35,000$ K with the ultraviolet line strength calculations at high gravity for solar composition of Henry, Shipman, and Wesemael (1984) yields an upper limit to the C IV and Si IV abundance from the non-appearance of photospheric

Fig. 1—A plot of relative flux versus velocity showing O I (1302), C II (1335) and Si II (1260), coadded on a common velocity scale for SWP 15900. Note the interstellar absorption feature slightly redshifted in the direction of the Hyades. The broad absorption trough centered at +260 km s⁻¹ is due to the K2V star's stellar wind absorbing white dwarf continuum (see text).

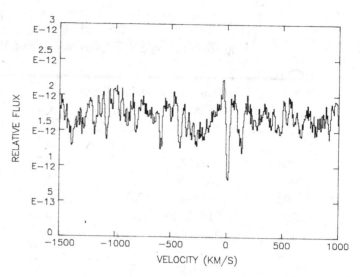

features to the detection limit of roughly 50 mÅ. The implied upper limits are (Si/H) $< 10^7$ and (C/H) $< 10^6$.

In all of the eight images in Table 1, weak interstellar lines invariably appeared at a slight redshift (+7 to +20 km s⁻¹), consistent with other studies of the line of sight to the Hyades (e.g., Gry *et al.* 1988). These were lowly-ionized species (C II, Si II, O I, N I) with implied column densities consistent with a fairly low density substrate of the local interstellar medium in that direction. A typical interstellar line is shown in the velocity coaddition plot of C II, Si II, and O I displayed in Figure 1 for SWP 15900.

In several images we detected the broad absorptions arising from the newly-discovered wind outflow emanating from the K2V star (Mullan *et al.* 1989). Note an example of this in Figure 1 (SWP 15900), where the broad feature at velocity −260 km s⁻¹ is the wind absorption in C II + Si II + O I, first noted by Bruhweiler and Sion (1986). These episodic wind features are also displayed in Figure 2, where the same cool ions have been coadded for SWP 32659. Here, the −500 km s⁻¹ phase-dependent wind component shows up clearly, consistent with the same preferred velocities noted by Mullan *et al.* (1989) in the eleven LWP high-resolution spectra in the regions of Fe II and Mg II h & k.

Strong evidence of the cool, very narrow, high-velocity circumbinary absorption at −1200 km s⁻¹ (Sion *et al.* 1989) and at −590 km s⁻¹ (Bruhweiler and Sion 1986) is present in those SWP images with the best signal to noise, an example of which is displayed in Figure 3 for SWP 15898. These features show up more strongly in the LWP images at the Fe II uv1 multiplet and in Mg II, as discussed by Sion *et al.* (1989). Both features may be associated with discrete mass ejection events associated with the white dwarf. The redshifted feature, longward-shifted by 0.5 to 0.9 Å, corresponding to orbital velocities of between 117 km s⁻¹ and 250 km s⁻¹ in

Fig. 2—The same ions as in Fig. 1, except for SWP 32659. The two absorption features near -500 km s^{-1} are real. The feature at -575 km s^{-1} is circumbinary gas (Bruhweiler and Sion 1986) and the feature at $+450$ km s^{-1} is due to orbital phase-dependent wind absorption (Mullan *et al.* 1989)

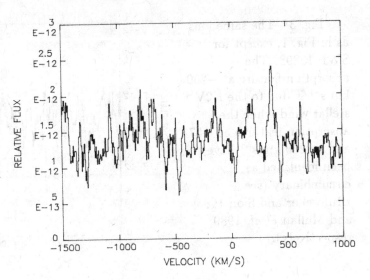

Figures 1, 2, and 3 is an interstellar feature artificially displaced in the coaddition process.

4 DOES THE PULSATING DA2 WHITE DWARF ACCRETE FROM ITS K2V COMPANION?

The detection of a cool wind emanating from the K2V star by Mullan *et al.* (1989) and the documented occurrence of large flares on the chromospherically-active K2V star (cf. DeCampli and Baliunas 1983; Young *et al.* 1983), both suggest the possibility that the DA2 star may accrete substantial amounts of material from its Roche-lobe detached companion. It is even possible that accretion may play a role in the 555-second *EXOSAT* X-ray (Jensen *et al.* 1986) and optical (Robinson *et al.* 1988) oscillations, be they due either to rotational modulation of X-ray-opaque magnetic accretion poles or to non-radial g-mode pulsations.

Stellar wind accretion would seem unavoidable if one applies the Bondi-Hoyle (1944) theory to wind accretion in the system. The characteristic impact parameter for the K2V wind material to be captured by the white dwarf is simply (in the absence of a magnetic field) the Bondi-Hoyle accretion radius given by $R_a = 2GM_{wd}/V_{rel}^2$, where $V_{rel} = V_w + V_{orb}$.

We adopt the following parameters for the V471 Tauri system: orbital separation $a = 1.15 \times 10^{11}$ cm (Jensen *et al.* 1986), orbital period 12.5 hours, $V_{orb} = 160$ km s^{-1}, $M_{wd} = 0.7\,M_\odot$, and radius of the white dwarf $R_{wd} = 7 \times 10^8$ cm. The fraction of wind material gravitationally captured by the white dwarf is then expressed as $\dot{M}_{acc} = (\pi R_a^2/4\pi a^2)\,\dot{M}_w$.

If $\dot{M}_w = 2\text{--}4 \times 10^{11}\,M_\odot$ yr^{-1} and $V_w = 600$ km s^{-1} (Mullan *et al.* 1989), then the accretion radius R_a is found to be 2.33×10^{10} cm. The white dwarf should therefore

Fig. 3—The same ions as in Fig. 1, except for SWP 15898. The absorption feature at −250 km s^{-1} is due to the K2V stellar wind while the absorption features at −575 and −1200 km s^{-1} have been identified as circumbinary (see Bruhweiler and Sion 1986 and Mullan *et al.* 1989, respectively).

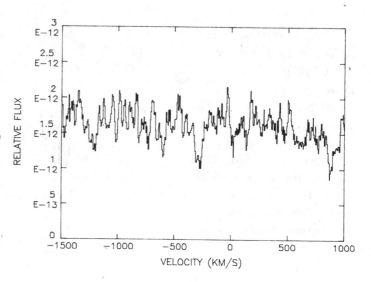

be accreting at the rate $\dot{M} = 4\text{--}9 \times 10^{-13}\,M_\odot\ \text{yr}^{-1}$. If the white dwarf has been accreting at this rate over its cooling age since emergence from the common envelope ($\tau_{\text{cool}} = 2.5 \times 10^{7}$ years), then it would be expected to have accreted a hydrogen layer mass of roughly $10^{-5}\,M_\odot$, sufficient perhaps to trigger a thermonuclear runaway. That this might possibly have occured is suggested by the discovery of high-velocity, cool, expanding gas in the line of sight to this system (Bruhweiler and Sion 1986; Sion *et al.* 1989). The major question we address here briefly, is: if the white dwarf is in fact accreting at the above rate, why is there no direct spectroscopic evidence of photospheric accretion in either low-resolution or high-resolution *IUE* spectra where the white dwarf dominates the light, virtually uncontaminated by the K star?

Several possible answers are worth noting: (1) Smearing out of the lines by the possibly rapid rotation of the white dwarf could be reponsible but there is no *a priori* reason why it should be rotating any faster than post-AGB field white dwarfs (upper limit $V \sin i < 65$ km s^{-1}), since it is not necessarily tidally locked to the orbital period. However, putative accretion torques could have spun it up. (2) The diffusion timescale for accreted metals in a similar DA2 star is on the order of a fraction of a year (Pelletier *et al.* 1986), which would mean rapid sinking to spectroscopic invisibility (although if the accretion is episodic as expected according to the activity cycle of the K2 star and the nature of its mass loss, it would seem probable that in all of the *IUE* coverage, some features would have been detected before diffusing out). (3) The possibility also exists that the ram pressure of the wind is insufficient to penetrate the magnetosphere of the white dwarf so that virtually no material is accreted, a suggestion due to Lamb (1989). For the sake of brevity, the quantitative exploration of these possibilities will be explored elsewhere.

5 CONCLUSIONS

The principal and potentially most important conclusion of our investigation is the apparently complete absence of line features that can be attributed to the photosphere of the DA2 white dwarf. This result, in the face of a large mass outflow from the K2 V star and the existence of X-ray and optical pulsations in which accretion may be playing a role, presents an enigma. Is the white dwarf rotating so rapidly that weak metal lines would escape detection? Is accreted material able to overcome and penetrate the putative magnetosphere of the (possibly) strongly-magnetic rotating white dwarf? Or is the diffusion timescale of a DA star of this temperature so short that the accreted ions from the episodic wind/flare material sink out of sight at small optical depth, thus evading detection? These possibilities will be explored in a followup paper.

It is a pleasure to express our appreciation to Ms. Rosalie Ewald and the staff of the NASA Goddard RDAF for their prompt and competent assistance with data analysis. We gratefully acknowledge support of this work through NASA grant NAG5-343 and NSF grant AST88-02689 to Villanova University and NSF grant 87-20530 to the University of Delaware.

REFERENCES

Bruhweiler, F. C., and Sion, E. M. 1986, *Ap. J. (Letters)*, **304**, L21.

Bondi, H., and Hoyle, F. 1944, *M. N. R. A. S.*, **104**, 273.

Davidson, K., and Ostriker, J. P. 1973, *Ap. J.*, **179**, 585.

DeCampli, W., and Baliunas, S. 1979, *Ap. J.*, **230**, 815.

Guinan, E. F., Wacker, S., Baliunas, S., and Raymond, J. 1986, in *New Insights in Astrophysics*, ESA SP-263, p. 197.

Henry, R. B. C., Shipman, H. L., and Wesemael, F. 1986, *Ap. J. Suppl.*, **57**, 145.

Illarionov, A. F., and Sunyaev, R. A. 1975, *Astr. Ap.*, **39**, 85.

Jensen, K., Swank, J., Petre, R., Guinan, E., Sion, E., and Shipman, H. 1986, *Ap. J. (Letters)*, **309**, L27.

Lamb, F. K., Pethick, C. J., and Pines, D. 1973, *Ap. J.* **184**, 271.

Lamb, D. Q. 1989, private communication.

Mullan, D. J., Sion, E. M., Carpenter, K. G., and Bruhweiler, F. C. 1989, *Ap. J. (Letters)*, **339**, L33.

Pelletier, C., Fontaine, G., Wesemael, F., and Michaud, G. 1986, *Ap. J. Suppl.*, **307**, 242.

Sion, E. M., and Starrfield, S. G. 1984, *Ap. J.*, **286**, 760.

Sion, E. M., Bruhweiler, F. C., Mullan, D. J., and Carpenter, K. G. 1989, *Ap. J. (Letters)*, **341**, L17.

Young, A., *et al.* 1983, *Ap. J.* **267**, 655.

WHT Spectroscopy of Globular Cluster CVs

G. Machin[1], P. J. Callanan[1], P. A. Charles[2], and T. Naylor[3]

[1]Department of Astrophysics, Oxford University, Keble Road, Oxford OX1 3RH
[2]Royal Greenwich Observatory, Apartado 321, 38780 Santa Cruz de La Palma,
 Tenerife, Canary Islands
[3]Institute of Astronomy, Madingley Road, Cambridge CB3 OHA

1 INTRODUCTION

Many active binaries are thought to be formed in the dense cores of globular clusters either by two-body tidal capture (Fabian *et al.* 1975) or three-body encounters (Hills 1976). These can take the form of either neutron star or white dwarf binaries.

There is a wealth of observational evidence in support of neutron star binaries being common in some globular cluster cores. This comes from both X-ray (Hertz and Grindlay 1983) and radio observations (e.g., Lyne *et al.* 1988).

Detailed calculations predict that large numbers of white dwarf binaries are formed by tidal capture in the cores of globular clusters (Verbunt and Meylan 1988). However, as yet only two candidate globular cluster dwarf novae (DN) have been discovered optically: V101 in M5 (Margon *et al.* 1981) and V4 in M30 (Margon and Downes 1983). Also, Richer and Fahlman (1988) have found six UV objects in M71 and suggested that they might be CVs.

Here we present WHT Faint Object Spectrograph (FOS-2) (Allington-Smith *et al.* 1989) spectroscopy of the two brightest CV candidates in M71 and the known globular cluster DN candidates M5, V101 and M30, V4.

2 DATA REDUCTION

All the spectra were reduced using the on-line FOS reduction software at the WHT and reanalyzed using the optimal CCD spectral extraction procedure of Horne (1986). Synthetic *B* and *V* magnitudes were obtained by folding the flux-calibrated spectrum through Johnson filter response curves for each object.

3 RESULTS

3.1 The CV candidates in M71

We show in Figure 1 spectra of the two brightest candidate objects in M71. Both the colors and spectra strongly suggest that the objects are sdB stars (Machin *et al.* 1990).

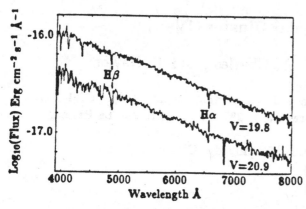

Fig. 1—Spectra of two of the CV candidates in M71. Star 1 $[V = 19.8; (B - V) = 0.08]$ has Balmer absorption lines. Star 2 $[V = 20.9; (B - V) = 0.01]$ has $H\alpha$ and $H\beta$ and possible He II $\lambda 4686$Å in absorption. The blueward edge ($\lambda 6840$Å) of the B band is still visible. These spectra are not like those of CVs (e.g., Fig. 2).

The fainter objects found by Richer and Fahlman (1988) are too faint ($M_V \geq +8.7$) to be sdB stars and members of M71. It IS more likely that the majority of these objects are actually field white dwarfs. Using the luminosity function of Downes (1986), we estimate that ~ 3 white dwarfs should be observable in the field of M71.

We scaled the calculations of Verbunt and Meylan (1988) to the core parameters of M71 and estimate only ~ 0.3 CVs would be formed in M71.

3.2 V4 in M30

In Figure 2 we show the mean spectrum of M30, V4.

Fig. 2—The quiescent spectrum of M30, V4 ($V \sim 19$). The spectrum shows strong emission lines of $H\alpha$ to $H\delta$ as well as He I $\lambda 6678$Å and $\lambda 5875$Å. The quiescent color of V4 $[(B - V) = 0.1]$ implies that there is very little flux contribution from the secondary star.

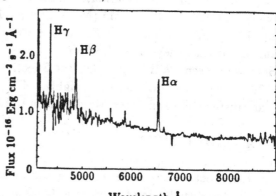

The period-color relation of Echevarria and Jones (1984) for field DN implies that the orbital period of V4 is ≤ 5 hrs.

If V4 is a cluster member then its absolute quiescent magnitude ($M_V = +4.6$) is inconsistent with it being either a magnetic variable ($M_V = +10.7$) or a "low-$\dot{M}$" system ($M_V = +9.2$) (Patterson 1984). Also, the quiescent spectrum is unlike that of a magnetic variable (e.g., Wade and Ward 1985). Hence, if V4 is a cluster member, it must be a "high-$\dot{M}$" system.

V4 has a mean heliocentric-corrected radial velocity of -43 ± 34 km s^{-1} as derived from fits to the $H\alpha$ line. This compares to a cluster velocity of -175 km s^{-1}, suggesting that the object is not a cluster member (but we note for example the caveat of Naylor *et al.* 1988).

The object was observed in outburst by Rosino (1949) to be at $m_{pg} = +16.4$. Assuming cluster membership and $B - V \sim 0$ (with $B = m_{pg} + 0.11$) for DN in outburst, the outburst magnitude of V4 is $M_V \sim +2$. Using the relationship for field DN relating the outburst magnitude to the orbital period given by Warner (1987), we find $P_{orb} \sim 13$ hrs. This conflicts with the period inferred from the quiescent color. Also, using the relation found by Echevarria and Jones (1984) relating orbital period and the magnitude of the secondary, we find $M_V(\mathrm{sec}) = +3.2$ implying that the secondary should dominate the light from the system in quiescence, yet this is not observed.

These inconsistencies are most easily resolved if V4 is a foreground object. Assuming a mean $M_V(\mathrm{outburst}) = +4.3$ for "high-$\dot{M}$" systems (Patterson 1984), we estimate the true distance of V4 to be 2.8 kpc.

3.3 V101 in M5

We observed the variable M5 V101 in quiescence ($M_V = +6.5$), outburst ($M_V = +4.1$), and "bright outburst" ($M_V = +3.1$) states. The absolute magnitudes we derive are compatible with cluster membership and those of other known DN (Warner 1987). Figure 3 shows spectra of the three outburst states of M5 V101.

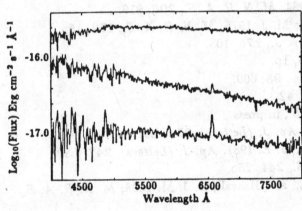

Fig. 3—M5, V101 was found in three distinct states during the course of the observations: quiescence ($V \sim 21.4$), outburst ($V \sim 19.0$), and "bright outburst" ($V \sim 18.0$). During quiescence Hα had $EW \sim 33$Å with Hβ and He I $\lambda 5875$Å in emission. In outburst only Hα in emission was visible. The "bright outburst" only occurred on one night when the star reddened significantly and Hα was in absorption.

V101's red color in quiescence [$(B - V)_0 \sim 0.7$] implies a long-period ($P_{orb} \geq 6$ hr) (Echevarria and Jones 1984). Our radial velocity measurements (Naylor et al. 1989) imply a $P_{orb} > 3$ hrs whilst Shara et al. (1987) estimate $P_{orb} \sim 11$ hrs. If the "bright outburst" magnitude ($M_V = +3.1$) is the outburst magnitude of V101 then this implies a long period ($P_{orb} \sim 10$ hrs). V101 is therefore above the period gap.

The low metallicity of the cluster coupled with the poor quality of our quiescent spectra prevents us from identifying spectral features in the secondary star.

V101 is not an AM Her system since neither its quiescent spectrum nor its magnitude are compatible with this hypothesis.

The "bright outburst" is very unusual. Although more observations are needed to confirm this behaviour, indirect confirmation of the "bright outburst" magnitude comes from Oosterhoff (1941) who found the object twice at $V \sim 17.5$.

If the secondary is on the main sequence then $M_V(\mathrm{sec}) = +4.7$ and the spectral type is G2 (Echevarria and Jones 1984). This is quite close to the turnoff point of M5 [$M_V(\mathrm{turnoff}) = +3.9$ to $+4.7$; Richer and Fahlman 1987] and hence the secondary could be evolved.

4 CONCLUSIONS

Spectroscopy of the two brightest CV candidates in M71 indicate that they are sdB stars. The remaining candidates are most likely to be field white dwarfs.

On the basis of the color, outburst magnitude, and radial velocity, we suggest it is unlikely that V4 is a member of M30 and is more likely to be a foreground object.

We find that M5 V101 lies above the period gap. M5 V101 must be regarded as the only DN whose properties are consistent with membership of a globular cluster.

REFERENCES

Allington-Smith, J. R., *et al.* 1989, *M. N. R. A. S.*, **238**, 603.

Downes, R. A. 1986, *Ap. J. Suppl.*, **61**, 569.

Echevarria, J., and Jones, D. H. P. 1984, *M. N. R. A. S.*, **206**, 919.

Fabian, A. C., Pringle, J. E., and Rees M. J. 1975, *M. N. R. A. S.*, **172**, 15p.

Hertz, P., and Grindlay, J. E. 1983, *Ap. J.*, **275**, 105.

Hills, J. G. 1976, *M. N. R. A. S.*, **175**, 1p.

Horne, K. D. 1986, *Pub. Astr.Soc. Pac.*, **98**, 609.

Lyne, A. G., *et al.* 1988, *Nature*, **332**, 42.

Machin, G., *et al.* 1990, *M. N. R. A. S.*, in press.

Margon, B., and Downes, R. A. 1983, *Ap. J. (Letters)*, **274**, L31.

Margon, B., Downes, R. A., and Gunn, J. E. 1981, *Ap. J. (Letters)*, **247**, L89.

Naylor, T., *et al.* 1989, *M. N. R. A. S.*, **241**, 25p.

Naylor, T., Charles, P. A., Drew, J. E., and Hassall, B. J. M. 1988, *M. N. R. A. S.*, **233**, 285.

Oosterhoff, P. Th. 1941, *Ann. Sterrwacht Leiden*, **17**, part 4.

Patterson, J. 1984. *Ap. J. Suppl.*, **54**, 443.

Richer, H. B., and Fahlman, G. G. 1987, *Ap. J.*, **316**, 189.

Richer, H. B., and Fahlman, G. G. 1988, *Ap. J.*, **325**, 218.

Rosino, L. 1949, *Mem. Italian Astr. Soc.*, **20**, 63.

Shara, M., Moffat, A. F. J., and Potter, M. 1987, *A. J.*, **94**, 357.

Verbunt, F., and Meylan, G. 1988, *Astr. Ap.*, **203**, 297.

Wade, R. A., and Ward, M. J. 1985, in *Interacting Binary Stars*, ed. J. E. Pringle and R. A. Wade (Cambridge: Cambridge University Press), p. 129.

Warner, B. 1987, *M. N. R. A. S.*, **227**, 23.

The First Survey of Probable Galactic Halo Cataclysmic Variables

Steve B. Howell[1] and Paula Szkody[2]

[1]Planetary Science Institute
[2]Department of Astronomy, University of Washington

1 INTRODUCTION

This paper is a short summary of the first observational survey of probable halo cataclysmic variables (CVs) (Howell and Szkody 1990). The motivation for this work came from recent results in the published literature. Howell (1982) found that only 14% of all cataloged CVs in the Variable Star Catalog were above ±30° Galactic latitude; Patterson (1984) listed all CVs with known periods at that time and out of 113 systems, only 9 were possible halo objects; and, finally, even large surveys like the PG survey (Green, Schmidt, and Liebert 1986) did not go nearly faint enough in apparent magnitude to contain all but a few probable halo CVs. The study of CVs has thus been severely limited to the Galactic disk.

We have isolated a sample of 84 CVs with Galactic latitudes $\gtrsim \pm 40°$. Very little was known about most of these stars so their distances were initially estimated using average M_V values. Warner (1987) showed that for CVs, the system inclination plays an important role in its M_V value. Even though the mean M_V of the nova and novalike classes is 4.5 and that for the dwarf nova is 7.5, the ranges are quite large: $M_V = 2\text{--}7$ for novae and novalikes and $M_V = 5\text{--}11$ for dwarf novae. We have calculated the *minimum* z distance (using ±40° Galactic latitude, which is our sample cutoff) that the sample stars would have for the M_V ranges given above. If we then assume that as many as half of the systems are inclined so as to have the lowest value of M_V, we find that *at most* only ∼ 20 systems from our sample would be less than 350 pc z distance. In fact, slightly less than half of our stars are near the 40° cutoff, so that realistically the number of possible sample members below a z distance of 350 pc is more likely to be *at most* ∼ 8.

The sample stars range in minimum magnitude from 15 to 22 and currently 35 systems have known orbital periods, over half of which have been found recently (Howell and Szkody 1988; Szkody *et al.* 1989; Howell *et al.* 1990). Figure 1 shows this information as a histogram.

We have discovered 3 new eclipsing systems (DV UMa, PG0134+070, AR Cnc), 3 new probable DQ Her systems (AL Com, AH Eri, CP Eri), and 3 new systems with orbital periods within the CV period gap (DV UMa, WX Cet, DM Dra). Other interesting systems include RZ Leo ($P = 1.7$ hrs) which has an orbital amplitude of ∼ 0.5 mag, RW UMi which is probably the faintest recovered old nova with a known

Fig. 1—Histogram of all the
stars contained in our sample
showing those with known orbital
periods. The single hatched area
is for non-magnetic CVs, while the
cross-hatched area shows the three
DQ Hers for which orbital periods
are known.

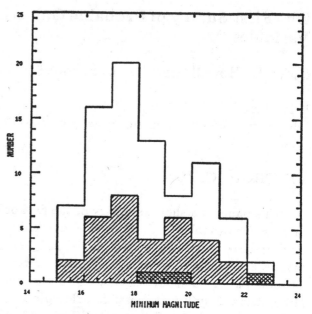

period orbital (117 min), and DM Dra which, so far, is our faintest system ($V = 21.7$)
with lightcurve information and a measured orbital period of 125 minutes.

Most of our information has been obtained via CCD photometry. Due to observational time constraints and our desire to collect much initial information quickly,
we are only complete in orbital period determination for systems with periodicities
of less than 4 hours unless there was a reason (e.g., an eclipse) for us to follow for a
longer time period. Approximately 70% of the observed systems showed a repeatable
photometric variation. Initial spectroscopy for a few of the most interesting systems
can be found in Mukai *et al.* (1990).

2 RESULTS AND DISCUSSION

Figures 2 and 3 show our initial results in terms of comparing the gross properties
of these probable halo objects to the disk CVs.

Figure 2 shows period histograms for our sample of probable halo stars (except
the DQ Her periods) and for a set of disk CVs taken from Patterson (1984). To
compensate for our observational bias toward periods less than 4 hours, we only
compare the number of systems of each kind below this period value. For the halo
objects, this represents a sample of 28 systems and for the disk stars it is a sample
of 49 systems.

While both sets of objects seem to have a preference for periods in the 1.5–2 hour
range, the halo CVs appear to have a higher percentage of systems below the gap
($16/21 = 76\%$) in comparison to the disk CVs ($28/49 = 57\%$). In addition, there is
a difference in the fraction of systems between 3–3.5 hours ($1/23 = 4\%$ for halo CVs
vs. $13/49 = 27\%$ for the disk CVs).

Figure 3 shows the relation between orbital period and outburst amplitude. The

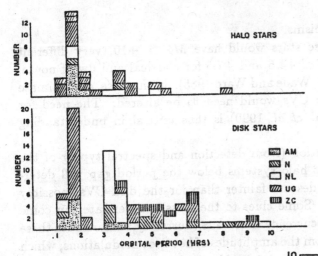

Fig. 2—Histogram of our sample stars with known orbital periods and a sample of disk CVs taken from Patterson (1984).

Fig. 3—The relation between orbital period and outburst amplitude for the sample dwarf novae stars. The halo systems with periods below the period gap have an average outburst amplitude of ~ 3 mag greater than the disk systems.

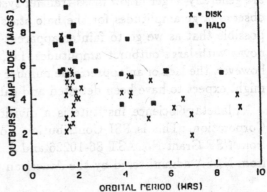

disk stars are Patterson's (1984) dwarf nova stars with known orbital periods while the halo stars are the dwarf nova objects from our sample. The number of halo systems with orbital periods below the period gap is only about half as many as for the disk systems, but it can clearly be seen that the mean outburst amplitude range for the two groups is different. The disk stars have $\overline{\Delta m} = 4.3 \pm 1.4$ ($\overline{\Delta m} = 4.1 \pm 1.1$ if the large-amplitude possibly odd disk system [WZ Sge; see Gilliland, Kemper, and Suntzeff 1986] is not used) while for the halo stars $\overline{\Delta m} = 6.9 \pm 1.7$. Above the gap, it also appears that the mean outburst amplitude may be larger for the halo stars, but there are too few data points to be statistically certain.

Two possibilities exist for interpreting the stars in our sample. They could be *bona fide* members of the Galactic halo, as we suspect, in which case some of them are extremely far away (up to 8 kpc!) or they could be intrinsically very faint systems (i.e., $M_V > 7.5$). In the former case, the halo CVs should be all of uniform age and older than their disk counterparts, should have formed under different initial conditions than the disk stars, and should consist of low-Z material. A different metal content could have an effect on the outburst amplitude, timescale, and duration of these systems (see Cannizzo, Shafter, and Wheeler 1988 and references therein). The large range in Δm observed for dwarf nova outbursts in the halo systems (see Fig. 3) is of interest. It may be an indication of a metallicity difference affecting the disk instability or superoutburst mechanisms.

If the latter is true, then these stars would have $M_v > +10$, very different from that of the accepted numbers of $+4.5$ and $+7.5$ for classical and dwarf novae, respectively (Kraft and Luyten 1965; Wade and Ward 1985). In this case, our estimate of space densities and lifetimes for CVs would need to be altered. The need for distance determination (e.g., Mukai *et al.* 1990) is thus critical in understanding these objects.

The correct distance determination (from detection and spectral typing of the secondary for example) of the eight halo systems below the period gap will determine if the quiescent disk magnitudes are fainter than for the disk CVs, possibly implying lower mass-transfer rates. Some clues to the mass-transfer rate can come from compilations of the outburst recurrence timescales (currently unknown for these poorly-studied faint systems) and from the amplitudes of hot spot modulations, which are generally larger in low mass-transfer systems. Alternatively, the fact that we only observe large amplitudes for the halo stars could be merely a selection effect. It is possible that as we go to fainter apparent magnitudes, we only discover the dwarf novae with large outburst amplitudes, i.e., ones which are easily detectable. Note, however, the lack of an appreciable number of large-amplitude disk systems than one might expect to have been detected and catalogued if they do indeed exist.

Planetary Science Institute is a division of Science Applications International Corporation. This is PSI Contribution No. 281. P. Szkody acknowledges support from NSF Grant No. AST 86-10226 and S. Howell acknowledges support by a grant from NASA administered by the American Astronomical Society.

REFERENCES

Cannizzo, J. K., Shafter, A. W., and Wheeler, J. C. 1988, *Ap. J.*, **333**, 227.

Gilliland, R. L., Kemper, E., and Suntzeff, N. 1986, *Ap. J.*, **301**, 252.

Green, R. F., Schmidt, M., and Liebert, J. 1986, *Ap. J. Suppl.*, **61**, 305.

Howell, S. B. 1982, *Pub. Astr. Soc. Pac.*, **94**, 969.

Howell, S. B., and Szkody, P. 1988, *Pub. Astr. Soc. Pac.*, **100**, 224.

Howell, S. B., and Szkody, P. 1990, *Ap. J.*, submitted.

Howell, S. B., Szkody, P., Kreidl, T. J., Mason, K. O., and Puchnarewicz, E. M. 1990, *Pub. Astr. Soc. Pac.*, submitted.

Kraft, R. J., and Layten, W. J. 1965, *Ap. J.*, **142**, 1041.

Mukai, K., *et al.*, 1990, *M. N. R. A. S.*, submitted.

Patterson, J. 1984, *Ap. J. Suppl.*, **54**, 443.

Szkody, P., Howell, S. B., Mateo, M., and Kreidl, T. J. 1989, *Pub. Astr. Soc. Pac.*, **101**, 899.

Wade, R. A., and Ward, M. J. 1985, in *Interacting Binary Stars*, ed. J. E. Pringle and R. A. Wade (Cambridge: Cambridge University Press), p. 129.

Warner, B. 1987, *M. N. R. A. S.*, **227**, 23.

Parameter Estimates of CVs

Eric M. Schlegel

Harvard-Smithsonian Center for Astrophysics and
NASA-GSFC/Universities Space Research Association

There exist about 6 to 8 CVs which are double-lined spectroscopic, eclipsing binaries. There are also many CV parameter estimators, usually used for the single-lined, non-eclipsing systems. A potentially useful approach is to test the estimators against the best data, namely, the 6 to 8 CV systems for which the best estimates of the parameters are available. The basic question to be answered, then, is *how* well do the "standard" parameter estimators do when compared to the best parameter values available? The motivation for this approach comes from Wade (1981).

The fewest assumptions possible have been made. As this work is based upon the published results of others, my assumptions will be the union of their assumptions. Two of the important ones made in the literature these days are that (*i*) the red dwarf fills its Roche lobe, and (*ii*) the bright spot is accurately located by a single-particle trajectory. I have also assumed that all the literature values are accurate and precise. This work is essentially a consistency check. Please note the effect of this assumption: if your favorite estimator does not do well as a predictor, it could be because your favorite estimator does not work, *or* because the "best" values are not correct.

There are several potential pitfalls with this approach. First, one must be very careful to note how each of the best values was obtained, to avoid choking noises and screams of "circular." Second, the intersection of the available data for the 6 to 8 systems and the required data often leaves only a few data points. This situation will improve as more CVs are studied in the infrared and the radial velocity curves of the secondary measured. For now, however, the results are quite tentative. A few representative results are presented in Tables 1 to 4. Table A lists the CV parameters collected from the literature.

Note that the error values quoted in Table A and in Tables 1 to 4 always refer to the last digit(s), regardless of the position of the decimal point. Finally, there remain several things still to be done. Other parameter estimators from the literature must be included. Plots of (predicted quantity − actual quantity) versus (*i*) actual quantity, (*ii*) q, (*iii*) system period, and (*iv*) CV type must be generated to see whether the estimators are systematically wrong. Figure 1 presents the results of Tables 1 and 2 in graphical form. The sense of "failure" of the predictors is clear. The failure is, of course, not a true failure, because the slope or offset of the predictors could be altered to bring them into agreement with the data. Furthermore, the disagreement

Table A
Table of Parameter Values for CVs

Parameter	V363 Aur	AC Cnc	Z Cha	EM Cyg	U Gem	IP Peg	OY Car	HT Cas
P (days)	0.3213	0.3005	0.0745	0.2909	0.1769	0.1582	0.0631	0.0736
P (hours)	7.71	7.21	1.79	6.98	4.25	3.79	1.51	1.77
K_{WD} (km s^{-1})	162±6	204±6	88±8	170±10	143±10	170±14	45±1	115±6
K_{RD} (km s^{-1})	181±5	165±9	430±16	135±3	285±15	293±11	440	?
R_{WD} (R$_\odot$)	?	?	0.0195±4	?	0.008±5	0.0068±11	0.0110±2	0.020±3
R_{RD} (R$_\odot$)	0.83±4	0.92±5	0.146	?	?	0.491±26	0.127±2	0.194
RD sp. type	G8/K0	G/K	M5.5V	?	M4.5V	?	?	?
i (deg)	70±2	72±3	81.7±2	63±1	69.7±7	79.3±9	83.3±2	76.4±6
$\delta\phi_{full}$	0.1	?	0.12	?	0.077	0.19±1	0.11	0.051
$\delta\phi_{WD}$	?	?	0.0534±9	?	?	?	0.0506±4	?
FWZI (km s^{-1})	1300	?	2100±50	?	?	3800	2700	2820
a (R$_\odot$)	2.3±1	2.3±1	0.634±5	?	1.71	1.47±5	0.608±3	0.659
M_{WD} (M$_\odot$)	0.86±8	0.82±13	0.544±12	0.57±8	1.12±13	1.09±10	0.685±11	0.53±7
M_{RD} (M$_\odot$)	0.77±4	1.02±14	0.081±3	0.76±8	0.53±6	0.64±9	0.070±2	0.19±2
q (M_{WD}/M_{RD})	1.12±4	0.81±8	6.6889±35	0.75±4	2.105±5	1.72±6	9.804±3	2.83
v_{disk} (km s^{-1})	400	334	600±25	390±15	?	511±114	750	600
R_{disk}/a	0.16±4	0.19±7	0.334±8	?	0.30±4	?	0.313	0.221
Sources	4	5	3,10,11,13	7	6,9,15	2	1,2,12	14,16

largely depends upon the high values of the intermediate-period points. This is not a particularly interesting case, nor a particularly strong demonstration of this inverse approach. Tests of the inclination predictors would be potentially more useful, but the sparseness of the data prevent such tests at this point. A more complete discussion will be presented elsewhere.

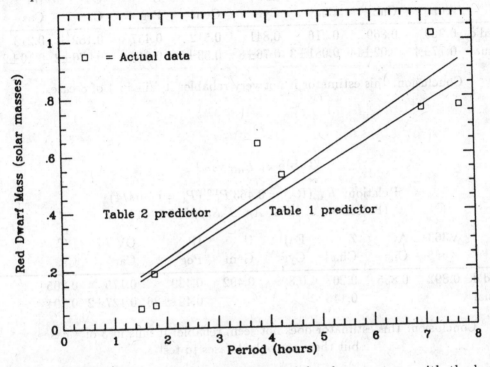

Fig. 1—Red dwarf mass versus orbital period for those systems with the best-measured system parameters.

Table 1: M_{RD} vs. P

Relation: $M_{RD} \sim 0.11\,P$ (hours)
(Faulkner, J. 1971, *Ap. J.*, **170**, L99.)

	V363 Aur	AC Cnc	Z Cha	EM Cyg	U Gem	IP Peg	OY Car	HT Cas
pred'd	0.848	0.793	0.197	0.768	0.468	0.417	0.166	0.195
actual	0.77±4	1.02±14	0.081±3	0.76±8	0.53±6	0.64±9	0.070±2	0.19±2

Conclusion: this estimator is not very reliable: it fails in 4 of 8 cases.

Table 2: M_{RD} vs. P

Relation: $M_{RD} \sim 0.482\, P_4^{1.0}$ ($P_4 = $ hours/4)
(Patterson, J. 1984, *Ap. J. Suppl.*, **54**, 443.)

	V363 Aur	AC Cnc	Z Cha	EM Cyg	U Gem	IP Peg	OY Car	HT Cas
pred'd	0.929	0.869	0.216	0.841	0.512	0.457	0.182	0.213
actual	0.77±4	1.02±14	0.081±3	0.76±8	0.53±6	0.64±9	0.070±2	0.19±2

Conclusion: this estimator is not very reliable: it fails in 4 of 8 cases.

Table 3: R_{RD} vs. P

Relation: $R_{RD}(R_\odot) = 0.463\, P_4^{1.0}$ ($P_4 = $ hours/4)
(Patterson, J. 1984, *Ap. J. Suppl.*, **54**, 443.)

	V363 Aur	AC Cnc	Z Cha	EM Cyg	U Gem	IP Peg	OY Car	HT Cas
pred'd	0.892	0.835	0.207	0.808	0.492	0.439	0.175	0.205
actual			0.146			0.491±26	0.127±2	0.194

Conclusion: this estimator does not seem reliable: it fails in 3 of 4 cases,
but there are only 4 cases to test.

Table 4: i via R_{RD}, R_{disk}, and $\delta\phi_{full}$

Relation: $R_{disk}/a = [\sin^2(180\delta\phi_{full}) + \cos^2 i]^{1/2} - R_{RD}/a$
(Longmore, A., *et al.* 1981, *M. N. R. A. S.*, **195**, 825.)

	V363 Aur	AC Cnc	Z Cha	EM Cyg	U Gem	IP Peg	OY Car	HT Cas
pred'd			80.2				80.5	68.7
actual	70±2	72±3	81.7±2	63±1	69.7±7	79.3±9	83.3±2	76.4±6

Conclusion: this estimator does not seem reliable,
but there are only 3 cases to test.

REFERENCES

1. Cook, M. 1985, *M. N. R. A. S.*, **215**, 211.
2. Hessman, F., Koester, D., Schoembs, R., and Barwig, H. 1989, *Astr. Ap.*, **213**, 167.
3. Marsh, T., 1988, *M. N. R. A. S.*, **231**, 1117.
4. Schlegel, E. M., Honeycutt, R. K., and Kaitchuck, R. H., 1986, *Ap. J.*, **307**, 760.
5. Schlegel, E. M., Kaitchuck, R. H., and Honeycutt, R. K., 1984, *Ap. J.*, **280**, 235.
6. Stover, R., 1981, *Ap. J.*, **248**, 684.
7. Stover, R., Robinson, E., and Nather, R., 1981, *Ap. J.*, **248**, 696.
8. Wade, R. 1981 in *Interacting Binaries*, ed. P. P. Eggleton and J. Pringle (Dordrecht: Reidel), p. 289.
9. Wade, R. 1981, *Ap. J.*, **246**, 215.
10. Wade, R., and Horne, K., 1988, *Ap. J.*, **324**, 411.
11. Wood, J., and Crawford, C. S., 1986, *M. N. R. A. S.*, **222**, 645.
12. Wood, J., Horne, K., Berriman, G., and Wade, R., 1989, *Ap. J.*, **341**, 974.
13. Wood, J., Horne, K., Berriman, G., Wade, R., O'Donoghue, D., and Warner, B. 1986, *M. N. R. A. S.*, **219**, 629.
14. Young, P., Schneider, D., and Shectman, S., 1981, *Ap. J.*, **245**, 1035.
15. Zhang, E-H., and Robinson, E. 1987, *Ap. J.*, **321**, 813.
16. Zhang, E-H., Robinson, E., and Nather, R. 1986, *Ap. J.*, **305**, 740.

Absolute Parameters of Cataclysmic Binaries

R. F. Webbink

Department of Astronomy, University of Illinois

1 INTRODUCTION

This contribution reports the preliminary results of a comprehensive survey of spectroscopic orbits and eclipse analyses of cataclysmic binaries, undertaken with the objective of re-deriving their masses and absolute dimensions using a uniform and consistent treatment of the data. For this purpose, empirical calibrations have been derived for several well-known relations between observable quantities and underlying physical parameters of these binaries: (1) Parameters characterizing emission-line profiles—velocity width at half- or mean-intensity (Shafter 1983), velocity separation of doubled emission peaks (e.g., Warner 1973), or rms line-widths (Williams 1983)— were empirically calibrated against a set of well-observed double-lined cataclysmic binaries (and Algol-type systems with double-peaked emission lines), and used to estimate mass ratios in a number of single-lined cataclysmic binaries. (2) This relation was used in part to calibrate an empirical mass-radius relation for the secondaries in cataclysmic binaries using 24 systems in which both component masses could be determined independently of any assumption regarding the main-sequence nature of the donor star. Surprisingly, the deduced relation replicates much more faithfully the features of theoretical lower main sequences than does a corresponding empirical lower main sequence based on classical spectroscopic and visual binaries.

With the adoption of this empirical main sequence, white dwarf masses could be estimated for a total of 84 systems. The Chandrasekhar limit was not invoked in constraining these mass estimates, but in fact only two systems (both estimates of lower quality) yielded nominal estimates exceeding this limit. Geometric mean masses are summarized according to orbital period and variable type in Table 1.

2 CONCLUSIONS

The lower main sequence as defined by the donor stars in cataclysmic binaries (see Fig. 1) , to the extent that it is uncomplicated by the effects of thermal disequilibrium resulting from the tidal mass loss process, provides strong evidence that the apparent discrepancy between theoretical lower main sequences and the empirical lower main sequence, as defined by observations of detached spectroscopic and visual

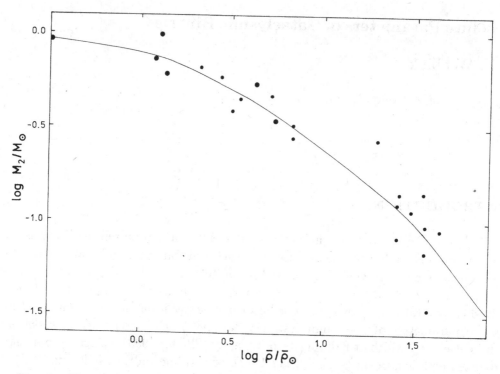

Fig. 1—The mass-density relation defined by donor stars in cataclysmic binaries. The choice of mean density as the abscissa is motivated by the fact that it
is determined, to within an extremely weak function of the mass ratio, solely by
the orbital period of the binary, which is a well-defined observable quantity. Observational errors are therefore almost completely confined to errors in the ordinate
(the mass of the donor star). Large points denote double-lined systems, small points
single-lined systems with indirect mass ratio determinations. The smooth curve is a
Gaussian-weighted running least-squares fit to the data.

binaries, may be an artifact of errors in the effective temperature scale for early-to-
middle M dwarfs. With the exception of the subdwarf eclipsing double-lined binary
CM Draconis, the empirical lower main sequence in the region of greatest discrepancy is defined entirely from visual binary data. The deduced radii of these dwarfs
depend entirely on the calibration of their effective temperature scale. With no direct
measurements of M-dwarf angular radii possible, and only CM Dra available among
eclipsing systems of known distance, this scale is in fact very poorly constrained.
A $\sim 10\%$ upward revision in effective temperatures around spectral type M0 would
resolve most of the alleged discrepancy between empirical and theoretical lower main
sequences. The small remaining mass difference between the CV lower main sequence
and theoretical main sequences is attributable to inadequacies in the usual application of the Roche approximation, i.e., equating of volume radius with that of a tidally
and rotationally undistorted star.

TABLE 1
WHITE DWARF MASSES[a]

	$P < 0.1$ day	$P > 0.1$ day	All P
Dwarf novae	0.50 ± 1.10 (15)	0.91 ± 0.08 (21)	0.74 ± 0.07 (36)
Classical novae	$\cdots$	0.91 ± 0.06 (8)	0.91 ± 0.06 (8)
Novalike systems	$\cdots$	0.74 ± 0.05 (22)	0.74 ± 0.05 (22)
Magnetic CVs	0.70 ± 0.10 (12)	0.79 ± 0.11 (9)	0.73 ± 0.07 (21)
All systems	0.61 ± 0.08 (26)	0.82 ± 0.04 (58)	0.74 ± 0.04 (84)

[a]Masses are in $M_\odot$. Standard errors of the mean are listed, followed by the number of systems in parentheses. SW UMa, V1500 Cyg, and DQ Her have been counted both as magnetic variables and as dwarf nova and classical novae, respectively.

Among the long-period CVs, marginally significant differences in mean white dwarf masses appear to exist between CVs of different types, with dwarf novae and novae having higher mean masses than novalike and magnetic variables. Among the long-period CVs of any one type, however, there is no evident systematic trend in mean white dwarf mass with orbital period. The trend which appears to exist in Figure 2 is a result of the different orbital period distributions of the sample systems of different types, with the novalike and magnetic variables, in particular, congregating toward the short-period end, just above the period gap. The estimates of low mass ($\lesssim 0.45\,M_\odot$, the core helium ignition mass among low-mass giants) that might contain helium white dwarfs are data of low quality, and of questionable significance.

Among the short-period CVs, the white dwarf mass distribution appears possibly bimodal. The same massive white dwarf contingent is present as is found among long-period CVs, including several systems (UZ For, WW Hor, and VV Pup) with white dwarf masses approaching the Chandrasekhar limit. This provides evidence (though not proof) that CVs do indeed evolve across the period gap. However, there exist in addition a number of systems, both dwarf novae and magnetic variables, which appear to contain very low-mass (helium) white dwarfs, and which have no obvious counterparts among long-period CVs. Such systems are expected, theoretically, to be found only among short-period CVs.

Contrary to the hypothesis of Hameury et al. (1988), the AM Her systems within the period spike at 0.08 day, immediately below the period gap, show clear evidence of a wide dispersion in white dwarf masses. These are estimates of high quality, being based on direct measurements of the radial velocity amplitudes of the donor M dwarfs, as determined from adsorption lines detected in the near-infrared, or from narrow emission-line components (attributed to the heated face of the secondary) detected in optical spectra.

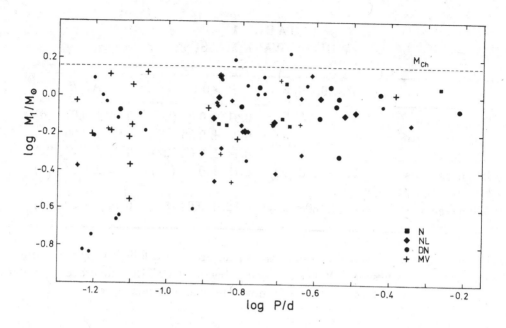

Fig. 2—The distribution of white dwarf masses versus orbital period, according to the type of cataclysmic variable (N – nova; NV – novalike; DN – dwarf nova; MV – magnetic cataclysmic), for the 84 systems for which such a determination required, at most, the assumption of a mass-radius relation for the donor star. Large symbols denote systems for which velocity amplitudes of the donor stars have been measured directly (the mass of the white dwarfs being most sensitive to this quantity). Smaller symbols denote systems with indirect mass ratio determinations. The Chandrasekhar limit is marked by the dashed line.

This work was supported in part by NSF grant AST 86-16992 to the University of Illinois.

REFERENCES

Hameury, J. M., King, A. R., Lasota, J. P., and Ritter, H. 1988, *M. N. R. A. S.*, **231**, 535.
Shafter, A. W. 1983, Ph.D. thesis, University of California at Los Angeles.
Warner, B. 1973, *M. N. R. A. S.*, **162**, 189.
Williams, G. 1983, *Ap. J. Suppl.*, **53**, 523.

Cosmic Dispersion on the Lower Main Sequence – Consequences for Mass Estimates in Cataclysmic Variables

Richard A. Wade

Steward Observatory, University of Arizona

1 INTRODUCTION

In the business of interacting binary stars, we are fond of using the "Roche lobe filling constraint," along with a "main-sequence mass-radius relation," to find the mass of the cool secondary star. This allows us to find solutions for binary systems that are, by classical criteria, underconstrained.

Do the secondary stars actually fill their Roche lobes? Most of the time, we answer "yes" with high confidence, although occasionally the suggestion is made that mass can be transferred (by means of prominence activity, for example) from a star that underfills its Roche lobe.

A power-law form, $R = \beta M^\alpha$, is usually adopted for the main-sequence mean mass-radius relation, and empirical constants are taken from, e.g., Lacy (1977), Echevarria (1983), or Patterson (1984). The empirical relation is not well determined, being based on rather few points, most of which have rather large error bars. With the mean relation poorly known, it follows that the dispersion about the mean is also poorly known. If the dispersion happens to be large by some criterion, we need to take it into account in assessing the likely errors of our mass estimates.

2 IS THERE A DISPERSION IN THE MAIN-SEQUENCE MASS-RADIUS DIAGRAM?

Such a dispersion can arise from the difference in radii of ZAMS and TAMS stars (for $M \sim 1\,M_\odot$; see Patterson 1984), from stars not yet having reached the ZAMS ($M \sim 0.1\,M_\odot$), or from composition or other differences among stars (at any mass). That such a dispersion among low-mass stars may exist is suggested by Greenstein (1989) who shows for late-type dwarfs in the field that at a given *color* there is a dispersion in *absolute magnitude* of ± 0.58 mag. If this were due solely to a spread in radii, the dispersion in radius would be 30%. If even half of the variance in magnitude is due to *radius* variations at a fixed *mass*, i.e., $\sigma(R) = 0.20R$, the consequences for the interacting binary business are sobering to contemplate. First, it means that many more determinations of masses and radii must be carried out for late-type field dwarfs before the *mean* mass-radius relation can be known with good precision. Second, it means that using the Roche constraint to help determine the stellar masses in any *individual* interacting binary system carries with it the penalty of a large uncertainty in the resulting masses, since it cannot be known whether the secondary star lies above or below the mean relation, or by how much.

(I do not claim to show that a significant dispersion in radius at fixed mass does exist; rather I treat such a case hypothetically, using 20% as a benchmark value for the size of the dispersion. This question deserves further discussion. Nevertheless, I think it is unwise to assume that there is *no* dispersion among the secondary stars in interacting binaries. As an example, some models for the synchronization torque in AM Her binaries require the dissipation to occur within the secondary star. For a re-synchronizing system like V1500 Cyg (Schmidt 1990), the dissipation power is similar in magnitude to the secondary star's intrinsic luminosity. The structure of this star might therefore differ from that of an unperturbed star.)

3 HOW SERIOUS ARE THE IMPLICATIONS OF SUCH A DISPERSION FOR OUR USE OF THE ROCHE CONSTRAINT?

I now illustrate the size of this uncertainty for three cases of interest. For ease of illustration I use a linear proportionality of radius and mass,

$$R = c_2 M \tag{1}$$

and I use the Paczyński (1971) approximation of the Roche lobe mean radius,

$$\frac{R_2}{a} = c_3 \left(\frac{M_2}{M_1 + M_2} \right)^{1/3} \tag{2}$$

where a is the semimajor axis,

$$a^3 = c_1^3 (M_1 + M_2) P^2 \tag{3}$$

from Kepler's third law. These equations can be combined to give the mass of the secondary star,

$$M_2 = C^{3/2} P \tag{4}$$

where $C = c_1 c_3 / c_2$. We immediately see that when equation (4) is used to find M_2, the fractional uncertainty in M_2 is 1.5 × the fractional uncertainty in c_2. (The physical constant c_1 is "exact". Equation (2) can be made exact for the masses of interest in any particular case if the "constant" $c_3 \sim 0.462$ is allowed to vary by less than 2%, a negligible amount.)

Of more interest is how the uncertainty in determining M_1, the mass of the white dwarf, depends on the uncertainty in c_2. The three cases of interest are when K_1 and K_2 are known but $\sin i$ is not, or when $\sin i$ is known (from a sufficiently deep eclipse) and either K_1 or K_2 is known but not both.

Case 1. Both velocity curves known, $\sin i$ unknown. We have

$$M_1 = \left(\frac{K_1}{K_2}\right) M_2 = \left(\frac{K_1}{K_2}\right) C^{3/2} P$$

and therefore

$$\left[\frac{\sigma(M_1)}{M_1}\right]^2 = \left[\frac{\sigma(K_1)}{K_1}\right]^2 + \left[\frac{\sigma(K_2)}{K_2}\right]^2 + \left[\frac{3}{2}\frac{\sigma(c_2)}{c_2}\right]^2 \tag{5}$$

so that, if c_2 is uncertain by 20%, M_1 must be uncertain by at least 30%.

Case 2. K_1 known, $\sin i$ known. The mass function equation

$$\frac{PK_1^3}{2\pi G} = \frac{M_2 \sin^3 i}{(1+q_{12})^2}$$

can be solved for $q_{12} = M_1/M_2$, using equation (4), and M_1 then follows:

$$M_1 = (2\pi G)^{1/2} \left(\frac{\sin i}{K_1}\right)^{3/2} PC^{9/4} - PC^{3/2}.$$

This expression leads to

$$\left[\frac{\sigma(M_1)}{M_1}\right]^2 = \left[\frac{3}{2}\frac{1+q_{12}}{q_{12}}\right]^2 \left[\left\{\frac{\sigma(\sin i)}{\sin i}\right\}^2 + \left\{\frac{\sigma(K_1)}{K_1}\right\}^2\right] + \left[\left(\frac{9}{4} + \frac{3}{4q_{12}}\right)\frac{\sigma(c_2)}{c_2}\right]^2$$

or

$$\frac{\sigma(M_1)}{M_1} \geq 2.25 \frac{\sigma(c_2)}{c_2} \tag{6}$$

where equality in equation (6) holds if $M_1 \gg M_2$ and there are no uncertainties (!) in K_1 or $\sin i$. Thus an uncertainty of 20% in c_2 implies a mandatory uncertainty of more than 45% in M_1.

Case 3. K_2 known, $\sin i$ known. The mass function equation and equation (4) lead to a cubic equation for M_1. To find the dependences of M_1 on uncertainties in C, K_2, and $\sin i$, do a linear perturbation expansion of the cubic equation around its solution, in terms of the error quantities δM_1, δC, δK_2, and $\delta(\sin i)$. Combining errors in quadrature one finds, to first order,

$$\left[\frac{\sigma(M_1)}{M_1}\right]^2 = \left[3\frac{1+q_{12}}{3+q_{12}}\right]^2 \left[\left\{\frac{\sigma(\sin i)}{\sin i}\right\}^2 + \left\{\frac{\sigma(K_2)}{K_2}\right\}^2\right] + \left[\frac{3}{3+q_{12}}\frac{\sigma(c_2)}{c_2}\right]^2. \tag{7}$$

For 20% uncertainties in c_2 and no other uncertainties, the uncertainty in M_1 varies between 5% and 15% as q_{12} varies from 10 to 1.

Other cases (K_1 known but $\sin i$ unknown because there is no eclipse, for example) cannot be solved without imposing additional constraints, such as the desire not to have the white dwarf mass exceed the Chandrasekhar limit. In the case of an eclipsing double-lined spectroscopic binary, of course, the Roche constraint is unnecessary, although it can serve as a check. Finally, I have not considered some "non-classical" cases, in which additional constraints are derived from the gas stream or bright spot position or from the velocity widths of lines.

4 SUMMARY

Variations in the radii of low-mass "main-sequence" stars at a given mass are expected to occur, and have possibly been observed (although indirectly). These variations arise in field stars due to differences in composition, age, and possibly other factors such as rotation rate. Additional effects in close binary systems, such as dissipative forces in some but not all cases, could increase the magnitude of any such dispersion. The consequences are that (1) the "mean" mass-radius relation that we rely on must remain uncertain until many more accurate masses and radii have been measured for field stars, and (2) since we do not know enough about the composition, age, and evolutionary state of the secondary stars in individual interacting binaries (to say nothing of irradiation-induced or mass-transfer-induced radius changes), we cannot know where the secondary stars lie with respect to the mean relation. When the Roche constraint is used to help find the masses for the component stars in these binary systems, this uncertainty *alone* gives rise (as outlined above) to errors in the masses as follows:

If there is a dispersion of 20% in radius among late-type dwarfs of a given mass, then for the simple mass-radius relation and the cases considered here, the uncertainty in the derived mass of the secondary star is 30% and the uncertainty in the derived mass of the white dwarf primary star is between 5% and more than 45%, depending on the mass ratio and what other information has been used in the mass solution. This is *in addition to* the uncertainty introduced by errors in measuring K_1, K_2, or the like. The most favorable case in which the Roche constraint is used is that of an eclipsing system for which K_2 has been measured. The more common cases are much worse.

I happily acknowledge conversations with K. D. Horne, J. R. Thorstensen, and G. D. Schmidt, and support from National Science Foundation grant AST-8818069.

REFERENCES

Echevarria, J. 1983, *Rev. Mexicana Astr. Ap.*, **8**, 109.
Greenstein, J. L. 1989, *Pub. Astr. Soc. Pac.*, **101**, 787.
Lacy, C. H. 1977, *Ap. J. Suppl.*, **34**, 479.
Paczyński, B. 1971, *Ann. Rev. Astr. Ap.*, **9**, 183.
Patterson, J. 1984, *Ap. J. Suppl.*, **54**, 443.
Schmidt, G. D. 1990, this volume.

Spin Period Variations of Intermediate Polar Emission Lines

Coel Hellier and Keith O. Mason

Mullard Space Science Laboratory, University College London,
Holmbury St. Mary, Dorking, Surrey, RH5 6NT U. K.

ABSTRACT

Spin-phase resolved spectroscopy of intermediate polars reveals the motions of the magnetically dominated emission regions and so provides more information than simple photometry. We find that the emission lines show velocity variations with the spin cycle and that the lines are bluest at flux maximum. We attribute the motion to the quasi-radial infall of material onto the magnetic poles of the white dwarf, implying that the upper pole points away from the observer at flux maximum. A simple model, invoking the variable aspect of an optically thick accretion curtain of material as it sweeps round with the spin cycle, can explain the spin modulations of the optical continuum, the emission lines, and the X-ray flux, providing a natural explanation of their similar pulse profiles and phasings. The mechanism for the optical spin pulse is not efficient at low inclinations.

1 THE SPIN PERIOD IN INTERMEDIATE POLARS

The intermediate polar (IP) subclass of cataclysmic variable is distinguished by quasi-sinusoidal modulations of the X-ray and optical light curves which are believed to signify the spin of a highly magnetic white dwarf. Although a magnetic field dominating the later stages of accretion provides the basic explanation for spin-pulsed emission, there is as yet no consensus over the modulation mechanism in either the X-ray or the optical band. Indeed, there is not even a consensus over the basic accretion geometry. In the conventional picture, material emanating from the secondary first circularizes into an accretion disk which is then magnetically disrupted nearer the white dwarf. Inside the magnetosphere, material corotates with the magnetic field while accreting along the field lines onto the magnetic poles. Alternatively, as proposed by Hameury, King, and Lasota (1986), the magnetic field may prevent any disk from forming. In this case the accretion stream from the secondary impacts directly onto the white dwarf's magnetosphere, producing a complex mixture of corotating material stripped from the accretion stream together with large blobs of material which have not yet been penetrated by the field and so continue on a ballistic trajectory, possibly as far as the white dwarf surface.

An early attempt to explain the X-ray light curves of IPs (King and Shaviv 1986) proposed that the polar region over which accretion occurred was a large fraction ($\sim 50\%$) of the white dwarf surface. With an offset magnetic axis, part of the region would disappear over the white dwarf limb as the star rotated. This mechanism would, to first order, produce an energy-independent modulation at the spin period. However, the observations show that in all known IPs the modulation is markedly deeper at low energies (e.g., Norton and Watson 1989). Additionally, Rosen, Mason, and Córdova (1988) concluded that with the geometry of a disrupted disk, the accreting fraction is unlikely to exceed $\sim 1\%$ of the white dwarf surface. The major accretion shock will occur above the surface, producing a tall, thin X-ray-emitting accretion 'curtain' whose optical depth is least through its sides. They suggested that the major modulation mechanism is photoelectric absorption, with flux minimum occurring when we look down the accretion column (magnetic pole pointing towards us) and the optical depth in the line of sight is greatest. Buckley and Túohy (1989) reached similar conclusions from a study of TX Col (1H0542–407) but also proposed that electron scattering could contribute significantly to beaming the X-ray radiation through the sides of the accretion column; this mechanism may dominate at higher energies where photoelectric absorption is inefficient.

While prominent spin-period modulations are always seen in IP X-ray light curves, this is not universally so in the optical continuum. An optical spin pulse is seen in some objects, such as EX Hya and FO Aqr, but in others, e.g., V1223 Sgr and TX Col, the optical light curve is modulated at the beat period between the spin and orbital periods while the spin period is not seen. TV Col shows no optical modulation at either spin or beat frequencies, while in AO Psc the beat pulse is normally larger but at times has a reduced amplitude, allowing the spin pulse to dominate. The beat-period modulation is simply explained; the X-ray radiation, which takes the form of a fan beam rotating with the spin period, may be reprocessed by non-axisymmetric structure fixed in the binary frame (e.g., the secondary star or a bulge on the disk) producing optical emission pulsed at the beat period.

Warner (1985) offered two explanations of the spin-pulsed optical emission. First, a direct component could arise from the white dwarf surface and be modulated by occultation. Second, X-ray reprocessing by axisymmetric disk structure could produce a spin modulation if there were a front-back asymmetry. For example, the rear of a concave disk and its inner face at the magnetosphere would both be preferentially viewed. In the original model, with the accretion pole pointing towards us at spin maximum, this pulse would be anti-phased with the X-ray pulse and the direct optical pulse. Observations show, though, that in general the optical and X-ray spin pulses of IPs are in phase.

IP spectra show strong, Doppler-broadened hydrogen and helium emission lines. We have undertaken a program of time-resolved spectroscopy of IPs to discover the motions of the accreting regions and so gain more information than can be derived from simple photometry. Here we present an overview of our results; fuller reports

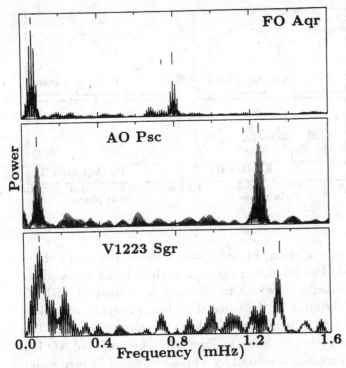

Fig. 1—Fourier transforms of the V/R ratio of the He II λ 4686 line. Tick marks indicate the orbital, spin, and beat (smaller ticks) frequencies. All show the major variations at the orbital and spin periods regardless of whether the spin or beat period dominates photometrically.

on individual stars are given in Hellier *et al.* (1987) on EX Hya (Paper 1); Hellier, Mason, and Cropper (1990) on FO Aqr (Paper 2); and Hellier, Cropper, and Mason (1990) on AO Psc (Paper 3).

2 SPIN-PERIOD VARIATIONS OF THE EMISSION LINES

Penning (1985) first reported that IP emission lines show radial velocity variations with the spin period. We confirm this with Fourier transforms of the V/R ratio of the emission lines in FO Aqr, AO Psc, and V1223 Sgr (Fig. 1). We have calculated the V/R ratio as the ratio of the equivalent widths of the line on either side of the rest wavelength. This measure is thus sensitive to both line profile changes and radial velocity motion of the line (as the lines are a complex sum of several components, a distinction between these effects is not appropriate when considering the line as a whole). Note that the dominant V/R change is at the spin period in all the systems, regardless of whether a spin-period modulation is seen photometrically. In EX Hya (Paper 1), we found a V/R variation in the emission lines at a level well below that in the above three systems; it was detected above the noise only by restricting the V/R calculation to the high-velocity line wings which originate nearest the white dwarf and are most affected by the magnetic field.

Fig. 2—V/R ra-
tio variations over the
spin period. All show
maximum blueness at
phase 1 (defined as X-
ray and optical flux
maximum).

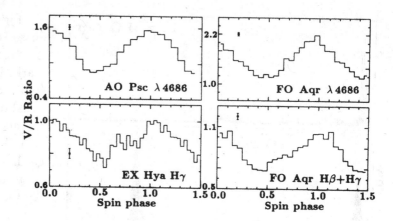

To investigate the velocity modulation, Figure 2 shows the V/R ratios of lines
from EX Hya, FO Aqr, and AO Psc folded on the spin cycles. In all cases phase
0 is optical photometric spin maximum which in all cases is coincident with X-
ray spin maximum. The V/R variation in the three stars has a roughly sinusoidal
profile phased with maximum V/R (maximum blueshift) at spin maximum. The
line profiles folded over the spin cycle are shown in Figure 3. The profile of AO Psc
He II $\lambda\,4686$ shows that the V/R motion is caused by a broad 'S-wave' feature which
is bluest at spin maximum. We measure the broad S-wave to have a FWHM of
450 ± 50 km s^{-1}, a semi-amplitude of 260 ± 30 km s^{-1} with maximum blueshift at
phase 0.95 ± 0.03 and mean velocity -15 ± 20 km s^{-1} (Paper 3).

The FO Aqr He II $\lambda\,4686$ line shows a similar S-wave with the same phasing, although there is the additional complexity of equivalent width changes. The Balmer
lines show parts of the same S-wave, but it is not prominent throughout the cycle
and is least visible when reddest. This results in a bulge on the blue wing of the
emission line at phase 1 but no corresponding bulge on the red wing at phase 0.5.
The Balmer line profiles of EX Hya show a bulge at phase 1 on both the blue and
the red wings, producing a total line width much greater than at spin minimum.

3 MODELS OF THE SPIN-MODULATED EMISSION

Penning (1985) proposed a reprocessing model for the spin-modulated line emission. He suggested that the X-ray beam swept round, illuminating different regions
of the disk, producing line emission with the local Keplerian velocity, and so a velocity variation. However when the X-ray beam points along the line of sight (at
spin maximum), the reprocessing would occur in material travelling across the line
of sight. Hence, at spin maximum we would expect to see zero velocity rather than
the observed maximum blueshift.

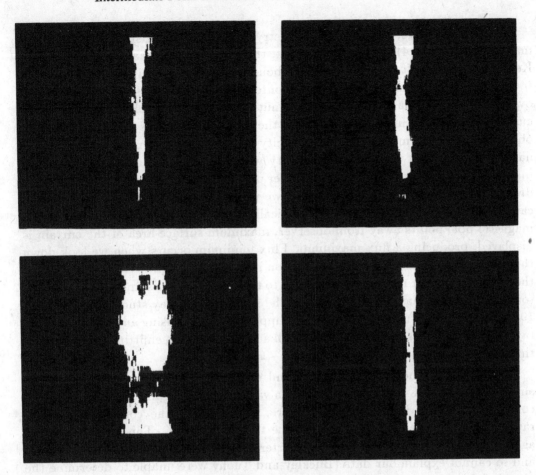

Fig. 3—Spin-cycle variations of intermediate polar emission lines showing; AO Psc He II λ 4686 (*top left*), FO Aqr He II λ 4686 (*top right*), FO Aqr Hγ (*bottom right*), and EX Hya Hγ (*bottom left*). The data are normalized to the continuum so that the greyscaling depicts quasi-equivalent width changes. In all cases phase 1 corresponds to optical and X-ray flux maximum. The data show prominent S-waves, phased with maximum blueness at spin maximum, which we attribute to material falling onto the magnetic poles. We believe the symmetric structure in EX Hya results from viewing both poles. Note that the structure near the center of the EX Hya Hγ line is due to an orbital cycle S-wave which does not smear out when folded on the spin cycle due to the near commensurability of the two periods (Paper 1). Similarly, the apparant redness of the FO Aqr Hγ line results from the smearing out of a blueshifted absorption S-wave on the orbital cycle (Paper 2).

As an alternative, consider the geometry of an accretion disk disrupted by the magnetic field (Fig. 4). While in the disk, the accreting material circulates at the Keplerian velocity. It then encounters a boundary layer where its velocity changes to quasi-radial infall along the field lines onto the magnetic poles. Accretion occurs over a range of azimuth, forming an azimuthally-extended, arch-shaped accretion curtain whose footprint is an arc around the magnetic pole. As the phasing of the observed variation implies that the velocity is parallel to the magnetic axis, it is natural to associate the observed velocity with the infall. The blueshift at spin maximum implies that the dominant upper pole is then on the far side of the white dwarf. Hence in Paper 1 (on EX Hya) we proposed that the spin modulation is caused by the varying aspect of the optically thick accretion curtain. When the magnetic pole points away from observer, maximum surface area of the curtain is displayed, producing a flux maximum. Flux minimum occurs when we look down the accretion column with most absorption in the line of sight. This model placed the magnetic pole at the opposite phasing to that previously assumed (pole pointing toward us at flux maximum). However, the subsequent X-ray studies (Rosen *et al.* 1988; Buckley and Tuohy 1989) provide support for this phasing and explain the X-ray modulation in a similar way, with a tall accretion column emitting preferentially through its sides.

Buckley and Tuohy (1989) also found spin-period velocity variations in the emission lines of TX Col. They proposed a variant model, again with magnetically-controlled material in the magnetosphere, but invoking the corotation velocity rather than the infall velocity to produce the modulation. This, though, would again produce maximum blueshift a quarter of a cycle away from flux maximum, and so cannot explain our data (Buckley and Tuohy were unable to determine the phase of their data relative to the X-ray pulse).

Our 'self-absorption' model, proposed for EX Hya, is directly applicable to FO Aqr and AO Psc. The large S-wave motion is expected when viewing one pole predominantly. We can also explain some of the differences between the systems by considering the shape of the arch-shaped accretion curtain. The material leaving the disk initially has a low infall velocity as it climbs to the top of the arch, but will rapidly accelerate as it approaches the white dwarf. The actual velocities will

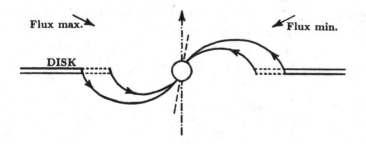

Fig. 4—Edge-on illustration of the accretion geometry resulting from a magnetically-disrupted disk.

depend on the radius at which the disk is disrupted; for illustration, with a system inclination of 60° and a 0.8 M$_\odot$ white dwarf, the measured velocity of the AO Psc S-wave (260 km s^{-1}) corresponds to the increase in radial velocity during freefall from a distance of 2.0×10^{10} cm to 1.2×10^{10} cm. The S-wave in AO Psc has no net velocity shift, implying that the flux-weighted mean velocity of the emitting material is parallel to the disk plane, presumably at the apex of the arch. We would expect to view this region equally well at all spin phases and, as seen from Figure 3, the S-wave shows little variation in its intensity through the cycle.

The FO Aqr S-wave extends to higher velocities than that of AO Psc and is more prominent on the blue side than the red. The line fluxes are also modulated with the spin cycle (Paper 2). We suggest that in this star the emission centroid is nearer the white dwarf, past the apex of the arch. This region is readily visible at spin maximum when the pole points away from the observer and the S-wave is bluest (see Fig. 4). However, as FO Aqr is at a high inclination ($\sim 70°$, Hellier et al. 1989), when the pole points towards us the inner regions would be obscured by material further out, reducing the S-wave while it is red. To test these ideas, measurement of the S-wave γ velocity may locate the emission region along the arch. Additionally, at large radii we would expect the infall velocity to be accompanied by a substantial co-rotation velocity, providing further information.

In EX Hya, we see both extended blue and red wings at flux maximum. In Paper 1 we suggested that we were seeing material accreting onto both magnetic poles (also proposed by Kaitchuck et al. 1987). This would produce two mirror-image S-waves. Partial obscuration of the lower pole by material in the disk could result in a stronger upper pole, creating an asymmetry and so the observed V/R variation with an amplitude much lower than in FO Aqr and AO Psc. One might expect a second increase in width at phase 0.5 when the two S-waves are again at maximum velocity, whereas the data show a width minimum at this phase. It may be that the S-waves are invisible at this phase due to the same effect which reduces the red part of the single S-wave in FO Aqr. The high velocity of EX Hya's line wings (± 3500 km s^{-1}) implies that the emission must originate within at most a few white dwarf radii (Paper 1). At a high inclination (EX Hya is eclipsing), these inner regions would be almost totally obscured near flux minimum by material rising out of the plane near the magnetosphere. In addition, the lower pole could disappear behind the white dwarf at this phase, removing the blue part of its S-wave and enhancing the V/R asymmetry.

4 DISCUSSION

We have proposed a model for the spin-period variations of the emission lines in IPs which has applicability to several systems. The main ideas, in which the spin modulation is caused by the varying aspect of an optically thick accretion curtain of material as it rotates with the magnetic field, are similar to those now used to

Table 1: Intermediate polar pulse amplitudes. Optical values are peak-to-trough as a fraction of the mean. X-ray values are 2–4 keV, from Norton and Watson (1989).

	Optical spin amp. (%)	Optical beat amp. (%)	X-ray spin amp. (%)	Inclination	Refs.
EX Hya	40	...	26	78°	1
FO Aqr	43	17	84	70°	2,3
BG CMi	45	...	45	?	4
AO Psc	3	6	78	~60°	5,6
V1223 Sgr	...	15	30	<40°	7
TX Col	...	14	24	~25°	8
TV Col	...	...	31	?	9

(1) Hellier *et al.* 1987 (2) Patterson and Steiner 1983 (3) Hellier *et al.* 1989 (4) McHardy *et al.* 1987 (5) van der Woerd *et al.* 1984 (6) Paper 3 (7) Watts *et al.* 1985 (8) Buckley and Tuohy 1989 (9) Barrett *et al.* 1989.

explain the X-ray spin modulation (Rosen *et al.* 1988; Buckley and Tuohy 1989; see also Mason, Rosen, and Hellier 1988 for a discussion of the X-ray model and its relation to our optical model). The similarity of the production mechanisms explains why the optical and X-ray modulations are always in phase and why they have similar pulse profiles. This model applies both to systems with a prominent optical photometric spin pulse and those without, such as V1223 Sgr and AO Psc — all show similar spin-period emission line velocity variations.

The model does not exclude a contribution to the optical spin pulse from re-processed X-rays in a manner similar to that proposed by Warner (1985). The hard X-rays from the accretion column, which are least affected by absorption, would preferentially illuminate the side of the disk that the pole pointed towards. Thus, with a front-back asymmetry so that rear of the disk was preferentially viewed, an additional pulse would be observed peaking when the pole pointed away from us. In this model, where X-ray spin maximum occurs with the upper pole pointing away from us, the reprocessed optical pulse is in phase with the direct X-ray and optical pulses from the accretion curtain.

Finally, we consider why some IPs show an optical spin pulse while others don't. As both of the above mechanisms for an optical spin modulation would be most efficient at high inclinations, this could simply be an aspect effect. In a face-on system, there would be no front-back disk asymmetry while our view of the accretion curtain would be similar at all spin phases. In Table 1 we give the spin and beat pulse amplitudes of well-studied IPs together with the best estimate of the inclination. This confirms that those showing a prominent optical spin pulse tend to be at high inclination (although the inclinations are reliable only for EX Hya and FO Aqr, where eclipses have been seen). There is no obvious trend in the amplitudes of the beat periods with inclination; these are probably related to the vertical structure in the accretion disk rather than the system aspect. Table 1 suggests that where the beat period dominates the optical light curves, it is due to the absence of a spin modulation rather than a large beat period amplitude. The X-ray spin amplitudes show some evidence of a similar correlation (note that as we see X-ray emission from both poles of EX Hya, it may not follow the trend). As the X-ray emission is produced nearer the white dwarf, a low inclination has a smaller effect on the variation in our view of the emission region over the spin cycle; this may explain why we still see an X-ray modulation at low inclinations but no optical spin pulse.

CH acknowledges receipt of an SERC Fellowship while KOM is supported by the Royal Society, London.

REFERENCES

Barrett, P., O'Donoghue, D., and Warner, B., 1988, *M. N. R. A. S.*, **233**, 759.

Buckley, D. A. H., and Tuohy, I. R. 1989, *Ap. J.*, **344**, 376.

Hameury, J.-M., King, A. R., and Lasota, J.-P. 1986, *M. N. R. A. S.*, **218**, 695.

Hellier, C., Cropper, M. S., and Mason, K. O. 1990, in preparation (Paper 3).

Hellier, C., Mason, K. O., and Cropper, M. S., 1989, *M. N. R. A. S.*, **237**, 39p.

Hellier, C., Mason, K. O., and Cropper, M. S., 1990, in press (Paper 2).

Hellier, C., Mason, K. O., Rosen, S. R., and Córdova, F. A. 1987, *M. N. R. A. S.*, **228**, 463 (Paper 1).

Kaitchuck, R. H., Hantzios, P. A., Kakaletris, P, Honeycutt, R. K., and Schlegel, E. M. 1987, *Ap. J.*, **317**, 765.

King, A. R., and Shaviv, G. 1984, *M. N. R. A. S.*, **211**, 883.

Mason, K. O., Rosen, S. R., and Hellier, C. 1988, *Adv. Sp. Res.*, **8**, #2, 293.

McHardy, I. M., Pye, J. P., Fairall, A. P., and Menzies, J. W. 1987, *M. N. R. A. S.*, **225**, 355.

Norton, A. J., and Watson, M. G. 1989, *M. N. R. A. S.*, **237**, 853.

Patterson, J., and Steiner, J. E. 1983, *Ap. J. (Letters)*, **264**, L61.

Penning, W. R. 1985, *Ap. J.*, **289**, 300.

Rosen, S. R., Mason, K. O., and Córdova, F. A. 1988, *M. N. R. A. S.*, **231**, 549.

Warner, B. 1985, in *Cataclysmic Variables and Low-Mass X-ray Binaries*, ed. D. Q. Lamb and J. Patterson (Dordrecht: Reidel), p. 269.

Watts, D. J., Giles, A. B., Greenhill, J. G., Hill, K., and Bailey, J. 1985, *M. N. R. A. S.*, **215**, 83.

van der Woerd, H., de Kool, M., and van Paradijs, J. 1984, *Astr. Ap.*, **131**, 137.

Synthetic Spectra of Accretion Disks in DQ Her Binaries

Christopher W. Mauche[1], Guy S. Miller[1], John C. Raymond[2], and Frederick K. Lamb[3]

[1]Space Astronomy and Astrophysics Group, Los Alamos National Laboratory
[2]Harvard-Smithsonian Center for Astrophysics
[3]Physics Department, University of Illinois at Urbana-Champaign

1 INTRODUCTION

Disk accretion by a strongly magnetic compact star differs fundamentally from disk accretion by a nonmagnetic star (see Lamb 1989). The interaction of the disk plasma with the stellar magnetic field drives electrical currents between the disk and the star. The resulting torque acts to bring the disk plasma and the star into corotation. At large radii, the gravitational and inertial stresses on the disk plasma greatly exceed the magnetic stress, and the angular velocity in the disk is nearly Keplerian. However, if the stellar magnetic field is strong enough, at some radius the rapidly increasing magnetic stress will dominate the other stresses, ending the Keplerian disk flow before it reaches the stellar surface. Here we call a star with such a strong magnetic field "strongly magnetic." Plasma approaching such a star leaves the disk near its inner edge and flows through the magnetosphere to the stellar surface, perhaps forming accretion columns above the magnetic poles.

The interaction of the disk plasma with the stellar magnetic field causes angular momentum to flow between the disk and the star. This flow is carried by the magnetic field, which is wound up by the shear between the relatively dense disk plasma, which orbits the star with the local Keplerian angular velocity, and the relatively tenuous magnetospheric plasma, which rotates with the angular velocity of the star. In a quasi-steady flow, the pitch of the magnetic field is determined by competition between the shear, which tends to increase it, and dissipation of the electrical currents in the disk and magnetosphere, which tends to decrease it. Thus, an accretion disk around a strongly magnetic star differs in two important ways from a disk around a weakly magnetic star. First, the disk ends before it reaches the stellar surface. Second, the local heating rate within the disk is altered, both by changes in the rate of viscous energy dissipation caused by the angular momentum flows between the disk and the star and by the heating caused by dissipation of electrical currents flowing within the disk. Both the termination of the disk flow above the stellar surface and the changes in the local heating rate within the disk should profoundly affect the spectrum of the radiation emitted by the disk.

This effect should be observable in the optical and ultraviolet spectra of DQ Her stars, which have magnetic fields strong enough to end the disk flow before it reaches the stellar surface (Bianchini and Sabadin 1983; Wu *et al.* 1989). This expectation is supported by Verbunt's (1987) tabulation of the (de-reddened) ultraviolet spectral indices of a large sample of cataclysmic variable (CV) spectra from the *IUE* archive: the median wavelength spectral index (α, defined by $f_\lambda \propto \lambda^{-\alpha}$) of the 1460–2880 Å ultraviolet continua of DQ Her stars is 1.3, which is the lowest of *all* the tabulated groups of "high accretion rate" disk-accreting CVs. The median ultraviolet spectral index of all other groups is 2.0.

Here we explore the effect of the stellar magnetic field on the spectrum of the disk by comparing synthesized disk spectra to the observed ultraviolet spectrum of the well-studied DQ Her binary GK Per. We use the parameterized disk-magnetosphere interaction models of Miller and Lamb (1990; hereafter ML) to calculate the radius of the inner edge of the disk and the local heating rate within the disk as a function of radius. With appropriate choices of parameters, these models can describe the models of Ghosh and Lamb (1979*a*, *b*; hereafter GL) and Wang (1987). Once the local heating rate is determined, we assign an effective temperature to each of a series of disk annuli. The disk spectrum is then calculated by summing the flux contributed by each annulus, assuming that each annulus has the same spectrum as a main-sequence star with the assigned effective temperature.

2 CALCULATION OF DISK STRUCTURE

The inner edge of the Keplerian disk occurs where the magnetic stress becomes large enough to remove the angular momentum of the disk flow in a radial distance small compared to the distance to the accreting object (see Lamb 1989). Thus, the inner edge of the disk has nothing to do with pressure balance or the radial velocity in the disk, but is determined by the rate at which angular momentum is removed from the disk by the magnetic stress. The differential rate at which the magnetic stress removes angular momentum from the disk is

$$\frac{dN_{\mathrm{mag}}}{dr} = -r^2 B_\phi B_z \,, \tag{1}$$

where B_ϕ and B_z are respectively the azimuthal and vertical components of the magnetic field evaluated on a surface above the disk flow. ML show that the radius R_0 of the inner edge of the disk is close to the radius given implicitly by the relation

$$c_0 \dot{M} \Omega_K(R_0) = B_\phi(R_0) B_z(R_0) R_0 \,, \tag{2}$$

with $c_0 \approx 0.5$. Here $\dot{M}$ is the mass-accretion rate and $\Omega_K(r) = \sqrt{GM_*/r^3}$ is the local Keplerian angular velocity.

As noted in the Introduction, the disk is heated both by viscous energy dissipation and by dissipation of electrical currents. As is usual for thin disks, we assume that the energy dissipated at each radius is radiated locally. Then the differential luminosity of the disk at radius r is given by the radial derivatives of (1) the radial flux of gravitational potential and kinetic energy, (2) the radial energy flux carried by the viscous stress, and (3) the rate of energy injection by the magnetic stress. Thus,

$$\frac{dL}{dr} = \dot{M}\frac{d}{dr}\left(\Phi_{\text{grav}} + \tfrac{1}{2}r^2\Omega_K^2\right) + \frac{d}{dr}(\Omega_K N_{\text{visc}}) - \Omega_*\frac{dN_{\text{mag}}}{dr} \tag{3a}$$

$$= -N_{\text{visc}}\frac{d\Omega_K}{dr} - [\Omega_* - \Omega_K(r)]\frac{dN_{\text{mag}}}{dr}, \tag{3b}$$

where Φ_{grav} is the gravitational potential energy, N_{visc} is the angular momentum flux carried radially outward through the disk by the viscous stress, and Ω_* is the stellar angular velocity. In expression $(3b)$, the various contributions to dL/dr have been rearranged to display explicitly the luminosity contributed by viscous energy dissipation and by dissipation of electrical currents.

The angular momentum flux carried by the viscous stress can be evaluated by using the fact that in a steady flow, the angular momentum entering and leaving any annulus of the disk must balance. Integrating the resulting differential equation from the inner edge of the disk, one obtains

$$N_{\text{visc}}(r) = \dot{M}\left[r^2\Omega_K(r) - R_0^2\Omega_K(R_0)\right] - \int_{R_0}^{r} \frac{dN_{\text{mag}}}{dr'}\,dr'. \tag{4}$$

The magnetic stress is determined by $B_\phi(r)$ and $B_z(r)$. ML consider two different prescriptions for $B_z(r)$, which give two sets of models. In the first set, $B_z(r)$ is assumed to be given by the z-component of an unscreened, dipolar stellar magnetic field, that is

$$B_z(r) = -\frac{\mu}{r^3}, \tag{5}$$

where μ is the dipole moment. In the second set of models, screening of the dipolar field by currents leaving the disk at R_0 is taken into account. As a result,

$$B_z(r) = -\frac{\mu}{r^3}\frac{B_z^2(R_0)}{B_z^2(R_0) + B_\phi^2(R_0)}. \tag{6}$$

In both sets of models, ML assume that (1) the interaction of the disk plasma with the magnetospheric field sweeps the field forward inside the disk corotation point (where the angular velocity of the disk plasma exceeds that of the star) and backward outside it (where the angular velocity is less than that of the star) and (2) the magnitude of the azimuthal component $B_\phi(r)$ is proportional to $B_z(r)$ times a power of r/R_A, where the Alfvén radius $R_A \equiv \mu^{4/7}(GM_*)^{-1/7}\dot{M}^{-2/7}$ is the usual length formed from μ, GM_*, and $\dot{M}$. Consistent with these assumptions, they write

$$B_\phi = g_0 B_z \left(\frac{|B_z|r^3}{\mu}\right)^{-4q/7} \left(\frac{r}{R_A}\right)^q \left|\frac{\Omega_* - \Omega_K}{\Omega_K}\right|^p \text{sign}(\Omega_* - \Omega_K). \tag{7}$$

The factor $|B_z|r^3/\mu$ is equal to unity in the models that assume an unscreened poloidal field. In the models that include screening, this factor effectively scales the radius

R_A, which is defined in terms of the unscreened field, bringing it into accord with the local poloidal field strength.

These models allow one to study the observational consequences of various radial variations of the magnetic pitch by considering different exponents p and q and different values of the "coupling constant" g_0. They can also be used to explore some of the effects of screening of the poloidal magnetic field.

3 CALCULATION OF DISK SPECTRA

We assume that the differential luminosity of the disk around a strongly magnetic star is given by equation (3). For comparison, we assume that the differential luminosity of the disk around a weakly magnetic star is given by

$$\frac{dL}{dr} = \frac{3}{2} \frac{GM_*\dot{M}}{r^2} \left[1 - \left(\frac{R_*}{r}\right)^{1/2} \right] , \tag{8}$$

which is appropriate if the star is rotating slowly. Once the differential luminosity is determined, we compute the effective temperature using the relation

$$\sigma T_{\text{eff}}^4 \equiv \frac{1}{4\pi r} \frac{dL}{dr} , \tag{9}$$

where σ is the Stefan-Boltzmann constant.

We assume that each annulus of the disk radiates the same spectrum as a main-sequence star with the effective temperature given by equation (9). The spectra have been chosen from a library of 46 de-reddened main-sequence stellar spectra ranging in temperature from 3,100 K to 47,000 K for spectral types M5 to O5. The vacuum ultraviolet ($\lambda\lambda \sim 1150$–3200 Å) spectra in the library were taken from the *IUE* spectral atlas of Wu *et al.* (1984) while the optical ($\lambda\lambda \sim 3500$–7500 Å) spectra were taken from the KPNO library of Jacoby, Hunter, and Christian (1984).

These elemental spectra have been synthesized into a disk spectrum using a computer code developed by S. Kenyon (Cannizzo and Kenyon 1987; Kenyon 1989). This code is similar to the one developed by Wade (1984). Like Wade, we assume that the disk flow is everywhere optically thick, physically thin, and steady. We approximate the continuous variation of the disk temperature with radius by a grid of temperatures chosen to match the temperatures of library spectra. An approximate spectrum of the complete disk is then constructed by summing the library spectra that correspond to the temperatures in the grid, weighting each one by the area of the corresponding annulus. In order to compare the spectra of disks around strongly and weakly magnetic stars, we synthesize the spectra of both types of disks for every choice of parameters.

4 RESULTS

For the sake of brevity, we consider here only the well-studied DQ Her binary GK Per; a more complete discussion will be published elsewhere. The "observables" for this source are: *IUE* spectra SWP 13497 and LWP 10143 and FES measurements of GK Per in outburst, the distance ($d = 470$ pc), the inclination ($i < 73°$), and the reddening ($E_{B-V} = 0.3$ mag), which we used to produce Figure 1; the orbital period ($P_{orb} = 47.9$ hr) and the mass ratio ($q = M_*/M_{sec} = 3.6$), which imply an accretion disk outer radius $R_d \approx 2 \times 10^{11}$ cm; and the mass ($M_* = 0.9\,M_\odot$) and spin period ($P_{spin} = 351$ s) of the white dwarf, which imply that the corotation radius is $R_c = (GM_*/\Omega_*^2)^{1/3} = 7.5 \times 10^9$ cm. We focus our attention on outburst spectra because they are least affected by emission from the secondary, the white dwarf, and the hot spot where the accretion stream strikes the outer disk. In comparing our synthesized disk spectra to observed spectra, we assume that contributions to the observed spectra from the boundary layer at the inner disk edge and from the accretion columns and stellar surface are negligible.

In order to define a model with a weakly magnetic white dwarf, one must specify M_*, R_*, R_d, and $\dot{M}$. If the white dwarf is strongly magnetic and rotating at its equilibrium spin rate, one must also specify P_{spin} and g_0, p, and q [cf. eq. (7)]. Finally, if the white dwarf is strongly magnetic but is not rotating at its equilibrium spin rate, one must also specify μ. In the two latter types of models, the white dwarf spin rate can be characterized by the fastness parameter $\omega_s \equiv \Omega_*/\Omega_K(R_0)$.

The fastness parameter at which the torque on the star vanishes is called the critical fastness and is denoted ω_c. Although the fastness of GK Per in outburst is not necessarily the equilibrium fastness, as illustrative examples we restrict ourselves to the consideration of models for which $\omega_s = \omega_c$. ML found that an unscreened model with $g_0 = 5$, $p = 1.0$, and $q = 1.2$ (which implies $\omega_c = 0.28$ and $R_0 = 0.42\,R_c$) was consistent with the $\dot{P}(L)$ relation found by Parmar *et al.* (1989) and power spectrum data obtained by Angelini, Stella, and Parmar (1989) from observations of EXO 2030+375. (However, these parameters are subject to large uncertainties because the orbital elements of this source are poorly known.) We shall refer to this choice of parameters as strongly magnetic white dwarf model A. The ML parameters that reproduce the magnetic stresses of the GL physical model are $p = 0.5$ and $q = 0.86$, which give $\omega_c = 0.74$. We refer to this choice of parameters as model B. The value of ω_c originally quoted by GL for their model was 0.35; this lower value was the result of an algebraic error (see Lamb 1989). Using $\omega_c = 0.74$, the radius of the inner edge of the disk is $R_0 = 0.82\,R_c$ rather than $R_0 = 0.50\,R_c$.

Figure 2 compares the predictions of three different models of the accretion disk in GK Per: the weakly magnetic white dwarf model and the strongly magnetic white dwarf models A and B. All three models assume a mass-accretion rate $\dot{M} = 10^{-7.5}\,M_\odot\,yr^{-1}$. The effects of injection and extraction of energy by the magnetic stress are clearly seen in both models A and B, for which the magnetic field strengths are found to be 3.3 MG and 7.5 MG, respectively. The magnetic stress extracts energy from the disk in the interval between R_0 and R_c but injects energy into the disk at radii larger than R_c. Because model A extends to smaller radii than model B, its maximum temperature is higher. However, the maximum temperatures

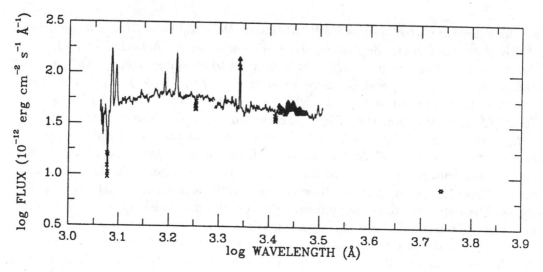

Fig. 1—The dereddened optical through ultraviolet outburst spectral distribution of GK Per, scaled to a distance of 100 pc and corrected for its inclination ($f_\lambda \leftarrow f_\lambda \times (470/100)^2/\frac{2}{5}(1+\frac{3}{2}\cos 70°)$). The spectral index $\alpha \equiv -\log(f_{1460}/f_{2880})/\log(1460/2880)$ is ≈ 1.0, although the spectrum is poorly fit by a power law. The various geometric symbols represent various data quality flags: extrapolated ITF pixels (*triangles*), saturated pixels (*filled triangle*), and reseaux (*crosses*).

of both strongly magnetic white dwarf models are significantly less than the maximum temperature of the weakly magnetic white dwarf model. The peak temperature of model A is $\approx 50,000$ K, compared to the $\approx 125,000$ K peak temperature of the weakly magnetic white dwarf model; the peak temperature of model B is $\approx 20,000$ K, less than half that of model A. According to Figure 1, the flux distribution of GK Per during outburst peaks at ≈ 1600 Å, implying a maximum temperature in the disk of $\sim 18,000$ K. Thus, the reduction of the peak temperature that occurs when the white dwarf is strongly magnetic is comparable in size to the reduction needed to explain the spectrum during outburst.

Turning to the resulting disk spectra, the 1460–2880 Å ultraviolet spectral index changes from 2.5 to 2.2 to 2.0 as one goes from the weakly magnetic white dwarf disk to the strongly magnetic model A disk and then to the model B disk. Although the spectra of all three models match the observed spectrum poorly (cf. Fig. 1), this decrease of ≈ 0.5 in the ultraviolet spectral index is comparable to the difference of ~ 0.7 between the ultraviolet spectral indices of Verbunt's (1987) sample of *IUE* spectra of DQ Her binaries and the spectral indices of all other "high accretion rate" CVs, indicating that the magnitude of the effect is correct. The spectral index of the Balmer continuum in stars is greater than the spectral index of a blackbody with the same temperature because radiative transfer processes redistribute flux from shortward of the Balmer and Lyman edges to longward of them. As a result, disk

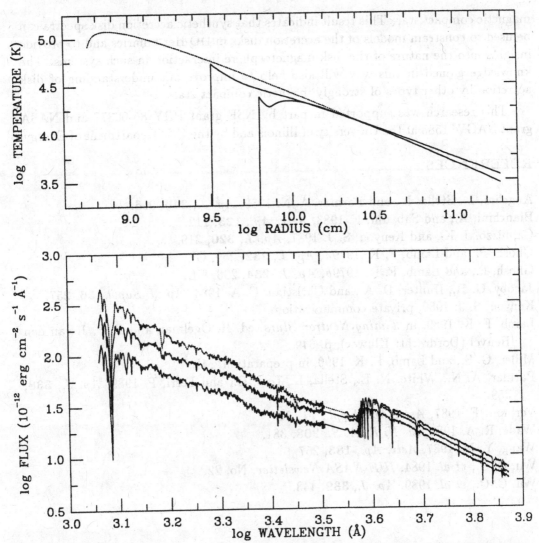

Fig. 2—Disk temperature versus radius curves (top panel) and the resulting optical through ultraviolet spectral distributions (bottom panel) for three models of GK Per: (1) the weakly magnetic white dwarf model (*upper, dotted curves*), (2) the strongly magnetic white dwarf model *A* (*middle, solid curves*), and (3) the strongly magnetic white dwarf model *B* (*lower, solid curves*).

spectra synthesized from blackbody spectra will be flatter than the spectra shown in Figure 2. The construction of synthetic disk spectra based on blackbody spectra is currently under way.

In summary, we find that it is possible to reproduce both the level and spectral index of the observed optical through ultraviolet spectrum of GK Per in outburst for certain choices of the parameters in the ML description of disk accretion by a strongly magnetic compact star. This result indicates that synthetic accretion disk spectra can be used to constrain models of the accretion disks in DQ Her binaries and to provide insights into the nature of the disk-magnetosphere interaction in such systems. The knowledge gained in this way will also help to improve our understanding of disk accretion by other types of strongly magnetic compact stars.

This research was supported in part by NSF grant PHY 86-00377 and NASA grant NAGW 1583 at the University of Illinois and by the U.S. Department of Energy.

REFERENCES

Angelini, L., Stella, L., and Parmar, A. N. 1989, *Ap. J.*, **346**, 906.
Bianchini, A., and Sabadin, F. 1983, *Astr. Ap.*, **125**, 112.
Cannizzo, J. K., and Kenyon, S. J. 1987, *Ap. J.*, **320**, 319.
Ghosh, P., and Lamb, F. K. 1979*a*, *Ap. J.*, **232**, 259, GL.
Ghosh, P., and Lamb, F. K. 1979*b*, *Ap. J.*, **234**, 296, GL.
Jacoby, G. H., Hunter, D. A., and Christian, C. A. 1984, *Ap. J. Suppl.*, **56**, 257.
Kenyon, S. J. 1989, private communication.
Lamb, F. K. 1989, in *Timing Neutron Stars*, ed. H. Ögelman and E. P. J. van den Heuvel (Dordrecht: Kluwer), p. 649.
Miller, G. S., and Lamb, F. K. 1990, in preparation, ML.
Parmar, A. N., White, N. E., Stella, L., Izzo, C., and Ferri, P. 1989, *Ap. J.*, **338**, 359.
Verbunt, F. 1987, *Astr. Ap. Suppl.*, **71**, 339.
Wade, R. A. 1984, *M. N. R. A. S.*, **208**, 381.
Wang, Y.-M. 1987, *Astr. Ap.*, **183**, 257.
Wu, C.-C., *et al.* 1984, *IUE NASA Newsletter*, No. 22.
Wu, C.-C., *et al.* 1989, *Ap. J.*, **339**, 443.

Spin-Up in BG CMi = 3A0729+103

Joseph Patterson

Columbia University

1 INTRODUCTION

McHardy *et al.* (1984) identified the high-latitude X-ray source 3A0729+103 with a 15th magnitude cataclysmic variable, subsequently named BG Canis Minoris. They reported coherent optical pulses with a period of 913 seconds, maintaining a constant phase over the 90-day baseline of observations. This established the star's membership in the DQ Herculis class, in which a rapid periodicity is presumed to arise from the rapid rotation of an underlying magnetic white dwarf. West, Berriman, and Schmidt (1987) then found 1.7% circular polarization in the infrared (1.25 μm), which, combined with the lack of strong linear polarization, constitutes impressive evidence that the light originates in a region of strong magnetic field ($\sim 10^7$ G). Although it is still somewhat surprising that the polarized flux does not show the periodicity (to within present limits), it is fair to say that this observation has at last revealed the critical signature of a magnetic field, establishing the basic soundness of the magnetic rotator model of DQ Her stars.

I have been accumulating high-speed photometry of DQ Her stars for the past 13 years, in an effort to track their rotational period changes. I added BG CMi to the program in 1982, and have been following it ever since. Here I report on the detection of a period decrease.

2 LIGHT CURVE AND PULSE TIMINGS

Most of the observations were obtained in blue or ultraviolet light, at the Cassegrain focus of the #2 0.9-m telescope of Kitt Peak National Observatory. Integration times ranged from 1 to 30 seconds. The only exceptions were the runs in September and December 1989, which were obtained with the Lick Observatory high-speed photometer (Stover and Allen 1987) on the Anna Nickel 1.0-m telescope. The latter instrument was used without filters, yielding an effective wavelength of ~5500 Å. Because the star's variability is a broadband phenomenon, no dependence of pulse timings on wavelength was expected, and none was found. Figure 1 shows a long light curve of the star, exhibiting the characteristic 3.2 hour orbital hump, the coherent 913 s variation, and erratic flickering. In Table 1 I present the complete log of

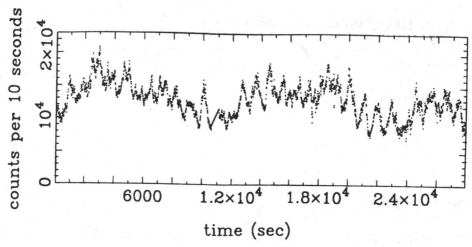

Fig. 1—A 7.8 hour light curve of BG CMi obtained in blue light on 4 Feb 1989, showing the 3.2 hour orbital humps, the 15 minute periodicity, and random flickering.

observations.

The last column of Table 1 gives the mean pulse arrival time for each night, obtained by fitting a 913.48 s sinusoid to the light curve. No barycentric or dynamical-time corrections were applied; these are of no consequence for such a long period.

The distribution of observations over time is not ideal. The cycle count can be regarded as established with certainty only in those intervals with densely spaced observations, namely: (a) the 90-day baseline covered in McHardy et al.'s first paper (1984); and (b) the February 1988–December 1989 data of Table 1. These two intervals yield periods of 913.480 ± 0.002 and 913.492 ± 0.001 seconds, respectively.

At face value, this suggests a period increase. But McHardy et al.'s quoted uncertainty of 0.002 s implies a flexure in phase of only 0.02 cycles over the 90-day baseline, and this seems much too optimistic. Considering the distribution of their observations, I would estimate their true 1σ error to be 3–4 times larger—in which case the period would be consistent with the 1988–9 value. However, even after this more generous assignment of uncertainty, I find that shortly after McHardy et al.'s initial intensive coverage, the phase began to depart significantly from their ephemeris. This is shown in the lower frame of Figure 2, which is an O−C diagram relative to McHardy et al.'s best-fit period. They did not publish explicit pulse timings, so I have represented their data as a shaded rectangle, assuming a typical scatter in the timings of 0.05 cycles. By February 1984 the phase appears to have slipped a full 1.5 cycles, implying a substantially longer period. While the O−C diagram gives a rather poor fit to McHardy et al.'s data (that is, it intersects the rectangle at a surprisingly steep angle), it corresponds to the only cycle count that makes any sense at all. I conclude that the period during 1982–1984 is reasonably well-established, and is 913.501 ± 0.002 seconds.

TABLE 1
LOG OF OBSERVATIONS

UT Date	Start	Duration	JD$_\odot$ maximum (2440000+)
21 Nov 1982	9:56:46	0^{h}35^m	5294.9340
10 Mar 1983		2 02	5403.6872
16 Mar 1983	4:05:23	2 13	5409.7145
7 May 1983	3:32:26	0 37	5461.6594
6 Oct 1983	9:32:09	1 10	5613.9207
10 Oct 1983	11:47:20	0 31	5618.0012
15 Nov 1985	12:11:25	0 30	6385.0131
25 Feb 1988	3:37:56	5 20	7216.6653
27 Feb 1988	5:20:42	3 20	7218.7266
29 Feb 1988	2:58:58	2 11	7220.6301
19 May 1988	3:35:18	0 36	7300.6550
20 May 1988	3:53:08	0 35	7301.6704
10 Nov 1988	11:01:20	1 34	7475.9636
11 Nov 1988	7:37:22	2 33	7476.8202
14 Nov 1988	8:00:04	3 53	7479.9717
16 Nov 1988	9:40:01	1 50	7481.9580
3 Feb 1989	4:28:09	2 17	7560.6950
4 Feb 1989	2:36:26	7 47	7561.6142
11 Feb 1989	6:18:03	3 36	7568.8347
12 Feb 1989	6:39:30	2 30	7569.8503
22 Sep 1989	11:30:30	1 30	7791.9866
23 Sep 1989	11:39:44	1 21	7793.0108
24 Sep 1989	11:31:30	1 35	7794.0145
25 Sep 1989	11:37:45	1 34	7795.0200
2 Nov 1989	9:45:07	1 35	7832.9229
4 Nov 1989	9:59:07	2 31	7834.9842
8 Nov 1989	10:08:07	2 22	7838.9591
4 Dec 1989	8:31:15	2 23	7864.8948

During 1984–7, the star was less diligently observed. Six timings are available during this interval, but there is no period that fits all of them. When the data are this sparse, a single bad timing can contaminate the whole study, so I have chosen to ignore this interval. The cycle count can probably be recovered if timings are continued through about 1991.

During 1988–9 I observed the star more frequently, attempting to exclude the normal aliases that afflict such period studies (1 day, 1 month, 1 year). These timings confirm the basic cycle count of McHardy *et al.*, and establish a period of 913.492 ±

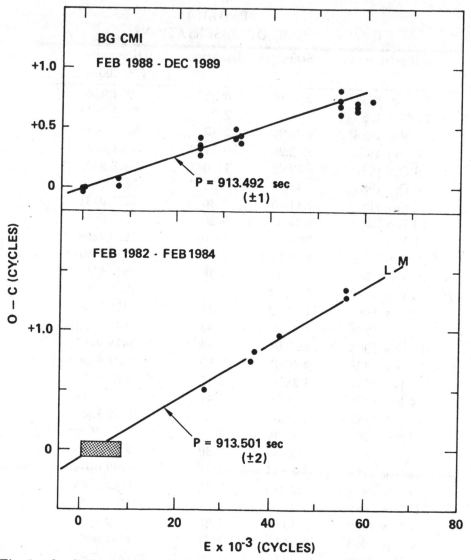

Fig. 2—O−C diagrams for the two intervals of densely spaced observations, relative to a test period of 913.48 s. The best-fit straight lines are labelled with their corresponding periods. The shaded rectangle and "L" represent timings from McHardy *et al.* (1984, 1987); "M" is from Mateo (1985); filled circles are from Table 1.

0.001 seconds. This implies a period decrease of 0.0018 (±4) s yr^{-1}.

3 INTERPRETATION

Most of the well-studied DQ Her stars have shown changes in their rotation periods, if observed sufficiently often to establish a cycle count over a few years. In their study of DQ Herculis itself, Herbst, Hesser, and Ostriker (1974) attributed the observed period decrease to the spin-up torques exerted by the accreting matter, and found that this explanation sufficed for an accretion rate of $\sim 10^{-8}\,M_\odot\,\mathrm{yr}^{-1}$, comparable to the rate expected if the luminosity derives entirely from accretion. At the time it seemed like spin-up measurements might be a promising way to derive the accretion rates and/or the physics of the disk/star interaction—if only other DQ Her stars could be found.

Well, now we have found a fistful of DQ Her stars, but are we any closer to realizing that promise?

On the theoretical side, Ghosh and Lamb (1979) have developed a model for explaining period changes in accreting magnetic compact stars, which includes spin-down torques from distant field lines dragged about by slowly moving gas in the disk, as well as the spinup torques exerted on the star by accreting matter. In this theory, the ratio $\omega_s = \Omega_*/\Omega_{\mathrm{Kep}}$, where Ω_* and Ω_{Kep} are respectively the rotation frequency of the star and the matter at the inner radius of the disk, determines whether the spinup or spindown torques will dominate. If the $\dot{P}$ we have measured in BG CMi is representative, then the star would be classified as a "slow rotator" (because it is secularly spinning up). In this case the matter torques dominate and we can write

$$\dot{P} = -9.3 \times 10^{-4}\,\mathrm{s\ yr^{-1}}\,\mu_{33}^{2/7} L_{34}^{6/7}, \tag{1}$$

where μ_{33} and L_{34} are respectively the star's magnetic moment and accretion luminosity in units of 10^{33} G cm^3 and 10^{34} erg s^{-1}. This can alternatively be written as

$$\dot{P} = -2.4 \times 10^{-3}\,\mathrm{s\ yr^{-1}}\,B_7^{2/7} M_1^{6/7} \dot{M}_{17}^{6/7}, \tag{2}$$

where B_7, M_1, and $\dot{M}_{17}$ are the star's magnetic field in units of 10^7 G, mass in solar masses, and accretion rate in units of 10^{17} g s^{-1}.

Can this theory account for the observed spinup in BG CMi? Comparing the behavior of the polarized light to that of AM Her, West, Berriman, and Schmidt (1987) estimated $B = 5$–10 MG in BG CMi. A value much lower than this would fail to produce polarized light, and a value much higher would dominate the accretion flow everywhere and prevent the formation of a disk. White dwarf masses are pretty reliably in the range 0.5–$1.0\,M_\odot$ (Shipman 1979), and those in CVs are not remarkably different (Shafter 1984). The accretion rate must be regarded as poorly known, because no reliable distance-finding techniques are available for stars not showing the spectrum of the secondary. But if we use the $\dot{M}$-orbital period relation of Patterson (1984), we estimate an accretion rate of 2×10^{16} g s^{-1}. Putting these three factors in equation (2), we expect $\dot{P} = 0.0004$ s yr^{-1}. Since the scatter in the empirical $\dot{M}$-P_{orb} relation alone is about a factor of 5, I certainly regard this as satisfactory agreement.

Nevertheless, a census of the eight "well-studied" DQ Her stars (i.e., having ephemerides established over long baselines) is downright confusing: 3 stars are spinning up, 2 are spinning down, 1 is going both ways, and 2 others have strong upper limits on period changes ($|\dot{P}| < 10^{-13}$ s s^{-1}). The Ghosh and Lamb theory is sufficiently flexible to accommodate such behavior, but only if we are willing to assign values of ω_s in an *ad hoc* manner. In order to claim that we understand the physics (not the *real* physics, just the torques) of the disk-star interaction, we need to find *independent* evidence of a star's status as a fast or slow rotator. This is an ambitious but not a hopeless task; it could be done, for example, by long-term studies of $\dot{P}$ versus L (Ghosh and Lamb 1979), or by careful studies of the emission line profiles which might reveal the innermost extent of the disk. For the latter case, the discovery of another DQ Her system showing *bona fide* deep eclipses would be extremely promising.

In the meantime, I promise to keep digging away in the trenches to keep track of cycle counts. I can't say that I think we're on the verge of *solving* these interesting problems, though; for me it is more a matter of individual torment:

> One day on my yacht,
> A plague I have caught,
> To discover P-dot.
>
> In time it did seem
> It would take a great team
> To unravel the crazes
> Of cycles and phases.
>
> 'Gainst this I have fought,
> Lest it all come to nought.
> Now I've got a P-dot,
> But hardly a thought,
> For my brain's turned to rot.

REFERENCES

Ghosh, P., and Lamb, F. K. 1979, *Ap. J.*, **234**, 296.

Herbst, W., Hesser, J. E., and Ostriker, J. P. 1974, *Ap. J.*, **193**, 679.

Mateo, M. 1985, in *Proceedings of the Ninth North American Workshop on Cataclysmic Variables*, ed. P. Szkody (Seattle: University of Washington), p. 80.

McHardy, I. M., Pye, J. P., Fairall, A. P., Warner, B., Cropper, M., and Allen, S. 1984, *M. N. R. A. S.*, **210**, 663.

McHardy, I. M., Pye, J. P., Fairall, A. P., and Menzies, J. W. 1987, *M. N. R. A. S.*, **225**, 355.

Patterson, J. 1984, *Ap. J. Suppl.*, **54**, 443.

Shafter, A. W. 1984, Ph.D. thesis, Univ. Calif. at Los Angeles.

Shipman, H. L. 1979, *Ap. J.*, **228**, 240.

Stover, R. J., and Allen, S. L. 1987, *Pub. Astr. Soc. Pac.*, **99**, 877.

West, S. C., Berriman, G., and Schmidt, G. D. 1987, *Ap. J. (Letters)*, **322**, L35.

A GINGA Observation of FO Aqr

A. J. Norton[1], M. G. Watson[2], A. R. King[3], I. M. McHardy[4] and
H. Lehto[4]

[1]High Energy Astrophysics Group, Physics Department,
University of Southampton, Southampton SO9 5NH, U. K.
[2]X-Ray Astronomy Group, Department of Physics and Astronomy,
University of Leicester, Leicester LE1 7RH, U. K.
[3]Astronomy Group, Department of Physics and Astronomy,
University of Leicester, Leicester LE1 7RH, U. K.
[4]Department of Astrophysics, South Parks Road, Oxford OX1 3RQ, U. K.

1 INTRODUCTION

The Intermediate Polar (IP) FO Aqr has an orbital period, $P_{orb} = 4.85$ hours
and a white dwarf spin period, $P_{spin} = 1254.45$ seconds (both values recently refined
by Osborne and Mukai 1989). Orbital modulation has previously been detected in
optical photometry and spectroscopy and spin modulation in both X-ray and optical
studies. Modulation at the beat period between the orbital and spin periods has
also been seen in optical data. Details of these observations may be found in the
recent publications by Osborne and Mukai (1989), Chiappetti *et al.* (1989), Hellier,
Mason, and Cropper (1989), and references therein. In common with other IPs, the
EXOSAT observations of FO Aqr showed the X-ray flux to be strongly modulated
at the white dwarf spin period, with a decreasing modulation depth at increasing
energies (Cook, Watson and McHardy 1984; Chiappetti *et al.* 1989; Norton and
Watson 1989, hereafter Paper 1). The X-ray spectrum of FO Aqr was found to be
flat and highly absorbed with a strong iron Kα emission line (Paper 1).

2 DETAILS OF THE OBSERVATION

FO Aqr was observed by the *GINGA* Large Area Counter (Turner *et al.* 1989)
for $\sim 115 \times 10^3$ seconds beginning at 16:56 UT on 29 October 1988. As usual with
GINGA, a large amount of this time was lost as a result of Earth occultation and
passage of the satellite through the South Atlantic Anomaly, leaving $\sim 46.5 \times 10^3$
seconds of useful data. This was accumulated in 48 energy channels spanning $\sim$ 2–
30 keV with a time resolution of 16 seconds. All or part of 92 spin cycles and 7
orbital cycles of FO Aqr were observed, although coverage of the orbital cycle was
limited to the phase bands: $\phi_{orb} \sim 0.05$–0.30, $\phi_{orb} \sim 0.40$–0.65, and $\phi_{orb} \sim 0.75$–0.95.
This arose because the orbital period of *GINGA* is $\sim$ one-third that of FO Aqr and
similar regions of the FO Aqr orbit were lost on each satellite orbit. Background

subtraction was performed using methods described by Hayashida *et al.* (1989). The mean flux for FO Aqr during this observation was 2.9×10^{-11} erg cm^{-2} s^{-1} (2–10 keV), slightly fainter than the *EXOSAT* measurements of 3.4×10^{-11} erg cm^{-2} s^{-1} (2–10 keV).

3 TIME SERIES ANALYSIS

The power spectrum of the data was calculated and cleaned to remove structure arising as a result of the *GINGA* orbit. The resulting spectrum (Fig. 1) shows the strongest peaks at the known frequencies of modulation in FO Aqr, namely the orbital frequency, $\Omega = 1/P_{orb}$ and the spin frequency, $\omega = 1/P_{spin}$. The third strongest peak coincides with the beat frequency, $\omega - \Omega$, and weaker peaks at $\omega + \Omega$ and $\omega \pm 2\Omega$ are also present. Finally, we see harmonics of the spin and orbital frequencies, indicating both modulations to be non-sinusoidal. We note that analysis of the *EXOSAT* data (Chiappetti *et al.* 1989) revealed a similar structure, but without the beat frequency, $\omega - \Omega$.

Fig. 1—Cleaned power spectrum of the 2–20 keV light curve of FO Aqr.

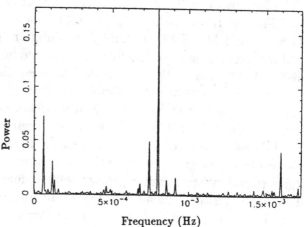

3.1 Orbital Modulation

The energy-resolved data folded on the orbital ephemeris of Osborne and Mukai (1989) are shown in Figure 2. This ephemeris defines phase zero at minimum optical flux. The limited phase coverage of the *GINGA* data is readily seen, as is a strong modulation of the X-ray flux at the orbital period. Fitting these folds with sinusoids yields modulation depths of: $49 \pm 3\%$ (2–4 keV), $46 \pm 2\%$ (4–6 keV), $36 \pm 2\%$ (6–10 keV), and $27 \pm 4\%$ (10–20 keV), defined as peak-to-peak amplitude divided by maximum flux. In each case the minimum of the sinusoid is at $\phi_{orb} = 0.89 \pm 0.02$. Since the modulation depth decreases with increasing energy, photoelectric absorption by material fixed in the orbital frame must be responsible, at least in part, for this effect. If the modulation is caused solely by absorption and electron scattering, an additional column density of several $\times 10^{23}$ cm^{-2} is needed to produce a depth of 27% in the 10–20 keV band (see Paper 1). However, a column of this size would imply a modulation depth of 100% at energies below 4 keV. This is not seen and suggests that a 'soft' diluting component is present, such as leakage through a patchy absorber or a genuinely unmodulated flux component. Alternatively, a fraction of the orbital modulation may be caused by an energy-independent effect, arising when material impacts the white dwarf at a region fixed in orbital phase which is then occulted by the body of the white dwarf (Hameury, King, and Lasota 1986).

On re-examining the *EXOSAT* light curves, we found that they also show a minimum flux at orbital phase $\phi_{orb} \sim 0.8$–0.9 (see Fig. 3, this is Observation C from Paper 1). Hence, this is a stable feature of the system and not an artifact of the *GINGA* coverage. The *EXOSAT* data shown span only a single orbital cycle (albeit with complete coverage) and so the individual pulses are not smeared out, as occurs to some extent in the *GINGA* data.

Fig. 2—Energy-resolved *GINGA* light curve of FO Aqr folded at the orbital period.

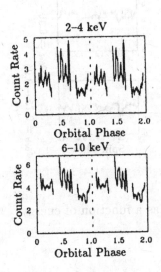
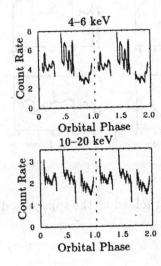
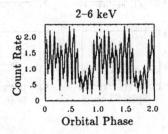

Fig. 3—*EXOSAT* light curve of FO Aqr folded at the orbital period.

3.2 Spin Modulation

We examined the pulse profile of FO Aqr as a function of energy *and* as a function of orbital phase. The resulting 12 independent folded data sets are shown in Figure 4. These profiles have arbitrary (though consistent) phase zeros, are repeated over two cycles for clarity, and have the same vertical axis scales for all 3 profiles at a given energy, to facilitate comparison. Using the quadratic ephemeris of Osborne and Mukai (1989), we note that the times of optical and X-ray pulse maximum agree within the estimated errors.

It is apparent that the pulse profile undergoes significant changes in amplitude and in shape as a function of both energy and orbital phase. In particular, the orbital minimum coincides with the virtual disappearance of the pulsations. Folded hardness ratios show that the spin modulation must be due, at least in part, to photoelectric absorption since the light curve becomes significantly harder at pulse minimum. The pulse profile itself may be described in terms of two components: a smooth sinusoidal modulation and a narrow 'notch' at $\phi_{spin} \sim 0.4$. This is the best evidence yet seen in any IP for a structured emission region, or regions, at the white dwarf surface. Previous pulse profiles have all been essentially sinusoidal with only hints of further structure (see Paper 1). The profiles presented here prove that in FO Aqr the emission region cannot be a single disk, ring, or arc-shaped polecap.

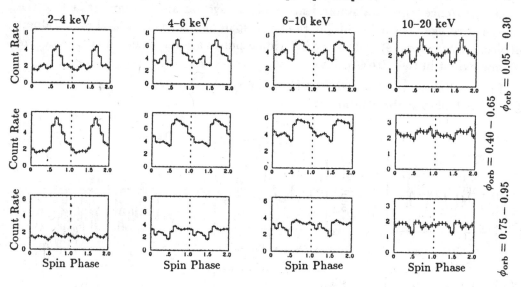

Fig. 4—Light curve of FO Aqr folded at the spin period, as a function of energy and orbital phase.

3.3 Beat Period Modulation

The beat period is defined as $P_{beat} = (1/P_{spin} - 1/P_{orb})^{-1}$ and is therefore the rotation period of the white dwarf in the reference frame of the binary. Optical photometric modulations at this period are commonly seen in IPs and are interpreted as emission from a reprocessing site which is fixed in the orbital frame of the system. The only other IP to show a beat period modulation in its X-ray light curve is H0542-407 (Buckley and Tuohy 1989), where an X-ray beat pulse is assumed to arise as a result of reflection from the secondary star. In FO Aqr, the X-ray spin pulse obviously undergoes changes as a function of orbital phase. Hence, a peak in the power spectrum at the beat frequency would naturally be expected to occur, without the addition of an explicit 'X-ray beat pulse'.

Osborne and Mukai (1989) found that the optical white dwarf rotation and beat period pulses occur simultaneously at orbital phase $\phi_{orb} \sim 0.8$. They suggest this must be the phase of inferior conjunction of the reprocessing site, where the accretion stream from the secondary hits material circulating the white dwarf. We can now add that at this orbital phase, the X-ray spin pulse virtually disappears and the X-ray flux is a minimum.

4 SPECTRAL ANALYSIS

Modelling the X-ray spectra of IPs has been problematical in the past, due to confusion over the nature of the model to use and the fact that the spectra undergo changes as a function of orbital and/or spin phase (see Paper 1 and Norton, Watson and King 1990). This remains a problem and for now we note the following results. The phase-averaged spectrum of FO Aqr is badly fit by simple power-law or thermal bremsstrahlung plus homogenous absorber models ($\chi_\nu^2 = 17$ and 20, respectively); the major failing of these models being that a large excess flux below ~ 4 keV is apparent in the spectrum. The fit is significantly improved by the inclusion of a partial-covering absorber or a second thermal bremsstrahlung continuum ($\chi_\nu^2 = 3.5$ and 3.7, respectively). In the partial-covering model, the bremsstrahlung temperature is high ($kT \sim 50$ keV) and the column density to the covered fraction (covering $\sim 80\%$) is $\sim 2 \times 10^{23}$ cm^{-2}. In the two-continuum model, one component has high temperature and a high column density ($kT \sim 90$ keV; $N_H \sim 2 \times 10^{23}$ cm^{-2}), while the other has low temperature and a low column density ($kT \sim 4$ keV; $N_H \sim 2 \times 10^{21}$ cm^{-2}). Resolving the spectrum as a function of orbital and/or spin phase shows that the column density varies significantly, reaching maximum values at orbital/spin minimum. The only well-constrained and model-independent feature of the spectrum is an emission line at 6.4 keV, presumably due to Kα fluorescence in cold iron, with an equivalent width of ~ 550 eV.

A more detailed discussion of the results from this observation will be presented elsewhere.

REFERENCES

Buckley, D. A. H., and Tuohy, I. R. 1989, *Ap. J.*, **344**, 376.

Chiappetti, L., *et al.* 1989, *Ap. J.*, **342**, 493.

Cook, M. C., Watson, M. G., and McHardy, I. M. 1984, *M. N. R. A. S.*, **210**, 7p.

Hameury, J-M., King, A. R., and Lasota, J-P. 1986, *M. N. R. A. S.*, **218**, 695.

Hayashida, K., *et al.* 1989, *Pub. Astr. Soc. Japan*, **41**, 373.

Hellier, C., Mason, K. O., and Cropper, M. 1989, *M. N. R. A. S.*, in press.

Norton, A. J., and Watson, M. G. 1989, *M. N. R. A. S.*, **237**, 853 (Paper 1).

Norton, A. J., Watson, M. G., and King, A. R. 1990, in preparation.

Osborne, J. P., and Mukai, K. 1989, *M. N. R. A. S.*, **238**, 1233.

Turner, M. J. L., *et al.* 1989, *Pub. Astr. Soc. Japan*, **41**, 345.

Soft X-ray Emission from AE Aquarii

Julian P. Osborne

EXOSAT Observatory, Astrophysics Division, Space Science Dept. of ESA
Current Address: X-ray Astronomy Group, Univ. of Leicester, LE1 7RH, U. K.

1 INTRODUCTION

Patterson (1979) detected a coherent 33.0767 second photometric period from AE Aqr which was shown to originate from the white dwarf. A short X-ray observation with *Einstein* showed AE Aqr to have a temperature of $> 10^6$ K (> 0.1 keV), and to have transient 30% modulation at and close to a period of 33 seconds (Patterson *et al.* 1980). This was taken as confirmation of Patterson's original oblique magnetic rotator model of AE Aqr.

Here, I use data from a short *EXOSAT* observation to show that the contrast between the intermediate polars and the shorter-period DQ Her systems is strengthened by the low temperature of the X-ray emission from AE Aqr. This temperature is too low to be due to an accretion shock cooled primarily by thermal bremsstrahlung, such as occurs in the intermediate polars.

2 OBSERVATIONS

EXOSAT observed AE Aqr between 1984 April 8 23:26 and April 9 03:32 UT. The low-energy telescope CMA (LE1; de Korte *et al.* 1981) was only switched on between 01:29 and 02:52, the 3000 lexan filter (0.05–2.0 keV) was used throughout. The ME detectors (1–50 keV; Turner, Smith, and Zimmermann 1981) were operated with one half on the source and the other monitoring the background, no array swap was performed. The background was stable in both ME and LE instruments during the observations.

The ME spectrum was accumulated over the entire observation. The background spectrum was taken from the slews to and from the target and was normalized to the source spectrum using channels 50–60, where no source flux could occur. AE Aqr was not detected. The 2σ upper limit to the half 1 count rate in channels 6–21 is 0.078 counts s^{-1}, corresponding to a 1–5 keV flux limit of 4.2×10^{-12} erg cm^{-2} s^{-1} for an unabsorbed 1 keV thermal spectrum (Mewe, Gronenschild, and van den Oord 1985).

AE Aqr was clearly detected in the LE with a count rate of 0.0465 ± 0.0037 counts s^{-1}, corrected for vignetting and dead time. In 4 minute bins, the light curve is flat, being well fit by a constant intensity.

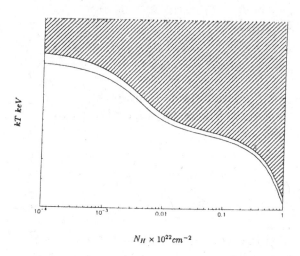

$kT\ keV$

$N_H \times 10^{22} cm^{-2}$

Fig. 1—The 90% and 99% confidence upper limits to the temperature of the emitting gas in AE Aqr from the combined ME and LE data, the excluded region is shaded. Optically thin thermal emission according to the model of Mewe, Gronenschild, and van den Oord (1985) is assumed. Both axes are logarithmic.

3 MODULATION AT THE 33 SECOND PERIOD?

The LE light curve was folded on the 33.0767 second optical period. A fit for constant intensity gave $\chi^2/\nu = 15.7/98$, corresponding to a 92% confidence rejection of the constant intensity hypothesis. To further investigate the reality of this modulation, a power spectrum of the LE data was calculated following Leahy et al. (1983). A single FFT was created from the continuous 1 second binned time series. No clear signal at the 33 second period was seen, nor at any of the harmonics claimed by Patterson et al. (1980) in their analysis of the Einstein data. The modulation seen at the 33 second period, while possibly significant, is not in excess of that seen at other periods and thus is probably a result of real noise in the data.

4 THE EMISSION TEMPERATURE

The detection by EXOSAT of AE Aqr at low energies, but not at high energies, allows an upper limit to be placed on the emission temperature. Model optically thin thermal line and continuum spectra from Mewe, Gronenschild, and van den Oord (1985) absorbed according to Morrison and McCammon (1983) were folded through the LE and ME response matrices for comparison with the observed count rates. Allowing the emission measure to vary, a grid of χ^2 was calculated as a function of temperature and absorbing column. Figure 1 shows a plot of this grid with contours at $\chi^2/\nu = 23.5/16$ and $\chi^2/\nu = 32.0/16$, corresponding to the 90% and 99% confidence levels of the fit. It shows that the maximum temperature of optically thin emission from AE Aqr allowed by the data is 1.8 keV. The luminosity is not well constrained; for a temperature of 0.5 keV, an absorbing column of 3×10^{20} cm², and a distance of 90 pc (Warner 1987; Berriman 1987), a bolometric luminosity of 3×10^{31} erg s⁻¹ is implied. This technique was also used to derive a limit to the temperature of possible blackbody emission from AE Aqr. The 99% confidence upper limit in this case is

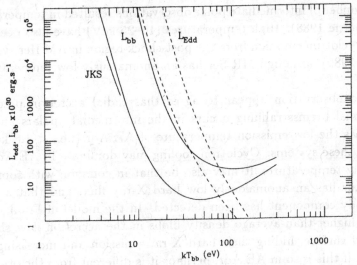

Fig. 2—Solid curves: the blackbody luminosity implied by the LE count rate, $N_{\rm H} = 3 \times 10^{20}$ cm^{-2} (*right*) and $N_{\rm H} = 3 \times 10^{19}$ cm^{-2} (*left*). Dashed curves: the respective Eddington limits derived from the fractional emitting area. If $L_{\rm bb} < L_{\rm Edd}$ and $F < 1.0$, then $6 < kT_{\rm bb} < 70$ eV. The line labelled JKS shows the parameter range from the UV line ratio study of Jameson, King, and Sherrington (1980).

0.26 keV.

Figure 2 shows the bolometric luminosity required of a blackbody emitter to produce the observed LE count rate. The column density to AE Aqr can be estimated as $N_{\rm H} \approx 3 \times 10^{20}$ cm^{-2} from Paresche (1984). This value and 3×10^{19} cm^{-2} are used in Figure 2. Consideration of the emitting area fraction, F, of the white dwarf implied assuming a distance of 90 pc and $R_{\rm wd} = 6 \times 10^8$ cm (from $M_{\rm wd} = 0.94\,{\rm M}_\odot$ [Patterson 1979] and the mass-radius relationship of Nauenberg 1972) enables the local Eddington luminosity to be calculated, $L_{\rm Edd} \approx 1.3 \times 10^{38}\, F\,(M/M_\odot)$ erg s^{-1}. If we require the emitting area to be less than that of the white dwarf and the luminous regions not to be brighter than the Eddington limit, then $6 < kT_{\rm bb} < 70$ eV.

5 DISCUSSION

The temperature of an accretion shock on a radially-accreting white dwarf is given by the kinetic energy of the free-falling gas at the shock. For a typical white dwarf, an observed temperature of ~ 7–60 keV is anticipated for pure thermal bremsstrahlung cooling except at the highest luminosities (Imamura and Durisen 1983). This temperature may be reduced if an accretion disk extends close to the white dwarf, although this is only significant if the disk reaches within $\sim R_{\rm wd}$ of the surface. However, post-shock cyclotron cooling can also reduce the observed X-ray temperature.

Measured temperatures of radially-accreting white dwarfs range from 8 keV (EX Hya: Rosen, Mason, and Córdova 1988) to 31 keV (AM Her: Rothschild *et al.*

1981), and a number of intermediate polar observations resulted in temperature limits > 10 keV (Osborne 1988). High temperatures (10–25 keV) have also been found by modelling the cyclotron emission from the post-shock region in AM Her systems (e.g., Cropper *et al.* 1989), although MR Ser has an anomalously low temperature of ~ 5 keV.

Theory and observation appear to agree that radial accretion predominantly cooled by thermal bremsstrahlung occurs in the intermediate polars and AM Her systems, whereas the low emission temperature of AE Aqr (i.e., < 1.8 keV) distinguishes it from these systems. Cyclotron cooling may dominate in AE Aqr, quenching a high X-ray temperature. It may also be that in common with some AM Her systems, AE Aqr has an anomalously low hard X-ray flux, and that a soft X-ray pseudo-blackbody component has been detected. In the model of Frank, King, and Lasota (1988), higher-than-average density blobs in the accretion flow shock below the white dwarf surface, hiding their hard X-ray emission and increasing the total soft X-ray flux. If this is so in AE Aqr, perhaps it is different from the other systems because the high accretion rate (implied by the long orbital period of AE Aqr), or the short spin period of the white dwarf, enhances the formation of these high density blobs.

REFERENCES

Berriman, G. 1987, *Astr. Ap. Suppl.*, **68**, 41.
Cropper, M., *et al.* 1989, *M. N. R. A. S.*, **236**, 29p.
de Korte, P. A. J., *et al.* 1981, *Space Sci. Rev.*, **30**, 495.
Frank, J., King, A. R., and Lasota, J. P. 1988, *M. N. R. A. S.*, **202**, 183.
Imamura, J. N., and Durisen R. H. 1983, *Ap. J.*, **268**, 291.
Jameson, R. F., King, A. R., and Sherrington, M. R. 1980, *M. N. R. A. S.*, **191**, 559.
Leahy, D. A., *et al.* 1983, *Ap. J.*, **266**, 160.
Mewe, R., Gronenschild, E. H. B. M., and van den Oord, G. H. 1985, *Astr. Ap. Suppl.*, **62**, 197.
Morrison, R., and McCammon, D. 1983, *Ap. J.*, **270**, 119.
Nauenberg, M. 1972, *Ap. J.*, **175**, 417.
Osborne, J. P. 1988, *Mem. Soc. Astr. Italiana*, **59**, 117.
Paresche, F. 1984, *A. J.*, **89**, 1022.
Patterson, J. 1979, *Ap. J.*, **234**, 978.
Patterson, J., Branch, D., Chincarini, G., and Robinson, E. L. 1980, *Ap. J. (Letters)*, **240**, L133.
Rosen, S. R., Mason, K. O., and Córdova, F. A. 1988, *M. N. R. A. S.*, **231**, 549.
Rothschild, R. E., *et al.* 1981, *Ap. J.*, **250**, 723.
Turner, M. J. L., Smith, A., and Zimmermann, H. U. 1981, *Space Sci. Rev.*, **30**, 513.
Warner, B. 1987, *M. N. R. A. S.*, **227**, 23.

The Pulse-Timing Orbit and the Emission Line Orbit of the White Dwarf in AE Aquarii

Edward L. Robinson, Allen W. Shafter, and S. Balachandran

McDonald Observatory and Department of Astronomy, University of Texas

ABSTRACT

The orbital motion of the white dwarf in AE Aqr can be estimated in two ways, from the emission-line radial velocity curve and from the pulse arrival times of the 33 s pulses in the light curve. We find that the pulse-timing orbit is not well fit by a sine curve and also suffers a 60° phase lag. The emission-line orbit is stable in amplitude, it is well fit by a sine curve, and its phase lag is not measurably different from zero. Contrary to our expectations, therefore, the emission-line orbit is much less distorted than the pulse-timing orbit. The emission-line orbit is the best estimator of the true orbital motion of the white dwarf.

1 INTRODUCTION

The spectrum of the bright ($V \sim 10.0$–12.5) novalike variable AE Aqr has an absorption-line component with a K5 V spectral type that comes from the late-type star in the system, and complex emission lines of H, He I, and Ca II from the gas disk around the white dwarf. Radial velocity measurements by Chincarini and Walker (1981b) yielded $K_{emis} = 135 \pm 5$ km s^{-1} for the emission-line orbit and $K_{K5V} = 160 \pm 1$ km s^{-1} for the orbit of the K5 V star.

The light curve of AE Aqr has large-amplitude flickering similar to the flickering seen in other cataclysmic variables, but it also has intervals of up to a few hours when the flickering disappears and the light curve is quiescent. During the intervals of quiescence and occasionally during the flares, a low-amplitude (0.002–0.01 mag) coherent modulation with a period of 33 s can be detected in the light curve (Patterson 1979). The pulse shape has two nearly equal peaks so that the harmonic at 16.5 s usually has a larger amplitude than the fundamental at 33 s. Because the 33 s pulse is highly coherent and because the X-ray flux from AE Aqr is also pulsed with a period of 33 s, it is generally accepted that the pulses are caused by rotation of the white dwarf, which is magnetized and accreting mass transferred from the K5 V star (Patterson, Branch, Chincarini, and Robinson 1980).

The 33 s pulses can be used to measure a pulse-timing orbit for the white dwarf if they come directly from the white dwarf and are not reprocessed anywhere else

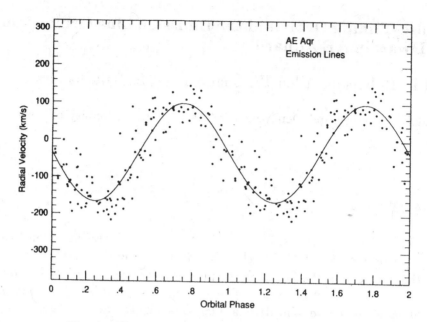

Fig. 1—The emission-line radial velocity curve.

in the system. Previous measurements of the pulse-timing orbit indicated that it is nearly circular and close to 180° out of phase from the spectroscopic orbit of the K star (Patterson 1979). These results suggested that the pulse-timing orbit gives an unbiased measurement of the true orbit of the white dwarf. The emission-line orbit was thought to give a biased measurement of the true orbit but only because the emission-line orbits in other cataclysmic variables are often severely distorted, not because of direct evidence that it is distorted in AE Aqr. We have remeasured both the pulse-timing and the emission-line orbits of the white dwarf in order to compare the two methods for measuring the orbit of the white dwarf more precisely.

2 OBSERVATIONS

We obtained new high-speed photometry of AE Aqr at McDonald Observatory during the 1982 and 1983 observing seasons and we obtained new spectroscopic observations of AE Aqr at McDonald Observatory in September and October of 1985. Both sets of data were reduced and analyzed using our standard methods (e.g., Kepler, Robinson, Nather, and McGraw 1982; Shafter, Szkody, and Thorstensen 1986).

Feldt and Chincarini (1980) revised the orbital period given by Chincarini and Walker (1981b) using only absorption-line radial velocities, on the grounds that the absorption-line velocities comprise a larger and more accurate set of data. They found

$$T_o = \mathrm{JD}_\odot 2439030.827(\pm 3) + 0.4116579E \,, \tag{1}$$

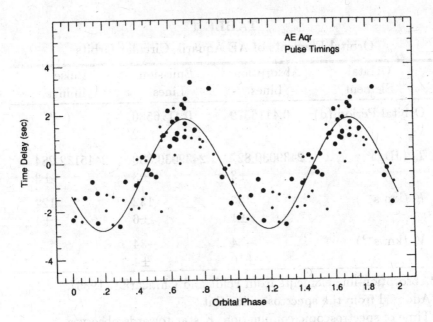

Fig. 2—The pulse-timing orbit for the white dwarf.

where T_o is the time of spectroscopic conjunction with the K5 V star towards the observer. We derived the ephemeris for the emission-line orbit by combining our emission-line velocities with the emission-line velocities listed in Chincarini and Walker (1981a). The ephemeris is

$$T_o = JD_\odot 2439030.830(\pm 3) + 0.4116580(\pm 2)E . \qquad (2)$$

Several alias periods also fit the emission-line velocities but we are confident that we have chosen the correct orbital period because it gives the formal best fit to the data and because it agrees with the period found by Feldt and Chincarini (1980). The radial velocity curve for the emission lines is shown in Figure 1. Both our velocity measurements and those of Chincarini and Walker (1981a) have been plotted in the figure. The amplitude of the radial velocity curve is 132 ± 6 km s^{-1} and its eccentricity is zero to within the measurement error.

Based on our new photometry, the ephemeris for the 16.5 s component of the 33 s modulation is

$$T_{max} = BJED2445171.999844(\pm 1) + 0.000191416425(\pm 1)E . \qquad (3)$$

The period in this ephemeris differs significantly from the period derived by Patterson (1979) because he used heliocentric times while we have used barycentric times. The pulse-timing orbit for the white dwarf was derived by folding the (O−C) diagram on the following ephemeris, in which the orbital period is the emission-line period and T_o is arbitrary:

$$T_o = JD_\odot 2445172.000 + 0.4116580E . \qquad (4)$$

TABLE 1

Orbital Elements of AE Aquarii, Circular Orbits

Orbital Element	Absorption[a] Lines	Emission Lines	Pulse Timings
Orbital Period (d)	0.4116579	0.4116580 ±2	(−)[b]
T_o (JD$_\odot$)[c]	2439030.827 ±3	2439030.830 ±3	2445172.284 ±3
K (km s^{-1})	160 ±1	132 ±6	122 ±4
V_o (km s^{-1})	−64 ±10	−34 ±4	

[a]Absorption-line elements from Feldt and Chincarini (1980).
[b]Adopted from the spectroscopic orbit.
[c]Time of spectroscopic conjunction, K star towards observer.

The (O−C) diagram for the pulse-timing orbit is shown in Figure 2. The best fitting circular orbit yields a semi-amplitude of 2.30 ± 0.07 s, which is equivalent to $K_{pulse} = 122 \pm 4$ km s^{-1}, and a revised T_o of JD$_\odot$2445172.284 ± 0.003. The pulse orbit is clearly distorted and is not well represented by a circular orbit. A summary of the orbital elements is shown in Table 1.

3 DISCUSSION AND CONCLUSIONS

To be reliable, a method for measuring the orbital motion of the white dwarf must at least give repeatable results, the measured orbit must be circular, and it must be 180° out of phase with the orbit of the K5 V star. Chincarini and Walker (1981*b*) measured $K_{emis} = 135 \pm 5$ km s^{-1} for the emission-line orbit and our new measurements give $K_{emis} = 132 \pm 6$ km s^{-1}. Patterson (1979) measured an amplitude of $K_{pulse} = 127 \pm 5$ km s^{-1} for the pulse-timing orbit and our new measurements give $K_{pulse} = 122 \pm 4$ km s^{-1}. Therefore, both the emission lines and the pulse timings give repeatable results for the amplitudes of the orbital motion. The eccentricity of the emission-line orbit is zero to within the measurement error. The pulse-timing orbit is, however, visibly distorted from a circular orbit.

Equations (1) and (2) show that the phase lag between the emission line and absorption line orbits is negligible, $\Delta\phi = 3° \pm 4°$. We updated T_o for the absorption-line orbit to the same epoch as the pulse-timing orbit using the best orbital period, 0.4116580 day, finding $T_o = $ JD$_\odot$2445172.353. Comparing this to T_o for the pulse-timing orbit, we find that the pulse timing orbit is out of phase by $\Delta\phi = 60° \pm 3°$, pulse orbit too early. Patterson (1979) found that the phase lag of the pulse-timing orbit was much smaller and that it was consistent with no phase lag. We have recalculated the phase lag for Patterson's pulse-timing orbit using the new orbital period and find $\Delta\phi = 64°$, in close agreement with the phase lag for our data, the change being entirely due to the improved orbital period. The pulse-timing orbit must be severely distorted to produce a phase lag this large.

Since the pulse orbit is distorted from a sine curve and since its phase lag is so large, we conclude that the pulse orbit is not a reliable estimator of the true orbital motion of the white dwarf in AE Aqr. The emission-line orbit is stable in amplitude, it is not significantly distorted from a sine curve, and it is also precisely in phase with the expected orbit of the white dwarf. Contrary to our expectations, therefore, there is no reason not to accept the emission-line orbit as a reliable estimator of the true orbit of the white dwarf. Adopting the spectroscopic orbital elements, we find

$$M_K \sin^3 i = 0.48 \pm 0.03 \, M_\odot \tag{5}$$

$$M_{wd} \sin^3 i = 0.58 \pm 0.04 \, M_\odot \tag{6}$$

$$q = M_{wd}/M_K = 1.21 \pm 0.05 . \tag{7}$$

This research was supported in part by NSF grant AST-8704382.

REFERENCES

Chincarini, G., and Walker, M. F. 1981a, ESO Preprint No. 134.

Chincarini, G., and Walker, M. F. 1981b, Astr. Ap., **104**, 24.

Feldt, A. N., and Chincarini, G. 1980, Pub. Astr. Soc. Pac., **92**, 528.

Kepler, S. O., Robinson, E. L., Nather, R. E., and McGraw, J. T. 1982, Ap. J., **254**, 676.

Patterson, J. 1979, Ap. J., **234**, 978.

Patterson, J., Branch, D., Chincarini, G., and Robinson, E. L. 1980, Ap. J. (Letters), **240**, L133.

Shafter, A. W., Szkody, P., and Thorstensen, J. R. 1986, Ap. J., **308**, 765.

The X-ray Eclipse in EX Hya: Emission from Two Poles

Keith O. Mason and Simon R. Rosen

Mullard Space Science Laboratory, University College London, U. K.

1 INTRODUCTION

EX Hya is an important member of the intermediate polar subgroup of cataclysmic variables because it is the only such system known where the red dwarf companion occults the white dwarf star. This affords us the opportunity to probe the X-ray emission region of such a system in unprecedented detail. EX Hya has a 98-minute orbital (eclipse) period, while the white dwarf star in the binary rotates with a 67-minute period. It was observed extensively using the *EXOSAT* satellite (Córdova, Mason, and Kahn 1985; Rosen, Mason, and Córdova 1988 [hereafter RMC]; Beuermann and Osborne 1988; see also Mason, Rosen, and Hellier 1988). In Section 2 we summarize knowledge of the system gained from the *EXOSAT* data. We then describe new observations made using the *GINGA* satellite which have improved sensitivity, particularly at the higher X-ray energies. Based on these data we can conclude that we see X-ray emission from two regions on the white dwarf, only one of which is eclipsed. We associate the two emission regions with the two poles of a dipole magnetic field, one above the orbital plane and one below.

2 EXOSAT OBSERVATIONS

The principal results of the *EXOSAT* observations of EX Hya can be summarized as follows:

• The amplitude of the 67-minute rotational modulation decreases rapidly with increasing energy between 0.5 and 6 keV.

• The 98-minute X-ray eclipse is partial, and its depth increases with increasing energy between 0.5 and 6 keV.

• The depth of the 98-minute X-ray eclipse is a function of 67-minute phase, being greatest at maximum flux.

The behaviour of the 67-minute modulation depth with energy is reminiscent of the effect of photoelectric absorption, and was explained as such by RMC. In their model, the accretion disk in EX Hya is disrupted by the magnetic field of the white dwarf and accreting gas then flows down field lines to form a narrow arc-shaped accretion region around whichever of the two magnetic poles is nearest to the point

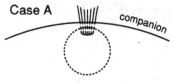

mid-eclipse: 67 min phase 0.5

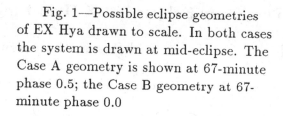

Fig. 1—Possible eclipse geometries of EX Hya drawn to scale. In both cases the system is drawn at mid-eclipse. The Case A geometry is shown at 67-minute phase 0.5; the Case B geometry at 67-minute phase 0.0

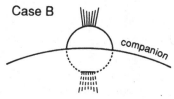

mid-eclipse: 64 min phase 0.0

of disruption. Absorption in the 'accretion curtain' causes the flux modulation in this model, which was also adopted by Beuermann and Osborne (1988), and a feature of this geometry is that the maximum X-ray flux is seen when the magnetic polar axis above the orbital plane points *away* from the observer. This geometry is supported by optical spectroscopic observations of EX Hya (Hellier *et al.* 1987; see also Hellier and Mason 1990).

RMC considered various explanations for the X-ray eclipse behaviour of EX Hya. They were left with two possible accretion geometries which were consistent with the *EXOSAT* data, referred to as Case A and Case B (see Fig. 1). In Case A only one emitting polar region is visible (the one above the orbital plane). The X-ray emission from this pole is significantly extended above the white dwarf's surface ($\sim 1\,R_{\mathrm{WD}}$) and only the lowest portions of this accretion column are occulted by the companion. The observed energy dependence of the eclipse then implies that the hottest emission regions must be closest to the white dwarf's surface. In Case B two emitting poles are visible, one above the orbital plane and one below, and the partial X-ray eclipse occurs because only the pole below the disk plane is occulted. In this model the energy dependence of the eclipse depth arises because radiation from the lower pole is absorbed by material in the inner part of the accretion disk, or material trapped in the magnetosphere of the white dwarf. In Case B, emission from the lower pole will be modulated with the 67-minute period because of occultation by the white dwarf, in phase with the absorption modulation of the upper pole.

An important point to note is that the Case A and Case B geometries predict different behaviour at energies above 6 keV where the *EXOSAT* observations were not sensitive. If Case A is applicable, the depth of eclipse should approach 100% at high energies where the hottest emission regions close to the white dwarf's surface dominate. Also, the dependence of eclipse depth on 67-minute phase should dis-

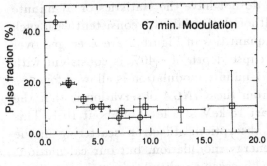

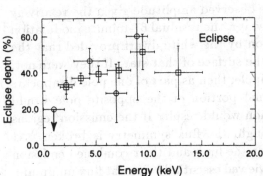

Fig. 2—Combined *GINGA*
(squares) and *EXOSAT* (circles)
data showing (above) the pulse
fraction of the 67-minute modulation as a function of energy
and (below) the maximum eclipse
depth as a function of energy.

appear at high energies where the accretion column is transparent to photoelectric
absorption. In Case B, however, the eclipse depth should be no more than 50%
at maximum in the 67-minute cycle. This assumes that both polar regions emit
intrinsically the same X-ray flux, and is the limit when all the lower pole is occulted
(Fig. 1). At the same time, in Case B the depth of eclipse should continue to be a
function of 67-minute phase at high energies, since the emission from the eclipsed
pole will be modulated by the body of the white dwarf as it rotates.

3 GINGA OBSERVATIONS

EX Hya was observed using the *GINGA* observatory for about 16 hours on 1988
June 16. Useful data were obtained on 8 eclipses. A full account of the observation
and data analysis is contained in Rosen *et al.* (1990).

The amplitude of the 67-minute modulation and the depth of the eclipse as a
function of energy have been extracted from the *GINGA* data in exactly the same
way as from the *EXOSAT* data. In Figure 2 we plot these quantities as a function
of energy, combining the *EXOSAT* and *GINGA* results. From this we see that the
eclipse is significantly detected in the *GINGA* data even in the 10.5–15.0 keV band,
and that the mean eclipse depth tends to an approximately constant value of ~40%
above about 3 keV. This is strong support for the Case B geometry which predicts
that the eclipse depth should approach 50% at high energies at the maximum in the
67-minute cycle. Case B also predicts that the depth of eclipse should be a function

of 67-minute phase at high energies. We do not have the statistics or 67-minute phase coverage to verify this prediction, although the data are consistent with such behaviour. Note that since the observed quantities in Figure 2 are averaged over several 67-minute phases, the measured eclipse depth of ~40% is consistent with the predicted value of 50% when a small 67-minute modulation is allowed for.

Another important point to emerge from the *GINGA* observation is that the 67-minute modulation persists up to at least 15 keV at a level of about 10%. This is clearly inconsistent with photoelectric absorption. Electron scattering in the accretion column could result in a high-energy modulation, but our calculations suggest that it is difficult to reproduce the observed amplitude given the relatively low accretion rate inferred in EX Hya. However, the residual 67-minute modulation can be explained as the result of occultation by the white dwarf provided that the emission regions are slightly raised above the surface of that star. If they were not, and the geometry is that of a symmetric dipole, then as part of one pole disappears over the limb of the white dwarf, an equal portion of the opposite pole would appear over the limb and no net modulation would result. If the emission regions are slightly raised above the surface, though, the flux symmetry is broken. At maximum light, both polar regions are near the limb and their combined emission exceeds that from the upper pole, which is viewed essentially alone at flux minimum.

4 CONCLUSIONS

The *GINGA* observations favor a Case B geometry where emission from polar regions both above and below the orbital plane are visible in EX Hya. Only emission from the polar region below the plane is occulted by the red dwarf companion. A ~10% modulation with the 67-minute period at energies up to at least 15 keV can be attributed to occultation by the white dwarf, if the polar emission regions stand slightly above the white dwarf's surface.

KOM thanks the Royal Society of London for support.

REFERENCES

Beuermann, K., and Osborne, J. O. 1988, *Astr. Ap.*, **189**, 128.

Córdova, F. A., Mason, K. O., and Kahn, S. M. 1985, *M. N. R. A. S.*, **212**, 447.

Hellier, C., and Mason, K. O. 1990, this volume.

Hellier, C., Mason, K. O., Rosen, S. R., and Córdova, F. A. 1987, *M. N. R. A. S.*, **228**, 463.

Mason, K. O., Rosen, S. R., and Hellier, C. 1988, *Adv. Space Res.*, **8**, No. 2-3, p. 293.

Rosen, S. R., Mason, K. O., and Córdova, F. A. 1988, *M. N. R. A. S.*, **231**, 549.

Rosen, S. R., Mason, K. O., Mukai, K., and Williams, O. R., 1990, *M. N. R. A. S.*, in press.

Phase-Dependent UV Observations of EX Hya [1]

Joachim Krautter and Werner Buchholz

Landessternwarte, Königstuhl, D-6900 Heidelberg, F. R. G.

EX Hya is a cataclysmic binary and belongs to the group of intermediate polars. Hence, the accretion onto the white dwarf occurs via an accretion column along the magnetic field lines. EX Hya is well studied in both the optical and X-ray spectral ranges (see, e.g., Beuermann and Osborne 1985; Hellier, Mason, and Córdova 1987; Rosen, Mason, and Córdova 1988). An orbital period of $P = 98$ min has been found for EX Hya. A second cycle with a period of 67 min, which is seen as a quasi-sinusoidal modulation in both the optical and UV spectral range, is interpreted as the rotation period of the white dwarf. Partial and narrow eclipses were found for EX Hya in both the optical and X-ray spectral range. The depth of the eclipses varies with the 67-min cycle, being highest at the maximum of the 67-min cycle.

Besides EX Hya, only one other eclipsing intermediate polar (DQ Her) is known. Hence, EX Hya offers an excellent opportunity to study the accretion geometry in this class of objects. Several models for the accretion geometry have been proposed, but it is still unknown what is really going on. A powerful tool to study the hot emission line-emitting regions close to the white dwarf are observations in the ultraviolet spectral range. In this note we describe phase-dependent UV observations of EX Hya.

Phase-dependent UV observations of the intermediate polar EX Hya were carried out with *IUE* on two occasions in 1984 and 1985 using the facilities of the ESA-Villafranca tracking station. In addition, we used spectra from the *IUE* data archive which were taken in 1982 and 1984 at the NASA Goddard Space Flight Center by F. Córdova and K. Mason. A total of 31 SWP and 28 LWP (LWR) images were obtained. Since 12 multiple exposures were taken in both the short- and long-wavelength range, a total of 43 and 40 spectra, respectively, were available for evaluation. 5 SWP spectra obtained at VILSPA were trailed around orbital phase $\Phi_{Orb} = 0.00$. All spectra were taken through the large aperture in the low-resolution mode. Since the *IUE* standard reduction does not work for multiple exposures, all spectra were reduced using our own reduction procedure. Typical exposure times are about 15 minutes. The observations have a reasonably good coverage in the phase space of the 98 min versus 67 min periods. For the evaluation of the results, the spectra were binned in phase steps $\Delta\phi = 0.1$ for both the 98-min and 67-min cycle

---

[1] Based on observations collected at the Villafranca Satellite Tracking Station of the European Space Agency by the *International Ultraviolet Explorer*.

with a mean number/bin of about 4 spectra.

The strongest emission lines found in the UV spectral range are N V $\lambda1240$, Si IV $\lambda1398$, C IV $\lambda1550$, and Mg II $\lambda2800$. Both continuum and line fluxes show a sinusoidal modulation with the 67-min period. The continuum modulation is slightly higher at longer wavelengths; the amplitude of the flux variation is about 30 percent in the short-wavelength region (1225–1950 Å) and 38 percent in the long-wavelength region (1950–3200 Å). The line fluxes show a stronger variation than the continuum flux; the flux variations of N V, Si IV, C IV, and Mg II are about 50 percent. No dependence of continuum and line fluxes with the orbital period could be found.

The five spectra which were obtained by trailing around the eclipse ($\Phi_{\text{Orb}} = 0.00$) have a high time resolution of about one minute. However, due to the low signal-to-noise ratio, only C IV $\lambda1550$ could be used for further evaluation. Eclipses are visible in both the blue and red wings of the C IV emission line; they first show up in the blue wing and about four minutes later in the red wing. Significant differences of the eclipse behaviour with respect to the 67-min period are found. Around $\Phi_{67} \simeq 0.60$, the blue wings are much weaker than the red wings, whereas around $\Phi_{67} \simeq 0.15$, the blue and red wings are of comparable strength.

This behaviour strongly supports case B of the model for the eclipse geometry by Rosen, Mason, and Córdova (1988). In their case B, only a partial eclipse of the accretion stream below the disk plane takes place, whereas in case A the white dwarf and the accretion stream below the disk plane are totally eclipsed. For $\Phi_{67} \simeq 0.60$ one expects for both case A and case B a strong red and weak blue wing. However, for $\Phi_{67} \simeq 0.00$ case A predicts a strong blue and weak red wing, contrary to case B, where the blue and the red wing should be of comparable strength.

The total UV luminosity (1200–3200 Å $L_{\text{UV}} = (2.6 \pm 0.4) \times 10^{32}$ erg s^{-1}) is slightly higher than the luminosity in the soft X-ray range (1.5–10 keV) reported by Rosen, Mason and Córdova. This suggests that the mass-transfer rate might be higher than the value of $\dot{M} = 3 \times 10^{15}$ g s^{-1} as derived by Rosen, Mason and Córdova from the soft X-ray luminosity.

A more complete description of these results will be submitted to *Astronomy and Astrophysics*.

REFERENCES

Beuermann, K., Osborne, J. O. 1985, *Space Sci. Rev.*, **40**, 117.

Hellier, C., Mason, K. O., and Córdova, F. A. 1987, *M. N. R. A. S.*, **228**, 463.

Rosen, S. R., Mason, K. O., and Córdova, F. A. 1988, *M. N. R. A. S.*, **231**, 549.

Observations and Models of the Intermediate Polar TX Col

David A. H. Buckley[1] and Ian R. Tuohy[2]

[1]Department of Astronomy, University of Cape Town, South Africa
[2]British Aerospace Australia, P. O. Box 180, Salisbury, South Australia 5108

1 INTRODUCTION

TX Col (1H0542-407) is an X-ray-selected CV, found during the systematic *HEAO-1* optical identification program (e.g., Remillard 1985; Remillard *et al.* 1986; Tuohy *et al.* 1986; Buckley 1989). It was discovered to be an intermediate polar (IP) on the basis of the ~ 1920 s soft X-ray period found with *EXOSAT*. Furthermore, the harder (> 2 keV) X-rays were found to be modulated at a significantly longer period of ~ 2100 s (Tuohy *et al.* 1986). This system has subsequently been intensively studied at optical (Buckley and Tuohy 1989) and UV wavelengths (Mouchet *et al.* 1990).

2 RADIAL VELOCITIES AND MASS FUNCTION

Observations on 1985 November 17 and 18 using the AAT confirmed two periods in the radial velocities: one at 5.7 hr, with a K-velocity ~ 50 km s^{-1}, and a second at 0.53 hr (~ 1910 s), with $K \sim 20$–30 km s^{-1} (see Figure 1). The latter period is consistent with the X-ray-determined white dwarf spin period (1920 ± 20 s), while the longer period is presumed to be the orbital period of the system. A Gaussian convolution analysis (e.g., Schneider and Young 1980) showed that there was no significant difference between the core and wing radial velocity parameters (γ, K, and ϕ). However, the further into the line wings the velocity was measured, the less obvious was the shorter-period (1910 s) variation.

The mass function for the system, based on $K = 51 \pm 3$ km s^{-1} and $P_{orb} = 5.72$ hr, is $3.3 \pm 0.6 \times 10^{-3}$ M$_\odot$. Assuming an empirical secondary star mass of 0.57–0.63 M$_\odot$, corresponding to the 5.7 hr period (Faulkner 1971; Patterson 1984), leads us to the conclusion that the system inclination is $< 25°$. In Figure 2 we present a diagnostic mass diagram with constant inclination contours.

3 LINE PROFILES AND THE SPIN MODULATION

We found that an adequate fit to the Balmer emission lines could be achieved using two Gaussians with FWHMs of ~ 800 and 1200 km s^{-1}. In the case of the

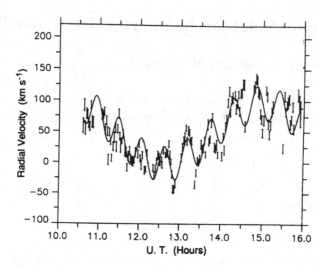

Fig. 1—Radial velocity variations of the Balmer lines for TX Col determined by cross correlations.

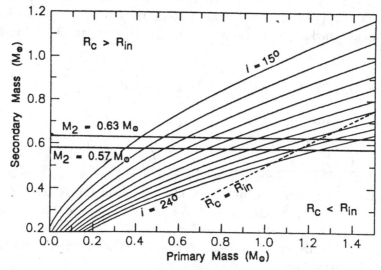

Fig. 2—Diagnostic diagram for TX Col based on the derived mass function.

He II λ4686 line, a single component fit was sufficient. This line was also seen to be more strongly modulated at the white dwarf spin period.

One explanation proposed by Penning (1985) for the spin period velocity variations in intermediate polars is a spot on the accretion disk, illuminated by an X-ray beam from the rotating white dwarf. This model has difficulties in explaining the very low velocity amplitudes observed (~ 20 km s^{-1} in TX Col and < 100 km s^{-1} in the sample discussed by Penning), since the velocity of material in this illuminated spot is essentially Keplerian. An alternative model which we propose has material corotating with the white dwarf's magnetosphere at or near the magnetospheric radius (Buckley and Tuohy 1989). For TX Col the distance from the white dwarf of this corotating emission region varies from ~ 10 to 2 white dwarf radii for masses 1.4 to $0.3\,M_\odot$, using the observed 1911 s spin period and the inclination from Figure 2.

Clearly, if there are truncated accretion disks in these systems, the condition $R_c \leq R_{mag} \leq R_{inner}$ has to hold, where R_c is the distance of the line emitting plasma corotating in the magnetosphere, R_{mag} is the magnetospheric radius, and R_{inner} is the inner radius of the accretion disk.

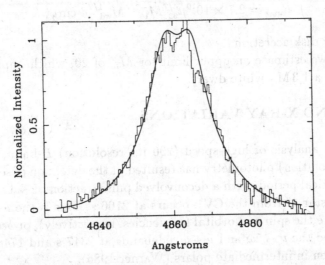

Fig. 3—Model accretion disk profile (for $i = 25°$, $M_{wd} = 1.3\,M_\odot$, $R_{inner} = 7\,R_{wd}$, $R_{outer} = 195\,R_{wd}$, $\alpha = 2.2$, $\sigma = 0.2$) plotted with the mean Hβ broad component for TX Col.

4 THE ACCRETION DISK

The observed second, broad component of the Balmer lines (and possibly He I) in TX Col led us to investigate whether this emission is possibly produced in a partially intact accretion disk. An initial estimate of the inner radius of such a disk was derived based on the observed FWZI of this broad component (~ 2800 km s^{-1}), which is naïvely equated to the maximum Keplerian velocity attained in the disk (i.e., at R_{inner}). Subsequently, we fitted the standard power-law Keplerian disk profile (Horne and Marsh 1986) to the mean broad component. A consistent solution, namely one in which i and M_{wd} were in agreement with the dynamical solution, was obtained and is shown in Figure 3. The solution is also consistent with the condition $R_c \leq R_{inner}$, with $R_{inner} \sim R_c \sim 1.5 \times 10^9$ cm, or $\sim 7\,R_{wd}$ for a $1.3\,M_\odot$ white dwarf. Furthermore, evidence for a disk bright spot is found from a careful analysis of the orbitally binned spectra. For the He II $\lambda 4686$ line, an asymmetry of the profile is occasionally seen, possibly due to the superposition of a narrower component to the line. This occurs at the two phases corresponding to the expected maximum approach and recession velocity of a disk bright spot.

5 MAGNETIC FIELD STRENGTH

Equating the inner radius of the accretion disk with the magnetospheric radius leads to an estimate of the magnetic moment, using the standard expression for R_{mag} (e.g., King, Frank, and Ritter 1985):

$$R_{\mathrm{mag}} = 2.7 \times 10^{10} \mu_{33}^{4/7} \dot{M}_{16}^{-2/7} M_{\mathrm{wd}}^{-1/7} \phi \; \mathrm{cm} \,, \tag{1}$$

where $\phi = 0.5$ for disk accretion.

For TX Col we estimate an upper limit for $\dot{M}_{16}$ of 20, which implies $\mu_{33} \sim 0.1$, or $B \sim 4$ MG for a $1.3 \, \mathrm{M}_\odot$ white dwarf.

6 OPTICAL AND X-RAY VARIATIONS

A time series analysis of high-speed (200 ms resolution) B-band, and $UBVRI$ (typically 40 s resolution) photometry has resulted in the detection of several periods. The dominant optical period, with a deconvolved pulse fraction of $\sim 17\%$ in B (there is a $V \sim 16$ F5V star 4″ from the CV), occurs at 2106 s. This is the $\nu - \omega$ sideband (where ν and ω are the spin and orbital frequencies, respectively), or *beat* period. We have also identified the $\nu - 2\omega$ and $\nu + \omega$ sidebands at 2347 s and 1748 s, which are expected to be seen in intermediate polars (Warner 1986).

The usual interpretation of optical beat periods in IPs is reprocessing of X-rays in a fixed region in the binary frame of reference; the inward face of the secondary star (e.g., Patterson and Price 1981) and/or the disk bright spot (e.g., Hassall *et al.* 1981; Wickramasinghe, Stobie, and Bessell 1982). Reprocessing in the axisymmetric disk could lead to a *spin* period optical variation if the inclination were high enough to produce a front-back asymmetry due to disk curvature (e.g., Cropper 1986). We note that, interestingly, all of the low inclination ($i < \sim 30°$) IPs (V1223 Sgr, AO Psc, and TX Col) have their dominant optical periods at the reprocessing *beat* period, which is consistent with the above picture: disks seen nearly face-on would show no asymmetry.

One of the more interesting features of TX Col is that the hard (> 2 keV) X-rays are *also* modulated at the 2106 s optical period. We have subsequently investigated whether these hard X-ray photons might be reflected in the same region (i.e., the hot spot and secondary) where the softer X-rays are photoabsorbed and reprocessed into optical light. Compton reflection in the atmospheres of stars has been investigated by a number of authors (e.g., Milgrom and Salpeter 1975; Felsteiner and Opher 1976). Albedos as high as 0.4 may be applicable for X-rays above ~ 4 keV. Our conclusions are that, providing the hard X-rays are fan-beamed preferentially in a direction parallel to the disk plane, the hard X-ray and optical pulse fractions can be explained. We have invoked the accretion models of Imamura and Durisen (1983), and find that for sufficiently small fractional accretion area ($f \sim 0.005$), the expected high *effective* accretion luminosity (L_{acc}/f) of $\sim 4 \times 10^{37}$ erg s^{-1} (assuming $\dot{M}_{16} \sim 20$) will result in a significant difference between the electron scattering opacities parallel and perpendicular to the accretion column (i.e., to the magnetic field lines). In fact, the ratio of perpendicular to parallel hard X-ray intensities is ~ 10, sufficient to produce the required fan beam. The low value of f is consistent with accretion at the footprints of tilted (~ 10–$20°$) dipole field lines which intersect the disk and trap material over a small radial extent $\Delta r/r \sim \nabla B/B \sim R_{inner}/3$.

The optical reprocessing was modelled assuming a Shakura-Sunyaev optically thick accretion disk, using the formulae for the temperature, disk thickness and opacity given by Frank, King, and Raine (1985). A circular hot spot (Bath *et al.* 1974) was added, and a Roche lobe-filling M0V secondary star of $T_{eff} \sim 3800$ K. The effective temperatures of small area elements on the disk, hot spot, and secondary were accordingly modified, depending on the effective X-ray flux and albedo. The observed optical $UBVRI$ pulse fractions were all close to the model predictions. Furthermore, the spectral shape of the model closely matched the observed optical continuum and agreed with the mean $UBVRIJHK$ magnitudes for TX Col (Buckley and Tuohy 1989).

7 DISCUSSION AND CONCLUSION

We have shown that TX Col is an established intermediate polar, with a spin period of ~ 1911 s and an orbital period of ~ 5.7 hr. It exhibits a number of periodicities both in the optical and X-ray, which can be understood in terms of X-ray reprocessing and reflection.

We estimate a distance for TX Col of ~ 500 pc, based on the observed IR magnitudes, and those expected for a Roche lobe-filling M0V secondary (i.e., Bailey 1981). This is also in agreement with predicted fluxes for our illuminated disk/hot spot/secondary model, and also the accretion arc model, which requires a high effective X-ray luminosity. This model also predicts a large, but absorbed, soft X-ray (< 2 keV) luminosity. Furthermore, there is a quantitative agreement between our accretion disk parameters, and those derived by Mouchet *et al.* (1990), based on *IUE* observations. We also note that larger distances for IPs than those usually adopted may help resolve the energy budget problems in the reprocessing models.

REFERENCES

Bailey, J. 1981, *M. N. R. A. S.*, **197**, 31.

Bath, G. T., Evans, W. D., Papaloizou, J., and Pringle, J. E. 1974, *M. N. R. A. S.*, **169**, 447.

Buckley, D. A. H. 1989, Ph.D. thesis, Australian National University.

Buckley, D. A. H., and Tuohy, I. R. 1989, *Ap. J.*, **344**, 376.

Cropper, M. 1986, *M. N. R. A. S.*, **222**, 225.

Faulkner, J. 1971, *Ap. J. (Letters)*, **170**, L99.

Felsteiner, J., and Opher, R. 1976, *Astr. Ap.*, **46**, 189.

Frank, J., King, A. R., and Raine, D. J. 1985, *Accretion Power in Astrophysics* (Cambridge: Cambridge University Press).

Hassall, B. J. M., *et al.* 1981, *M. N. R. A. S.*, **197**, 275.

Horne, K., and Marsh, T. R. 1986, *M. N. R. A. S.*, **218**, 761.

Imamura, J. N., and Durisen, R. H. 1983, *Ap. J.*, **268**, 291.

King, A. R., Frank, J., and Ritter, H. 1985, *M. N. R. A. S.*, **213**, 181.

Milgrom, M., and Salpeter, E. E. 1975, *Ap. J.*, **196**, 583.

Mouchet, M., Bonnet-Bidaud, J.-M., Buckley, D. A. H., and Tuohy, I. R. 1990, in preparation.

Patterson, J. 1984, *Ap. J. Suppl.*, **54**, 443.

Patterson, J., and Price, C. M. 1981, *Ap. J. (Letters)*, **243**, L83.

Penning, W. R. 1985, *Ap. J.*, **289**, 300.

Remillard, R. A. 1985, Ph.D. thesis, Massachusetts Inst. of Technology.

Remillard, R. A., *et al.* 1986, *Ap. J.*, **301**, 742.

Schneider, D. P., and Young, P. 1980, *Ap. J.*, **238**, 946.

Tuohy, I. R., Buckley, D. A. H., Remillard, R. A., Bradt, H. V., and Schwartz, D. A. 1986, *Ap. J.*, **311**, 275.

Warner, B. 1986, *M. N. R. A. S.*, **219**, 347.

Wickramasinghe, D. T., Stobie, R. S., and Bessell, M. S. 1982, *M. N. R. A. S.*, **200**, 605.

V795 Herculis (PG 1711+336): A New Intermediate Polar in the Period Gap

Allen W. Shafter[1], Edward L. Robinson[1], David Crampton[2], Brian Warner[3], and Richard M. Prestage[4]

[1]McDonald Observatory and
 Department of Astronomy, University of Texas at Austin
[2]Dominion Astrophysical Observatory
[3]Department of Astronomy, University of Cape Town and
 Department of Astronomy, Dartmouth College
[4]Steward Observatory, University of Arizona

1 INTRODUCTION

During their search for faint objects with ultraviolet excesses at high Galactic latitude, Green et al. (1982) identified V795 Her (PG 1711+336) as a possible cataclysmic variable based on its emission-line spectrum. Subsequent analysis of archival plates by Mironov, Moshkalev, and Shugarov (1983) showed that V795 Her exhibits slow brightness variations in the range $13.2 < B < 12.5$ with no evidence for dwarf nova eruptions. They also found a periodic variation in the light curve with an amplitude of about 0.2 mag and a period of 0.115883 day. Later observations by Baidak et al. (1985) suggested a slightly different value of 0.114488 day for the photometric period. Baidak et al. noted that if the photometric period is equal to the orbital period of the system, V795 Her would then be in the gap in the distribution of orbital periods of cataclysmic variables (Whyte and Eggleton 1980; Robinson 1983). An early attempt by Thorstensen (1986) to determine the orbital period from spectroscopic observations revealed possible radial velocity variations with a period of 0.6151 day. This result led Thorstensen to propose that V795 Her may be a member of the intermediate polar class of cataclysmic variables.

In 1985, we began a program of spectroscopic and photometric observations of V795 Her in order to firmly establish the orbital period and to resolve the discrepancies in the published photometric periodicities. A preliminary report of our work was published in Shafter et al. (1987) and a detailed report of our final results can be found in Shafter et al. (1990). Here, we present a brief summary of our final results.

2 OBSERVATIONS

Our photometric observations consist of high-speed photoelectric photometry obtained in two sets of 4 runs, one set in 1985 April and May and the other in 1986 June. The 1985 observations were obtained at McDonald Observatory with the 2.1-m telescope using the McDonald Observatory high-speed photometer (Nather 1973). The observations were made in unfiltered light with a blue-sensitive photomultiplier tube (an RCA 8850). The 1986 observations were obtained at McGraw Hill Observatory with the 1.5-m telescope using a standard photometer modified for high-speed photometry. The McGraw-Hill observations were made in unfiltered light with a red-sensitive RCA C31034A photomultiplier tube.

The spectroscopic observations were obtained at the Dominion Astrophysical Observatory (DAO) in 1985 and 1986 and at the McDonald Observatory in 1986 and 1988. These data were primarily obtained for a radial velocity study and cover the wavelength range of 4200–5000 Å. The details of our photometric and spectroscopic observations can be found in Shafter *et al.* (1990).

2.1 The Photometric Periodicity

As pointed out earlier, the published ephemerides for the photometric period of V795 Her are not in agreement. Mironov, Moshkalev, and Shugarov (1983) found $P = 0.11583$ day, while Baidak *et al.* (1985) found $P = 0.114488$ day. Very recently, Rosen *et al.* (1989) and Kaluzny (1989) found $P = 0.11575498$ day and $P = 0.1166728$ day, respectively. Our data show that none of these periods are correct.

By combining times of minima from the studies of Rosen *et al.* (1989) and Kaluzny (1989) with our timings, we find a uniquely-determined linear ephemeris:

$$T_{\min} = \mathrm{JD}_{\odot} 2446584.204 + 0.1164865\, E\,.$$

$$\pm 2 \qquad\quad \pm 4$$

This ephemeris is different from all previously published ephemerides. The previous ephemerides appear to be aliases of the true ephemeris, differing from the true ephemeris by one or more cycles in several months.

2.2 The Orbital Period

As part of our radial velocity study of V795 Her, we obtained 319 individual time-resolved spectra covering the Hγ, Hβ, and He II $\lambda 4686$ emission features. Radial velocities of the emission lines were extracted in the usual manner — by convolving their line profiles with a double Gaussian template (Shafter 1985; Shafter, Szkody, and Thorstensen 1986).

After measuring velocities for 319 individual spectra we searched for the orbital period using the Press and Rybicki (1989) algorithm for the Lomb-Scargle periodogram (Lomb 1976; Scargle 1982). A frequency of ~ 9.24 day^{-1} is strongly favored in our data. We adopt this as the true orbital frequency, noting that frequencies of ~ 10.23 and 8.23 day^{-1}, which correspond to one cycle per day aliases of the dominant frequency, cannot be ruled out absolutely. The dominant frequency corresponds to an orbital period of 0.1082468 day. We find no evidence for radial velocity variations at the period of 0.6151 day reported by Thorstensen (1986).

To find preliminary estimates of the remaining orbital elements, we fit the velocities with a sinusoid of the form:

$$V(t,s) = \gamma(s) - K(s) \sin\left[\frac{2\pi(t - T_o(s))}{P}\right],$$

where γ is the systemic velocity, K is the semi-amplitude of the emission line source, T_o is the time of conjunction, P ($= 0.1082648$ day) is the orbital period, and s is the Gaussian separation, which we initially took to be 1000 km s^{-1}. The systemic velocity was not reliable so we artificially shifted the mean of the velocities to zero. The fit yielded $K = 70 \pm 5$ km s^{-1} and $T_o =$ JD$_\odot$2447329.824 $\pm$ 0.002.

3 DISCUSSION

3.1 The Intermediate Polar Model

We have established that the the spectroscopic and photometric periods of V795 Her are different. This result is robust and would be unchanged even if we have chosen the incorrect aliases of either period. The preferred explanation for this behavior is that V795 Her is an intermediate polar. As for other intermediate polars, we identify the spectroscopic period with the orbital period and the photometric period with the spin of an accreting magnetic white dwarf. There is one interesting complication to the model: In V795 Her the photometric period is *longer* than the spectroscopic period. One expects that accretion torques should decrease the rotation period of the white dwarf until its period reaches an equilibrium value determine primarily by the mass-accretion rate and the strength of the white dwarf's magnetic field.

As it is difficult to understand how the white dwarf could be rotating with a lower angular frequency than the binary, we propose that the photometric period is the beat period between the rotation of the white dwarf and the orbital period. The rotation period of the white dwarf would then be 0.056108 day ($\sim$ 1.35 hr) for single-beam and 0.073907 day ($\sim$ 1.77 hr) for dipole-beam reprocessing. There is no evidence for either of these periods in our photometric or spectroscopic data. One or the other should, however, be present at X-ray wavelengths. In this regard it is surprising that *EXOSAT* observations by Rosen *et al.* (1989) did not reveal X-ray emission from this system. The 3σ upper limits to the X-ray fluxes were 1.2×10^{-12} erg cm^{-2} s^{-1} in the 0.02–2.5 keV band and 2.4×10^{-12} erg cm^{-2} s^{-1} in the 2.5–6 keV band.

These stringent upper limits are interesting because it is generally thought that intermediate polars are strong X-ray sources. The association of strong X-ray emission with intermediate polars may, however, be partly a selection effect: Most intermediate polars are first detected by their X-ray emission and later studied at optical wavelengths, guaranteeing that they are X-ray sources. A few exceptional systems have been identified as intermediate polars or DQ Her stars at optical wavelengths before they were observed at X-ray wavelengths, usually because they show highly-coherent periodic variations in their optical light curves. Limiting our list to the least controversial systems, we include DQ Her itself, AE Aqr, TT Ari, and now V795 Her among these systems. The X-ray emission from all these systems is weak or undetectable. The X-ray emission from AE Aqr has been detected and its properties are particularly instructive because it is pulsed at 33 s, demonstrating conclusively that AE Aqr is a magnetic accretor (Patterson 1979). Nevertheless, the X-ray luminosity of AE Aqr is low, 10^{31} erg s^{-1} in the 0.1–4.0 keV range, and the X-ray emission would not have been detected if it were not among the nearest of the cataclysmic variables.

It appears that magnetic accretion does not always result in detectable X-ray emission. Intermediate polars showing strong X-ray fluxes would then be only a subset of the entire class of magnetic accretors. We have no preferred explanation for the lack of strong X-ray emission from some intermediate polars but two possibilities seem worth investigating: The X-rays could be emitted in a tight beam that misses Earth or the X-ray spectrum could be extremely soft so that it is missed by the detectors or absorbed before reaching Earth.

3.2 The Evolutionary State of V795 Her

V795 Her is one of the few cataclysmic variables with an orbital period lying near the middle of the gap in the distribution of periods between roughly 2 and 3 hours. There may be two ways to account for the anomalous period of V795 Her. When the system was first formed in its pre-cataclysmic state its secondary star may have had an unusually low mass. If so, the orbital period at which the secondary star first established contact with its Roche lobe and began mass transfer could have been much shorter than normal, possibly less than 3 hours. V795 Her would then be the descendant of a system roughly like MT Ser (Green, Liebert, and Wesemael 1984).

Alternatively, V795 Her may not yet have entered the period gap. In their discussion of V Per, which has an orbital period nearly identical to V795 Her and may also be an intermediate polar, Shafter and Abbott (1989) noted that the secondary stars in systems with low rates of mass transfer are not driven as far out of thermal equilibrium as systems with high rates of mass transfer. The period gap may be much narrower for such systems. Binaries with low transfer rates should evolve to shorter orbital periods before their secondaries becomes fully convective, narrowing the gap from the upper end. Their secondaries may also shrink less in radius and re-establish contact at longer orbital periods, narrowing the gap from the lower end. We note, however, that the dispersion in the rates of mass transfer in cataclysmic variables above the gap appears to be small because the boundary of the gap is sharply defined. The evolutionary states of V Per and V795 Her are, therefore, most unusual.

4 CONCLUSIONS

The light curve of V795 Her varies with a period of 0.1164865 day. Its spectrum shows radial velocity variations with a *different* period of 0.1082648 day. We interpret V795 Her as an intermediate polar. We identify the 0.1082648 day spectroscopic period as the orbital period of the binary system and the 0.1164865 day photometric period as the beat period between the orbital period and the rotation period of the white dwarf. In this model the beating is caused by reprocessing of X-rays from the poles of the white dwarf in the atmosphere of the companion star. This interpretation predicts that the white dwarf has a rotational period of either 0.065108 day or 0.073907 day, depending on whether radiation is being reprocessed from one or both poles of the white dwarf.

This work has been supported in part by NSF grant AST 87-04382. BW thanks McGraw Hill Observatory for telescope time and the NSF for partial support under NSF grant AST85-15219 to G. Wegner. We thank Lu Wenxian for obtaining and reducing some of the spectra obtained at DAO. RMP acknowledges the receipt of a SERC/NATO fellowship during this work.

REFERENCES

Baidak, A. V., *et al.* 1985, *IBVS*, No. 2676.
Green, R. F., *et al.* 1982, *Pub. Astr. Soc. Pac.*, **94**, 560.
Greeen, R. G., Liebert, J., and Wesemael, F. 1984, *Ap. J.*, **280**, 177.
Kaluzny, J. 1989, preprint.
Lomb, N. R. 1976, *Ap. Sp. Sci.*, **39**, 447.
Mironov, A. V., Moshkalev, V. G., and Shugarov, S. Yu. 1983, *IBVS*, No. 2438.
Nather, R. E. 1973, *Vistas Astr.*, **15**, 91.
Patterson, J. 1979, *Ap. J.*, **234**, 978.
Press, W. H., and Rybicki, G. B. 1989, *Ap. J.*, **338**, 227.
Robinson, E. L. 1983, in *Cataclysmic Variables and Related Objects*, ed. M. Livio and G. Shaviv (Dordrecht: Reidel), p. 1.
Rosen, S. R., *et al.* 1989, preprint.
Scargle, J. D. 1982, *Ap. J.*, **263**, 835.
Shafter, A. W. 1985, in *Cataclysmic Variables and Low Mass X-Ray Binaries*, D. Q. Lamb and J. Patterson (Dordrecht: Reidel), p. 355.
Shafter, A. W., and Abbott, T. M. C. 1989, *Ap. J. (Letters)*, **339**, L75.
Shafter, A. W., *et al.* 1987, *Bull. AAS*, **19**, 1058.
Shafter, A. W., Szkody, P., and Thorstensen, J. R. 1986, *Ap. J.*, **308**, 765.
Shafter, A. W., *et al.* 1990, *Ap. J.*, in press.
Thorstensen, J. R. 1986, *A. J.*, **91**, 940.
Whyte, C. A., and Eggleton, P. 1980, *M. N. R. A. S.*, **190**, 801.

High-Speed Photometry of PG 1711+336 (V795 Her)

B. N. Ashoka[1], S. Seetha[1], T. M. K. Marar[1], V. N. Padmini[1],
K. Kasturirangan[1], U. R. Rao[1], J. C. Bhattacharyya[2], B. C. Bhatt[3],
D. C. Paliwal[3], and A. K. Pandey[3]

[1]ISRO Satellite Centre, Bangalore 560 017, India
[2]Indian Institute of Astrophysics, Bangalore 560 034, India
[3]UP State Observatory, Nainital 263 129, India

ABSTRACT

Fast photometric observations of PG 1711+336 are presented. The light curve shows a periodic modulation of ~2.8 hr. Although variations of ~0.15 mag, on timescales of ~10–12 min, are evident throughout the light curve, a search for periodicities in the range 500–1500 s did not reveal any coherent short period in the system. We also report that the photometric phase (0.3) of the relatively large-amplitude (0.2 mag) decrease in the light, lasting for ~10 to 15 min, coincides in phase with the decrease in equivalent width of a few Balmer lines determined by Rosen et al. (1989).

1 INTRODUCTION

Long-period light variations of PG1711+336 were first noticed by Miranov, Moshkalev, and Shugarov (1983b), who reported a photometric period of 0.115883 day in the system. Baidak et al. (1985) improved the ephemeris and refined the period to 0.114488 day. Rosen et al. (1989) arrived at two close periods of 0.11575498 and 0.11588066 days, while Shafter et al. (1990) have derived an ephemeris with a period of 0.1164865 day. Thorstensen (1986) observed a 14.8-hr radial velocity variation in the Hα emission line and suggested that 14.8 hr is the orbital period of the system and that the 2.8-hr modulation represents the spin period of the magnetized white dwarf. However, subsequent spectroscopic observations by Shafter et al. (1987) and Rosen et al. (1989) did not confirm the presence of the 14.8-hr period in the system. If the 2.8-hr period represents the true orbital period of the system, then further investigations of this interesting object are called for, as this period lies within the well-known period gap in the distribution of orbital periods of CVs.

2 OBSERVATIONS

High-speed photometric observations of PG 1711+336 were made with the 1-m telescope of the UP State Observatory at Nainital using a two-star photometer (Venkata Rao et al. 1989) on 7 and 10 April 1989. The star was observed in unfiltered light through the main channel using an uncooled RCA 8850 PMT and a diaphragm of 24 arcsec. The sky was clear and stable, as deduced from the data of the second channel in which star B of the sequence of Miranov, Moshkalev, and Shugarov (1983a) was monitored. On April 7 and April 10, ~1 hour and ~2.6 hours of data were collected, respectively, with an integration time of 5 s.

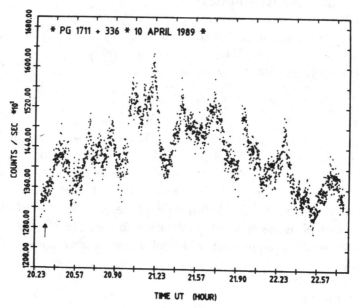

Fig. 1—Light curve of 10 April 1989 of PG 1711+336.

3 LONG-PERIOD LIGHT MODULATIONS

A visual inspection of the light curve shown in Figure 1 reveals light modulations on a timescale of 2.8 hr. The ephemeris of Rosen *et al.* (1989) did not fit with our observed minima, not only on 10 April 1989 (JD 2447627.3465), indicated by an arrow in Figure 1, but also with one of our previous observations on 11 May 1986 (JD 2446749.3592) (light curve not shown). We, however, note that our times of minima, as well as those of Rosen *et al.*, closely fit the ephemeris of Shafter (1990). (At the time of preparation of this paper, only the photometric ephemeris derived by Shafter *et al.* 1987 was known to us.) One can clearly discern a large-amplitude (~0.2 mag) decrease in light close to the photometric maximum in the 10 April data. Similar decreases were also noticed in some of the light curves of Rosen *et al.* and also in their combined light curve. Rosen *et al.* have noticed a decrease in the equivalent widths of the Hβ, Hγ, and Hδ lines at a photometric phase of 0.3. It is indeed remarkable that this phase coincides with that of the large-amplitude decrease seen in our 10 April data.

The maximum blueshift in the radial velocity curve using Shafter's ephemeris is at a photometric phase of 0.5 while the photometric maximum is around 0.3. Hence, the discrepancy in explaining the light modulations and radial velocity variations based on the hot spot model continues.

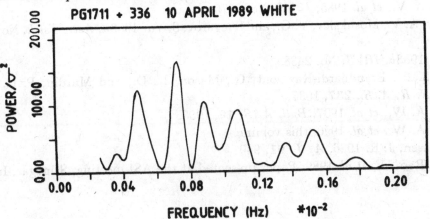

Fig. 2—Power spectrum of 10 April 1989 of PG 1711+336.

4 QUASI-PERIODIC LIGHT VARIATIONS

Superposed on the 2.8-hour light variations we notice structured oscillations throughout the 10 April 1989 light curve on timescales of 500–1500 s. These were also evident in our 11 May 1986 data (not shown). Visual inspection of the data indicates the presence of more than one period. The power spectrum of the April 10 data (Fig. 2) shows a maximum power around 1400 s, while the April 7 data shows a maximum around 560 s. The power spectrum also shows considerable power around 1136 s, 935 s, 735 s, and 658 s. To check the consistency of these periods, we subjected the data to a Discrete Fourier Transform analysis in parts. This did not show any consistency in the period, indicating that the oscillations are quasi-periodic in nature.

5 DISCUSSION

The 2.8-hour modulation in the light curve of PG 1711+336 is well established. The disagreement in the phases and shapes of light and radial velocity variations could indicate that each varies with a different period. The presence of flickering throughout the light curve indicates that they originate either from the disk or from the bright spot. The large-amplitude dip at the photometric maximum may be due to a geometric effect, as, for example, an occultation of the bright spot by the secondary star. Further study of the object is called for to resolve the true orbital period of the system.

REFERENCES

Baidak, A. V., *et al.* 1985, *IBVS*, No. 2676.

Miranov, A. V., Moshkalev, V. G., and Shugarov, S. Yu. 1983*a*, *Astr. Tsirk*, No. 1279, 6.

————. 1983*b*, *IBVS*, No. 2438.

Rosen, S. R., Branduardi-Raymont, G., Mason, K. O., and Murdin, P. G. 1989, *M. N. R. A. S.*, **237**, 1037.

Shafter, A. W., *et al.* 1987, *Bull. AAS*, **19**, 1058.

Shafter, A. W., *et al.* 1990, this volume.

Thorstensen, J. R. 1986, *A. J.*, **91**, 940.

Venkata Rao, G., *et al.* 1989, Paper presented at the ASI meeting, Srinagar, India.

Abundance Anomaly in Magnetic Cataclysmic Binaries

M. Mouchet[1], J. M. Bonnet-Bidaud[2], and J. M. Hameury[1]

[1]DAEC, Observatoire de Meudon, F-92190 Meudon
[2]Service d'Astrophysique, CEN Saclay, F-91191 Gif-sur-Yvette

ABSTRACT

In the course of the UV study of X-ray sources newly identified with magnetic variable systems, we have discovered two systems H0538+608 and H0542-407 for which the N V and C IV lines are radically different compared to the values typical of their class (strong N V/C IV ratio). Such an anomaly cannot be accounted for by present available photoionization models but might indicate non-cosmic abundances in the accreted matter from the companion whose external layers have been stripped off. This requires an initial mass of the secondary of ~ 1.1 M$_\odot$ for H0538+608 and much higher (~ 2.5 M$_\odot$) for H0542-407.

1 INTRODUCTION

The magnetic white dwarf binaries [the so-called AM Her systems or polars and the intermediate polars (IPs)] are semi-detached systems consisting of a magnetic white dwarf accreting matter from a low-mass late-type star. The main difference between these two types of systems is the rotational period of the white dwarf, which is synchronized with the orbital period for polars and spins faster for IPs. Both of them are strong X-ray and UV emitters. The UV spectra of these systems exhibit strong emission lines (N V, Si IV, C IV, and He II) superimposed on a 'blue' continuum, each class forming a homogeneous group as far as the line ratios and UV colours are concerned (Mouchet and Bonnet-Bidaud 1988).

However, our UV study of X-ray sources newly identified with magnetic binaries has revealed two systems H0538+608 and H0542-407 with unusual C IV and N V emission line strengths.

The X-ray source H0538+608 was identified as a polar by Remillard *et al.* (1986) on the basis of a circular polarized flux, a strong He II 4686Å emission line, and an inverted Balmer decrement. Recent optical observations (Mason *et al.* 1989) have shown rapid changes in the polarization behavior and an orbital period of 3.33 hr. The first published UV spectra (Bonnet-Bidaud and Mouchet 1987) revealed highly abnormal C IV and N V lines (see Fig. 1a).

H0542-407 is identified as an intermediate polar with a 32 min white dwarf rotation period (Tuohy *et al.* 1986) and a 5.7 hr orbital period (Buckley and Tuohy 1989). The UV spectrum also exhibits abnormal line intensities (see Fig. 1*b*), though they are less pronounced than for H0538+608 (Mouchet *et al.* 1989).

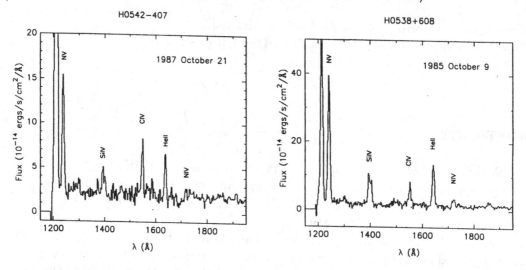

Fig. 1—*IUE* short-wavelength spectra of H0538+608 (*left*) and of H0542-407 (*right*). Note the strong N V line and the weak C IV line in both spectra.

In Figure 2 the line ratio N V/Si IV is plotted versus N V/C IV for polars and IPs. No obvious difference is detected between these classes; however, note that the values for H0538+608 and H0542-407 are markedly distinct. If we assume that the Si IV line intensity is normal, this implies a N V line ~ 3 times higher for both sources and a C IV line ~ 2 times lower for H0542-407 and ≳ 5 times for H0538+608.

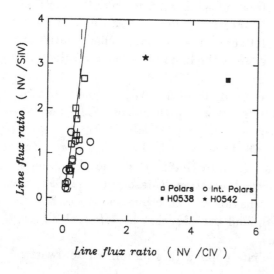

Fig. 2—Line intensity ratio (N V/ Si IV) versus (N V/C IV) for polars and IPs. The extreme values of H0538 +608 and H0542-407 are interpreted as an overabundance of nitrogen and a deficiency of carbon. Curves are line ratios predicted from collisional excitation with ion abundances resulting from photoionization by blackbody spectra with $T = 15$ eV (*full*) and 45 eV (*dashed*) (Kallman 1983).

2 INTERPRETATION

2.1 Orbital and Long-Term Variability?

The analysis of 13 *IUE* SWP spectra of the source H0538+608 shows that this abnormal line ratio is always present with a variation of the N V/C IV ratio of up to a factor 3, the C IV line being hardly detectable in some spectra. The optical level of the source decreased strongly between 1986 and 1987 (Mason *et al.* 1989), the UV flux also varied but by much less, while no change in the line ratios related with this continuum variation was observed.

On two occasions (Oct 85 and Aug 88), 3 spectra were obtained in two consecutive orbital cycles and at different phases: they still present abnormal line ratios. If the emission lines of different elements arise from distinct regions in the system, this rules out any strong occultation, along the orbital cycle, of a region responsible for one of the two lines.

We have obtained UV spectra of H0542-407 at 3 epochs (Apr 85, Oct 87, and Aug 88). The spectra obtained on Oct 87 and Aug 88 are quite similar, while that on Apr 85 exhibits a weak continuum but with typically anomalous line ratios.

2.2 Line Formation Processes?

The anomalous UV lines might result from a specific ionization state in the emitting region. However, if the abundances are cosmic, presently available photoionization models (Kallman 1983) fail to explain these unusual line intensities, while they account for the standard values observed in these magnetic systems (see Fig. 2).

2.3 Abundance Anomalies?

Another possible explanation for these line strengths is a CNO redistribution following a thermonuclear runaway at the surface of the white dwarf, as is observed in novae (Truran 1985). Neither of these two sources is known to have undergone a recent nova outburst. Large-slit spectra of H0538+608 obtained at the 3.6-meter CFH telescope failed to reveal any nebular line arising from a residual shell.

The most promising interpretation of these abnormal values of the emission line strengths is that they result from non-cosmic abundances in the matter accreted from a companion whose external layers have been stripped off by mass transfer. This is supported by the absence of any large variability on a secular timescale in the line intensities.

We have computed the abundance ratio N/C, using a double polytrope code, as a function of the observed orbital period, depending of the initial secondary mass $M_{2,i}$ (see Fig. 3). We assume that mass transfer starts 3×10^8 yr after the formation of the system. The primary mass is 1.4 $M_\odot$, so that mass transfer is always stable,

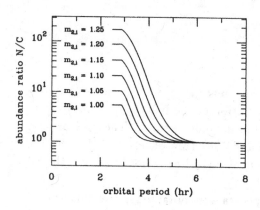

Fig. 3—The abundance ratio N/C normalized to cosmic values expressed as a function of observed orbital period, for different values of the initial secondary mass $M_{2,i}$.

and all the accreted mass is assumed to be ejected during nova explosions, so that M_1 remains constant. The abundance profiles $X(M)$ just prior to the onset of mass transfer have been computed using a detailed stellar code. The surface abundances vary with time according to:

$$\frac{dX_{\text{surf}}}{dt} = \frac{d(M_2 - M_{\text{conv}})}{dt} \times \frac{X(M_{\text{rad}}) - X_{\text{surf}}}{M_{\text{conv}}},$$

where M_{conv} is the mass of the convective envelope and $X(M_{\text{rad}})$ is the abundance at the top of the radiative core, which is given by the initial abundance profile.

It is seen that a significant deviation from the normal value of the ratio N/C can be produced for periods less than 5 hr and for initial secondary masses larger than about 1 $M_\odot$. In the case of H0538+608 ($P = 3.3$ hr), the secondary must have been more massive than ~ 1.1 $M_\odot$ to account for the observed N V/C IV ratio. If the primary is less massive than 1.1–1.2 $M_\odot$, there must have been an episode of unstable mass transfer with the formation and ejection of a common envelope. In the case of H0542-407 ($P = 5.7$ hr), the secondary must have been initially more massive than about 2.5 $M_\odot$; in that case, dynamical mass transfer has certainly occurred.

REFERENCES

Bonnet-Bidaud, J. M., and Mouchet, M. 1987, *Astr. Ap.*, **188**, 89.

Buckley, D. A. H., and Tuohy, I. R. 1989, *Ap. J.*, **344**, 376.

Kallman, T. R. 1983, *Ap. J.*, **243**, 911.

Mason, P. A., Liebert, J., and Schmidt, G. D. 1989, *Ap. J.*, submitted.

Mouchet, M., and Bonnet-Bidaud, J. M. 1988, in *A Decade of UV Astronomy with the IUE Satellite*, ESA SP-281, p. 271.

Mouchet, M., Bonnet-Bidaud, J. M., Buckley, D., Tuohy, I. R. 1989, in preparation.

Remillard, R. A., *et al.* 1986, *Ap. J. (Letters)*, **302**, L11.

Truran, J. 1985, in *Production and Distribution of CNO Elements*, ESO Colloquium, p. 211.

Tuohy, I. R., *et al.* 1986, *Ap. J.*, **311**, 275.

New DQ Hers

Paula Szkody[1], Peter Garnavich[1], Steve Howell[2], and T. Kii[3]

[1]Department of Astronomy, University of Washington
[2]Planetary Science Institute
[3]Institute of Space and Astronautical Science, Japan

1 INTRODUCTION

During the past few years, a variety of observational programs have led to the discovery of properties of several cataclysmic variables that are suggestive of a magnetic DQ Her nature. We report here on the results for 4 systems (AH Eri, AL Com, S193, and V426 Oph). This includes optical photometry on all four, as well as *GINGA* X-ray results on V426 Oph.

From our perspective, the identifying traits of a DQ Her system include:

1) The presence of a pulse period in the optical that is less than a spectroscopically identified orbital period. The correct identification of this pulse period as the spin period must come from X-ray observations.

2) The presence of a relatively large X-ray flux. In general, the DQ Hers fit into a sequence of X-ray flux such that AM Her flux > DQ Her flux > dwarf novae flux.

3) The strength of He II λ4686. The validity of this parameter in the context of the above two will be discussed at the end of this paper.

2 RESULTS

2.1 AH Eri and AL Com

The data on these two dwarf novae present a convincing case for membership in the DQ Her category. For each system, three independent nights of CCD photometry (obtained on the KPNO 36- and 84-inch telescopes and the Lowell 72-inch telescope) reveal a strong peak in the power spectrum (Fig. 1) at a period of 42 min for each night. The peak-to-peak pulse modulation is 30–40% in the blue. Sample light curves are shown in Howell and Szkody (1988) and Szkody *et al.* (1989). Time-resolved spectra from the KPNO 4-m are being analyzed to determine the orbital periods.

2.2 S193

This object was identified by Downes and Keyes (1988) as a cataclysmic variable and the first photometry (Garnavich, Szkody, and Goldader 1988) revealed a highly-variable light curve with a periodicity at 19 min evident in most, but not all, the light curves. A 30-min periodicity was also often present. Preliminary Hα spectroscopy revealed the orbital period was between 3–4 hrs. At the current time, there are 9 nights of data obtained at the Manastash Ridge Observatory 30-inch telescope over a 2-yr period. Figure 2 shows a sample of the power spectra. During the two years, the mean optical brightness decreased from 12.1 to 13.3. Although there is still no conclusion about the stability of the 19-min timescale in terms of a rotational origin, it is apparent that some behavior at 19 min is still present during parts of each night. In addition, *IUE* and optical blue spectra have shown peculiarities in the strength of absorption at 1300 Å and in the He I lines. Further spectral analysis is proceeding in order to ascertain the true nature of this object.

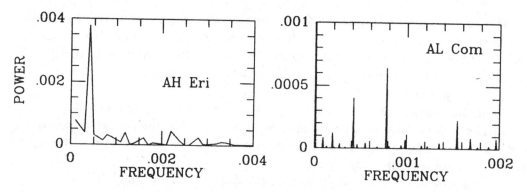

Fig. 1—The power spectra of AH Eri and AL Com.

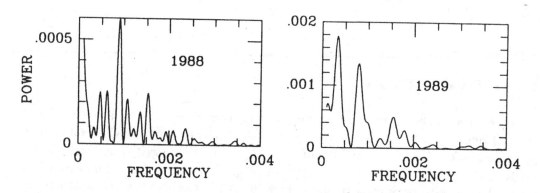

Fig. 2—Examples of the power spectra of S193 during 2 different years.

2.3 V426 Oph

Previous *EXOSAT* observations obtained at quiescence (Szkody 1986) led to a suggestion that this source might be modulated on a 1-hr timescale. Hessman (1988) has done a comprehensive spectroscopic study which showed V426 Oph to be a double-lined system with an orbital period of 6h 51 min with a K3 dwarf secondary at a distance of 200 pc.

The system was observed by *GINGA* for 4 days in Sept. 1988 when it was about one magnitude below outburst brightness (and one magnitude brighter than the previous *EXOSAT* observations). Simultaneous optical observations were obtained at KPNO during part of the X-ray observations. The timeline is shown below.

Sept. 14		Sept. 14 15		Sept. 16		Sept. 16 17
	12hr gap	X-ray 12.1hr	21hr gap	X-ray 3.7hr	5hr gap	X-ray 31.7hr
Opt 2.6hr		Opt 3.6hr		Opt 3.8hr		Opt 3.9hr

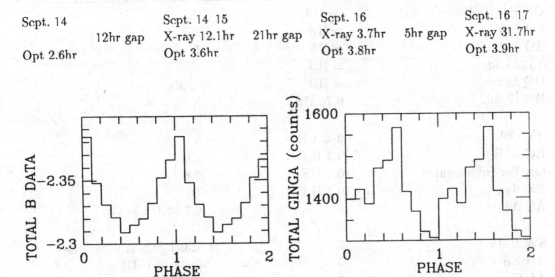

Fig. 3—The optical and X-ray data of V426 Oph folded on a period of 28 min and with the same starting epoch.

Sine-wave fitting and Fourier transform shows a period near 56 min in the first optical dataset but periods in the range of 25–33 min are evident in the remaining optical and the X-ray data. There does not appear to be one discrete constant period present in all the data, although excess power is present near 30 min. Folding the X-ray and optical data on a 28 min period and placing on the same epoch does yield a modulated curve (Fig. 3) but the relative phasings are opposite. Since this object was near outburst and other systems (EX Hya, SW UMa) are known to loose their pulse signature at outburst (Hellier *et al.* 1989*a*; Szkody, Osborne, and Hassall 1988), these data do not solve the DQ Her nature of this source. However, the lack of a spin period in quiescent spectra (Hellier *et al.* 1989*b*) argues against such an interpretation.

3 DISCUSSION OF HE II λ4686

During the last few years, it is common to find He II λ4686 strength associated with a magnetic nature for a system. This line is thought to arise from high-energy X-ray photons. In order to explore this line characteristic as an indicator of a DQ Her system, we have summarized in Table 1 the He II λ4686 strength (relative to Hβ)

TABLE 1
SUMMARY OF HE II AND X-RAY STRENGTHS

Object	He II λ4686	2–10 keV X-ray Flux ($\times 10^{-11}$ erg cm^{-2} s^{-1})
GK·Per (outburst)	$\gg$ Hβ	20
FO Aqr	$>$ Hβ	2.5
AO Psc	$=$ Hβ	5
V1223 Sqr	$=$ Hβ	6
DQ Her	$=$ Hβ	$<$ .01
H0542–407	0.75 Hβ	3
TV Col	0–0.4 Hβ	3
BG CMi	0.3 Hβ	1.6
GK Per (quiescence)	0.2 Hβ	0.8
EX Hya	0.1 Hβ	10
AE Aqr	0	0.7 (0.1–4 keV)
SW UMa	0	$<$ 0.7 (0.5 at 0.05–2 keV)
AH Eri	0	[0.08 (0.1–4)]
AL Com	0	[8 (0.5–2.8)]
S193	0.15	[2]
V426 Oph	0	8

and the available X-ray fluxes from *HEAO*, *EXOSAT*, and *GINGA* (fluxes with large uncertainties are given in brackets).

The table is ordered in groups with the first group all having very strong He II λ4686, the second group are confirmed DQ Hers having weak He II λ4686, and the third group are new or unconfirmed DQ Hers with basically no He II λ4686. It is evident that half of the DQ Hers have weak or no ionized helium. Other things to note are (*a*) 5/6 of the strong He sources in the first group are strong X-ray sources and (*b*) many of the strong X-ray sources that are weak He II λ4686 have outburst behavior (EX Hya, TV Col, AL Com). In addition, there are several novalikes with orbital periods between 3–4 hr (DW UMa, V1315 Aql, SW Sex) that have He II λ4686 comparable to Hβ but are not known to be X-ray sources nor to show any evidence for magnetic behavior. In lieu of the association of He II λ4686 with relatively constant disks (rather than those that might undergo instability), it may be that the ionized helium is a better indicator of accretion rate than magnetic field.

REFERENCES

Downes, R. A., and Keyes, C. D. 1988, *A. J.*, **96**, 777.

Garnavich, P., Szkody, P., and Goldader, J. 1988, *Bull. AAS*, **20**, 1020.

Hellier, C., *et al.* 1989*a*, *M. N. R. A. S.*, in press.

Hellier, C., O'Donoghue, D., Buckley, D., and Norton, A. 1989*b*, *M. N. R. A. S.*, in press.

Hessman, F. V. 1988, *Astr. Ap. Suppl.*, **72**, 515.

Howell, S. B., and Szkody, P. 1988, *Pub. Astr. Soc. Pac.*, **100**, 224.

Szkody, P. 1986, *Ap. J. (Letters)*, **301**, L29.

Szkody, P., Osborne, J., and Hassall, B. J. M. 1988, *Ap. J.*, **328**, 243.

Szkody, P., Howell, S. B., Mateo, M., and Kreidl, T. 1989, *Pub. Astr. Soc. Pac.*, **101**, 899.

Cyclotron Humps in AM Her Systems

Mark Cropper and Keith Mason

Mullard Space Science Laboratory, University College London, U. K.

1 INTRODUCTION

The strength of the magnetic field on the white dwarf primary star in AM Her systems sets them apart from other cataclysmic variables (CVs). This statement has, of course, to be put in the context of other system parameters, such as binary separation and white dwarf mass which also have an effect; nevertheless it was recognized very early on that determining the strength of the magnetic field would be one of the most important measurements which could be made for these systems.

The earliest estimates of the field strength were made by Tapia (1977) in his discovery paper of AM Her. These were based on the premise that the cyclotron frequency would have to occur at optical wavelengths, requiring field strengths of the order of 10^8 Gauss. Recent field strength determinations have centered around using the spacing and position of the cyclotron harmonic peaks seen at low amplitudes in the spectra of some systems, and on the position of photospheric Zeeman absorption features seen when the system is in one of its episodic low states and the mass-transfer rate is substantially reduced (see Cropper 1990 for a review of these methods and of AM Her systems in general).

2 THE SURVEY

Until recently, the occurrence of cyclotron features in AM Her systems was thought to be extremely rare. Prominent humps had been seen in VV Pup by Visvanathan and Wickramasinghe (1979), but there were no further firm reports for almost ten years. There were a number of reasons for this: the principal reason was that most spectroscopy was concentrated at blue wavelengths in order to derive radial velocity information from the many lines available there. Cyclotron humps are broader features that are masked by the prominent emission lines and often washed out in the reduction process. In addition, the wavelength coverage generally obtained was small, observations being optimized for resolution. We therefore began a survey of all AM Her systems using 4-m class telescopes with high-efficiency spectrographs providing wide wavelength coverage from 4000 to 10000 Å. Most of our observations were made on the 4.2-m William Herschel Telescope, but the most southerly were made with the Anglo-Australian Telescope. We obtained spectra on every confirmed AM Her system, except Grus-V1 and V834 Cen which was in a faint state. The observations were made between 1988 February and 1989 January.

The results are as follows: we have observed cyclotron humps in seven AM Her systems, five of which, DP Leo, MR Ser, 1E1048.5+5241, AN UMa, and H0538+608, were new detections (Cropper et al. 1989; Cropper, Mason, and Mukai 1990). An example of the cyclotron hump pattern seen in H0538+608 can be seen in Figure 1. The field strengths we determined ranged from 47 MG down to 25 MG. Our observations of VV Pup and EXO033319−2554.2 confirm the existence of two sets of humps as reported by Wickramasinghe, Ferrario, and Bailey (1989) in the latter case and by Beuermann, Thomas, and Schwope (1988) and Ferrario and Wickramasinghe (1989) in the former. We have placed upper limits on the existence of humps for EXO032957−2606.9 (Cropper et al. 1990). We obtained only limited phase coverage of EXO023432−5232.3 and V1500 Cyg, but no cyclotron features are visible in the spectra we have. Analysis of our results is continuing, and we believe that we have a reasonable chance of detecting cyclotron humps in other systems, particularly in BL Hyi. In some cases, for example DP Leo and MR Ser, we have followed up our original single-exposure spectra with phase-resolved spectra covering a full orbit.

Fig. 1—The spectrum of H0538+608 plotted against wavenumber in cm^{-1} at the top. Beneath this is a synthesized spectrum with the residual cyclotron spectrum in the third plot. This is binned and smoothed to produce the cyclotron hump spectrum shown at the bottom. Tick marks indicate the position of the harmonic peaks predicted in Cropper et al. (1989). From Cropper et al. (1989).

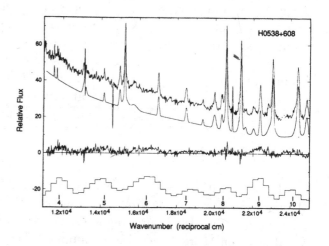

3 THE CASE OF DP LEO

Cropper et al. (1989) determined a magnetic field for this eclipsing system of 44 MG from a single long-exposure spectrum centered on the rise to the bright phase. This spectrum was dominated by a prominent hump at 6300 Å. Because DP Leo is an eclipsing system in which the cyclotron-emitting region is seen over a range of viewing angles, and it was clear that the system would be important for investigating the behaviour of the cyclotron humps with viewing angle, we made further observations of the system covering almost two orbits. We found that the humps were most prominent on the rise to and fall from the bright phase when the accretion region is on the limb. However, we found that humps are also visible during the *faint* phase integrations; these probably arise in a flow to a different accretion region and which may also be responsible for the small levels of circular polarization seen during the faint phase by Biermann *et al.* (1985) (although, see Brainerd 1990). During the rise to the bright phase, where we obtained observations with sufficient time resolution, the harmonic redward of that at 6300 Å became progressively more prominent as the system brightness rose until maximum was reached, after which all humps disappeared. The 6300 Å hump was still the most prominent cyclotron feature at all phases. Despite the change in viewing angle, none of the humps was seen to move in wavelength as a function of phase.

During the rise to maximum brightness, narrow absorption features become evident (see Fig. 2); these can be identified as Hα Zeeman absorptions in a field of 29 ± 2 MG. Following Wickramasinghe, Tuohy, and Visvanathan (1987) who saw the same features at similar viewing angles during periods of normal accretion rate in V834 Cen, we identify their origin as a halo around the main X-ray emitting accretion core. The features are more prominent on the rise to maximum than on the fall from maximum, indicating that the more of the halo is seen against the emission core on the rise to than on the fall from the bright phase, and that the halo is therefore probably asymmetric with respect to the core.

Fig. 2—The averaged spectrum of DP Leo for phase 0.75 (on the rise to the bright phase). The smooth curve fitted to the three prominent cyclotron humps are Gaussians on a linear continuum; these were fitted to locate the peak emission wavelength. Below the data are the predicted positions for the Zeeman components in a 29 MG field.

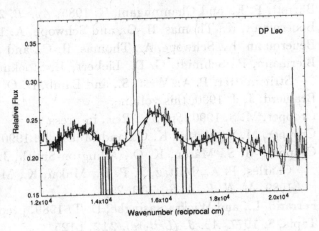

Taking the narrow Zeeman features into account, we fitted broad Gaussians to the cyclotron humps to determine their centroids (Fig. 2). We found that their wavenumbers ($1/\lambda$) are similar to those obtained in Cropper *et al.* (1989). Applying the approximation given in Barrett and Chanmugam (1985) to determine the field strength, we calculated a similar field, 48 MG, to that determined by Cropper *et al.* (1989). However, using the results from physically more appropriate calculations tabulated in Wickramasinghe (1988), the prominent harmonic at 6300 Å is identified as harmonic 6 rather than 5, and the field strength estimate must then be reduced to 32 ± 1 MG. We note that this field strength is very similar to that deduced from the Zeeman absorption features.

We believe that DP Leo is the first system in which both Zeeman absorption features and cyclotron humps have been identified (see also Beuermann *et al.* 1990), although with the benefit of hindsight, cyclotron features were seen but not recognized in V834 Cen by Wickramasinghe *et al.* (1987).

A surprising result of our observations is that the cyclotron features do not move with phase, despite the range of viewing angles at which the emission region is seen and contrary to theoretical prediction (see, for example, Wickramasinghe 1988). The inconsistency can be accommodated if it is assumed that it is mostly that part of the emission region seen successively on the limb of the white dwarf and (therefore perpendicularly to the line of sight) which contributes most heavily to the cyclotron features.

Acknowledgement: We are grateful to PATT for the allocation of telescope time for the survey. KOM acknowledges the support of a Royal Society Fellowship.

REFERENCES

Barrett, P. E., and Chanmugam, G. 1985, *Ap. J.*, **298**, 743.

Beuermann, K., Thomas, H.-C., and Schwope, A. 1988, *Astr. Ap.*, **195**, L15.

Beuermann, K., Schwope, A., Thomas, H.-C., and Jordan, S. 1990, this volume.

Biermann, P., Schmidt, G. D., Liebert, J., Stockman, H. S., Tapia, S., Kühr, H., Strittmatter, P. A., West, S., and Lamb, D. Q. 1985, *Ap. J.*, **293**, 303.

Brainerd, J. J. 1990, this volume.

Cropper, M. S. 1990, *Sp. Sci. Rev.*, in press.

Cropper, M. S., Mason, K. O., and Mukai, K. 1990, *M. N. R. A. S.*, in press.

Cropper, M. S., Mason, K. O., Allington-Smith, J. R., Branduardi-Raymont, G., Charles, P. A., Mittaz, J. P. D., Mukai, K., Murdin, P. G., and Smale, A. P. 1989, *M. N. R. A. S.*, **236**, 29p.

Ferrario, L., and Wickramasinghe, D. T. 1989, preprint.

Tapia, S. 1977, *Ap. J. (Letters)*, **212**, L125.

Visvanathan, N., and Wickramasinghe, D. T. 1979, *Nature*, **291**, 47.

Wickramasinghe, D. T. 1988, in *Polarized Radiation of Circumstellar Origin*, ed. G. V. Coyne, *et al.* (Vatican City State: Vatican Press), p. 199.

Wickramasinghe, D. T., Ferrario, L., and Bailey, J. A. 1989, preprint.

Wickramasinghe, D. T., Tuohy, I. R., and Visvanathan, N. 1987, *Ap. J.*, **318**, 326.

Cyclotron Features in VV Pup and UZ For

Jeremy Bailey[1], D. T. Wickramasinghe[2], and Lilia Ferrario[3]

[1]Joint Astronomy Centre, Hawaii
[2]Mathematics Department, Australian National University
[3]Astronomy Department, University of Leicester

1 INTRODUCTION

VV Puppis was the first AM Herculis system to be recognized as having accretion onto both poles of the magnetic white dwarf. Observations by Liebert and Stockman (1979) showed negative circular polarization during the faint phase of the orbital cycle, as well as positive circular polarization during the bright phase when the main emission region is visible. It was also the first system in which resolvable cyclotron harmonics were detected (Visvanathan and Wickramasinghe 1979), from which a magnetic field of 32 MG for the main emission region was derived.

We here present spectropolarimetric observations of VV Pup which clearly show two sets of cyclotron harmonics which can be attributed to the main and secondary poles, and allow the magnetic fields of both poles to be determined. A more detailed account of this work is given by Wickramasinghe, Ferrario, and Bailey (1989). We have also obtained similar observations of UZ For (EXO 033319-2554.2) which also shows strong cyclotron emission features.

2 OBSERVATIONS

The observations were obtained using the 3.9-m Anglo-Australian Telescope with the Pockels Cell Spectropolarimeter using the RGO Spectrograph and IPCS as detector (McLean *et al.* 1984). A series of 500-second exposures of VV Pup were obtained covering one cycle on 1988 Nov 29 and two cycles on 1988 Nov 30. Similar observations of UZ For covering a little over one cycle were also obtained on each night. For analysis, we have coadded spectra obtained during the faint and bright phases of the cycle to obtain representative spectra. We see little evidence for any significant changes in the spectrum over the duration of each bright or faint phase.

3 DISCUSSION

3.1 VV Pup

Figure 1 shows data obtained during the faint phase of VV Pup. The polarization data are here presented in the form of polarized flux. The spectra clearly show the presence of four broad emission features at wavelengths of 3600 Å, 4200 Å, 5100 Å, and 6700 Å. The 3600 Å feature is confused with the Balmer jump in the intensity spectrum, but shows up clearly in the polarized flux spectrum.

These features can be fitted with cyclotron models as shown in Figure 1. The two models shown are simple constant temperature models with $T_e = 20$ keV, $B = 57$ MG, and $\Lambda = 32$ (*solid line*), and $T_e = 5$ keV, $B = 54.6$ MG and $\Lambda = 1,100$ (*dashed line*). Although a reasonable fit to the intensity spectrum can be obtained, it proves impossible to simultaneously fit the intensity and polarization with this type of model. It seems likely that more sophisticated models involving a range of physical conditions through the emitting region will be required.

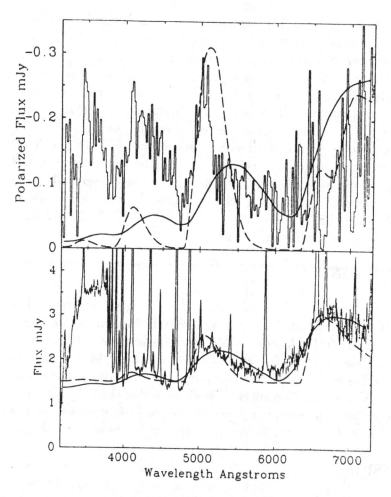

Fig. 1—Faint phase total and polarized flux spectra of VV Pup obtained on 1988 Nov 30. The lines are cyclotron model fits as described in the text.

The bright phase data show more complex structure. Although the polarization is generally positive, there is a dip in polarization at about 5100 Å. This suggests that the faint phase cyclotron harmonics, which show a strong negative polarization peak at 5100 Å are still present during the bright phase, and the complex structure we observe is the result of the combination of two sets of cyclotron harmonics. We therefore subtracted the faint phase data from the bright phase data to obtain the result shown in Figure 2. It can be seen that the 5100 Å feature is removed by this procedure, and the remaining intensity spectrum is similar to that originally observed by Visvanathan and Wickramasinghe (1979). The model fit shown in Figure 2 has $T_e = 10$ keV, $B = 30.5$ MG, and $\Lambda = 6 \times 10^5$.

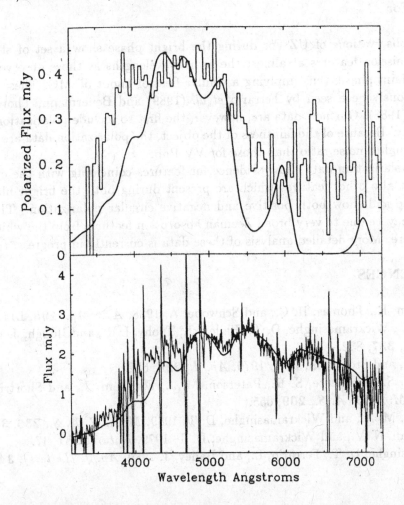

Fig. 2—Bright phase total and polarized flux spectra of VV Pup obtained on 1988 Nov 29 with the corresponding faint phase data subtracted. The line is a cyclotron model fit as described in the text.

The fact that the faint phase harmonics seem to be visible during the bright phase with little change in strength, implies that the viewing angle to the secondary emission region does not change much with phase and therefore the region is close to the rotational pole of the white dwarf. Its colatitude is likely to be less than 10°. The colatitude of the main pole is thought to be about 145° (Meggit and Wickramasinghe 1989). Thus, the two emission regions are not opposite each other, but could be located on the footpoints of a closed field line. The difference in magnetic field measured for the two emission regions can be explained with a dipole which is displaced by $0.1 R_{wd}$ in the direction of the dipole axis.

3.2 UZ For

The observations of UZ For during the bright phase show a set of strong cyclotron emission features at almost the same wavelengths as those observed in the VV Pup faint phase, thus implying a similar field of about 55 MG. These features had previously been seen by Ferrario *et al.* (1989) and Beuermann, Thomas, and Schwope (1988). Our new data are, however, the first to include polarization as well as intensity. Because of the faintness of the object, the polarization data are of much poorer signal-to-noise ratio than those for VV Pup.

The polarization data show evidence for features coinciding with the cyclotron peaks, but also other features which are present during both the bright phase and faint phase and show both positive and negative circular polarization. The latter features may be due to very broad Zeeman absorption features from the white dwarf photosphere. More detailed analysis of these data is currently in progress.

REFERENCES

Beuermann, K., Thomas, H. C., and Schwope, A. 1988, *Astr. Ap.*, **195**, L15.

Ferrario, L., Wickramasinghe, D. T., Bailey, J., Tuohy, I. R., and Hough, J. H. 1989, *Ap. J.*, **337**, 832.

Liebert, J., and Stockman, H. S. 1979, *Ap. J.*, **229**, 652.

McLean, I. S., Heathcote, S. R., Paterson, M. J., Fordham, J., and Shortridge, K. 1984, *M. N. R. A. S.*, **209**, 655.

Meggitt, S. M. A., and Wickramasinghe, D. T. 1989, *M. N. R. A. S.*, **236**, 31.

Visvanathan, N. V., and Wickramasinghe, D. T. 1979, *Nature*, **281**, 47.

Wickramasinghe, D. T., Ferrario, L., and Bailey, J. 1989, *Ap. J. (Letters)*, **342**, L35.

Cyclotron and Zeeman Spectroscopy of V834 Cen[*]

Klaus Beuermann[1,2], Axel D. Schwope[1], Hans-Christoph Thomas[3], and Stefan Jordan[4]

[1]Institut für Astronomie und Astrophysik, TU Berlin, F. R. G.
[2]MPI für Extraterrestrische Physik, Garching, F. R. G.
[3]MPI für Astrophysik, Garching, F. R. G.
[4]Institut für Theoretische Physik, Universität Kiel, F. R. G.

1 OBSERVATIONS

The observational properties of AM Herculis binaries depend strongly on the strength and orientation of the magnetic field in the emission region on the accreting white dwarf. The measurements of cyclotron harmonics and Zeeman splitting are the principal methods to infer the strength of the field (Cropper 1989a).

We performed low-resolution phase-resolved spectrophotometry of V834 Cen in its high state (1986 March 1 and 2) and in the low state (1989 January 26 and 29) when accretion had ceased or decreased to a very low level (Beuermann, Thomas, and Schwope 1989). Full orbital coverage was available in both cases. Figure 1 shows the V-band light curves phased on the post-1985 high-state timings T_o of minimum light (Cropper 1989, private communication). At $\phi = 0$, we are looking down the accretion funnel and cyclotron beaming in the high state is most pronounced. In the low state, we find no evidence for remnant cyclotron radiation and conclude that the light maximum at $\phi = 0$ must be due to increased photospheric emission of the hot former accretion region. Independent low-state spectrophotometry of V834 Cen was performed by Mason, Puchnarewicz, and Murdin (1989).

2 HIGH-STATE CYCLOTRON AND ZEEMAN SPECTROSCOPY

Wickramasinghe, Tuohy, and Visvanathan (1987, hereafter WTV) first observed broad intensity maxima and minima in the high-state spectrum of V834 Cen. They identified a dip at 6050 Å as the Zeeman σ^- component of Hα. They also considered the interpretation of three broad humps as cyclotron emission features but finally rejected this possibility because the broad lines did not seem to change in wavelength with viewing angle as predicted by cyclotron theory.

Using our high-state observations, we are able to disentangle cyclotron and Zeeman features and to measure the shifts in wavelength as a function of orbital phase. Figure 2 depicts the observed spectrum at $\phi = 0.5$ (top) and the same spectrum after removal of the strong emission lines, subtraction of the low-state spectrum, and divi-

---

[*] Based on observations with the ESO/MPI 2.2-m telescope at La Silla, Chile, in MPI time.

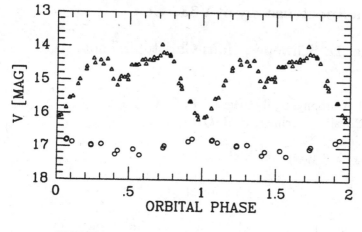

Fig. 1—V-band light curves of V834 Cen in the high state (triangles) and in the low state (circles).

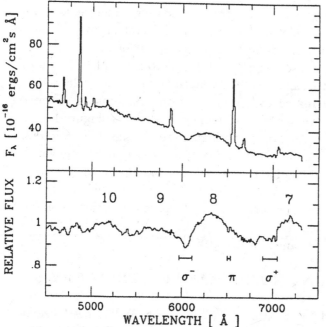

Fig. 2—High-state spectrum at $\phi \simeq 0.5$ (top). Same after removal of the emission lines and division by a smooth fit to the continuum (bottom). Cyclotron harmonics and Zeeman components of Hα are indicated.

sion by a smooth fit to the remaining continuum (bottom). The final identifications of the cyclotron harmonics and Zeeman dips are indicated. Cyclotron harmonics are discernible between $\phi \simeq 0.2$ and $\phi \simeq 0.8$. They are almost stationary at $\phi \simeq 0.4$–0.6, but outside this range display the rapid shift in wavelength expected from theory. In order to follow the features, we displayed their relative intensity as a function of wavelength and time (phase) on the screen of an image processing unit and determined the positions of cyclotron maxima and Zeeman minima interactively with the cursor. We found that the eye is very potent in following the weak features. Figure 3 shows the positions (with errors) of cyclotron maxima measured in spectra of three orbital cycles (horizontal bars). Also shown are the positions of Zeeman minima measured in the phased spectra (open circles).

We determined the field strength from comparison of the observed cyclotron

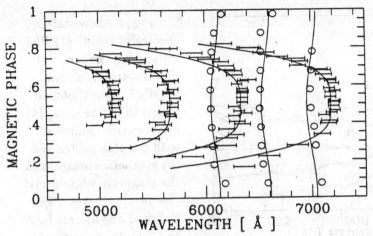

Fig. 3—Observed wavelengths of cyclotron maxima (bars) and Zeeman dips (circles). Solid curves are predicted positions (see text).

maxima with those predicted by cyclotron theory for an isothermal plasma. The best fit was obtained by identifying the observed maxima as the 7th to 10th harmonic in a field of 23.5 MG (for $kT = 10$ keV). Alternative fits, starting with the 6th or 8th harmonic, were significantly less successful.

The shift expected for the positions of the maxima depends on the variation of the viewing angle of the accreting field line with orbital phase and thereby on the inclination i and the angle δ between the field line and the rotation axis. The solid lines in Figure 3 were calculated for $i = 50°$ and $\delta = 40°$. The standstill in the position of the cyclotron maxima between $\phi = 0.4$ and 0.6 requires $i + \delta = 90° \pm 5°$. The angles determined from polarization measurements (Cropper 1989a, b), $i + \delta \simeq 70°$, are not consistent with the observed motion of the cyclotron lines. This need not be a contradiction because the weak cyclotron humps may originate from a subsection of the extended emission region.

The position and width of the Zeeman σ^- component of Hα at $\phi = 0.5$ implies $B \simeq 23$ MG with a dispersion of less than 2 MG. Surprisingly, the Zeeman dips appear all redshifted as one approaches $\phi = 0$, suggesting that absorption occurs in matter which freefalls along the field lines. The relevant solid curves in Figure 3 were calculated for $i = 50°$, $\delta = 40°$, and $v = 4000$ km s^{-1}. The near equality of the field strengths derived from cyclotron harmonics and Zeeman dips indicates that the absorbing cool gas is located very close to the emitting hot plasma, as noted already by WTV. Near $\phi = 0$, the Zeeman dips widen and absorption may extend to some height $h \simeq 0.05 R_{wd}$ above the surface of the white dwarf.

3 LOW-STATE ZEEMAN SPECTROSCOPY

In the low state, the observed flux distribution can be accounted for by contributions from the dM5–6.5 late-type star and the white dwarf with its non-uniform temperature distribution (Fig. 4). After correction for the red star, the Zeeman absorption components of Hα are clearly visible. The σ^- component of Hα is located

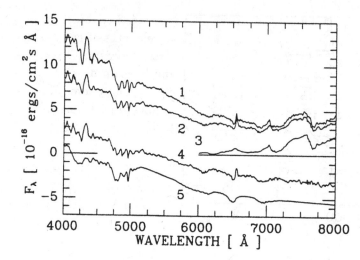

Fig. 4—Low-state spectra of V834. (1) Photometric maximum, (2) photometric minimum, (3) dM5⁺ star Gl866 adjusted to fit the data, (4) photometric minimum with red star subtracted, (5) synthetic spectrum for the magnetic white dwarf (see text). Curves 4 and 5 shifted downwards by 5 and 7 units, respectively.

at nearly the same position as in the high state, implying a mean field strength of $\sim$ 22 MG. The observed absorption features are probably a mixture produced both in the cool photosphere of the white dwarf and in the former accretion region.

As a first step to a more detailed interpretation, we compare the observed spectrum at photometric minimum with the synthetic spectrum calculated for a magnetic white dwarf which emits uniformly over its surface and has a dipolar field geometry. The spectrum shown in Figure 4 is for $T_{\text{eff}} = 15{,}000$ K, $\log g = 8$, a polar field strength $B_{\text{p}} = 35$ MG, and a viewing angle relative to the magnetic axis of 40°. The agreement is satisfactory. For details of the calculation see Jordan (1989, and a forthcoming paper). Considering our knowledge of the geometry of V834 Cen (Cropper 1989a, b), a polar field strength of 35 MG is difficult to accommodate. If the accretion region with $B \simeq 23.5$ MG is located $\sim 30°$ from the rotation axis, the polar field strength would be $B_{\text{p}} = 26$ MG for an aligned dipole. However, the model of a uniformly-emitting white dwarf is likely to be too simple in the present case. The obvious next step is to construct synthetic spectra for a magnetic white dwarf with an extended hot spot of the type presented by WTV.

REFERENCES

Beuermann, K., Thomas, H.-C., and Schwope, A. 1989, *IAU Circ.*, No. 4775.
Cropper, M. 1989a, *Space Sci. Rev.*, in press.
Cropper, M. 1989b, *M. N. R. A. S.*, **236**, 935.
Jordan, S. 1989, in IAU Coll. No. 114, *White Dwarfs*, ed. G. Wegner (Berlin: Springer), p. 333, see also Jordan, S. 1988, Ph.D. thesis, University of Kiel.
Mason, K. O., Puchnarewicz, E. M., Murdin, P. G. 1989, *IAU Circ.*, No. 4717.
Wickramasinghe, D. T., Tuohy, I., and Visvanathan, H. 1987, *Ap. J.*, **318**, 326 (WTV).

Observations of Bowen Fluorescence in AM Her Stars

Jonathan Schachter[1], Alexei V. Filippenko[2,3], Steven M. Kahn[1,2,4], and Frits B. S. Paerels[1,4,5]

[1]Department of Physics and Space Sciences Laboratory,
 University of California, Berkeley
[2]Department of Astronomy, University of California, Berkeley
[3]Presidential Young Investigator
[4]Also affiliated with Laboratory for Experimental Astrophysics,
 Lawrence Livermore National Laboratory
[5]Also affiliated with Institute of Geophysics and Planetary Physics,
 Lawrence Livermore National Laboratory

1 INTRODUCTION

The near-ultraviolet (UV) Bowen fluorescence (BF) lines are radiative cascades from the O III 2p3d level, the strongest at $\lambda\lambda 3133, 3444$ (see Schachter, Filippenko, and Kahn 1989 for a Grotrian diagram). These lines are produced because of the near-coincidence in wavelengths of He II Ly α ($\lambda 303.782$) and an O III $2p^2$–2p3d transition ($\lambda 303.799$). The BF *yield* (hereafter, y_{HeO}), defined as the probability of producing an O III cascade per incident He II Ly α photon, can be a sensitive function of the line-center opacity in He II Ly α (hereafter, τ_L; Kallman and McCray 1980). The yield may be measured directly from observed near-UV line fluxes, via $y_{HeO} \propto I_{3133}/I_{3204}$, where I_{3204} is the intensity of He II Pa β, a measure of the He II Ly α pumping rate (e.g., case B recombination). The quantity I_{3133} is a measure of the total cascade intensity (Saraph and Seaton 1980).

The quantity y_{HeO} can be large (≈ 0.1–1) in accretion-powered sources (Kallman and McCray 1980; Deguchi 1985). Because BF lines are uniquely sensitive to conditions in regions strongly ionized by the X-ray continuum in these systems, we recently initiated an observational program to quantify the role of this process more fully. The observations, performed at the Lick 3-m Shane telescope, cover a wide spectral range (≈ 3100 Å to 1 μm); extensive wavelength coverage was necessary in order to obtain diagnostics of the reddening, temperatures, densities, and opacities.

We first analysed BF in the low mass X-ray binary Sco X-1 (Schachter, Filippenko, and Kahn 1989), and more recently in a sample of Seyfert galaxy nuclei (Schachter, Filippenko, and Kahn 1990). Currently, we are extending this work to include BF in AM Her stars. These cases may be more complicated because of optical depth effects in the He II line spectrum (Stockman *et al.* 1977). Detailed knowledge of the physical conditions is therefore required to understand the Bowen emission. We discuss the physical constraints imposed by the observations after a brief overview of the data.

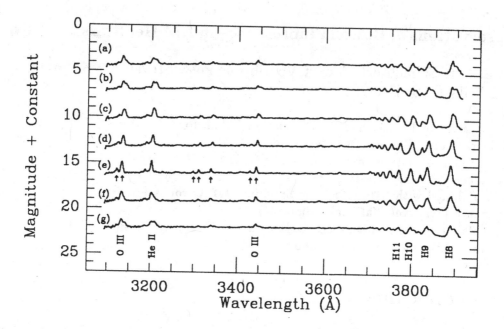

Fig. 1—Spectra of AM Her over the range 3100–3900 Å obtained 16 September 1988 UT at these orbital phases: (*a*) −0.08, (*b*) 0.08, (*c*) 0.24, (*d*) 0.44, (*e*) 0.59, (*f*) 0.76, and (*g*) 0.92. Constant shifts (in magnitudes) added to data are (*a*) −9, (*b*) −6, (*c*) −3, (*d*) 0, (*e*) +3, (*f*) +6, and (*g*) +9. Seven O III Bowen lines (*arrows*) are seen in (*e*): O III λλ 3123, 3133, 3299, 3312, 3341, 3429, 3444. The two strongest Bowen lines (O III λλ3133, 3444), He II λ3204, and several high-order H I Balmer lines, present in all the spectra, are labelled in (*g*).

2 NEW RESULTS

In Figure 1, we present a set of spectra of AM Her, obtained over a single binary period on one observing night. The new data demonstrate the presence of at least seven BF lines in AM Her, the strongest at λλ3133, 3444, as in Figure 1(*e*). The emission-line spectrum shortward of the Balmer jump ($\approx$ 3647 Å) is completely dominated by Bowen lines and by He II λ3204.

An example of the wide spectral coverage of AM Her in our data is presented in Figure 2. The Balmer jump is very prominent, and the Balmer decrement is relatively flat, as reported by Stockman *et al.* (1977). The new features present in our data include strong H I Paschen emission lines, and the Paschen jump in emission. Since the Paschen lines are detected to high order (Pa 13), we expect that the Paschen decrement is considerably flatter than case B.

We have also discovered the λλ3133, 3444 BF lines in EF Eri (not shown), suggesting that BF is a general property of AM Her stars. The Balmer decrement is again relatively flat compared with case B, although the Balmer jump in EF Eri is not as strong as in AM Her. Because this object is relatively faint (approximately

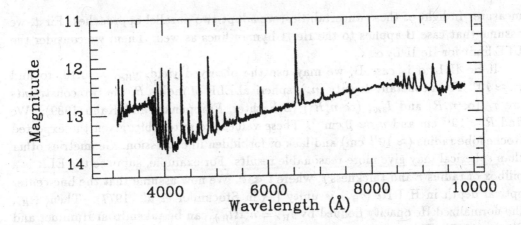

Fig. 2—Spectrum of AM Her (taken 17 July 1988 UT) from the atmospheric limit (≈ 3100 Å) to 1 μm, showing strong H I Balmer series and high-order Paschen series emission lines. The most prominent Paschen line is Pa 8 ($\lambda 9847$). The strongest line in the entire spectrum is Hα. Note also the prominent Balmer and Paschen continuum emission, and the large size of the Balmer jump.

2 magnitudes fainter than AM Her), the remainder of the discussion focuses on AM Her.

3 DISCUSSION

The electron temperature (T_e) may be calculated from the slope of the Balmer continuum (e.g., Osterbrock 1989, Appendix 1). Assuming the observed continuum shortward of the Balmer jump is due only to Balmer recombination emission gives $T_e < 20,000$ K. Subtracting an estimate of the contribution of the white dwarf photosphere (Heise and Verbunt 1988) yields $T_e \gtrsim 14,000$ K. These temperatures are, strictly, appropriate only to the H I emission-line region (ELR; see below). The slope and amplitude of the continuum shortward of the Balmer jump are relatively constant over the orbit. However, the emission lines are variable, broad, and often blended. They have a complex structure (at least two narrow components; Fig. 1). Thus, the observational test that the He II lines are produced by case B recombination, and the subsequent calculation of y_{HeO} are both difficult to perform.

If we take the value of I_{3204}/I_{4686} (He II Pa β / He II Pa α) estimated from observations, we conclude that the He II Paschen lines are produced by case B recombination (at the derived temperature). Nevertheless, it is possible that the spectrum of lower-series He II lines (Balmer, Lyman) is significantly affected by optical depth and collisional processes, as has been suggested for the H I Balmer lines (Stockman *et al.* 1977). The predicted value of He II Pa β / He II Ly α could then differ significantly from case B; this would complicate the calculation of the Bowen yield and emission measure. In light of these uncertainties, we adopt two parallel approaches. First, we assume that case B applies to the He II Lyman lines as well. Then, we consider the LTE limit for He II Ly α.

If He II Ly α is case B, we may use the observed yield, $y_{\text{HeO}} \approx 0.16$, to find $\tau_L \approx 10^3$. For a constant density (n_e), spherical ELR of radius R, the two constraints are τ_L ($\propto n_e R$) and I_{3204} ($\propto n_e^2 R^3$) (Schachter, Filippenko, and Kahn 1989). We find $R \approx 10^{17}$ cm and $n_e \approx 1$ cm^{-3}. These values are untenable, given the expected Roche lobe radius ($\approx 10^{11}$ cm) and lack of forbidden line emission. Geometries other than spherical may give more reasonable results. For example, suppose the ELR is a pillbox of radius r and thickness t, where $t \ll r$. We now assume that the line-center optical depth in H I He (τ_{He}) is unity (as in Stockman *et al.* 1977). Then, Ω_{He}, the normalized He opacity defined by $\tau_{\text{He}} = n_e^2 t \Omega_{\text{He}}$, can be taken from Hummer and Storey (1987). This approach gives $n_e \approx 10^{14}$ cm^{-3}, $r \approx 10^9$ cm, and $t \approx 10^4$ cm. The derived density is consistent with earlier results, while the emitting radius is clearly within the Roche lobe (Stockman *et al.* 1977); nevertheless, the inferred thickness seems a bit too low. Of course, this argument assumes that the H I and He II lines come from the same region.

If, instead, optical depth effects and collisions are important, then $I_{\text{He II Ly}\alpha}$ approaches its LTE value — $B_{304}(T_e)$, the blackbody function at 304 Å. In this limit, the pumping rate of the O III 2p3d level is given by $b_{2p^2, 2p3d}[B_{304}(T_e)]$, where b is the Einstein b coefficient. Therefore, $I_{3133} \propto n_{\text{O III}} R^3 b_{2p^2, 2p3d}[B_{304}(T_e)]$, where $n_{\text{O III}}$ is the number density of O III, and spherical geometry has been assumed. Additional constraints come from the observed value of the continuum at the Balmer jump ($\propto n_e^2 R^3$), and from the estimated system size. We find $T_e \approx 17,000$–$20,000$ K, $R \approx 6 \times 10^9$–10^{11} cm, and $n_e \approx 4 \times 10^{11}$–$9 \times 10^{13}$ cm^{-3}. The estimates are somewhat more reasonable than those derived above.

This work was supported by grants from the California Space Institute and the NASA Innovative Research Fund.

REFERENCES

Deguchi, S. 1985, *Ap. J.*, **291**, 492; **303**, 901.

Heise, J., and Verbunt, F. 1988, *Astr. Ap.*, **189**, 870.

Kallman, T. R., and McCray, R. 1980, *Ap. J.*, **242**, 615.

Hummer, D. G., and Storey, P. J. 1987, *M. N. R. A. S.*, **224**, 801.

Osterbrock, D. E. 1989, *Astrophysics of Gaseous Nebulae and Active Galactic Nuclei* (Mill Valley: University Science Books).

Saraph, H. E., and Seaton, M. J. 1980, *M. N. R. A. S.*, **193**, 617.

Schachter, J., Filippenko, A. V., and Kahn, S. M. 1989, *Ap. J.*, **340**, 1049.

Schachter, J., Filippenko, A. V., and Kahn, S. M. 1990, *Ap. J.*, submitted.

Stockman, H. S., Schmidt, G. D., Angel, J. R. P., Liebert, J., Tapia, S., and Beaver, E. A. 1977, *Ap. J.*, **217**, 815.

A Multiple-Component Line Ratio Study of AM Herculis

K. Mukai and M. L. Edgar

Mullard Space Science Laboratory, University College London,
Holmbury St. Mary, Dorking, Surrey RH5 6NT, U. K.

ABSTRACT

The emission lines of AM Herculis are known to have two components. We present some preliminary results from our high-resolution spectroscopy of this system, particularly at and beyond the Balmer limit, in an effort to constrain the density and temperature of the emitting regions. We find a new upper limit to the density of the narrow line-emitting region and that our dataset may enable us to perform a spatially-resolved study of the accretion stream (the broad line-emitting region).

1 EMISSION LINES IN AM HER SYSTEMS

Emission lines of AM Her systems have been studied extensively over the years (Mukai 1988, and references therein; see Cropper 1989 for a general review of AM Her systems). This is largely due to the complex line profiles reflecting the geometry of the accretion stream from the L_1 point of the secondary to the magnetic white dwarf.

The study of two helium lines in AM Her by Greenstein et al. (1977) has established the presence of a broad and a narrow component, the latter of which was proposed (and later confirmed by Young and Schneider 1979) to be from the heated face of the secondary. Schneider and Young (1980a, b) have successfully modelled the broad component in EF Eri, AN UMa, and VV Pup as originating in a curved stream confined by the magnetic field. Subsequent studies of these and other AM Her-type systems followed the same path but often found the emission line profiles to be far more complex than the two-component picture found by Greenstein et al. (1977) (e.g., 4 components found in V834 Cen by Rosen, Mason, and Córdova 1987).

These studies have mostly concentrated on the geometry of the accretion stream, a fascinating subject in itself. Our work is a natural extension of such studies and aims to study the density, temperature, and volume of the line-emitting regions. This work is still in an early stage; however, some interesting results have already emerged (Section 3). Also, the data quality may allow us to perform a spatially-resolved study of the accretion stream (Section 4).

2 OBSERVATION

We obtained data using the 2.5-m Isaac Newton Telescope at the Roque de los Muchachos Observatory (La Palma) with the Intermediate Dispersion Spectrograph on three nights in July 1989. We observed more than 1 orbital cycle (3.1 hr) of AM Herculis in 4 different wavelength ranges. The integration time for each individual integration was kept at 100 s to prevent orbital motion smearing out the sharp features. The IPCS detector was used except for the Hα region, for which we used a GEC CCD camera. The journal of the observations is given in Table 1.

TABLE 1
JOURNAL OF OBSERVATIONS

Date	λ-range	Res.	Phase
20 Jul	3500–4015	0.85Å	0.65– 1.67
1989	4460–4955	0.75Å	2.04– 2.54
21 Jul	6470–6675	0.73Å	8.29– 9.29
1989	4490–4985	0.73Å	9.81–10.38
22 Jul	3015–4060	1.8 Å	15.95–17.01
1989	4490–4985	0.75Å	17.39–18.04

The original two-dimensional data have been optimally reduced to produce wavelength- and flux-calibrated one-dimensional spectra. The first two nights were photometric; on the third night, we had a cloudless sky but with some dust, which added a uniform and stable component to the extinction. The average spectrum of AM Her during our observation is shown in Figure 1. The continuum levels at different wavelength ranges show a reasonable (but not perfect) match, which should be taken as an indication of the spectrophotometric quality of the data.

According to recent I. A. U. Circulars, AM Herculis was probably 0.5–1.0 mag fainter than the nominal bright state during our observations, although these measurements were not exactly simultaneous with our spectroscopy. Thus, it is possible that the conclusions from this dataset may not apply to this system at all epochs. Furthermore, we will assume that there were no gross changes during the three nights of our observations; we can confirm this to some degree by comparing our Hβ–He II λ4686 region data from different nights, although this is subject to the usual limitations of narrow-slit spectrophotometry.

3 CONSTRAINTS ON THE DENSITY

Stockman *et al.* (1977) used the higher members of the Balmer series to derive a constraint of the density of the line-emitting region (note that this is before

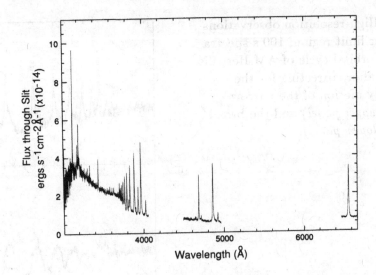

Fig. 1—Average spectrum of AM Herculis during our observations. The three regions shown are (1) around Hα, (2) Hβ–He II λ4686 region, and (3) ~ 4000 Å down to the atmospheric cut-off.

Greenstein *et al.* 1977). They observed H13 and H14 resolved in their spectrum of AM Her; for Stark broadening not to blend these lines, $n_e \lesssim 2 \times 10^{14}$ cm^{-3} in the emission region. However, we now know that there are (at least) two components. Which component did Stockman *et al.* observe?

We specifically observed the Balmer limit region at high resolution to answer this question. First, we calculated the correct phasing and amplitude of the sharp emission-line component using the He II λ4686 data. We then applied the velocity shift appropriate to each 100 s integration, then summed all the spectra. The result is shown in Figure 2 (upper panel), in which we observe H18 clearly resolved from H19. The FWHM of the narrow component is of the order of 100 km s^{-1} (after deconvolving the instrumental resolution) both in He II λ4686 and in the reasonably contamination-free Balmer lines such as H14, H16, and H18 (within uncertainties). For this amount of Doppler broadening and zero Stark broadening, we expect to be able to detect up to ~ H25 in the high signal-to-noise ratio limit.

If the observed limit of detectability is due to Stark broadening, then we obtain the number density of ~ 10^{13} cm^{-3} in the narrow-emission region (i.e., the chromosphere on the heated face of the secondary). However, the summed spectrum suggests that we may be limited by the low signal-to-noise ratio due to the weakness of the lines. If this is the case, the density could be considerably lower.

We have also calculated the broad line radial velocity variation, shifted the individual spectra and calculated the sum (lower panel, Fig. 2). We can clearly see H13 resolved from H14. However, the interpretation is not as straightforward as in

Fig. 2—High-resolution observations of the Balmer limit region; 100 s spectra taken over 1 orbital cycle of AM Herculis are summed after correcting for the radial velocity motion of the narrow component (*upper panel*) and the broad component (*lower panel*).

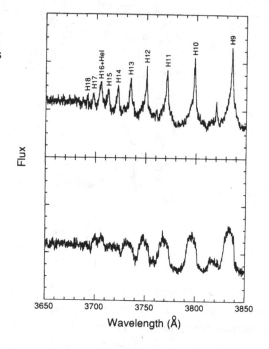

the case of the narrow emission line. This is because the Doppler width of the broad component (together with the smeared-out narrow component) is sufficient (FWHM ~ 900 km s^{-1}) to ensure that higher members of the Balmer series merge. Therefore, we are unable to put a secure lower limit on the density of the broad component region. There is some indication in the data, however, that the wings of Balmer lines start to merge around H12 (the 'continuum' between H11 and H12 appears to be higher than that expected from extrapolation from longer wavelengths); if confirmed, this would indicate a density as high as ~ 3×10^{14} cm^{-3}. In any case, the detection of H13 in Stockman *et al.*'s data cannot be used to infer the conditions in the broad line-emission region.

In the above discussion, we have assumed a simple two-component (a broad and a narrow) structure for the emission lines in AM Her. From the preliminary analysis of the line profiles, we estimate that, although the true structure can be more complicated, the bulk of the flux is contained in the two components.

4 LINE PROFILES AND RATIOS

The other important clue which Stockman *et. al.* used is the line flux ratios (Balmer decrement). It is clear from Figure 1 that the ratio of Hα to Hβ is inverted relative to the Case B calculations, for which the ratio should be 2.8 (for $T \sim$ 10,000 K). Any uncertainties in the absolute flux level are far too small to account for this inverted Balmer decrement. Stockman *et al.* interpreted this phenomenon as due to some lower members of the Balmer series being optically thick. The ratio of Hα through to Hδ in their data is consistent with a Planckian function for $T \sim 10,000$ K (although the final value adopted by Stockman *et al.* actually came from the Balmer continuum). Our superior quality data allows, in principle, a similar study of different components of the emission lines.

We present, in Figure 3, the line profiles of Hβ and H8 at three representative phases. The profile of H8 at $\phi = 0.02$ shows a clear two-component structure; the profile of Hβ is more complex, and may be interpreted as the doubling of the narrow component. At this phase, the stream is nearly pointing towards Earth and the stream near the primary is presented to the observer most favorably, causing the asymmetry seen in the broad component.

Fig. 3—Profiles of Hβ and H8 at three representative orbital phases.

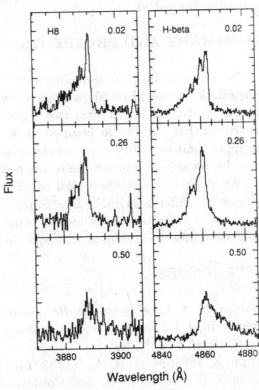

We have attempted some profile fitting, which shows that the Hα to Hβ ratio of the narrow component is less extreme than that of the broad component, but is still different from the Case B prediction. We therefore conclude that the narrow component is probably formed in an optically thick (to the lower members of Balmer series) region, but the physical condition is significantly different from that in the broad-emission line region.

We also note that the width of the broad component seems to vary with phase. In particular, at maximum redshift ($\phi \sim 0.5$), the line appears to be broadest in Hβ while it is at its narrowest in H8. We note that H8, which is supposed to be optically thin, shows some evidence of flux modulation around the orbital cycle, with the minimum at about $\phi = 0.5$. If confirmed, we may have to revise our picture of simple division between optically thick and thin lines. Perhaps the line is optically thin across the stream whereas the optical depth along the stream is significant.

As for Hβ, the maximum line width occurs at the extremes of its radial velocity curve, while the flux is maximum at quadrature. This is entirely consistent with the expectation for an optically thick line from a stream. The stream presents maximum area to the observers at quadrature while the velocity dispersion is at its minimum; at the extreme velocities, the area is at a minimum and the velocity dispersion is greatest.

5 SUMMARY AND PROSPECTS

We have obtained high-quality spectra of AM Her with the aim of deriving constraints on the physical parameters in the line-emitting regions, taking into account that there are at least two regions. At this early stage, we have already discovered that the narrow component is visible at least up to H18, which places an upper limit of $\sim 10^{13}$ cm^{-3} on the number density in the emitting region. We may also be able to improve the estimate for the broad emission-line region.

We observe line profile (radial velocity and flux) variations around the orbital cycle, which differ from line to line. Since different parts of the stream are presented favorably at different phases and velocities, it may enable us to study the density gradient within the stream as it is squeezed by the magnetic field of the primary.

REFERENCES

Cropper, M. S. 1989, *Space Sci. Rev.*, in press.

Greenstein, J. L., Sargent, W. L., Boroson, T. A., and Boksenberg, A. 1977, *Ap. J. (Letters)*, **218**, L121.

Mukai, K. 1988, *M. N. R. A. S.*, **232**, 175.

Rosen, S. R., Mason, K. O., and Córdova, F. A. 1987, *M. N. R. A. S.*, **224**, 987.

Schneider, D. P., and Young, P. 1980a, *Ap. J.*, **238**, 946.

Schneider, D. P., and Young, P. 1980b, *Ap. J.*, **240**, 871.

Stockman, H. S., Schmidt, G. D., Angel, J. R., Liebert, J., Tapia, S., and Beaver, E. 1977, *Ap. J.*, **217**, 815.

Young, P., and Schneider, D. P. 1979, *Ap. J.*, **230**, 502.

Rapid Optical Variability in AM Her Objects: QPOs and Flares

Stefan Larsson

University of Tromsø

1 INTRODUCTION

AM Her systems consist of a magnetic white dwarf accreting matter from a red dwarf companion. The strong magnetic field directs the accretion flow towards the magnetic pole(s), where a strong shock is formed. The radiation emitted from this accretion column shows variability on a large range of timescales. Inhomogeneities in the accretion flow are believed to be responsible for flickering on timescales down to seconds. In a few of the AM Her objects the optical emission has also been found to show quasi-periodic oscillations (QPOs) with periods in the range of 1–3 seconds. Such QPOs were first seen in V834 Cen (E1405-451) and AN UMa (Middleditch 1982) and then later in EF Eri and VV Pup (Larsson 1987b, 1989).

Non-simultaneous color observations of the QPOs in V834 Cen (Larsson 1985) and AN UMa (Imamura and Steiman-Cameron 1986), suggested a color similar to that of the cyclotron spectrum for these objects. This together with the eclipses in VV Pup localizes the oscillations to the accretion column. Only in one previous study were observations made simultaneously in two wavelength bands (Larsson 1987a). Although the data were noisy, they indicated that, at least for V834 Cen, the QPOs with higher oscillation frequencies were bluer than the lower-frequency ones.

Here we report preliminary results from simultaneous four channel ($UBVR$) fast photometry of AN UMa, made with the Stiening photometer at the 2.7-m telescope at McDonald Observatory.

2 A MULTIWAVELENGTH STUDY OF THE QPOS IN AN UMA

A 40-minute data sequence obtained on December 31, 1986, which showed strong QPOs, was selected for this first analysis. Power spectra were calculated and averaged along the lines described in Larsson (1989). Resulting power spectra are shown in Figure 1. The rms pulsed fraction in the U, B, V, and R bands are $3.6^{+0.3}_{-0.3}\%$, $4.0^{+0.3}_{-0.3}\%$, $2.6^{+0.6}_{-0.7}\%$, and $1.7^{+0.5}_{-0.6}\%$, respectively. The blue color is consistent with the identification of the oscillating light source with the cyclotron emission. Most striking, however, is the differences in QPO frequency distribution between the four bands. The peak of the power excess shifts gradually from 0.5 Hz in R to 0.7 Hz in U.

Fig. 1—Power spectra for AN UMa, calculated from simultaneous observations in the U, B, V, and R bands. The peak QPO frequency increases gradually from 0.5 Hz in R to 0.7 Hz in U.

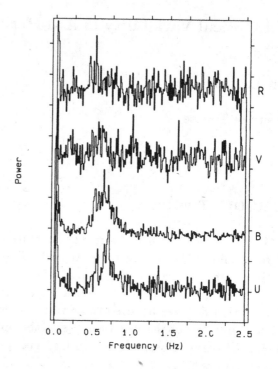

Fig. 2—The cross correlation function for the U and B band observations of AN UMa. It is found that the QPOs in the B band lag the ones in the U band by about 50 milliseconds.

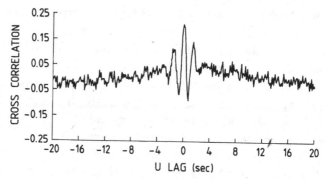

To further investigate the wavelength dependence of QPO properties, we also calculated the cross correlation between the various wavelength bands. Since the signal-to-noise ratio is highest for the U and B bands, we limit ourselves here to a discussion of the correlation between those two bands. The cross correlation function (CCF), shown in Figure 2, reveals clearly the presence of correlated oscillations with a low degree of coherence. In addition to the coherence, there are two other properties associated with the QPOs that show up in the CCF. One is an asymmetry in the damping of the oscillation in the CCF. Such an asymmetry could be produced by a systematic color variation during the 'lifetime' of an oscillation. A similar asymmetry is seen in the CCF (of soft and hard X-rays) for the QPOs in low mass X-ray binaries, in which case it is attributed to oscillations in 'softening shots' (van der Klis 1988). A second feature of the CCF is a small negative phase shift for the oscillation. From the position of the first minima on each side of the central peak, the shift is found to be about -50 milliseconds, which is just half of the 100 millisecond time resolution. The time shift implies that the oscillations in the B band efficiently lag the ones in the U band by a few percent of the oscillation period. A shift of this size could easily be produced by a difference in pulse shape for the two wavelength bands.

What, then, are the implications of these results for models of the oscillations? The fact that the higher-frequency QPOs are bluer than the lower-frequency ones has a straightforward interpretation in the oscillating shock models (e.g., Chanmugam, Langer, and Shaviv 1985). An increase in accretion flow density will shift the peak of the cyclotron spectrum to higher radiation frequency and also decrease the cooling timescale below the shock, which increases the oscillation frequency. One has to keep in mind that differences in thickness of the radiating plasma region also affect the cyclotron spectrum. However, since eclipses of accretion columns show that the emitting region is more extended in red than in blue, it seems likely that the red emission is more dominated by lower-density regions than is the case for the blue part of the cyclotron emission. A bluer color for the higher-frequency QPOs is therefore expected for shock height oscillations. For Alfvén wave resonances, on the other hand, the opposite dependence is expected, since the Alfvén velocity is proportional to the inverse square root of the density. We therefore conclude that the observed correlation supports an oscillating shock model rather than one based on Alfvén waves.

3 V834 CEN IN A FLARING STATE

The optical flickering in AM Herculis has been modeled as shot noise by Panek (1980). The shots would have a pulse duration of 70–90 seconds with many overlaping pulses at any particular time. A typical example of flickering in an AM Her object, V834 Cen, is shown at the top of Figure 3. During an observation of the same object in January 1987, the character of the flickering had changed dramatically as seen in the lower part of Figure 3. In this case the flickering is dominated by clearly-distinguishable high-amplitude flares. The object remained in this flaring state over the 5 nights of observation. Since the flaring was strong enough to be seen directly with the guiding camera, we can rule out the kind of errors (such as a second star on the diaphragm edge) that otherwise might affect the high-speed photometry.

Power spectra of the data show a steepening of the low-frequency noise component from a ν^{-1} power law in the earlier observations to ν^{-2} in the flaring state. During an observation two months later (March 1987), flares were still seen but they were now less prominent. The power spectrum slope was also in between those of the previous nonflaring and flaring states.

It is interesting to note that the phasing of the light curve minima suggests a possible correlation between the flaring activity and the phase jitter of the minimum. During the January 1987 observation, the minima were late by about 4–5 minutes,

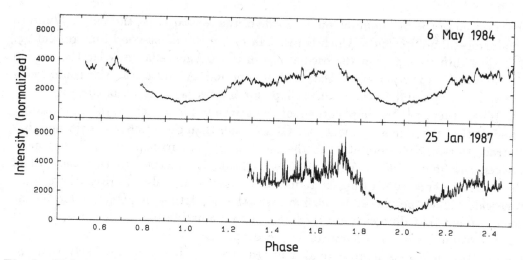

Fig. 3—Light curves for V834 Cen showing the usual flickering behaviour (top), and the flaring state seen in January 1987 (bottom). Both observations were made in white light (4000Å–9000Å), with the 1.5-m Danish telescope at ESO.

which is the largest delay in all of our data. The phase jitter is likely to be produced by variations in the penetration depth of the accretion stream. Compared to the ordinary situation, the flaring state would then correspond to accretion of fewer and brighter (in cyclotron emission) blobs that are threaded by the field further away from the white dwarf. To move the threading region outwards requires either a lower stream density (lower ram pressure) or that the threading process is more efficient.

A detailed study of the distribution in flare length and amplitude might provide information about the blob spectrum and hence about the fragmentation process.

Acknowledgement: I want to thank E. Robinson, K. Horne, and R. Stiening for making the observations of AN UMa at McDonald Observatory.

REFERENCES

Chanmugam, G., Langer, S. H., and Shaviv, G. 1985, *Ap. J. (Letters)*, **299**, L87.
Imamura, J. N., and Steiman-Cameron, T. 1986, *Ap. J.*, **311**, 786.
Larsson, S. 1985, *Astr. Ap.*, **145**, L1.
Larsson, S. 1987a, *Ap. Sp. Sci.*, **130**, 187.
Larsson, S. 1987b, *Astr. Ap.*, **181**, L15.
Larsson, S. 1989, *Astr. Ap.*, **217**, 146.
Middleditch, J. 1982, *Ap. J. (Letters)*, **257**, L71.
Panek, R. J. 1980, *Ap. J.*, **241**, 1077.
van der Klis, M. 1988, *Adv. Space Res.*, **8**, #2, 383.

Absorption Dips in EF Eri and the Properties of the Accretion Stream

M. G. Watson[1], R. F. Jameson[2], A. R. King[2], and K. O. Mason[3]

[1]X-ray Astronomy Group, Dept. of Physics and Astronomy,
 University of Leicester
[2]Astronomy Group, Dept. of Physics and Astronomy, University of Leicester
[3]Mullard Space Science Laboratory, University College London

ABSTRACT

We present preliminary results from a simultaneous X-ray and infrared (IR) study of the absorption dips in the AM Her system EF Eri. We discuss the structure seen in the dips and the implications for the properties of the accretion stream. We use the X-ray and IR measurements to place strong constraints on the density and radius of the stream, and hence on the accretion rate onto the white dwarf.

1 INTRODUCTION

Narrow absorption dips occurring at a fixed orbital phase are a common feature of the soft X-ray light curves of AM Her systems (e.g., Mason 1985). In some cases the dips also appear at near IR or optical wavelengths. These dips are believed to result from occultations of the emission region on the white dwarf by parts of the accretion stream from the secondary at distances $\sim 10^{10}$ cm from the white dwarf (King and Williams 1985). Such dips are inevitable in systems which have their emitting pole in the same hemisphere as the observer, providing the colatitude of the pole is less than the system inclination. Photoelectric absorption in the stream causes the X-ray dips, as is demonstrated by their dependence on X-ray energy, while free-free absorption is the likely cause of the near IR and optical dips. Since the photoelectric and free-free optical depths for a given path length through a stream of density N vary as N and N^2, respectively, a comparison of the dip profiles in the X-ray and IR can be used to measure directly the density profile across the stream, and hence to determine other parameters of the stream (Watson et al. 1989).

Amongst AM Her systems, EF Eri has the most prominent dips in both its X-ray and IR light curves (e.g., Watson et al. 1989). Here we report preliminary results from a detailed study of EF Eri made with X-ray coverage from *GINGA* together with J-, K-, and V-band photometry from UKIRT which extends our earlier work on this object.

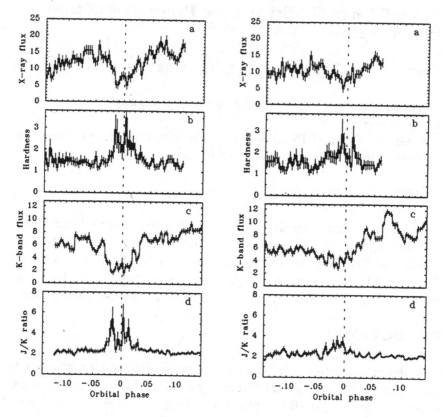

Fig. 1—Detailed X-ray and IR light curves for two dips (*left:* dip 17, *right:* dip 18) plotted as a function of orbital phase (phase zero is the nominal dip center). (*a*) X-ray light curve (2–3.5 keV); (*b*) X-ray hardness ratio = (3.5–6 keV flux)/(2–3.5 keV flux); (*c*) *K*-band IR light curve; and (*d*) *J/K* IR flux ratio.

2 OBSERVATIONS

The observations of EF Eri discussed here were made with the *GINGA* Large Area Counters (LAC: ∼ 2–50 keV, Turner *et al.* 1989) and with UKIRT between 1988 November 30 and December 2. The X-ray observations span 36 cycles of the 81-minute orbital period of EF Eri, and provide coverage of 8 absorption dips. The simultaneous IR (*J*- and *K*-band) and optical (*V*-band) observations of EF Eri were obtained using the "two-banger" and VISPHOT photometers on UKIRT, giving coverage of 6 complete orbital cycles split between two nights of observing. The *V*-band data will be presented in a later publication. Completely simultaneous X-ray and IR observations were obtained for two consecutive dips on Dec 2.

3 RESULTS

3.1 Overall Dip Properties

Absorption dips are evident in each orbital cycle for which we have X-ray and/or IR coverage. The dips show considerable variability from cycle to cycle and considerable structure within each dip, as can be seen in Figure 1 which shows the detailed X-ray and IR light curves for 2 dips. The residual flux at the bottom of the dip (expressed as a fraction of the non-dip flux, i.e., $I_{\rm dip}/I_0$) varies between 0.15 and 0.45 in the lowest X-ray energy band (2–3.5 keV), between 0.3 and 0.7 in the J band, and between 0.15 and 0.5 in the K band. The range of effective optical depths $\tau = -\ln(I_{\rm dip}/I_0)$ implied is ~ 0.5–2, although for the IR measurements these values increase when dilution by the secondary, etc. is taken into account. The full phase-width is $\Delta\phi \approx 0.05$ in the IR and $\Delta\phi \approx 0.05$–0.07 in the X-ray data ($\Delta\phi$ in orbital cycles). The average dip parameters obtained here are comparable to those determined in our earlier study of EF Eri (Watson *et al.* 1989), but the variability observed is greater, as we would expect, given the increased number of dips covered.

3.2 Dip Structure

None of the dips observed shows anything like a smooth profile in either the X-ray or IR data (e.g., Fig. 1). The typical structure within each dip consists of flare-like features lasting 50–100 seconds (0.01–0.02 cycles) in which the flux varies by as much as a factor of two from a smooth profile (a smooth profile is expected for occultation by any stream which has a symmetrical density profile). This dip structure is in fact very similar to the flickering that characterizes other parts of the X-ray and IR light curves, but can be distinguished from the general flickering by the behaviour of both the IR colors and the X-ray hardness ratio. Both indicators are strongly correlated with the dip structure, but show a much weaker correlation with the general flickering outside the dips (see Fig. 1). This implies that the dip structure is caused by fluctuations in the absorption, whereas the other flickering corresponds to genuine changes in the accretion-driven luminosity.

Fluctuations in the absorption can be produced by changes in either the stream density or the stream thickness, but changing the density will have a more dramatic effect on the IR flux (since the IR optical depth scales as N^2). The larger absorption variations apparent in the IR strongly suggest that the dip structure is primarily caused by density fluctuations. Assuming that this is the case, the apparent density fluctuations could correspond to: (*a*) true density structure across the stream; (*b*) temporal fluctuations in the average stream density. This ambiguity arises because the line of sight crosses the stream with an effective velocity which is very much smaller than the flow velocity along the stream. If (*b*) were true this would immediately imply fluctuations in the instantaneous $\dot{M}$ in the stream which should

Fig. 2—Spectral fitting results for
the *GINGA* observations of dip 17. (*a*)
X-ray light curve (2–3.5 keV, cf. Fig. 1);
(*b*) normalization of the X-ray spectrum;
(*c*) fitted column density N_H in units of
10^{21} cm^{-2}.

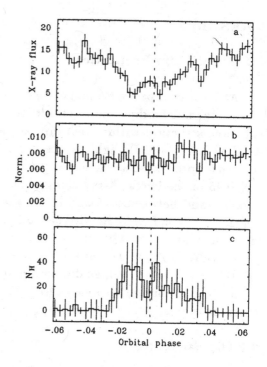

manifest themselves in changes in the emitted X-ray and IR luminosity (by up to a
factor 2) occurring with a time delay $\Delta t \sim 50$ seconds, corresponding to the free-fall
time from the region where the stream crosses the line of sight. Such variations are
not apparent in the higher energy X-ray data within the dips (where absorption
is unimportant). This points to explanation (*a*) being correct, although (*b*) might
still be valid if the $\dot{M}$ variations resulted in a smeared-out luminosity changes.

3.3 Dip Column Densities from X-ray Data

We have used the *GINGA* X-ray spectra to measure the absorbing column
density within the dips. As an example, Figure 2 shows the results obtained from
fitting simple model spectra (a $kT = 18$ keV bremsstrahlung plus photoelectric
absorption) to the time-resolved *GINGA* data for dip 17 (see Fig. 1). The results
show that the spectral normalization is essentially constant, whilst the column
density shows considerable variations. This is consistent with the dips being due
entirely to photoelectric absorption. The measured column density varies from
$N_H < 5 \times 10^{21}$ cm^{-2} to $N_H \approx 4 \times 10^{22}$ cm^{-2} across the dip. Analysis of other dips
gives column densities in the same range.

3.4 Stream Properties from Modelling the X-ray and IR Data

In our earlier study of EF Eri (Watson *et al.* 1989), we found that the profile

Fig. 3—X-ray flux (2–3.5 keV) versus IR K-band flux for dip 17. The curves shown are for the following values of N_0 (in units of 10^{13} cm^{-3}) and $y_{\max}$ (in units of 10^9 cm): (a) (0.3, 13) (*lowest curve*); (b) (0.5, 8); (c) (0.7, 6); (d) (1, 4) (*highest curve*).

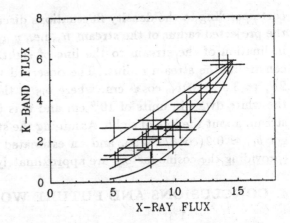

of dips observed simultaneously in X-rays and in the K band could be adequately modelled with a cylindrical stream which had a dense core and rarer envelope. Limits could be placed on the stream density, but uncertainties in the amount of dilution (from the secondary, etc.) and the nature of the absorption in the IR (i.e., whether it was in fact free-free absorption as has been commonly assumed), prevented more definite estimates of the stream parameters. We now have better quality observations in several IR/optical bands, and coverage of many more dips, and should be able to resolve these uncertainties with the present data. Such a detailed analysis is in progress. Here we present simple modelling of the data from one dip which indicates the power of this technique.

The correlation between the X-ray and IR (K-band) fluxes for dip 17 is shown in Figure 3. Points on this graph should be given by the following equations (see Watson *et al.* 1989 for further details):

$$I_X = I_{X,0} \exp(-\sigma N y) \tag{1}$$

and

$$I_K = I_{K,0} \exp(-\kappa N^2 y) \tag{2}$$

where $I_{X,0}$ and $I_{K,0}$ are the X-ray and K-band fluxes outside the dips, σ is the effective photoelectric cross-section for the X-ray energy range selected, κ is the effective free-free absorption coefficient for the K band, and N and y are the density and path length through the stream at any particular instant. (Note that the points in Figure 3 have been corrected for the likely IR dilution by the secondary, etc. — but see Section 4).

The locus of points in Figure 3 thus constrains how N and y vary through the dip. The curves shown correspond to simple models with different constant stream densities N_0 and maximum path lengths $y_{\max}$ through the stream, constrained to give the observed maximum column density (Section 3.3), i.e., $N_{H,\max} = N_0 y_{\max} \lesssim 4 \times 10^{22}$ cm^{-2}. The best fit is achieved with $y_{\max} \approx 6 \times 10^8$ cm and density $N_0 \approx 7 \times 10^{13}$ cm^{-3}, although there is of course considerable scatter about these curves corresponding to the density fluctuations discussed above. The value of y_{max} gives the projected radius of the stream R_s, i.e., $y_{max} = 2R_s \sec\theta$, where θ is the effective inclination of the stream to the line of sight. The phase-width of the dips also constrains the stream radius. The observed value $\Delta\phi \approx 0.05$ (Section 3.1) gives $2R_s \approx 3.5 \times 10^9 b_{10} \cos\psi$ cm, where b_{10} is the distance between the stream and the white dwarf in units of 10^{10} cm and ψ is the effective angle of rotation of the stream about the line of sight. Assuming the stream is roughly cylindrical, we thus get $b_{10} \approx 0.2\,(\cos\theta/\cos\psi)$, and an estimated accretion rate $\dot{M} \approx 3 \times 10^{16}$ g s^{-1} (providing the cosine factors are approximately equal).

4. CONCLUSIONS AND FUTURE WORK

Our X-ray/IR study of the dips in EF Eri demonstrates the power of this technique for making direct estimates of the basic properties of the accretion stream which are only poorly determined by other methods. The presence of significant structure in the dip profiles, hinted at in earlier observations, is amply confirmed in these new data. This structure points to the existence of density fluctuations *across* the stream, although we cannot presently rule out the possibility that these are caused by density (i.e., $\dot{M}$) variations *along* the stream.

Preliminary analysis of the two-color IR data in the dips indicates two unanticipated results. First, the IR dilution at the bottom of the dips (i.e., the contribution to the total IR flux which is not absorbed in the dips) seems to vary by a factor of 2 between nights. This would suggest that there is a significant accretion-powered component to the diluting flux since such a variation in the secondary seem unlikely. Second, the ratio between the optical depths in the J and K bands appears to be inconsistent with the value expected for free-free absorption, pointing to other forms of opacity in the stream. Further work on these problems is in progress.

REFERENCES

King, A. R., and Williams, G. 1985, *M. N. R. A. S.*, **215**, 1p.
Mason, K. O. 1985, *Space Sci. Rev.*, **40**, 99.
Turner, M. J. L. T., *et al.* 1989, *Pub. Astr. Soc. Japan*, **41**, 345.
Watson, M. G., King, A. R., Jones, M. H., and Motch, C. 1989, *M. N. R. A. S.*, **237**, 299.

H0538+608: An X-ray and Optical Study of an Unusual AM Her-Type System

A. Silber[1], R. Remillard[1], H. Bradt[1], M. Ishida[2], and T. Ohashi[2]

[1]Massachusetts Institute of Technology
[2]University of Tokyo

1 INTRODUCTION

H0538+608 is an AM Her-type cataclysmic variable that has shown unusual behavior since its identification (Remillard *et al.* 1986). In the discovery paper, H0538+608 was shown to have circularly-polarized light, hard X-ray emission, and an optical light curve that varied with a period of about 3.1 hours. The optical polarization data displayed evidence of chaos, as opposed to the highly-repeatable variation in other AM Her systems. Further questions came from the discovery of an anomalous *IUE* spectrum that could be explained by an excess of nitrogen at the expense of carbon, relative to solar abundances (Bonnet-Bidaud and Mouchet 1987). When H0538+608 was observed with *EXOSAT* (Shrader *et al.* 1988), aperiodic changes were seen in the light curve. Eclipses of the low-energy X-rays were detected with a period of 3.30 ± 0.02 hours. These eclipses can be explained by an occultation of the hard X-ray emission region by the accretion column, such as those seen in EF Eri (Patterson, Williams, and Hiltner 1981). It was also found that H0538+608 lacked the low-energy blackbody component that is found in all other AM Her systems.

2 OBSERVATIONS AND ANALYSIS

A program of concurrent X-ray and optical observations was carried out during 1988 Feb 8–10. H0538+608 was observed with the Japanese X-ray observatory *GINGA* (Makino and the *Ginga* team 1987) for 3 days. During this time, relative CCD photometry with a time resolution of 30 s and spectroscopy (5750–7000 Å) with a resolution of 6 Å and an integration time of 5 min were obtained at the Michigan-Darmouth-M.I.T. Observatory.

Figure 1 shows concurrent optical (4000–6000 Å) relative CCD photometry and X-ray (1.2–11.6 keV) light curves. The optical light curve shows aperiodic flickering superposed on an orbital variation, which changes each night. On Feb 8 and 9 the folded light curve (Fig. 2) is double-peaked with the separation in phase between the peaks being 0.3 and 0.4, respectively, based on a period of 3.33 hours. The shape of the optical light curve evolves significantly on a timescale of 1 day. On Feb 10, the light curve is single-peaked with an average luminosity of twice that of the previous days.

Variation is seen in the X-ray light curve over a wide variety of timescales. However, when the X-ray light curve is folded over the nominal period of 3.33 hours, the only persistent feature is a drop in intensity of 50% at (arbitrary) phase 0.55 on Feb 8 and 9 (Fig. 2). This dip occurs at a phase when the optical emission is high, ruling out an eclipse by the secondary. The most straightforward explanation is that the X-ray emission region is occulted by the white dwarf. On Feb 10, when the optical and X-ray intensities have significantly increased, this X-ray dip is no longer seen. The emission region must have changed its geometry such that it no longer passed over the limb of the white dwarf.

Power density spectra (PDS) were calculated for each night of the optical data (Fig. 2). On Feb 9 and 10, excess power was detected at 31.5 and 36.2 min, respectively. No similar peak was seen in the data for Feb 8. The pulsation appears to be coherent on a timescale of 8 hours. There is no corresponding peak in the X-ray PDS. In a search for short-period (2–20 min) quasi-periodic oscillations (QPOs), the PDS from 51-min segments were summed. No QPOs of this type were detected. Though noncoherent pulsations have been seen in AM Her systems (e.g., EF Eri at 6 min; Patterson, Williams, and Hiltner 1981), none have shown periods as long as 30 min. For short-term variability, cross correlation analysis of our optical and X-ray data show a high degree of correlation with a timelag of 0 ± 24 s.

The X-ray spectrum is well fit to a thermal bremsstrahlung spectrum with an Fe line at 6.6 keV absorbed by a column of cool gas. The temperature of the bremsstrahlung region was 30 keV from Feb 7, 18:00 until Feb 9, 12:00 (UT). After Feb 9, 12:00, when the X-ray self eclipse disappeared, the temperature dropped to 25 keV.

TABLE 1
PERIODS OF H0538+608

X-ray light curve (1.8–18.6 keV)	3.333 ± 0.027 hr
Optical Photometry	3.30 ± 0.03
Low-Energy X-ray Eclipse (Shrader *et al.* 1988)	3.30 ± 0.03
Optical Polarization (Mason, Liebert, and Schmidt 1989)	3.331 ± 0.015
Optical Spectral lines (Hα):	
Velocity of the narrow comp.	3.37 ± 0.03
Velocity of the broad comp.	3.34 ± 0.02
Equiv. width of narrow comp.	3.32 ± 0.02
Equiv. width of broad comp.	3.30 ± 0.02

One explanation for the rapidly-evolving light curve is asynchronous rotation. We searched for periods in the optical and X-ray light curves and in the optical

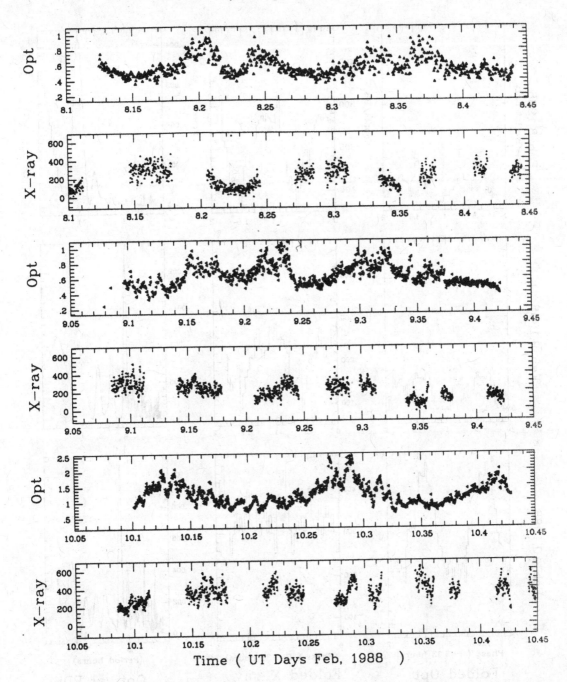

Fig. 1—Concurrent optical and X-ray light curves of H0538+608. The units of the y axis are relative intensity (optical) and counts per 16 s (X-ray).

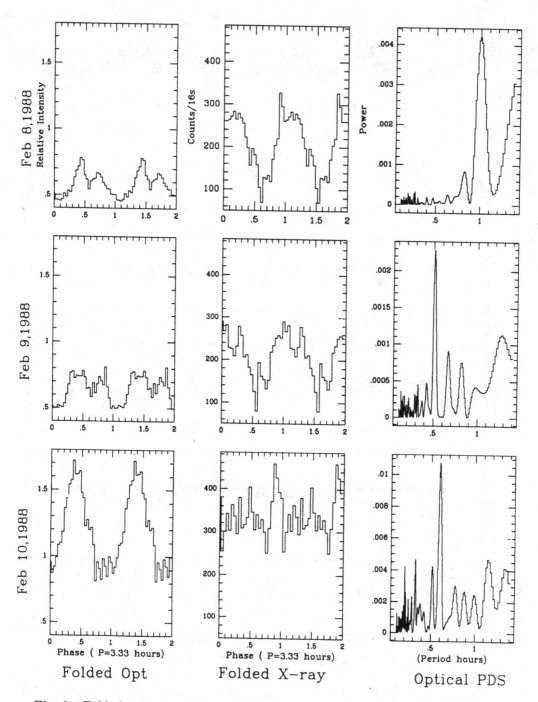

Fig. 2—Folded optical and X-ray light curves and optical power density spectra of H0538+608.

spectroscopy. The spectral search was carried out by separating the Hα line into narrow and broad components. It is believed that the narrow component comes from the surface of the companion and hence would track the orbital motion. The broad component light curves and polarization would track the white dwarf spin period. However, all searches for a second period have been unsuccessful with all detected periods consistent with 3.33 hours (Table 1). Nonetheless, asynchronous rotation cannot be ruled out.

3 CONCLUSION

This study confirms the uniqueness of H0538+608 among the AM Her systems. The possibility of a QPO at about 30 min and clearly-evolving optical and X-ray light curves are new features to be added to the list that makes H0538+608 an unusual system. Even after the intensive study that has been undertaken by us and others, it cannot be said with confidence whether or not this system is rotating synchronously. Dramatic changes in the accretion rate may explain the considerable changes on the third night of observations. Why the accretion geometry changes with time remains unknown. Further studies based on these data and new observations with a longer timespan are clearly needed.

REFERENCES

Bonnet-Bidaud, J. M., and Mouchet, M. 1987, *Astr. Ap.*, **188**, 89.

Makino, M., and the *Ginga* team. 1987, *Ap. Letters Comm.*, **25**, 223.

Mason, P. A., Liebert, J., and Schmidt, G. D. 1989, *Ap. J.*, **346**, 941.

Patterson, J., Williams, G., and Hiltner, W. A. 1981, *Ap. J.*, **245**, 618.

Remillard, R. A., Bradt, H., McClintock, J. E. , Patterson, J., Roberts, W., Schwartz, D. A., and Tapia, S. 1986, *Ap. J. (Letters)*, **302**, L11.

Shrader, C. R., Remillard, R., Silber, A., McClintock, J. E., and Lamb, D. Q. 1988, in *A Decade of UV Astronomy with the IUE Satellite*, ESA SP-281, Vol. 1, p. 137.

Re-Locking of an AM Herculis Binary

Gary D. Schmidt

Steward Observatory, University of Arizona

1 INTRODUCTION

An AM Herculis binary differs from its non-magnetic counterpart in two important qualitative respects: *a*) a strong magnetic field on the primary star funnels the accreting gas onto generally one or two localized accretion shocks near the white dwarf's magnetic pole(s), where X-ray bremsstrahlung and strongly polarized optical/infrared cyclotron emission arise, and *b*) the white dwarf spin and binary orbital motions are locked in a rigid corotating geometry.

The reality of these two properties is well-known, but the reason for their intimate coexistence is not. For example, in summarizing the physics of magnetic CVs, Lamb and Melia (1988; see also Lamb 1988) point out that asynchronous magnetic systems (DQ Her binaries) ought to be found in both disk-accreting and column-accreting geometries, depending on the size of the Alvfén radius relative to the circularization radius of the incident stream. The two situations have important distinctions: In a disk-accreting geometry, the azimuthally-symmetric accretion disk would feed a variety of magnetic field lines, producing large accretion caps or arcs on the stellar surface. The resulting emission would be expected to vary smoothly through the rotational cycle, and the net polarization would be small due to both geometric cancellation and dilution by emission from the disk. With dilution largely absent in the case of stream-accretion, polarization and light curves should be more dramatic and exhibit qualitative changes through the spin/orbit lap cycle as the columns rearrange to accommodate the continuously changing magnetic geometry.

Curiously, most, and perhaps all, of the known DQ Her systems appear to possess disks, a conclusion consistent with their weak or nonexistent circular polarization (e.g., Berriman 1988), while the strongly-polarized AM Hers are all synchronously rotating. To date, the only CV found to have substantial optical polarization modulated on a period *different* than its orbital cycle is the recent nova V1500 Cyg. Its photometric record and current near-equality of periods imply that it, too, was a locked AM Her binary prior to outburst in 1975, and that synchronism was broken as a result of angular momentum transfer between the stellar components during the common-envelope phase of the nova (Stockman, Schmidt, and Lamb 1988; hereafter Paper 1).

As the sole example of a near-synchronous magnetic system, V1500 Cyg is of unique importance for studying the locking processes which operate in close magnetic binaries. In contrast to the DQ Her systems, rotation of the white dwarf in V1500 Cyg is available *directly* through its polarization curves. Moreover, the inference of synchronization prior to 1975 implies that it will relock before its next nova outburst, suggesting a synchronization timescale sufficiently short ($< 10^{3-4}$ yrs) that it might be directly measurable. This paper describes a detection of spin-down in V1500 Cyg and implications of the rate of period change for the likelihood of identifying other AM Her systems in the process of formation.

2 THE ROTATIONAL EPHEMERIS OF V1500 CYG

V1500 Cyg is the host binary of the spectacular Nova Cyg 1975, an outburst of such magnitude and brilliance that it ranks as the best-studied nova event in the history of astronomy. Although the system failed to exhibit evidence of magnetism in polarimetry obtained in 1978, today—more than 3 mag fainter—cyclotron emission is manifested as a net circular polarization of $V \approx 1.5\%$, modulated on a period 3.5 min (1.8%) *shorter* than the 3 hr 21 min orbital cycle (Paper 1).

We have continued a program of systematic monitoring of V1500 Cyg since our discovery of circular polarization more than $2\frac{1}{2}$ years ago. Due to its faintness ($V \approx$ 17–18.5) and modest polarization (cf. a typical AM Her), data were acquired in "white" light ($\lambda\lambda \sim 3200$–8600 Å) using the 2.3 m telescope of Steward Observatory. As in Paper 1, circularly polarized *flux* was chosen for analysis in order to avoid the aliasing effects of variable dilution by the heated inner face of the orbiting companion star. In all, nine epochs of polarimetry are now available, spaced at intervals of 2–6 months, and on each occasion strongly-modulated polarization is apparent. All epochs include a polarization zero-crossing, and all but two span at least one rotational cycle (four epochs include more than one night).

The uncertainty for each datum of polarized flux was calculated from photon statistics (for the degree of polarization) plus an allowance for errors in the photometry. A sine wave was fit by least-squares to the data string at each epoch to best estimate the time of positive zero-crossing of polarized flux and its uncertainty. Assuming a constant period, these timings can be expressed as a best-fit ephemeris for the rotation of the magnetic primary:

$$HJD = 2446875.971 + 0.137164E.$$
$$\pm \qquad 0.004 \pm 0.000001$$

The O−C diagram for this fit is shown in Figure 1 as the distribution of points about the dashed line. It is obvious that a simple linear ephemeris is a rather poor fit to the timings. Specifically, as the observational baseline has lengthened, both the derived mean period as well as the reduced χ_ν^2 index have increased. By Sept 1989 the latter stood at a value of 6.7 ($\nu = 7$), corresponding to a probability for random occurrence of less than 10^{-7}. A quadratic relation, on the other hand, provides the excellent fit indicated by the dot-dash line in Figure 1. The χ_ν^2 index for this is 0.80 and the F-test assigns a probability for significance of a second-order term at 0.9996. This fit is described by the equation

$$HJD = 2446876.008 + 0.137142E + 3.0 \times 10^{-9}E^2 ,$$
$$\pm \qquad 0.007 \pm 0.000004 \quad \pm 0.5$$

corresponding to a rate of change of period $\dot{P}_{rot} = +4.4\,(\pm0.7) \times 10^{-8}$: the rotation period is gradually lengthening. The timescale for synchronization with the orbital cycle is thus $\tau_s = (P_{orb} - P_{rot})/\dot{P}_{rot} = 150 \pm 25$ yrs.

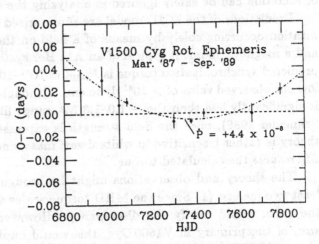

Fig. 1—Residuals of the timings of positive zero-crossing of polarized flux relative to the best-fit linear ephemeris. The parabolic curve, which represents a quadratic fit to the data, corresponds to a spin-down rate $\dot{P}_{rot} = +4.4 \times 10^{-8}$.

3 THE SYNCHRONIZATION TORQUE AND LOCKING OF MAGNETIC CVS

All theories for synchronizing the primaries in AM Her systems involve the stressing of magnetic field lines which connect the stellar components. As Lamb and Melia (1988; hereafter LM) recently summarized, proposed ideas tend to divide into two classes according to the dissipation mechanism involved: (1) a magneto-static interaction, in which the time-varying field experienced by the companion is dissipated by ohmic or turbulent processes within the star; or (2) the driving of field-aligned currents between the two stars and the ensuing dissipation of energy through MHD instabilities in the connecting region. Formulation of the latter mechanism is sufficiently explicit in LM (see also Lamb et al. 1983) to permit direct comparison with our observed spin-down in V1500 Cyg, and it is used here by example.

The mere presence of a polarized cyclotron component ensures that the white dwarf is accreting. Thus, $\dot{P}_{rot}$ measures the difference between the braking action of the synchronization torque, N_{sync}, and the spin-up effect of accretion, N_{acc}:

$$N_{obs} = N_{sync} - N_{acc} = I\dot{\omega}_{rot} = \frac{-2\pi I \dot{P}_{rot}}{P_{rot}^2} .$$

With a moment of inertia $I \approx 10^{50}$ g cm^2, $N_{obs} \approx 2 \times 10^{35}$ dyne cm, and an error bar of 30–40% covers the uncertainties in both $\dot{P}_{rot}$ and I.

For a slowly-rotating, strongly-magnetic system, the accretion torque is roughly $N_{acc} \approx \dot{M} \, l_{R1}$, where $\dot{M}$ is the mass accretion rate and l_{R1} the angular momentum per unit mass at the L1 point. For a "typical" CV at this orbital period with $\dot{M} \approx 3 \times 10^{-10}$ M$_\odot$ yr^{-1} (Patterson 1984), the accretion torque, $N_{acc} \approx 3 \times 10^{34}$ dyne cm, is nearly an order of magnitude weaker than the value observed. In fact, V1500 Cyg appears to accrete at a rate significantly below the norm both now as well as prior to outburst (Patterson 1984; Paper 1). We conclude that the effects of accretion can be safely ignored in analyzing the spin-down of this system.

Predictions of the MHD model are summarized in Figure 10 of LM for synchronization occurring solely by means of a field on the primary. For $P_{orb} = 3.35$ hrs and a magnetic moment typical of an AM Her system ($\mu_1 \approx 10^{33}$–10^{34} G cm^3), the predicted synchronization torque is $N_{MHD} \approx 10^{32}$–10^{34} dyne cm, which falls well below our observed value of $> 10^{35}$ dyne cm. Although LM's assumed 0.7 M$_\odot$ primary is significantly less than the ~ 1.0–1.3 M$_\odot$ range likely for V1500 Cyg (Horne and Schneider 1988), once the field strength is expressed as a magnetic moment, the theory is rather insensitive to white dwarf mass, and in any case a larger value for M_1 *reduces* the calculated torque.

The theory and observations might be brought into better agreement by at least two means: (1) Since the MHD torque scales as μ_1^2, a moment which is $\sim 5\times$ the norm would rectify the discrepancy. However, given the large mass (small size) of the primary in V1500 Cyg, this would imply a surface field strength well above 100 MG, making this system much more strongly magnetic than any of the known synchronized systems. Spectropolarimetry by Schmidt and Stockman (1989) suggests that the shape of the cyclotron continuum in V1500 Cyg is actually typical of AM Hers, so it likely has a similar surface field strength. (2) The existence of a magnetic companion would have the same effect. In this case, $N_{MHD} \propto \mu_1\mu_2$, and $\mu_2 \gtrsim 2 \times 10^{34}$ G cm^3 is required to synchronize. The implied (dipolar) surface field is nearly 1 kG. At the same time, we note that the MHD theory as developed thus far fails by a factor of ~ 2 in providing torque sufficient to synchronize *any* system longward of the period gap. Thus, as with all of the mechanisms proposed to date, specific predictions must be regarded with considerable uncertainty (LM).

The dissipated rotational energy amounts to a luminosity $L_{rot} = I\omega_{rot}\dot{\omega}_{rot} \approx 10^{32}$ erg s^{-1}. If, as ohmic dissipation schemes require, it is expended in heating the secondary, the input is comparable to the normal luminosity of a 0.3 M$_\odot$ main-sequence star. The process therefore provides an alternative to irradiation by the nova remnant (Patterson 1984; Paper 1) for the hot ($\sim 10,000$ K) inner hemisphere of the companion of V1500 Cyg. The mechanism of Lamb *et al.* (1983) implies that the energy is dissipated in MHD turbulence within the magnetosphere. Here, as they propose, effects include particle acceleration. Such is the explanation Book-binder and Lamb (1987) have proposed to account for their detection of radio emission from the asynchronous (DQ Her-type) system AE Aqr. We point out here that the dissipated rotational energy of V1500 Cyg is nearly a *million* times larger than the radio luminosity of AE Aqr. In the context of the MHD picture, the difference is explicable by V1500 Cyg's presumably larger magnetic moment, its smaller binary separation, and the inefficiency of acceleration processes. Obviously, a deep radio image of V1500 Cyg might prove very interesting.

Synchronization of a DQ Her into an AM Her system proceeds by a process identical with that currently occurring in V1500 Cyg. In the former case, the white dwarf must spin down from a (much faster) rate determined by details of magnetic coupling into the accreting gas. This is an extremely difficult problem which involves the presence or absence of a disk. Again, we adopt the arguments of LM, who suggest a ratio of spin to binary frequency $2 < \omega_{rot}/\omega_{orb} < 50$ *at synchronization*. Scaling N_{obs} with binary period according to the MHD prescription, we find that the synchronization torque ought to dominate the accretion torque near $P_{orb} = 4.5$ hrs and locking should require 2×10^5 to 5×10^6 yrs. Assuming a lifetime of, say, 10^9 yrs as a strongly-polarized system, the implied probability of detecting an AM Her *in the process of synchronizing for the first time* is roughly 0.0002–0.005. It is therefore no surprise that, among the 16 polarized systems identified to date, only the recent nova V1500 Cyg exists in this intriguing state.

The author is happy to acknowledge a continuing collaboration with Pete Stockman and Don Lamb on this and other aspects of the physics of magnetic accretion binaries. Financial support was provided by NSF grant AST 86-19296.

REFERENCES

Berriman, G. 1988, in *Polarized Radiation of Circumstellar Origin*, ed. G. V. Coyne, *et al.* (Vatican City State: Vatican Observatory), p. 281.

Bookbinder, J. A., and Lamb, D. Q. 1987, *Ap. J. (Letters)*, **323**, L131.

Horne, K., and Schneider, D. P. 1989, *Ap. J.*, **343**, 888.

Lamb, D. Q. 1988, in *Polarized Radiation of Circumstellar Origin*, ed. G. V. Coyne, *et al.* (Vatican City State: Vatican Observatory), p. 151.

Lamb, D. Q., and Melia, F. 1988, in *Polarized Radiation of Circumstellar Origin*, ed. G. V. Coyne, *et al.* (Vatican City State: Vatican Observatory), p. 45 (LM).

Lamb, F. K., Aly, J.-J., Cook, M., and Lamb, D. Q. 1983, *Ap. J. (Letters)*, **274**, L71.

Patterson, J. 1984, *Ap. J. Suppl.*, **54**, 443.

Schmidt, G. D., and Stockman, H. S. 1989, in preparation.

Stockman, H. S., Schmidt, G. D., and Lamb, D. Q. 1988, *Ap. J.*, **332**, 282 (Paper 1).

The Structure of the Optically Thick Boundary Layer in Cataclysmic Binaries

Wilhelm Kley

Institut für Astronomie und Astrophysik der Universität München,
Scheinerstr. 1, D-8000 München 80, F. R. G.

ABSTRACT

Two-dimensional numerical calculations solving the radiation hydrodynamics equations to investigate the structure of the boundary layer (BL) of accretion disks in binary systems have been performed. Different values of the mass and the rotation rate of the white dwarf (WD) and the viscosity coefficient have been considered. The results are briefly described.

1 INTRODUCTION

The boundary layer (BL) is the innermost part of the accretion disk, where the disk grazes the central star, a white dwarf (WD) in cataclysmic variables (CV). There frictional interaction decelerates the disk material to the stellar rotational velocity. Knowing the detailed structure of this BL is of great importance for accretion disk physics because, depending on the rotation of the central star, up to one half of the total available accretion energy can be released here. The emitting area is much smaller than the surface of the accretion disk so that the dissipated energy will be radiated away predominantly in the high energy part of the spectrum. Simple models of the BL (Pringle 1977; Pringle and Savonije 1979; Tylenda 1981) lead to the expectation of an optically thin BL emitting hard X-rays in the case of low mass flow $\dot{M}$ through the disk (the quiescent state) and an optically thick BL emitting soft X-rays in the case of high mass flow through the disk (outburst state). In fact, most of the CVs show hard X-ray emission during the quiescence and outburst state. Some systems (with $\dot{M}$, high M_{WD}, small distance) were found to emit soft X-rays during outburst. The spectra of those systems (U Gem and SS Cyg) confirm the view of an optically thick soft X-ray-emitting BL. Furthermore, the observed P Cygni line profiles of many CVs during outbursts point to a fast mass loss from these systems. The question of heating the WD and mixing of disk matter with the upper WD layers is very important for the theory of nova eruptions. All of these topics, the soft X-ray and FUV spectral distribution, the origin of the wind, the flow structure, and the mixing/heating processes with the WD are intimately connected to the structure of the optically thick BL. The investigations so far have basic shortcomings, such as a one-dimensional treatment or, in the case of two-dimensional calculations, an inadequate treatment of energy transport through radiation. For further details and the references see the review article by Shaviv (1987).

2 THE MODEL

Accordingly, we have developed a numerical algorithm which can treat time-dependent hydrodynamic flows simultaneously together with radiation processes. It is a mixed explicit/implicit method with the radiation transport treated in the flux limited diffusion approximation. The details of the method and the test calculations are described by Kley (1989a). The fixed model parameters used in the calculations are $T_{\text{eff}} = 1.5 \times 10^4$ K, $\dot{M} = 3 \times 10^{-8}$ $M_\odot$ yr^{-1}, which are typical (outburst) values in CVs. The other parameters used are

Name	Grid Size	M_{WD} ($M_\odot$)	ν (10^{15} cm^2 s^{-1})	Ω^*/Ω_K^*
LDTR3	61×61	1.0	1.0	0.0
LDTR9	61×61	1.0	1.0	0.333
crl2	85×85	1.0	1.0	0.0
crl3	85×85	1.0	4.0	0.0
crl5	85×85	0.6	0.0	0.0

Here Ω^* denotes the stellar rotation rate and Ω_K^* refers to the Keplerian value, i.e., in the second model the star rotates with one third of its breakup velocity. As initial condition in the radial direction we have taken a stellar atmosphere in hydrostatic equilibrium around the WD. Near to the equatorial plane we imposed a vertical equilibrium disk model. The calculations are done in spherical polar coordinates and the computational domain has the extent $R_{\text{WD}} \leq r \leq 1.5 R_{\text{WD}}$, and $0 \leq \alpha \leq \pi/2$, where r denotes the radial coordinate and α is the angular distance from the equator (i.e., $\alpha = \pi/2 - \theta$). The boundary conditions are described in Kley (1989a). The initial model is then evolved in time until a quasi-stationary state is reached.

3 RESULTS

Here we present some calculations performed with this program to study the structure of the interaction region of the inner accretion disk with the stellar surface layers in the case of a high mass-flow rate through the disk (i.e., the optically thick outburst case). Some of the results are described in more detail in Kley (1989b) Figure 1 shows in the central region the final density contours with velocity arrows superimposed. As can been seen from Figure 1, the matter flows in the disk towards the central star with the largest velocity at $r = 1.05$ R_{WD}. Nearly all the matter

Fig. 1—Final density contours with velocity arrows superimposed. Larger values of the labels at the contour lines indicate higher densities. The scaling is logarithmic with the values given in cgs units. The unit of time is given by the Keplerian rotational period at the surface of the star.

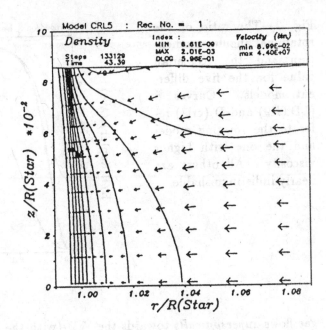

accumulates in an *equatorial belt* around the star. Only a very small amount of gas with low density is flowing towards higher stellar latitudes, where it partly leaves the computational domain. The mass loss through the outer boundary is generally less than 0.1% of the mass inflow rate. While the radial extent of the inner hydrostatic part is quite small, $< 0.004\,R_{WD}$, the BL extends in the vertical direction up to $z \lesssim 0.08$. This is essentially a consequence of the different pressure scale heights in the vertical and radial directions. In the BL, the energy increases strongly, and the whole region is a source of intense radiation. The difference of the comoving to the inertial flux is entirely negligible (at least in this case of accretion onto a WD). The hardness of the spectrum is determined partly by the area of the emitting region, which is given by viscous transition region. The radial extent can be estimated (using dimensional arguments) to be $\delta_{BL}/R_{WD}=\sqrt{\nu/h_K^*}$, where $h_K^* =\sqrt{GM_{WD}R_{WD}}$ is the specific angular momentum (at the stellar radius). This estimate is confirmed by the model calculations for the different parameters if one takes the inflection point of the angular velocity as reference. As can be seen in Figure 2, a four times larger viscosity coefficient leads to a doubling of δ_{BL}. The 0.6 $M_\odot$ and the 1.0 $M_\odot$ star have the same relative thickness because the product $M_{WD}R_{WD}$ is identical for these masses as follows from the mass-radius relationship for WDs. The stellar rotation does not significantly change the thickness: But the rotation of the WD reduces the amount of dissipated energy and subsequently the luminosity of the BL substantially. The dissipation in the transition zone is on average *half* the value of the non-rotating case, which is exactly expected on theoretical grounds. At the point of maximum dissipation (i.e., the inflection point, at $r \approx 1.05$ for the low-viscosity models) the

Fig. 2—The ratio of the rotational angular momentum and the Keplerian value for the five different models.   Curves B (LDTR9) and D (crl3) refer to the rotating model and the one with higher viscosity.   All others are nearly indistinguishable.

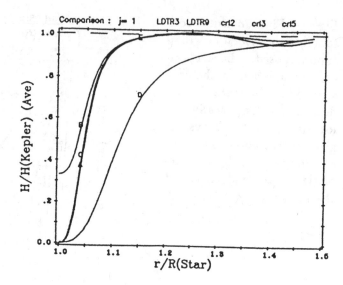

gas flows *supersonically* towards the WD (with the radial Mach number ≈ 6). This is a consequence of the large viscous information speed as implied by the viscosity coefficient ν.

4 OBSERVATIONAL CONSEQUENCES AND CONCLUSION

To estimate the observational properties of the BL we have applied a crude procedure to obtain a spectrum of the BL. To do so, first the surface of the disk-star interaction region was determined by integrating over the inverse mean free path until the value $\tau = 1$ was reached. To calculate the spectra, at each surface point the disk/BL is assumed to radiate a blackbody (BB) spectrum at the corresponding local radiation temperature $T_{rad}(\tau = 1)$.   Then, taking into account limb darkening, we integrated over the surface of the disk/BL-star under different inclinations to obtain a BB spectrum. To get an impression how our results compare with observations, we have superimposed the soft X-ray data of SS Cyg taken with *HEAO 1*. A much better fit to the observations was obtained in the case of the non-rotating star. To conclude, it seems possible (if the stellar parameters of the WD are known) to determine the rotation of the WD, and, last but not least, the mass inflow rate by comparing model calculations of this kind with observed spectra. For calculating nova outbursts, it is important to know in detail how the disk mixes with the stellar matter. It has to be noted that it is difficult to estimate the deepness of the penetration process of the disk into the star and the total fraction of matter that moves up to higher latitudes, because the real time covered is only a few dynamical timescales of the WD, which are of orders of magnitude smaller than the accretion timescales.

REFERENCES

Kley, W. 1989a, *Astr. Ap.*, **208**, 98.
Kley, W. 1989b, *Astr. Ap.*, **222**, 141.
Pringle, J. E. 1977, *M. N. R. A. S.*, **178**, 195.
Pringle, J. E., Savonije, G. J. 1979, *M. N. R. A. S.*, **197**, 777.
Shaviv, G. 1987, *Ap. Sp. Sci.*, **130**, 303.
Tylenda, R. 1981, *Acta Astr.*, **31**, 267.

Particle Simulation for Accretion Disks in Close Binary Systems

Florian Geyer, Heinz Herold, and Hanns Ruder

Lehr- und Forschungsbereich Theoretische Astrophysik Universität Tübingen, F. R. G.

1 INTRODUCTION

To model the hydrodynamics of accretion disks in close binary systems a pseudo particle method (Particle-In-Cell) is used, i.e., every particle represents a gas cell, which is still called a particle.

The original ideas of this work are from Hensler (Hensler 1982a, b, c). His formulation of the viscous interaction conserves neither the momentum nor the angular momentum, so we had to find a new formulation of the viscous interaction. Furthermore, we use more particles (10,000–40,000 instead of max. 2,000) to get a more exact picture of the dynamics. Because of these facts, we wrote a new code, also in order to speed up the calculations.

2 NUMERICAL METHOD

We formulate the equation of motion for each particle in a binary system as a restricted three-body problem in a corotating frame with pressure and viscosity:

$$\frac{d\boldsymbol{u}}{dt} = -\nabla\Phi_G - \boldsymbol{\Omega} \times (\boldsymbol{\Omega} \times \boldsymbol{r}) - 2\boldsymbol{\Omega} \times \boldsymbol{u} - \frac{1}{\rho}\nabla P + \nu\Delta\boldsymbol{u}.$$

To determine the pressure, we use the z-component of the equation of motion and the ideal gas equation and obtain the pressure at $z = 0$ as a function of the column density Σ and the temperature T (see Hensler 1982a, b):

$$P(x,y,0) = \rho(x,y,0)\frac{kT}{\mu m_{\mathrm{H}}} = \Sigma\sqrt{\frac{kTb}{2\pi\mu m_{\mathrm{H}}}} \quad \text{with} \quad b = -\frac{\partial^2}{\partial z^2}\Phi_G\big|_{z=0}.$$

In the first step, we integrate the equation of motion without the viscosity. As a particle mover we use the "Leap-Frog" scheme (described e.g., by Birdsall and Langdon 1985, p. 13).

In the second step, we apply the viscous interaction. Each particle is interacting with each other. Then the new velocity of particle i by interacting with particle k is given by:

$$\boldsymbol{u}'_i = \boldsymbol{u}_i + W(r_{ik})(\boldsymbol{u}_k - \boldsymbol{u}_i),$$

and also for particle k:

$$u'_k = u_k - W(r_{ik})(u_k - u_i).$$

This equation conserves only the momentum. To conserve the angular momentum simultaneously, we must shift the particles in the following way:

$$r'_i = r_i + \frac{W(r_{ik})}{1 - 2W(r_{ik})}(r_i - r_k), \qquad r'_k = r_k - \frac{W(r_{ik})}{1 - 2W(r_{ik})}(r_i - r_k).$$

The function W depends only on the distance of the two particles, and we choose the following linear function for W:

$$W(r_{ik}) = \gamma\left(1 - \frac{r_{ik}}{r_v}\right) \quad \text{for} \quad r_{ik} < r_v \qquad W(r_{ik}) = 0 \quad \text{for} \quad r_{ik} > r_v,$$

with r_{ik} being the distance between the two particles.

In this formulation we have two parameters for the viscous interaction: the viscosity radius r_v, which determines the range of the viscosity, and γ, which is the strength of the viscous interaction.

To get a relation between these two parameters and the kinematic viscosity ν in the equation of motion, we compute the total change in velocity of particle k after the viscous interaction:

$$\delta u_k = \sum_{l=1}^{n(k)} W(r_{kl})(u_k - u_l),$$

where $n(k)$ the number of particles which are in the viscosity range of particle k. If we assume that the local particle density $\Sigma_p(k)$ around particle k is approximately constant, we can write:

$$\Sigma_p(k) = \frac{n(k)}{\pi r_v^2}.$$

Now we expand u_l as a Taylor series around u_k up to second order, replace the sum over l by an integral over the viscosity area, and obtain:

$$\delta u_k = \frac{1}{2}\sum_{\alpha,\beta} \frac{\partial^2 u}{\partial x_\alpha \partial x_\beta}\Big|_k \int_0^{2\pi} \int_0^{r_v} \Sigma_p(k)W(r_{kl})(r_{kl})_\alpha(r_{kl})_\beta r_{kl}\,d\varphi\,dr_{kl} = \frac{1}{40}n(k)\gamma r_v^2 \triangle u.$$

So the kinematic viscosity is approximately given by:

$$\nu = \frac{1}{40}n(k)\frac{\gamma r_v^2}{\delta t},$$

where δt is the time step.

The energy δE, which is produced by the viscosity at each time step in a cell with the area δA, is assumed to be instantly released by radiation, leading to a local effective temperature of:

$$T_{\text{eff}} = \left(\frac{\delta E/2}{\delta A \cdot \delta t \cdot \sigma}\right)^{\frac{1}{4}}.$$

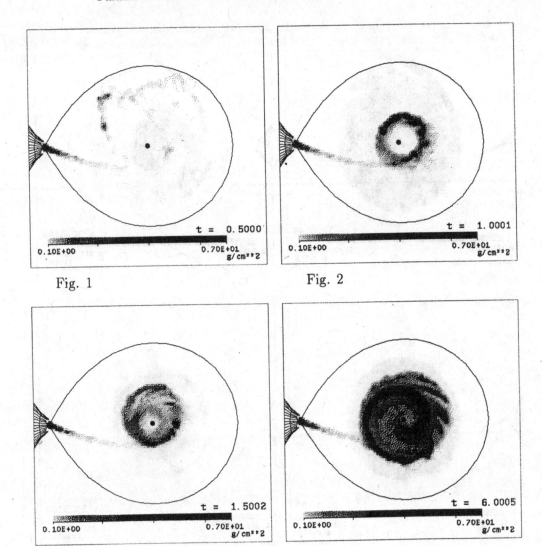

Fig. 1   Fig. 2

Fig. 3   Fig. 4

3 RESULTS

For two example calculations we use the following system parameters: primary: $M_1 = 1\,M_\odot$, secondary: $M_2 = 0.5\,M_\odot$, orbital period: $P_{orb} \simeq 0.2\,\text{day}$, mass transfer rate: $\dot{M} = 10^{-9}\,M_\odot\,\text{year}^{-1}$, radius of the primary: $R_a = 0.02\,R_\odot$, binary separation: $a = 1.647\,R_\odot$, mass of a "particle:" $M_{part} = 5.48 \times 10^{-16}\,M_\odot$. In Figures 1–6 the parameter of the viscosity are: radius of viscosity: $r_v = 0.005\,a$, strength of viscosity: $\gamma = 0.05$. Figures 1–5 show the temporal evolution of the column density (t in units of the orbital period), and Figure 6 shows the distribution of T_{eff}. In Figures 7–8 the parameters of the viscosity are: radius of viscosity: $r_v = 0.005\,a$, strength of viscosity: $\gamma = 0.01$. Figure 7 shows the column density and Figure 8 the distribution of T_{eff}.

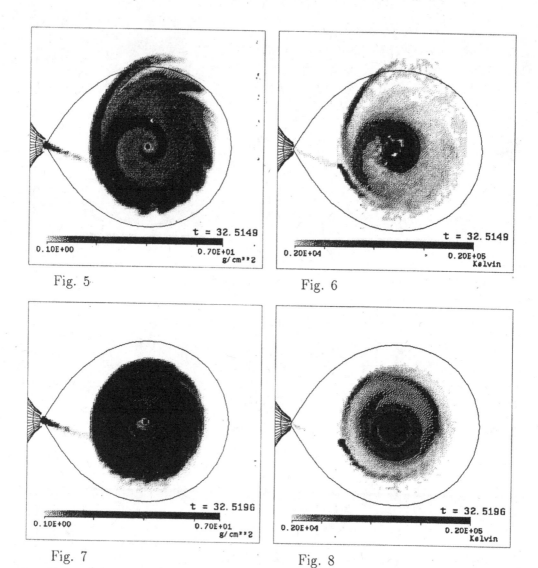

Fig. 5

Fig. 6

Fig. 7

Fig. 8

REFERENCES

Birdsall, C. K., and Langdon, A. B. 1985, *Plasma Physics via Computer Simulation* (New York: McGraw-Hill).

Hensler, G. 1982a, Ph.D. thesis, University of Göttingen.

Hensler, G. 1982b, *Astr. Ap.*, **114**, 309.

Hensler, G. 1982c, *Astr. Ap.*, **114**, 319.

Superhumps in Cataclysmic Variables

Robert Whitehurst and Andrew King

Astronomy Group,
University of Leicester,
Leicester LE1 7RH,
United Kingdom

ABSTRACT

We show that the structure of accretion disks can be dominated by the excitation of resonances in the accretion disk. In particular, we show that a near 3:1 commensurability with the binary orbit can explain the superhumps observed in superoutbursts of SU UMa dwarf novae. This resonance can only appear for mass ratios $M_2/M_1 \lesssim 0.25$–0.33: for larger mass ratios the available resonances are probably too weak to have observable effects. The 2:1 resonance may appear in CVs and LMXRBs with ultrashort periods, and give rise to similar phenomena.

1 INTRODUCTION

A class of dwarf novae, the SU UMa stars, from time to time undergo particularly long outbursts (superoutbursts). During these outbursts a photometric modulation of the optical light (the superhump) with a period a few percent longer than the orbital period always appears. Numerical studies of accretion disks with a particle simulation code (Whitehurst 1988) do give such a modulation provided that the disk is large enough and also that the ratio $q = M_2/M_1$ of secondary to primary (white dwarf) masses is less than ~ 0.25–0.33. The first requirement implicitly demands a high mass-overflow rate from the secondary: this will hold in reality if superoutbursts are mass transfer events, as is widely believed. The latter requirement, and indeed the physical origin of the superhumps, have up to now been unclear. Here we show that both are related to the presence of strong resonances between particle orbits in the disk and the motion of the secondary star.

2 RESONANCES

Resonance occurs in a disk when the frequency of radial motion of a particle orbit in the disk is commensurate with the angular frequency of the secondary star as seen by the particle. The former (epicyclic) frequency is $\Omega - \omega$, where Ω is the particle's mean angular frequency around the primary and ω the apsidal precession frequency of its orbit, both measured in a non-rotating frame. The latter (synodic) frequency is $\Omega - \Omega_{\text{orb}}$, where Ω_{orb} is the orbital frequency. Thus, resonance requires

$$k(\Omega - \omega) = j(\Omega - \Omega_{\text{orb}}), \tag{1}$$

where j and k are positive integers. (Cases with $k > 1$ only arise when orbital symmetry breaks down. Normally this is due to a small eccentricity of the binary orbit but is here due to the random radial excursions of material in the disk caused by the viscosity.) The apsidal precession rate is zero for periodic (self-closing) orbits in the disk and in any case much smaller than Ω and Ω_{orb} for aperiodic (rosette) orbits, so resonances appear close to commensurabilities of the form $j : j-k$ between particle and binary frequencies. The strength of a resonance is measured by its growth rate g, and goes as e^k, where e is the eccentricity of the particle orbit. Thus, the strongest resonances arise at the $j : j - 1$ commensurabilities, followed by $j : j - 2$, etc. Near a given commensurability, the orbits with the largest eccentricity compatible with stability resonate most strongly. Generally, these are aperiodic orbits with non-vanishing precession rates ω. Resonances thus act as filters picking out these extreme orbits provided they can be populated by perturbations.

Assuming that even quite extreme aperiodic orbits do not differ greatly from Keplerian circles, we may measure the typical radii of resonant orbits near the general $j : k$ commensurability as

$$R_{jk} = (GM_1/\Omega_{jk}^2)^{1/3}. \tag{2}$$

Using Kepler's law, the binary separation is

$$a = [G(M_1 + M_2)/\Omega_{\text{orb}}^2]^{1/3}, \tag{3}$$

so we find

$$R_{jk}/a = [(j - k)/j]^{2/3}(1 + q)^{-1/3}. \tag{4}$$

To see what resonances are possible for a given mass ratio, we compare this with the disk's tidal radius $R_T = 0.9R_{\text{Roche}}$, given by

$$R_T/a = 0.45/[0.6 + q^{2/3}\log_e(1 + q^{1/3})] \tag{5}$$

(Eggleton 1983), and ask when $R_{jk} < R_T$. It is readily apparent that the strongest ($k = 1$) resonances can only occur for very small mass ratios: $j = 2$, $k = 1$ requires $q \lesssim 0.025$, with still smaller ratios required for larger j. The 2:1 resonance is only possible in systems with mass ratios too extreme to be SU UMa dwarf novae but may be possible in some LMXRBs (see below). For a possible explanation of the superhumps in SU UMa stars we must turn to the next strongest ($k = 2$) resonances. The smallest radius is for $j = 3$, and is inside R_T for $q \lesssim 0.3$. Larger values of j require ratios at least as extreme as for the 2:1 resonance. Since larger k values give weaker resonances, this shows that the strongest resonance possible in the disks of

SU UMa stars is likely to be at $j = 3$, $k = 2$, and requires $q \lesssim 0.3$. There is good reason to believe such ratios to be typical in SU UMa systems, and the implied limit on the orbital period agrees well with their absence at periods $\gtrsim 3$ hr (Whitehurst 1988). Further, particles near the edge of the eccentric precessing disks producing superhumps in the numerical simulations (Whitehurst 1988) have periods very close to the 3:1 commensurability with the binary orbit. Clearly this offers a promising explanation for the superhump phenomenon, so we examine the detailed behaviour near the resonance.

3 SUPERHUMPS

The behaviour of *periodic* orbits ($\omega = 0$) near resonance has been thoroughly studied in the case of the restricted three-body problem by Henon (1965). Viewed in the reference frame corotating with the binary, resonances cause period-doubling of such orbits, that is, the orbit *circulates* twice. In the sidereal frame, simply-periodic orbits are thus seen to circulate *three* times before closing on themselves, resulting in rosette orbits whose periods are three times those of the simply-periodic orbits just inside the resonance, i.e., $6\pi/\Omega$. A similar result must hold for the aperiodic (precessing) orbits with nonzero ω which we know are the most likely to resonate in accretion disks. Since these have frequencies $\Omega = 3\Omega_{orb} - 2\omega$ just inside the resonance near the 3:1 commensurability [eq. (1)], the circulating orbits must have radial periods

$$P_{SH} = 6\pi/\Omega \approx P_{orb}(1 + 2\omega/3\Omega_{orb}), \qquad (6)$$

where $P_{orb} = 2\pi/\Omega_{orb}$ is the binary period and we have used the fact that $\omega \ll \Omega_{orb}$. The precession is prograde ($\omega > 0$), so the radial period of the resonant orbits is slightly longer than P_{orb}. We cannot calculate by exactly how much without knowing exactly which precessing orbit resonates. However, since ω must vanish in the limit of vanishing q, we expect the fractional difference between P_{SH} and P_{orb} to be proportional to q. We see that P_{SH} has precisely the properties of the observed superhump period.

For perfectly symmetrical orbits near the 3:1 commensurability, the growth rate of the disturbance is proportional to the square of both the eccentricity of the orbit e and the binary star e'. This gives a growth rate g of the form $g \propto e^2 e'^2$ (Greenberg 1984). However, in the simulations, $e' = 0$ and the growth of the resonance is due to the presence of the viscosity which breaks the symmetry. With a typical viscous scale length λ, the growth rate g is expressible as

$$g \propto \frac{e^2 \lambda^2}{a^2}, \qquad (7)$$

where a is the semi-major axis of the binary orbit.

In the numerical simulations halving the scale length λ had the effect of delaying the onset of the *precessing* disk from one hundred to five hundred orbital periods. In real systems the observed waiting time in superoutburst before the beginning of the superhump implies a viscosity equivalent to a turbulent viscosity with $\alpha \simeq 1$.

The arguments given above show compellingly that superhumps are probably a consequence of a near 3:1 resonance in the accretion disks of SU UMa systems as this produces a radial disturbance with the right period. The explicit form of the disturbance is not yet clear, however: it must involve only a subset of the resonant orbits since there could be no net effect if all phases of these orbits were equally populated. Further studies of the numerical simulations are needed to decide this question.

The orbital periods of SU UMa systems are $\lesssim 3$ hr, probably as a consequence of the mass-ratio condition $q \lesssim 0.3$, and cannot be less than the CV minimum period $\sim$ 80 min. No LMXRBs are known in this period range, so it is hardly surprising that superhumps have not been detected in them. As noted above, however, the much stronger 2:1 resonance can exist in systems with extreme mass ratios $q \lesssim 0.025$ and is likely to give rise to similar phenomena (Whitehurst and King 1989).

4 CONCLUSION

The work described here strongly suggests that the superhumps of SU UMa stars are the result of a 3:1 resonance in systems with mass-ratio $q \lesssim 0.3$, or a 2:1 resonance in ultrashort-period systems with $q \lesssim 0.025$, provided that the accretion disk grows out to fill most of the primary's Roche lobe. Given these simple criteria, it may be possible to use the presence or absence of superhumps as a way of discovering more about the underlying binary systems.

ARK and RW thank the Royal Society and the organizers of the Workshop for support in attending the CV Workshop and RW also acknowledges the UK SERC for a Research Associateship.

REFERENCES

Eggleton, P. P. 1983, *Ap. J.*, **268**, 368.
Greenberg, R. 1984, in *Saturn*, ed. T. Gehrels and M. Shapley Matthews (Tucson: University of Arizona Press), p. 593.
Henon, M. 1965, *Ann. d'Astr.*, **28**, 992.
King, A. R. 1988, *Quart. J. R. A. S.*, **29**, 1.
Whitehurst, R. 1988, *M. N. R. A. S.*, **232**, 35.
Whitehurst, R., and King, A. R. 1989, in *Proceedings of the ESLAB Symposium on X-ray Binaries, Bologna, 1989*, in press.

Accretion Disk Instabilities in A0620-00 and Related Systems

J. Craig Wheeler, Shin Mineshige, Min Huang, and S. W. Kim

Department of Astronomy, University of Texas at Austin

There has been considerable success in accounting for the properties of dwarf novae with models based on the thermal instability associated with the ionization of hydrogen (Smak 1984; Meyer and Meyer-Hofmeister 1984; Lin, Papaloizou, and Faulkner 1985; Mineshige and Osaki 1985; Cannizzo, Wheeler, and Polidan 1986 and references therein). Despite the success of these models in rather naturally reproducing typical burst recurrence times, amplitudes, and durations, and even such details as the optical-UV delay during rise to maximum (Mineshige 1988), there are still many problems to be explored. As an example, we cannot readily account for the dispersion in recurrence times in a single dwarf nova, nor among different systems (Cannizzo, Shafter, and Wheeler 1988; Cannizzo, this conference). There are a variety of reasons to continue and extend this effort to explore the application of accretion disk thermal instabilities. Study of a variety of time-dependent systems is the best way to put empirical bounds on the viscosity and thus to guide the development of a fundamental physical theory of the transport of angular momentum in accretion disks and related systems.

A0620-00 is an excellent laboratory with which to attempt to extend the accretion disk instability mechanism to X-ray transients. Its large outbursts separated by long intervals ($\sim$60 years) are intrinsically interesting. McClintock and Remillard (1986) have argued that the system contains a black hole which means that study of this system may yield new ways of probing black holes and may provide a link between dwarf novae on the stellar scale and active galactic nuclei on the galactic scale. Accretion disk instability models for A0620-00 have been studied by Huang and Wheeler (1989) and Mineshige and Wheeler (1989) and it remains a central target of ongoing studies.

The basic parameters in an accretion disk instability model are the mass of the central object, the (nominally constant) transfer rate from the companion, the inner and outer disk radii and a prescription for the viscosity. There is empirical evidence that a single value of the viscosity parameter α (Shakura and Sunyaev 1973) is not constant in dwarf novae but is lower in quiescence than outburst. There are physical arguments to suggest that it might vary as $(h/r)^n$ where n is some power (Meyer and Meyer-Hofmeister 1983; Vishniac and Diamond 1989). Here we adopt $\alpha = \alpha_o (h/r)^n$ where $n \sim 1$–2.

In the simplest models (without irradiation, magnetized central objects, etc.), one can still discriminate models with white dwarfs from more compact objects in some regimes. With fixed transfer rate and outer disk radius and a moderately large transfer rate such that the heating instability begins near the outer edge of the disk, there would be little difference between models of white dwarfs and neutron stars or black holes of identical gravitational mass. For lower accretion rates, such that the heating instability begins deeper in the disk, there will be systematic differences between white dwarf models with inner disk boundaries at $r \sim 10^8$–10^9 cm and neutron stars or black holes of the same mass but inner disk radii of 10^6–10^7 cm. The recurrence time will tend to be longer in the latter case. There is no immediate means by which an idealized neutron star and an idealized black hole of the same mass can be discriminated, although pursuit of means to do so in more realistic cases is a high priority. In practice, systems suspected to contain neutron stars and black holes tend to have larger orbital periods and hence larger disks than systems with white dwarf primaries. This factor alone gives larger recurrence times, all else being the same. Finally, for all else the same, larger-mass central objects will have longer viscous times in quiescence and hence longer recurrence times. This allows a way to discriminate black holes from neutron stars.

Huang and Wheeler (1989) and Mineshige and Wheeler (1989) showed that for appropriate parameters for A0620-00, the accretion disk heating instability naturally disrupts the quiescence. The burst magnitude, duration, and recurrence interval in both optical and soft X-rays (assumed proportional to the accretion rate through the inner edge of the disk) all correspond reasonably well with observations. Figure 1 gives a model reminiscent of A0620-00. It has $\dot{M} = 10^{15}$ g s^{-1}, $M_1 = 10\ M_\odot$, $\alpha = 10^2\ (h/r)^{1.5}$, $R_{\text{inner}} = 9 \times 10^6$ cm, and $R_{\text{outer}} = 3.2 \times 10^{10}$ cm. The 60-year interval between outbursts is consistent with a central object of $\sim 10\,M_\odot$, but not with one of $1\ M_\odot$, if the viscosity in the outer parts of the disk resemble that in dwarf nova disks under similar conditions. This is independent evidence that A0620-00 contains a black hole.

The models tend to give interbursts on somewhat shorter intervals which are several magnitudes dimmer as shown in Figure 1. It would be very interesting to check whether there is any evidence for such interbursts. Typical models in quiescence have a cold ($T \sim 3000$ K), optically thick disk which is very dim ($V \gtrsim 30^m$). This prediction is consistent with the photometric evidence for a dark, optically thick disk presented at this conference by Haswell, Robinson, and Horne (1990). The origin of the blue light in quiescence with a luminosity of about one-half that of the secondary is not easily explained by the model. The emission lines (Johnston, Kulkarni, and Oke 1989; Johnston and Kulkarni 1990) are also difficult to explain, but they are more apt to arise in a corona, not in an optically thin disk.

Aql X-1 is a prototypical soft X-ray transient. Priedhorsky and Terrell (1984) have presented *Vela* satellite data showing bursts at irregular intervals over a seven-year span. They argue that there may be a fundamental underlying period of $\sim$123

Fig. 1—The logarithm of the accretion rate through the inner edge of the disk, the V magnitude of the disk and the companion star (*dashed line*) and the logarithm of the total luminosity from the disk are given as a function of time for a model relevant to A0620-00.

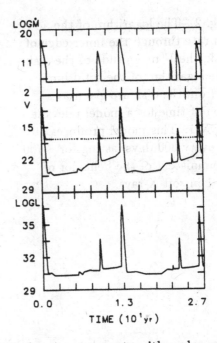

days in the sense that every large burst seems to be commensurate with such a period at intervals of 2 or 3 times the fundamental period. For appropriate parameters ($\dot{M}_T = 10^{16}$ g s^{-1}, $M_1 = 1\ M_\odot$, $\alpha = 10^{1.5}\ (h/r)^{1.5}$, $R_{\text{inner}} = 10^6$ cm, $R_{\text{outer}} = 10^{10.5}$ cm), a basic model gives principle bursts ~300 days apart, lasting for about a month, with smaller interbursts. Searching for parameters which match a certain burst interval, e.g., 123 days, has not been done pending the addition of the necessary effects of irradiation. Nevertheless, such calculations suggest that the intuitive idea that soft X-ray transients may be very similar to dwarf novae, but with a neutron star in place of the white dwarf, receives support from these quantitative models.

Priedhorsky, Terrell, and Holt (1983) give ~10 years worth of soft X-ray data from Cygnus X-1. They argue that there may be a regular pattern of minima about every three hundred days with a modulation in flux at about the 25 percent level. There was also a larger outburst (factor of $\gtrsim$ 2 in flux) beginning in late 1975 and lasting for ~3 months. Ling *et al.* (1987) report another such soft X-ray flare in 1980 which followed a rise in the hard X-ray flux. One might thus ask whether either of these temporal behaviors might be explained by an accretion disk thermal instability.

The answer is that either behavior might represent a disk instability, but such a physical process is unlikely to explain both simultaneously. Further work will be necessary to determine which, if either, of these phenomena might be associated with an accretion disk thermal instability. Figure 2 shows the results of a model chosen to fit the nominal parameters of Cygnus X-1. The model has $\dot{M} = 10^{17}$ g s^{-1}, $M_1 = 10\ M_\odot$, $\alpha = 10^{2.5}\ (h/r)^{1.5}$, $R_{\text{inner}} = 9 \times 10^6$ cm, and $R_{\text{outer}} = 10^{11}$ cm. This model gives outbursts every 300 days lasting ~100 days. The top panel is the accretion rate through R_{inner} which should be proportional to L_X. The second panel gives the V magnitude of the disk and shows that the optical modulation of the 9th magnitude

Fig. 2—The logarithm of the accretion rate through the inner edge of the disk, the V magnitude of the disk, and the logarithm of the total luminosity from the disk are given as a function of time for a model relevant to Cyg X-1. This model produces outbursts every 300 days lasting for $\sim$100 days reminiscent of the behavior of Cyg X-1 in soft X-rays.

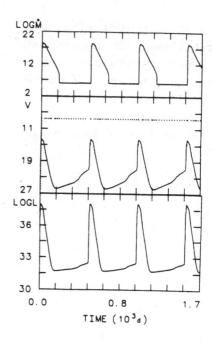

star (assumed to be at 2.5 kpc) should be negligible. The principal difference between this model and that for A0620-00 is the higher mass-transfer rate (10^{17} g s^{-1} vs. 10^{15} g s^{-1} for A0620-00).

Another nominal "Cyg X-1" model was run with the same parameters as in Figure 1, but with the coefficient of α reduced to $10^{2.0}$. This factor of three decrease in the scaling of the viscosity resulted in outbursts every $\sim$2.6 years. Clearly, fine-tuning could give outbursts 5 years apart as seen in the soft X-ray flares in 1975 and 1980. In this model, the V magnitude flux from the disk becomes $\sim$20 percent of the total light from the system. Limits on the change in optical luminosity during the soft X-ray flares could thus constrain the disk instability model for these outbursts.

In conclusion, the accretion disk instability model can provide a plausible explanation for the optical and soft X-ray observations of variable X-ray sources such as A0620-00, Aql X-1, and Cyg X-1. Most of the parameters of the model ($\dot{M}$, M_1, R_{outer}, R_{inner}) are constrained by observations of the specific systems. The prescription for viscosity is still a major physical uncertainty, but even it is constrained by observations of complementary systems, e.g., dwarf novae. There are thus encouraging reasons to pursue the investigation of this class of models for some X-ray transients. An important ingredient for deeper understanding of the model and its implications is the effect of external irradiation by the X-rays (Tuchman, Mineshige, and Wheeler 1990; Mineshige, Tuchman, and Wheeler 1990a, 1990b).

This research is supported in part by NSF Grant 8717166 and NASA Grant 7232. The computations were done with the resources of the University of Texas System Center For High Performance Computing.

REFERENCES

Cannizzo, J. K., Shafter, A. W., and Wheeler, J. C. 1988, *Ap. J.*, **333**, 227.

Cannizzo, J. K., Wheeler, J. C., and Polidan, R. S. 1986, *Ap. J.*, **301**, 634.

Haswell, C. A., Robinson, E. L, and Horne, K. D. 1990, this volume.

Huang, M., and Wheeler, J. C. 1989, *Ap. J.*, **343**, 229.

Johnston, H. M., and Kulkarni, S. R. 1990, this volume.

Johnston, H. M., Kulkarni, S. R., and Oke, J. S. 1989, *Ap. J.*, **345**, 492.

Lin, D. N. C., Papaloizou, J., and Faulkner, J. 1985, *M. N. R. A. S.*, **212**, 15.

Ling, J. C., Mahoney, W. A., Wheaton, W. A., and Jacobsen, A. S. 1987, *Ap. J. (Letters)*, **321**, L117.

Meyer, F., and Meyer-Hofmeister, E. 1983, *Astr. Ap.*, **128**, 420.

Meyer, F., and Meyer-Hofmeister, E. 1984, *Astr. Ap.*, **132**, 143.

McClintock, J. E., and Remillard, R. A. 1986, *Ap. J.*, **308**, 110.

Mineshige, S. 1988, *Astr. Ap.*, **190**, 72.

Mineshige, S., and Osaki, Y. 1985, *Pub. Astr. Soc. Japan*, **37**, 1.

Mineshige, S., Tuchman, Y., and Wheeler, J. C. 1990*a*, *Ap. J.*, submitted.

Mineshige, S., Tuchman, Y., and Wheeler, J. C. 1990*b*, this volume.

Mineshige, S., and Wheeler, J. C. 1989, *Ap. J.*, **343**, 241.

Priedhorsky, W. C., and Terrell, J. 1984, *Ap. J.*, **280**, 661.

Priedhorsky, W. C., Terrell, J., and Holt, S. S. 1983, *Ap. J.*, **270**, 233.

Shakura, N. I., and Sunyaev, R. A. 1973, *Astr. Ap.*, **24**, 337.

Smak, J. 1984, *Pub. Astr. Soc. Pac.*, **96**, 5.

Tuchman, Y., Mineshige, S., and Wheeler, J. C. 1990, *Ap. J.*, submitted.

Vishniac, E. T., and Diamond, P. 1989, *Ap. J.*, in press.

Structure and Evolution of Irradiated Accretion Disks

S. Mineshige[1,2], Y. Tuchman[1,3], and J. C. Wheeler[1]

[1]Department of Astronomy, University of Texas at Austin
[2]Institute for Fusion Studies, University of Texas at Austin
[3]Department of Physics, Hebrew University

1 INTRODUCTION

Soft X-ray transients exhibit repetitive eruptions similar to outbursts of dwarf novae (see the review by Priedhorsky and Holt 1987). Two competing models are proposed for outbursts of soft X-ray transients: irradiation instability of the secondary star (Hameury, King, and Lasota 1986) and the disk instability model (Cannizzo, Wheeler, and Ghosh 1985). It is now widely believed that the dwarf nova outbursts are caused by the disk thermal instability triggered by the recombination of the hydrogen and helium (reviewed by Smak 1984; see also Osaki 1974; Hōshi 1979; Meyer and Meyer-Hofmeister 1981). Following this idea, Huang and Wheeler (1989) and Mineshige and Wheeler (1989) have calculated time evolution of the thermally unstable disks in low-mass X-ray binaries (but without external heating sources) and reproduced the basic features of light curves of soft X-ray transients.

The key issue to both models is the magnitude of the irradiation. In order to trigger the mass-transfer burst instability, X-ray irradiation of the secondary surface during quiescence should be significant. On the other hand, if the external heating is strong enough to ionize the hydrogen and helium in the disk, then the dwarf-nova type instability will be suppressed (Meyer and Meyer-Hofmeister 1984). In the disk-instability model, the effect of irradiation should be small. We thus study the effects of external irradiation on the thermal stability of the disk.

2 STATIC STRUCTURE

Assuming that the irradiation flux is thermalized in the photosphere of the disk, we calculated the vertical structure of the disk by changing the surface boundary conditions. We found that the response of the disk structure to the irradiation is totally different for radiative disks and for convective disks. For radiative disks only the surface layer is heated by the irradiation, and there is practically no change in the internal structure. In the convective disks, however, the heat of irradiation can deeply penetrate into the disk, and thus the temperature increases at every depth.

Thermal equilibrium curves of irradiated accretion disks are displayed in Figure 1.

Fig. 1—Thermal equilibrium curves in the $(\log T_c, \log \Sigma)$ plane for different irradiation temperatures: $T_{\mathrm{irr}} = 0$, 3000, 4500, 6000, 7500, 9000, 10,500, and 12,000 K. The radius is $\log r(\mathrm{cm}) = 10.5$ and the viscosity parameter is $\alpha = 0.1$ (constant).

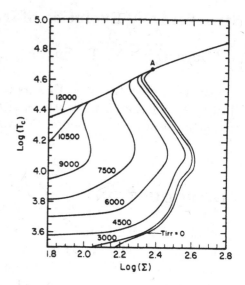

We take various values of the irradiation temperature, $T_{\mathrm{irr}} = 0$ (no irradiation), 3000, 4500, 6000, 7500, 9000, 10,500, and 12,000 K, where T_{irr} is defined in terms of the irradiation flux: $F_{\mathrm{irr}} \equiv \sigma T_{\mathrm{eff}}^4$. It is clear in Figure 1 that, for sufficiently strong irradiation with $T_{\mathrm{irr}} > 10,000$ K, the disk is completely stabilized against thermal instabilities of the sort invoked to explain dwarf novae (Meyer and Meyer-Hofmeister 1984). Even for moderately strong irradiation with $T_{\mathrm{irr}} \sim 6000$ K, however, there is still an unstable branch in the thermal equilibrium curve, although the difference in the temperature between the hot stable state and the cool stable state is less than that in non-irradiated disks. We can build a simple analytical model to understand these features. For details, see Tuchman, Mineshige, and Wheeler (1990).

3 DYNAMICAL EVOLUTION

As the next step, we calculated the non-linear evolution of an irradiated accretion torus using a one-dimensional dynamical code. It is shown that for moderately strong irradiation with $T_{\mathrm{irr}} \sim 6000$ K, the amplitude of the resultant light curves is smaller, the quiescent period is shorter, and the outburst duration is longer compared to those of non-irradiated models. If we vary F_{irr} in proportion to the mass accretion rate at the inner edge of the disk, $\dot{M}_a$, then we can reproduce light curves with a plateau in the decay from outbursts by introducing a time delay in the response of F_{irr} to the modulation in $\dot{M}_a$.

In Figure 2 we display the typical light curves of such models. Model parameters are the mass-transfer rate, $\dot{M} = 2 \times 10^{17}$ erg s^{-1}, the radius of the torus, $r_R = 10^{10.5}$ cm, the viscosity parameter $\alpha = 0.1$, and a time delay $\Delta t = 2.0$ d. The light curve in Figure 2 has a "shoulder" in the decline phase from outbursts, and is therefore very reminiscent of the light curves of black-hole candidate A0620-00 (Whelan *et al.* 1977)

Fig. 2—Time evolution of a one-zone torus. The irradiation is assumed to be proportional to the mass accretion rate but with a time delay of $\Delta t = 2.0$ d (*solid line*: T_c, *dotted line*: Σ, *dashed line*: T_{eff}, *dash-dot line*: T_{irr}).

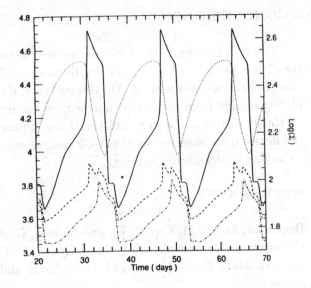

and of some soft X-ray transients which contain neutron stars (see compilations by Priedhorsky and Holt 1987). In this model, we assumed that the outer portions of the disk are shaded by the inner portions of the disk, thus $T_{\text{irr}} < 10^4$ K. This assumption is justified in the decay phase. If the shielding of the irradiation is inefficient, however, the irradiation is expected to be so strong that the dwarf-nova type thermal instability can be completely suppressed. We then require mechanisms to initiate the cooling instability. Wind mass loss (Begelman, McKee, and Shields 1983) or the irradiation instability may provide such mechanisms. For details, see Mineshige, Tuchman, and Wheeler (1990).

4 HARD X-RAY FLUX

Cyg X-1 displays a hard X-ray flux which is at minimum during soft X-ray maxima. In the disk instability model (see Wheeler *et al.* 1990) the cool disk ($T < 1$ keV) extends to the inner edge in quiescence and cannot produce a two-temperature radiation pressure dominated disk. In outburst, such a condition is possible, but would occur at small radii and simultaneous with the rise in soft X-rays, contrary to observations. Moreover, Liang and Dermer (1988), Makishima (1988), and Fabian *et al.* (1989), give arguments that the hard X-ray region is at $r \sim 10^8$ cm, much larger than the inner disk edge. A solution consistent with these conditions is a cold disk ($kT < \text{keV}$) plus a hot corona ($kT > 10$ keV) in quiescence at $r \sim 10^8$ cm. The corona could be formed by hard X-rays from the inner regions which inverse Compton scatter on soft photons from the cool disk. During disk outburst, the fatter disk may block the hard flux and prevent the formation of the corona. Similar phenomena may be at work in A0620-00 and in AGN.

5 CONCLUSIONS

The thermal equilibrium curves are very sensitive to the external irradiation. If the disk is exposed to the direct irradiation, the dwarf-nova type thermal instability can be effectively suppressed. On the contrary, if the outer parts of the disk are completely shaded from the X-ray radiation by the inner portions of the disk, the disk is thermally unstable and so exhibits outburst-quiescent limit cycles. The "shoulder" in the light curves observed in A0620-00 and other soft X-ray transients are interpreted to be due to irradiation heating.

REFERENCES

Begelman, M. C., McKee, C. F., and Shields, G. A. 1983, *Ap. J.*, **271**, 70.

Cannizzo, J. K., Wheeler, J. C., and Ghosh, P. 1985, in *Cataclysmic Variables and Low-Mass X-ray Binaries*, ed. D. Q. Lamb and J. Patterson (Dordrecht: Reidel), p. 307.

Fabian, A. C., Rees, M. J., Stella, L., and White, N. E. 1989, *M. N. R. A. S.*, **238**, 729.

Hameury, J. M., King, A. R., and Lasota, J. P. 1986, *Astr. Ap.*, **162**, 71.

Hōshi, R. 1979, *Progr. Theor. Phys.*, **61**, 1307.

Huang, M., and Wheeler, J. C. 1989, *Ap. J.*, **343**, 229.

Liang, E. P., and Dermer, C. D. 1988, *Ap. J. (Letters)*, **325**, L39.

Makishima, K. 1988, in *Physics of Neutron Stars and Black Holes*, ed. Y. Tanaka (Tokyo: Universal Academy Press), p. 175.

Meyer, F., and Meyer-Hofmeister, E. 1981, *Astr. Ap.*, **104**, L10.

Meyer, F., and Meyer-Hofmeister, E. 1984, *Astr. Ap.*, **140**, L35.

Mineshige, S., Tuchman, Y., and Wheeler, J. C. 1990, *Ap. J.*, submitted.

Mineshige, S., and Wheeler, J. C. 1989, *Ap. J.*, **343**, 241.

Osaki, Y. 1974, *Pub. Astr. Soc. Japan*, **26**, 429.

Priedhorsky, W. C., and Holt, S. S. 1987, *Space Sci. Rev.*, **45**, 291.

Smak, J. 1984, *Pub. Astr. Soc. Pac.*, **96**, 5.

Tuchman, Y., Mineshige, S., and Wheeler, J. C. 1990, *Ap. J.*, submitted.

Wheeler, J. C., Mineshige, S., Huang, M., and Kim, S. W. 1990, this volume.

Whelan, J. A. J., *et al.* 1977, *M. N. R. A. S.*, **180**, 657.

Emission Lines from an X-ray Illuminated Accretion Disk in Sco X-1

T. R. Kallman

Laboratory for High Energy Astrophysics,
NASA/Goddard Space Flight Center

1 INTRODUCTION

Emission lines are a defining characteristic of accreting binaries of all types. In the case of X-ray binaries it is likely that these features are formed as the result of reprocessing of X-rays by the accretion flow. This paper will discuss the formation of emission lines resulting from an illuminated accretion disk in the low mass X-ray binary Sco X-1. The use of these lines as diagnostics of the conditions in this gas and on the broad-band properties of the continuum will be discussed.

The emission lines from Sco X-1 can be separated into wavelength bands: 'hard' (2–10 keV) X-rays whose reprocessed emission is dominated by the iron K line, 'soft' (0.5–2 keV) X-rays, and ultraviolet (1200–3000 Å); we will discuss these in turn.

2 LINE FORMATION

2.1 Iron K Emission Lines

The energy of the iron K line, ε, has an energy in the range 6.4–7.1 keV depending on the ionization state of iron (cf., Hirano $et\ al.$ 1987). The line width, $\Delta\varepsilon$, depends on any of a variety of broadening mechanisms, including blending of multiple line components with differing intrinsic energies, virial motions of the emitting gas, Compton scattering, or general relativistic effects. The equivalent width, EW, depends on the amount of emitting material (parameterized by the emission measure at the appropriate temperature and ionization state, $EM = \int n^2 dV$, where n is the gas density and dV is the volume element) and the emission efficiency (we will make the 'coronal' assumption in calculating this quantity; e.g., Raymond and Smith 1977). The results from the $EXOSAT$ observatory compiled by White, Peacock, and Taylor (1985) and White $et\ al.$ (1986) show that for Sco X-1 the line centroid energy is $\varepsilon \simeq 6.7$ keV, the width (Full Width at Half Maximum) is FWHM $\simeq \langle \Delta\varepsilon \rangle \simeq 1$ keV, and the equivalent width is $EW \simeq 50$ eV.

These results imply that the dominant ionization state in the emitting gas is high (Fe XXIII or greater) and therefore that the temperature in the emitting region is $\geq 10^7$ K under a variety of plausible assumptions about the conditions in the emitting gas. From the limit on the line width we can derive an emission radius of $\leq 10^7$ cm if the broadening is due to virial motions, a Thomson depth ≥ 3 if the broadening is due to Compton scattering, or a radius $\leq 10^{6.5}$ cm if the broadening is due to gravitional effects.

2.2 Soft X-ray Lines

The wavelength range between 0.5–2 keV contains a large number of emission lines which are useful diagnostics of the emitting gas in X-ray binaries. The interpretation of soft X-ray line spectra is more complicated than that of the iron K lines owing to the much larger number of competing transitions and excitation mechanisms. However, the existing observations of these lines can be described as follows: line energies in the range 0.8–1.6 keV region due to ions including N VII, O VIII, Fe XVII–XVIII, and Fe XXIII–XXIV; line equivalent widths 1–10 eV; and line widths $\Delta\varepsilon/\varepsilon \leq 0.1$.

Attempts to fit the observed soft X-ray grating spectra with optically thin collisional equilibrium models (e.g., Raymond and Smith 1977) suggest that there is a broad distribution of ionization states and temperatures in the range 10^6–10^7 K in the emitting material, rather than a single set of conditions (Kallman, Vrtilek, and Kahn 1989), and emission measures $EM \sim 10^{58}$–10^{59} cm^{-3}. From the limit on the line width, we can derive an emission radius of $\geq 10^7$ cm if the broadening is due to virial motions, or a Thomson depth ≤ 12 if the broadening is due to Compton scattering.

2.3 UV Lines

UV emission lines from Sco X-1 have energies in the range 7–10 eV due to ions including He II, C IV, N IV, N V, O V, and Si IV; line equivalent widths 0.01–0.1 eV; and line widths $\Delta\varepsilon/\varepsilon \leq 0.01$ (Willis et al. 1980). The limit on the line width implies an emission radius of $\geq 10^{10}$ cm if the broadening is due to virial motions, or a Thomson depth ≤ 120 if the broadening is due to Compton scattering. Under the coronal assumption, the maximum available line emissivities for the UV lines are greater than for the X-ray lines by a factor ~ 10; the emission measures required to account for the line intensities are $\sim 10^{57}$–10^{58} cm^{-3}. The observed properties of the emission lines are summarized in Table 1, together with the inferred properties of the emitting regions.

3 INTERPRETATION

A likely site for the formation of the emission lines in Sco X-1 and other accretion-driven binaries is a corona above the accretion disk or stellar photosphere (Shakura and Sunyaev 1973; White and Holt 1982; Begelman, McKee and Shields 1983; Begel-

TABLE 1

Quantity	Fe K	Soft X-ray	UV
ε (keV)	6.6–6.7	0.8–1.5	0.007–0.009
$\Delta\varepsilon/\varepsilon$	~ 0.1	< 0.1	< 0.01
EW (keV)	$\simeq 0.05$	0.001–0.01	10^{-5}–10^{-4}
T (K)	$\sim 10^7$	10^6–10^7	$\sim 10^4$
emissivity (erg cm^3 s^{-1})	10^{-24}	10^{-24}	10^{-23}
EM (cm^{-3})	$\sim 10^{59}$	$\sim 10^{58}$	$\sim 10^{56}$
broadening:			
R_{virial} (cm)	$\leq 10^7$	$\geq 10^7$	$\geq 10^{10}$
τ_{Th}	3	< 12	< 120
R_{grav} (cm)	$\leq 10^{6.5}$	$\geq 10^{6.5}$	$\geq 10^7$
f	~ 0.1–1	≤ 0.1	$\simeq 10^{-2}$

man and McKee 1983). The conditions in an accretion disk corona (ADC) can be estimated under the assumption that the corona is supported by X-ray heating. If so, the temperature of the corona depends on the 'ionization parameter', Ξ, defined as the ratio of radiation pressure to gas pressure (cf., Krolik, McKee, and Tarter 1981). Crudely, the temperature T approaches the 'Compton temperature' $\sim 10^7$ K for $\Xi \geq \Xi_{\text{crit}} \sim 1$–10, and $T \sim 10^4$ K, the 'atomic temperature' for $\Xi \leq \Xi_{\text{crit}}$. The scale height, maximum density, vertical column density, and emission measure within a radius interval ΔR and height interval Δz are:

$$z_s = 8 \times 10^7 \text{ cm } R_9^{3/2} T_7^{1/2} \tag{1}$$

$$n_{\max} \simeq 1.9 \times 10^{17} \text{ cm}^{-2} R_9^{-2} L_{37} f T_7^{-1} \tag{2}$$

$$N_{\text{tot}} \simeq 1.5 \times 10^{25} \text{ cm}^{-2} L_{38} f R_9^{-1/2} T_7^{-1/2} \tag{3}$$

$$EM \simeq 3 \times 10^{60} \text{ cm}^{-3} \frac{L_{38}^2 f^2}{T_7^{3/2} R_{10}^{-1/2}} \frac{\Delta R}{R} \frac{\Delta z}{z}, \tag{4}$$

respectively, where R_{10} is the distance from the X-ray source in units of 10^{10} cm, L_{38} is the luminosity of the X-ray source in units of 10^{38} erg s^{-1}, T_7 is the coronal temperature in units of 10^7 K, and a mass of 1 M$_\odot$ has been assumed for the X-ray source. It is apparent that the column density of the corona increases with decreasing distance from the source, and that when $R \leq R_{\text{thick}} = 10^{11}$ cm $L_{38}^2 f^2 T_7^{-1}$ the Thomson depth through the corona exceeds unity. This buildup must ultimately be limited by the fact that X-rays will not penetrate to the base of a corona whose

Thomson depth, τ_{Th}, exceeds some critical value which we define as $\tau_{\text{Th crit}} \simeq 1\text{--}$
10. Therefore, we may treat the requirement $\tau_{\text{Th}} \leq \tau_{\text{Th crit}}$ as a second ("optically
thick") necessary condition for coronal equilibrium. When this condition applies, the
minimum ionization parameter in the corona may be shown to be

$$\Xi_{\text{min}} \simeq 100 \, L_{38} f R_9^{-1/2} T_7^{-1/2} \tau_{\text{Th crit}}^{-1} \, . \tag{5}$$

In the inner region of the disk where the optically thick criterion applies, the mean
ionization parameter may therefore be much greater than Ξ_{crit}, and it increases with
decreasing distance from the source.

The effects of geometry and the transfer of X-rays through the corona are combined in the factor $f = 4\pi R^2 F/L$, where F is the flux of X-rays incident on the
disk surface. $f = 1$ corresponds to unattenuated X-rays normally incident on the
disk. Inherent in the discussion so far has been the assumption that this flux proportionality constant, f, is known. Depending as it does on the detailed geometrical
relationship between the compact object and the disk, i.e., the disk flaring angle and
the size of the compact object, and on the effects of radiative transfer in the corona,
it is difficult to predict the value of this quantity theoretically (cf., London 1985).
Instead, we consider the value of f to be an important unknown in the theory of
X-ray illuminated disks, and seek to constrain it from the observations of emission
lines.

The estimates presented so far suggest that the corona can have conditions appropriate for the formation of the iron K line: temperature $\sim 10^7$ K, and emission
measure $\sim 10^{60}$ cm^{-3}. Further support for this conclusion comes from spectral modelling calculations of the dependence of emissivities of the K lines from the various
stages of iron on radial and vertical position in an ADC, embodying the ADC structure as summarized in the previous section along with detailed calculations of the
atomic physics affecting the line emission (including collisional, recombination, and
inner-shell fluorescence emission) (cf., Kallman and White 1988). These models show
that the requirement that the radius of the iron line emission be small is consistent
with both the line centroid energy, since the coronal ionization is greater at small
radii in the optically thick regime [eq. (5)], and with the line width as implied by
rotational broadening. The Thomson depth of the corona is never great enough to
provide the observed broadening of the iron lines by Compton scattering since the
X-rays required to excite the corona cannot penetrate to $\tau_{\text{Th}} \simeq 3$ where the broadening is sufficient. Broadening due to blending of multiple components results in a
line width of approximately 0.5 keV, and dominates over Compton broadening for
the ADC conditions, but is insufficient to account for the observed width. The line
blends are dominated by emission from ion stages Fe XXV and below, and so have
centroid energies in the range 6.2–6.7 keV. Thus, an ADC can produce the observed
iron K line properties if the emission occurs at small radii in the disk, $R \leq 10^8$ cm
where the corona is optically thick and so is highly ionized and where rotational
broadening is great enough to match that observed. In addition, the line emission

regions must shield the disk at larger radii from X-rays in order to avoid an excess of narrower, lower-energy line emission. This is because in the optically thin coronal region the line blend is receives a strong contribution from the 6.4 keV line which is not observed.

The last conclusion has potentially important consequences for models of the emission in wavelength bands other than the hard X-rays, since it implies that only a small fraction (≤ 0.1) of the flux from the X-ray source can be incident on the disk in the regions where these lines are emitted. This constraint is equivalent to requiring that $f \leq 0.1$ for $R \geq 10^8$ cm. Also included in Table 1 is the value of f required to produce the emission measure implied by equation (3) for the emission lines in the various wavelength bands assuming the radius values from the width constraints and $(\Delta R/R)(\Delta z/z) = 0.1$. This shows that neither the soft X-ray nor the UV lines require values of f in excess of 0.1. In fact, if the optically thin hypothesis is correct, the UV lines require $f \sim 10^{-2}$. The disk illumination becomes progressively weaker with increasing radius, from $f \sim 1$ at $R \leq 10^8$ cm (iron K lines) to $f \sim 0.1$ at $R \geq 10^8$ cm (soft X-ray lines), $f \sim 10^{-2}$ at $R \geq 10^{10}$ cm (UV lines). It is important to point out that even though the illumination is weak compared to the full diluted X-ray flux, the illumination is likely to dominate the local, viscous, release of energy in the disk in the outer, UV-emitting disk regions. For example, at a radius of 10^{10} cm, the ratio of local viscous flux to diluted X-ray flux is $F_{\text{viscous}}/F_{\text{X direct}} = 3GM\dot{M}/(2LR) = 4 \times 10^{-4} \dot{M}_{18} L_{38}^{-1} R_{10}^{-1}$.

REFERENCES

Begelman, M. C., McKee, C. F., and Shields, G. A. 1983, *Ap. J.*, **271**, 70.

Begelman, M. C., and McKee, C. F. 1983, *Ap. J.*, **271**, 89.

Hirano, T., Hayakawa, S., Nagase, F., Masai, K., and Mitsuda, K. 1987, *Pub. Astr. Soc. Japan*, **39**, 619.

Kallman, T., Vrtilek, S. D., and Kahn, S. M. 1989, *Ap. J.*, **345**, 498.

Kallman, T., and White, N. E. 1988, *Ap. J.*, **341**, 955.

Krolik, J. H., McKee, C. F., and Tarter, C. B. 1981, *Ap. J.*, **249**, 422.

London, R. 1985, in *Cataclysmic Variables and Low Mass X-ray Binaries*, ed. D. Q. Lamb and J. Patterson (Dordrecht: Reidel), p. 121.

Raymond J. C., and Smith, B. H. 1977, *Ap. J. Suppl.*, **35**, 419.

Shakura, N. I., and Sunyaev, R. A. 1973, *Astr. Ap.*, **24**, 337.

White, N. E., and Holt, S. S. 1982, *Ap. J.*, **257**, 318.

White, N. E., et al. 1986, *M. N. R. A. S.*, **218**, 129.

White, N. E., Peacock, A., and Taylor, B. G. 1985, *Ap. J.*, **296**, 475.

Willis, A., et al. 1980, *Ap. J.*, **237**, 596.

Numerical Synthesis of Wind-Formed Lines in Cataclysmic Variable Spectra

John A. Woods

Department of Astrophysics, University of Oxford,
Keble Road, Oxford OX1 3RH, U. K.

1 INTRODUCTION — OBSERVATIONAL EVIDENCE FOR WINDS

Mass loss in the form of a high-velocity wind is known to take place from cataclysmic variables with low white dwarf magnetic field strength and high $\dot{M}$, i.e., novalike variables and dwarf novae in outburst. Although signs of outflowing material have been detected in optical data (see, e.g., Marsh and Horne 1989), the principal observational evidence for this phenomenon comes from the C IV λ 1550, N V λ 1240, and Si IV λ 1400 resonance lines as observed with *IUE*. Two types of line feature are seen: Low and moderate ($\lesssim 60°$) inclination systems show broad blueshifted absorption profiles corresponding to terminal velocities of up to about 5000 km s^{-1}, sometimes with accompanying redshifted emission (e.g., RX And, see Fig. 1); whereas eclipsing systems show broad, asymmetric emission that is eclipsed to a much lesser degree than the continuum, implying an extended emitting region (e.g., OY Car, see Fig. 2). The current theoretical understanding of CV winds has been summarized by Drew (1989).

Fig. 1—Typical *IUE* spectrum of a low-inclination system (in this case RX And) showing blueshifted absorption features. The Si III λ 1300 feature is at its rest wavelength and is therefore not due to the wind. This spectrum is the mean of images SWP34808–SWP34812 (see Woods, Drew, and Verbunt 1989). The flux units are ergs cm^{-2} s^{-1} Å^{-1}.

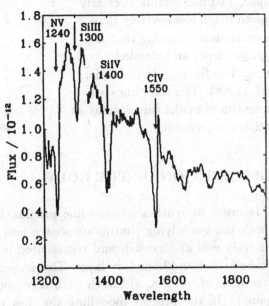

2 MOTIVATION BEHIND THE MONTE CARLO CODE

In view of several theoretical difficulties, not least the uncertainty as to the basic physics of the wind driving mechanism, the work here described aims to extend the approach of Drew (1987) and Mauche and Raymond (1987) in trying to interpret existing observations by comparison with synthesized spectra from simple wind models, without making any attempt to build a self-consistent model of the wind from first principles. A new code has been developed that solves the radiative transfer problem of resonance scattering in the wind by means of a Monte Carlo technique. Since this provides a means of solving the radiative transfer problem *exactly*, it is interesting in itself as a check on the accuracy of the approximate methods that have previously been applied to the problem. It also has a number of other desirable features: unlike codes using, e.g., the Sobolev escape probability method, it can accurately determine the emergent intensity around line centre; it copes well with complex geometries; it handles intrinsic doublets or multiplets accurately (all three principal wind lines are doublets); and it can use effectively any desired velocity law for the wind (the law need not be radial and monotonic).

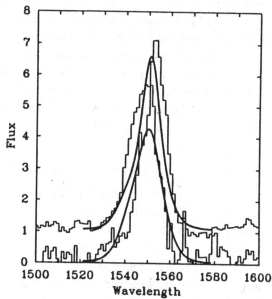

Fig. 2—*IUE* spectra of the high-inclination system OY Car showing the C IV λ 1550 emission line. The upper histogram is a mean out-of-eclipse spectrum, while the lower is the mean of several spectra taken during eclipse. The line feature is clearly eclipsed much less severely than the continuum, showing that it originates from an extended region. The figure is from Naylor *et al.* (1988). The solid lines are the results of model calculations as described in Section 3.

3 DESCRIPTION OF THE CODE

In order to synthesize wind line profiles, it is necessary to know the structure of both the underlying continuum source and the wind itself. The former is comparatively well understood, and represented in the code by a standard steady-state blackbody disk and boundary layer. This neglects the possibility of a disk component in the lines of interest, although such a component can be included trivially when required. In the present modelling the disk is assumed not to be limb-darkened, although future work will investigate the effects of relaxing this constraint. The occulting effects of disk, secondary, and white dwarf are determined using orbital parameters derived from optical studies of particular systems. The structure of the wind, on the other hand, is largely unknown. To specify a wind model for the purposes of the code, we require a velocity law, an ionization law, and the outflow geometry. While the code can cope with essentially any desired parameterization of these properties, the preliminary investigations carried out to date have followed Drew (1987) in adopting a linear velocity law, constant ionization, and a simple form of bipolarity in the outflow density.

Once the wind structure and the nature of the underlying continuum source and absorbing surfaces have been specified in this way, the predicted line profiles for a variety of inclination angles and orbital phases are synthesized by the Monte Carlo technique. This involves generating a representative sample of continuum photons and tracing their scattering history in this model wind.

Although the technique is by its nature statistical, the physics of the radiative transfer of each individual photon is handled exactly. Such considerations as the effects of doublet structure are correctly treated, this being particularly important in the interpretation of line profiles of Si IV $\lambda 1400$, where the doublet separation corresponds to some 2000 km s^{-1}. This introduces radiative coupling between parts of the wind that are widely separated in space, the effects of which have not previously been investigated for any geometry more complex than the stellar (spherically symmetric) case.

4 SOME EARLY RESULTS

The size of the parameter space that this code permits access to is enormous, and work to date has only explored a small fraction of it. However, some useful results have already been obtained. Of particular interest are models of two types of systems: those bright enough to have *IUE* high resolution spectra available, permitting detailed comparison of line profiles; and eclipsing systems where the secondary serves as a probe of the wind structure. An example of a model of the first type of system (in this case the novalike variable RW Sex) is shown in Figure 3, while the smooth curves shown in Figure 2 are in- and out-of-eclipse model spectra for OY Car. In this latter case very accurate parameters for the system geometry were taken from the optical study of Wood *et al.* (1989). None of these model spectra yet represents a formal fit to the data—they are included rather to indicate that there is promise of obtaining reasonable fits with the chosen class of model.

These early results broadly confirm the work of Drew (1987) and of Mauche and Raymond (1987) in that:

a) A slowly accelerated wind is required, not reaching terminal velocity before several tens of white dwarf radii;

b) Mass-loss rates of $\dot{M}_{\rm CIV} \simeq 10^{-10.5}$ M$_\odot$ yr^{-1} are needed (this being the mass-loss rate required assuming all carbon to be in the form of C IV);

c) The outflow for OY Car at least is fairly bipolar, although this conclusion is subject to an investigation of the effects of limb-darkening (see Mauche and Raymond 1987).

Fig. 3—High-resolution *IUE* spectrum (SWP 27089 from the *IUE* archive) of the novalike variable RW Sex showing the C IV λ 1550 line. The spectrum has been divided by a fitted continuum. The smooth curve is a model spectrum, as described in the text.

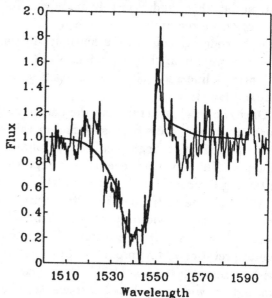

There are a number of interesting prospects for future investigation using this code. Obvious avenues to explore include deviations from spherical velocity laws (such as winds driven from the disk surface and/or with rotational components) and the departures from axial symmetry apparently required by observations of orbital variability in resonance line profiles such as those of YZ Cnc (Drew and Verbunt 1988) and SU UMa (Woods, Drew, and Verbunt 1989).

ACKNOWLEDGEMENT

JAW is in receipt of an SERC postgraduate studentship.

REFERENCES

Drew, J. E. 1987, *M. N. R. A. S.*, **224**, 595.
Drew, J. E. 1989, in *Proc. IAU Coll. No. 122, 'Physics of Classical Novae'*, in press.
Drew, J. E., and Verbunt, F. W. 1988, *M. N. R. A. S.*, **234**, 341.
Marsh, T. R., and Horne, K. 1989, preprint.
Mauche, C. W., and Raymond, J. C. 1987, *Ap. J.*, **323**, 690.
Naylor, T., *et al.* 1988, *M. N. R. A. S.*, **231**, 237.
Wood, J. H., Horne, K., Berriman, G., Wade, R. A. 1989, *Ap. J.*, **341**, 974.
Woods, J. A., Drew, J. E., and Verbunt, F. W., 1989, in preparation.

On the Stability Properties of White Dwarf Settling Solution Shocks

Michael T. Wolff[1], James N. Imamura[2], Kent S. Wood[3], and John H. Gardner[3]

[1]Universities Space Research Association, Columbia, Maryland
[2]University of Oregon, Eugene, Oregon
[3]Naval Research Laboratory, Washington, D.C.

1 INTRODUCTION

AM Her binaries consist of a magnetic white dwarf primary and a late-type, low-mass secondary The secondary overflows its Roche Lobe resulting in mass accretion onto the dwarf that is channeled by the dwarf's magnetic field toward one or both of the dwarf's magnetic poles. At a radius R_s a shock-front transition slows the plasma to subsonic velocity and heats it to X-ray-emitting temperatures. The post-shock plasma cools via optically thin bremsstrahlung, Compton cooling, and optically thick cyclotron cooling, and then merges with the dwarf's atmosphere. We refer to the structure consisting of the shock-front and the cooling region as a "radiative shock." Studies have revealed that radiative shocks are unstable to oscillations in the size of the cooling region under certain circumstances (Langer *et al.* 1982). Chevalier and Imamura (1982; see also Imamura *et al.* 1984) found that radiative shocks oscillate in a number of modes which they referred to, in order of increasing oscillation frequency, as the fundamental mode (F), the first overtone mode (1O), the second overtone mode (2O), and so on. Imamura (1985) showed that when the physics appropriate to accretion onto white dwarfs is included, all oscillation modes were stabilized for high accretion rates onto dwarfs more massive than 0.4 $M_\odot$, and that the 1O mode was unstable for dwarf masses less than 0.4 $M_\odot$ (see Wolff *et al.* 1989).

We model numerically the stability properties of radiative shocks as a function of white dwarf mass (M_*) and accretion rate ($\dot{M}$) with the stipulation that $\dot{M}$ be low enough to cause the shock height relative to the dwarf surface ($d_s = R_s/R_* - 1$, where R_* is the white dwarf radius) to be of the order or larger than the white dwarf radius (i.e., $d_s > 1$). Such shocks are referred to as "settling solutions" (Katz 1977).

2 NUMERICAL CONSIDERATIONS

We model the accretion flows using two hydrodynamic codes, "VEGA" and "FAS-TAR." These codes are one-dimensional, spherically symmetric, and one- (1T) and two-temperature (2T), respectively. We model 2T flows using the equations of mass and momentum conservation in conservative form, an equation for the electron internal energy, and an equation for the total (kinetic + internal + gravitational po-

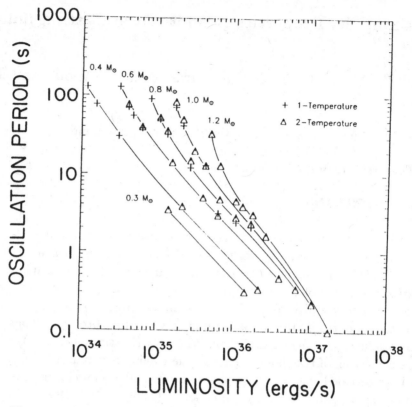

Fig. 1—The oscillation period of the F mode vs. the accretion luminosity, $L_{\rm acc}$. Some of the model points are taken from Wolff *et al.* (1989). The mass-accretion rate can be obtained from $L_{\rm acc}$. The curves are labeled according to the assumed M_*.

tential) energy. Further details regarding the 2T calculations can be found in Wolff *et al.* (1989). We model 1T flows using the continuity and momentum conservation equations as well as an internal energy equation. The gas physics in both the 1T and 2T models includes bremsstrahlung and Compton cooling, electron thermal conduction, and in the 2T models, electron-ion equilibration. This last physical effect is important in that it makes possible the accurate modeling of shocks where the white dwarf mass is greater than 1 $M_\odot$.

Our parameter range for the accretion luminosity ($L_{\rm acc} = GM_*\dot{M}/R_*$) and white dwarf mass can be represented as in Figure 1. Here the 1T models are the crosses and the 2T models are the triangles.

3 RESULTS FOR SETTLING SOLUTIONS

In the low $\dot{M}$ regime, the cooling in the immediate post-shock plasma is inefficient and the flows compressively heat before they are able to radiate significant amounts of energy. This leads to the shock-front having a radius larger than $2R_*$. We find

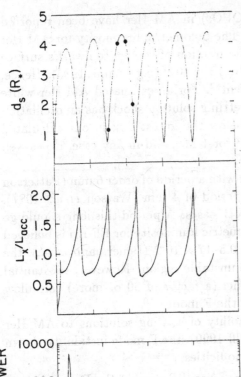

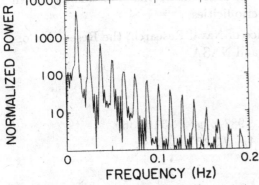

Fig. 2—The variation of the normalized shock height $d_s(t)$, and the normalized luminosity $L_x(t)/L_{acc}$ for a 2T model with $M_* = 1.0$ M$_\odot$ and $\dot{M} = 1.15 \times 10^{-3}\dot{M}_E$. The FPS of the full light curve is shown in the bottom panel.

that white dwarf settling solution radiative shocks oscillate in the F mode for flows onto white dwarfs with masses up to and including 1.0 M$_\odot$. The F mode is the only mode we isolate in the settling solution regime. Figure 2 shows a shock model with $M_* = 1.0$ M$_\odot$ and $\dot{M} = 1.15 \times 10^{-3}\dot{M}_E$, where $\dot{M}_E$ is the Eddington accretion rate. After an initial bounce the F mode oscillates in a limit cycle. This is in contrast to the high $\dot{M}$ regime where the F mode damps away and the 1O forms a limit cycle (Imamura 1985; Wolff et al. 1989). The F modes are unstable in our settling solution models for $L_{acc} < 1.2 \times 10^{34}$, 4.1×10^{34}, 1.1×10^{35}, and 1.7×10^{35} erg s^{-1}, for white masses 0.4, 0.6, 0.8, and 1.0 M$_\odot$, respectively. These L_{acc} values correspond to the mass accretion per unit area, $\dot{M}/A$, of 0.017, 0.047, 0.12, and 0.18 g s^{-1} cm^{-2}, respectively, for the above dwarf masses. In Figure 2 we also show the Fourier power spectrum (FPS) of the light curve for this model. The highest peak in the FPS corresponds to the F mode. The asymmetric profile of the luminosity pulses produced by a shock oscillating in the F mode results in a rich harmonic structure.

Hard X-ray quasi-periodic oscillations (QPOs) in AM Her have been reported with timescales of 400 s (Stella *et al.* 1986). The bolometric luminosity for AM Her is thought to be 10^{32}–10^{33} erg s^{-1}, with f (the fraction of the white dwarf's surface over which the accretion occurs) believed to be 10^{-3}–10^{-4} (e.g., Lamb 1985). Hence, the mass flux is in the range 0.1 to 10 g s^{-1} cm^{-2}. For an estimated AM Her white dwarf mass of 1.0 M$_\odot$ (see Imamura 1984) a settling solution shock has an oscillation period of 400 s with a mass accretion flux below the lowest flux in our calculated models, but it should be near 0.16 g s^{-1} cm^{-2} for 1 M$_\odot$, and in any case, it lies well inside the unstable region of parameter space.

EF Eri has exhibited X-ray QPO behavior with a period of order 6 min (Patterson *et al.* 1981; Beuermann *et al.* 1987) and a period of 4 min (Watson *et al.* 1987). The amplitudes of the QPO was 10–15% in both cases. A period this long should go with very low mass flux. However, as the bolometric luminosity for EF Eri is believed to be 3–7×10^{32} erg s^{-1}, with f in the range 0.5–17×10^{-5} (Beuermann *et al.*), the implied mass flux lies well above the F-mode unstable region. However, substantial variation of mass flux over the accretion funnel (a factor of 30 or more) can allow some portions of the funnel to be unstable to the F mode.

We conclude then that while the applicability of settling solutions to AM Her binaries is questionable (see Imamura and Wolff 1990), the F mode instability shown in our models can reproduce the observed periodicities.

This research was supported by the Office of Naval Research, the Research Corporation, the National Science Foundation, and NASA.

REFERENCES

Beuermann, K., Stella, L., and Patterson, J. 1987, *Ap. J.*, **316**, 360.
Chevalier, R. A., and Imamura, J. N. 1982, *Ap. J.*, **261**, 543.
Imamura, J. N. 1984, *Ap. J.*, **285**, 223.
Imamura, J. N. 1985, *Ap. J.*, **296**, 128.
Imamura, J. N., and Wolff, M. T. 1990, *Ap. J.*, in press.
Imamura, J. N., Wolff, M. T., and Durisen, R. H. 1984, *Ap. J.*, **276**, 667.
Katz, J. 1977, *Ap. J.*, **215**, 265.
Lamb, D. Q. 1985, in *Cataclysmic Variables and Low Mass X-ray Binaries*, ed. D. Q. Lamb and J. Patterson (Dordrecht: Reidel), p. 179.
Langer, S. H., Chanmugam, G., and Shaviv, G. 1982, *Ap. J.*, **258**, 289.
Patterson, J., Williams, G., and Hiltner, W. A. 1981, *Ap. J.*, **245**, 618.
Stella, L., Beuermann, K., and Patterson, J. 1986, *Ap. J.*, **306**, 225.
Watson, M., King, A. R., and Williams, G. 1987, *M. N. R. A. S.*, **226**, 867.
Wolff, M. T., Gardner, J. H., and Wood, K. S. 1989, *Ap. J.*, **346**, 833.

'Blob' Accretion in AM Herculis Systems

S. J. Litchfield

Astronomy Group,
Department of Physics and Astronomy,
The University,
Leicester LE1 7RH U. K.

1 INTRODUCTION

AM Herculis systems are short-period cataclysmic variables consisting of a magnetized white dwarf in synchronous lock with a late spectral-type companion from
which it is accreting. In such systems, the magnetic field of the white dwarf is of
sufficient strength to disrupt the formation of any accretion disk, forcing matter to
accrete along the field lines of the white dwarf. Furthermore, instabilities may result
in inhomogeneous accretion by means of discrete 'blobs' (e.g., Kuijpers and Pringle
1982; Frank *et al.* 1988). Whereas simple bremsstrahlung shock models fail to predict
the observed soft X-ray excess in such systems, blob accretion allows the possibility of
blobs penetrating deeply enough into the photosphere of the white dwarf to thermal-
ize their kinetic energy and produce soft X-ray emission. This heating should distort
the local photosphere, which may protrude sufficiently from the accreting surface to
produce the characteristic light curves of AM Herculis systems in their anomalous
state (Heise *et al.* 1985; Hameury and King 1988).

2 EQUATIONS AND SOLUTION

I have attempted to model the above situation of deeply-buried blobs in a white
dwarf atmosphere in steady-state conditions. This was done by numerically solving
the radiative diffusion approximation:

$$\nabla \cdot (\frac{1}{\kappa \rho} \nabla T^4) = 0 \,. \tag{1}$$

The opacities were assumed to be purely electron scattering, hence κ was taken as
constant, with ρ being given by applying hydrostatic equilibrium. I adopted the
following boundary conditions:

$$T^4 \to 3/4 T_o^4 (\tau + 2/3) \text{ as } \tau \to \pm\infty \,; r \to \infty \tag{2}$$

$$\text{and } \int_S \nabla T^4 \cdot d\mathbf{S} = 0, \, L_s \text{ on the } r = 0 \text{ boundary,} \tag{3}$$

where L_s is the luminosity of the buried heat source.

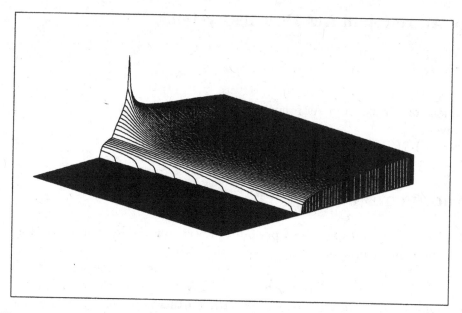

Fig. 1—Temperature distribution in a section through the white dwarf atmosphere shown by a surface plot. Temperature is plotted against radius and optical depth. The luminosity source is located at the point ($r = 0$, $\tau \approx 200$). Optical depth lies in the range $\tau \approx \pm 10^3$; negative values of τ have mostly imaginary temperatures and these are set to 0 in the figure. This particular model was for a 0.6 M$_\odot$ white dwarf with a blob luminosity of 10^{30} erg s^{-1}.

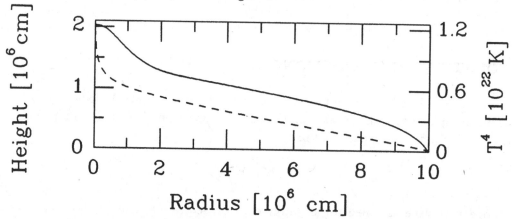

Fig. 2—Plot of both surface height (*solid line*) and T^4 (*dashed line*) against radius. The optical depth at the surface for any given radius is found by solving eq. (4) for the temperature distribution shown in Fig. 1. This is then converted to a physical height by integration. The values of T^4 at the surface are interpolated from the grid of temperatures, and from these are calculated the Planck intensities used in Figs. 3 and 4. Note that the 'hump' is far broader than it is high; this differs from the surface postulated by Heise *et al.* (1985).

Equations (1)–(3) can be expanded for some suitable choice of coordinates, e.g., cylindrical polars with assumed axisymmetry, and written in finite difference form to be solved by standard relaxation methods. However, this is typically prohibitively expensive in terms of computer processing time. Instead, a multi-grid algorithm was used to accelerate line relaxation (e.g., Brandt 1977). At optimum efficiency, this powerful technique can produce *at least* an order of magnitude increase in speed of solution. The iteration proceeded by assuming an initial temperature distribution; using this to give the values of ρ at each grid point; performing one multi-grid cycle; updating the densities once more, and so on until convergence. Once obtained, the solution was refined down to the region of interest, and the surface identified by solving:

$$\left| \frac{3\sigma}{4\kappa\rho} \nabla T^4 \right| = \sigma T^4 . \tag{4}$$

Fig. 3—Light curve obtained from the surface shown in Fig. 2. The flux is given in arbitrary units. The local intensities are given by $B_\nu(T)$, where T is the temperature at that radius. The bolometric light curve has a similar form.

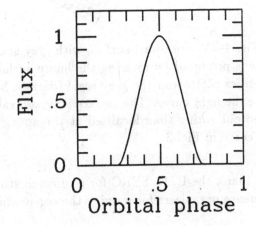

3 RESULTS AND SUMMARY

Preliminary results follow. Figure 1 shows the temperature distribution in the white dwarf atmosphere obtained from a 256 by 256 zoned model. The surface height and temperature variations with radius are shown in Figure 2, with the corresponding light curve shown in Figure 3. A three-dimensional representation of the surface with the intensities shown as a grey scale is given in Figure 4. The flux in the light curve plot was obtained by integrating Planck intensities, $B_\nu(T)$, filtered at a frequency of $\nu = 3.0 \times 10^{16}$ Hz (100 Å), over the hump surface. The light curve has a cosine form, indicating that the steady-state approximation is perhaps insufficient; or that the assumption of hydrostatic equilibrium is incorrect and some form of dynamic splash model is required.

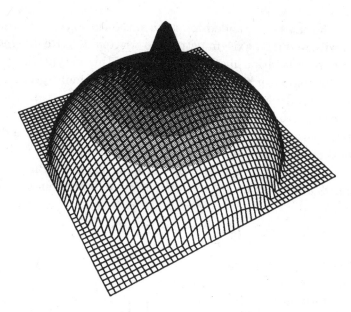

Fig. 4—White dwarf surface with grey scale intensities. This three-dimensional image is produced by rotating the hump surface in Fig. 2 through 2π radians. The intensities plotted on the grey scale (darkest being brightest) are those used to calculate the light curve. The surface is broad, rather than peaked, and the emission is spread out rather than localized at the apex. This explains the cosine form of the light curve in Fig. 3.

I thank the U. K. SERC for a research studentship and travel support, and the conference organizers for waiving the registration fee.

REFERENCES

Brandt, A. 1977, *Math. Comput.*, **31**, 333.
Frank, J., King, A. R., and Lasota, J.-P. 1988, *Astr. Ap.*, **193**, 113.
Hameury, J. M., and King, A. R. 1988, *M. N. R. A. S.*, **235**, 433.
Heise, J., *et al.* 1985, *Astr. Ap.*, **148**, L14.
Kuijpers, J., and Pringle, J. E. 1982, *Astr. Ap.*, **114**, L4.

Cyclotron Heating Above the AM Herculis Accretion Shock

J. J. Brainerd

The University of Chicago

1 INTRODUCTION

Research on cyclotron emission from a geometrically thin post-shock region dominates our community's efforts to understand the polarized optical emission produced by AM Herculis binaries (Meggitt and Wickramasinghe 1982; Barrett and Chanmugam 1984; Schmidt, Stockman, and Grandi 1986; Stockman and Lubenow 1987; Wickramasinghe and Ferrario 1988; Wu and Chanmugam 1988, 1989). Cyclotron emission from above the shock has received only slight attention (Chanmugam and Dulk 1981; Kylafis and Lamb 1982; Lamb 1985). In a recent paper (Brainerd 1989), I calculate the cyclotron emission from a hot pre-shock plasma flowing along magnetic field lines and conclude that such emission contributes to the polarized optical spectrum of AM Herculis binaries. In this paper, I show that cyclotron absorption and re-emission of shock radiation by the accretion stream immediately above the shock is inevitable.

2 PLASMA HEATING

Cyclotron absorption of blackbody radiation will heat a free-falling plasma if the blackbody temperature T_{bb} is larger than the plasma temperature T_e. These temperatures are in units of energy. The rate at which energy is absorbed per unit column length is given by equation (4) of Brainerd (1989), with the dimensionless temperatures converted to units of energy, the factor T_e set to T_{bb}, and the temperature in the function $K(T_e)$ set to a characteristic line width temperature T_{lw}. Figure 4 of Brainerd (1989) suggests that $K(T_{lw})$ behaves approximately as $K(T_{lw}) = 10^3 T_{lw}/m_e c^2$. The electron thermal energy per unit column length is $dE/dR = 4\pi r^2 n T_e = 2LT_e/m_p V_{ff}^3$, where r is the column radius, n is the electron number density, L is the accretion luminosity, m_p is the proton mass (I assume the plasma composition is fully-ionized hydrogen), and V_{ff} is the free-fall velocity. Dividing this by equation (4) of Brainerd (1989) gives the heating time scale t_h.

The temperature increases rapidly over a column length l only if t_h is much less than the flow time scale $t_{fl} = l/V_{ff}$. The shock geometry dictates a length scale of order the accretion column radius r, so $l = r = 2R_0 f^{1/2}$, where R_0 is the white

344   J. J. Brainerd

Fig. 1—Total (*solid*) and circularly polarized (*dotted*) flux plotted as a functions of frequency.

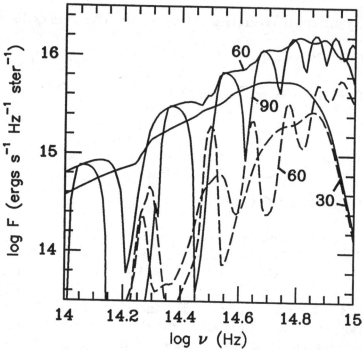

dwarf radius and $f > 10^{-4}$ is the fraction of the white dwarf surface covered by the accretion column. Setting $t_h \ll t_{fl}$ gives the condition on T_{bb} of

$$T_{bb} \gg \frac{\pi^2 m_e^3 c^5 L}{4 m_p e^3 G M R_0 B^3} \frac{T_e}{f K(T_{lw})} = 0.09 \frac{M_\odot L_{32} T_e}{M R_9 B_7^3 T_{lw} f} \, \text{keV} . \qquad (1)$$

The variables B and M are the surface magnetic field strength and the white dwarf mass. The luminosity L_{32} is in units of $10^{32} \, \text{ergs s}^{-1}$, the white dwarf radius R_9 is in units of $10^9 \, \text{cm}$, and the surface magnetic field strength B_7 is in units of $10^7 \, \text{G}$. When equation (1) holds, the accretion column comes into radiative equilibrium with the shock.

The term T_{lw} in equation (1) describes the influence of the cyclotron line structure on the heating rate. The temperature dependence is principally through line broadening from two sources: the electron thermal motion and the variation of the magnetic field. For absorption at angles much less than 90°, the thermal line width is of order $\Delta\nu/\nu \approx (T_e/m_e c^2)^{1/2}$. If the line width is from field variation, one has $\Delta\nu/\nu \approx \Delta B/B \approx \Delta R/3R$. Setting the radial variation equal to the column radius gives $\Delta R/R \approx 2 f^{1/2}$. The field variation line width temperature can therefore be defined as $T_{bf} = 4 f m_e c^2/9$. The actual line temperature T_{lw} is the maximum of T_e and T_{bf}, so one always has $T_{lw} \geq T_e$, with equality occurring when $T_e \geq 4 f m_e c^2/9 > 0.023 \, \text{keV}$.

Setting $T_{lw} = T_e$, $f = 10^{-3}$, and $B_7 = 2$, one finds that the accretion column achieves radiative equilibrium if $T_{bb} \gg 20 \, \text{keV}$. Increasing the field strength to $B_7 = 4$ lowers the condition to $T_{bb} \gg 2.6 \, \text{keV}$, which shows the extreme sensitivity of

Fig. 2—Linear (*upper half*) and circular (*lower half*) polarization fractions plotted as functions of frequency.

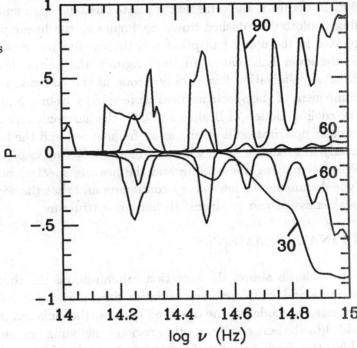

cyclotron heating on the magnetic field strength. Clearly, the temperature condition is met for a wide range of parameters characteristic of AM Herculis binaries. When these conditions are not met, one still finds that the plasma temperature has a small minimum value, because for any value of T_{bb}, one can find a plasma temperature $T_e \ll T_{lw} = T_{bf}$ that satisfies equation (1).

3 CYCLOTRON SPECTRUM AND POLARIZATION

To understand the type of spectrum produced through this process, I assume a temperature model for the plasma above the shock. Let the geometry of the problem determine the temperature so that the cyclotron emission into solid angle 4π at temperature T_e equals the cyclotron absorption over solid angle Ω of energy with temperature T_{bb}. For a height h above the white dwarf surface, the shock subtends a solid angle of $\Omega = 2\pi r^2/\sqrt{r^2 + h^2}\left(h + \sqrt{r^2 + h^2}\right)$. The plasma temperature is then $T_e = 2R_0^2 f T_{bb}/\sqrt{4R_0^2 f + h^2}\left(h + \sqrt{4R_0^2 f + h^2}\right)$. This temperature model assumes that the radiative transfer along the column length and the angle dependence of the line structure are unimportant.

Cyclotron spectra and polarization fractions, which are calculated in the manner outlined in Brainerd (1989), are plotted in Figures 1 and 2. These results are for a white dwarf mass and radius of $0.9\,M_\odot$ and 6.3×10^8 cm, an accretion luminosity of 10^{32} ergs s^{-1}, a surface field strength of 5×10^7 G, a blackbody temperature of $20\,keV$, and an accretion area of $f = 10^{-3}$. The emission angles are $30°$, $60°$, and

90°. In Figure 1, the total flux is plotted as solid lines and the circularly-polarized flux is plotted as dashed lines. In Figure 2, the linear polarization fractions are plotted in the upper half plane and the circular polarization fractions are plotted in the lower half plane. In these figures, the harmonic structure is strong, the circular polarization flux is only strong at the emission peak, and the low-energy component of the spectrum rises more rapidly than ν and less rapidly than ν^2. In the cooling models of Brainerd (1989), the harmonic structure is often absent, the circular polarization is strong at all frequencies, and the low-energy spectrum rises nearly as ν. In the cases where harmonic structure is present, the lines are broad. When combining the cooling and the heating spectra, one expects that the first type dominates the low-energy continuum and that the second type dominates the peak emission and produces the harmonic structure.

4 FINAL COMMENTS

As shown above, the accretion column above the shock heats to a high temperature and emits cyclotron radiation for conditions characteristic of AM Herculis systems. In modeling the observed spectra, the emission from this region must be added to the emission from other regions, including emission from a cooling region above the shock, emission from the post-shock region, and emission from the surface around the accretion region. This last unexplored source of optical emission is particularly intriguing: the large white dwarf surface area that absorbs radiation from the accretion column must emit low-temperature cyclotron radiation.

Yet unresolved is the influence of temperature and magnetic field variation over the width and the length of the accretion column on cyclotron radiation from both the accretion column and the underlying shock. The observations cannot be understood until this problem is solved.

REFERENCE

Barrett, P. E., and Chanmugam, G. 1984, *Ap. J.*, **278**, 298.

Brainerd, J. J. 1989, *Ap. J.*, **345**, 978.

Chanmugam, G., and Dulk, G. A. 1981, *Ap. J.*, **244**, 569.

Kylafis, N. D., and Lamb, D. Q. 1982, *Ap. J. Suppl.*, **48**, 239.

Lamb, D. Q. 1985, in *Cataclysmic Variables and Low-Mass X-Ray Binaries*, ed. D. Q. Lamb and J. Patterson (Dordrecht: Reidel), p. 179.

Meggitt, S. M. A., and Wickramasinghe, D. T. 1982, *M. N. R. A. S.*, **198**, 71.

Schmidt, G. D., Stockman, H. S., and Grandi, S. A. 1986, *Ap. J.*, **300**, 804.

Stockman, H. S., and Lubenow, A. F. 1987, *Ap. Sp. Sci.*, **131**, 607.

Wickramasinghe, D. T., and Ferrario, L. 1988, *M. N. R. A. S.*, **334**, 412.

Wu, K., and Chanmugam, G. 1988, *Ap. J.*, **331**, 861.

Wu, K., and Chanmugam, G. 1989, *Ap. J.*, **344**, 889.

The Linear Polarization from Asymmetric Accretion Shocks

G. Chanmugam and Kinwah Wu

Department of Physics and Astronomy,
Louisiana State University,
Baton Rouge, LA 70803

1 INTRODUCTION

A distinguishing characteristic of the AM Herculis binaries is the strong optical and infrared polarized radiation ($\sim 10\%$) they emit (see the review by Cropper 1989). This radiation is believed to be due to cyclotron radiation arising from the post-shock region in the accretion column. Linear polarization typically occurs in the form of a pulse per accretion pole and usually does so when the circular polarization changes sign (Tapia 1977). Nevertheless, in many cases only one such pulse is seen per orbital cycle instead of at both circular polarization crossing points when the magnetic field B is essentially perpendicular to the line of sight.

Models for the polarized radiation from the AM Her binaries have been presented in several papers, assuming a homogeneous emission region (e.g., Chanmugam and Dulk 1981; Barrett and Chanmugam 1984; Wickramasinghe and Meggitt 1985) or an inhomogeneous one (Wu and Chanmugam 1989; Wickramasinghe and Ferrario 1988; Brainerd 1989) where either the electron density $N(r)$ or temperature $T(r)$, or both, were assumed to vary. Here r is the distance from the symmetry axis of the accretion column. It is therefore surprising that the most basic problem of calculating the circularly- and linearly-polarized radiation from accretion shocks with a three-dimensional (3D) structure has only recently been performed for the first time (Wu and Chanmugam 1990). Here we focus on one aspect of these calculations: the linearly-polarized radiation emitted by an asymmetric accretion column. We show that the polarization light curves become asymmetric and that the linear polarization pulses have different heights and widths. We also show that for extremely asymmetric accretion, linear polarization pulses may occur only once per accretion pole per orbital cycle, as is often observed.

2 THE ASYMMETRIC ACCRETION MODEL

In calculating the structure of the shock we have assumed that the cooling is generally dominated by bremsstrahlung radiation. Since bremsstrahlung radiation is optically thin, the flow equations and the radiative transfer equations are decoupled and the problem is readily solvable. The plasma flow is assumed to be confined in magnetic tubes, and the 3D equations are therefore decoupled into independent sets of 1D equations. In each flow tube, the fluid elements pass through a shock, with the shock jump conditions. The temperature at the shock is $T \approx 3GM\mu/8Rk$, where M and R are the mass and radius of the white dwarf. With the physical conditions at the shock and the white dwarf surface ($T = 0$ K) as the boundary conditions, the flow equations can be then integrated (Aizu 1973; Chevalier and Imamura 1982), and the shock height, the flow velocity pattern, and the 3D temperature and electron number density structure in the post-shock region obtained. The optical/IR spectra and polarization can then be calculated by solving the radiative transfer equations for the case of large Faraday rotation assuming that cyclotron absorption (Robinson and Melrose 1984) is the dominant radiation process (Wu and Chanmugam 1990).

For asymmetric accretion rates across the accretion columns, the emission region will have an irregular shape (Fig. 1). The effective cross section of the accretion column is then no longer circular or ring-shape (Mukai 1988; Mason, Rosen, and Hellier 1988). Such a complex geometry will cause the effective optical depths of the emission region to vary dramatically as the orientation of the shock-heated region changes during an orbital cycle. Consequently, asymmetric polarization light curves with different shapes at different frequencies may be obtained. Many AM Her systems indeed have asymmetric light curves as, for example, EXO 033319-2554.2 (Berriman and Smith 1988) and polarization light curves with different shapes in different frequency bands (e.g., EF Eri, see Piirola, Reiz, and Coyne 1987), in support of this hypothesis.

All previous theoretical models with a simple geometry predicted two identical linear polarization pulses in one orbital period (Barrett and Chanmugam 1984). However, in most observations, pulses with different heights and widths (Cropper 1986) are seen while sometimes only one per accretion pole per orbital cycle is seen (Cropper 1985). The pulse asymmetry may be explained by assuming that the emission region is offset from the pole so that the field lines are tilted (Wickramasinghe 1989). Here, we point out that asymmetric accretion can also produce such an asymmetry so that the actual polarization light curves will be due to a combination of these effects. To illustrate the effects due to asymmetric accretion we consider such a case (see Fig. 1).

Fig. 1—The cyclotron emission region with an asymmetric accretion rate. A and B are two viewing directions which are perpendicular to the field.

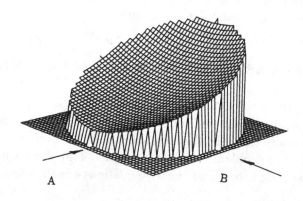

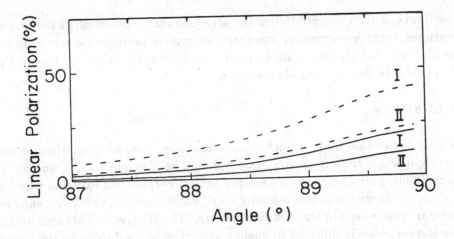

Fig. 2—The linear polarization due to two accretion columns, each of radius $r_0 = 10^7$ cm, are plotted against the viewing angle for $s = 8$. The dashed curves correspond to the viewing direction A and the solid curves to B. Model I corresponds to an asymmetric column where the peak in the electron number density is shifted by $0.7r_0$ while model II to a less asymmetric column with a shift of $0.2r_0$.

For simplicity, consider only one active accretion column, in the form of a cylinder of radius $r_0 = 10^7$ cm outside which there is no accretion, and assume it to be off the rotating axis. We take $N(\mathbf{r}) = 2 \times 10^{16} \exp(-|\mathbf{r} - \mathbf{r}_s|/0.6r_0) \, \mathrm{cm}^{-3}$, where $\mathbf{r}$ is the displacement from the central axis of the accretion column and $\mathbf{r}_s$ is the shift of the peak in the electron number distribution on a plane perpendicular to the central axis. Figure 2 shows results for two asymmetric accretion rates I and II where the latter is less asymmetric than the first. As the white dwarf rotates during the orbital cycle, there are two positions where the line of sight is perpendicular to the magnetic field line so that linear polarization pulses are expected to be observed. Since the accretion rate across the accretion column is asymmetric, the optical depths are different when the emission region is viewed at these positions. The corresponding linear polarization for the two viewing directions A and B are shown in Figure 2 for a harmonic number $s = 8$, indicating different heights and widths for the two polarization pulses. However, the polarization does not change significantly with the angle θ between the line of sight and the magnetic field when viewed from B, implying a very small polarization pulse. Such a small pulse can easily be contaminated by background fluxes and remain undetected. Therefore, only one polarization pulse is seen in an orbital cycle at this frequency.

Furthermore, we emphasize that if only one or two linear polarization pulses are observed in an orbital cycle, one cannot immediately conclude whether it has one or two active accretion poles by merely counting the number of pulses (Meggitt and Wickramasinghe 1989). For the cases with two linear polarization pulses, both pulses can arise from one accretion pole when the accretion rate is essentially axi-symmetric or, sometimes, from two accretion poles with extremely asymmetric accretion rates. If more than two polarization pulses are observed (Piirola, Reiz, and Coyne 1987), then it is probably due to two-pole accretion.

CONCLUSIONS

We have shown that for asymmetric accretion rates, the linear polarization pulses due to asymmetric accretion columns have different heights and widths, and for some frequency bands, only one pulse is seen per orbital cycle. This explains why linear polarization pulses with different sizes are seen, while sometimes only one such pulse in an orbital cycle is seen in the AM Her systems. The relative heights and widths of the polarization pulses in different frequency bands can be used to probe the accretion rate inhomogeneities and degree of asymmetry of the accretion column.

This research was supported by NSF grant AST-8822954.

REFERENCES

Aizu, K. 1973, *Prog. Theoret. Phys.*, **49**, 1184.

Barrett, P. E., and Chanmugam, G. 1984, *Ap. J.*, **278**, 298.

Berriman, G., and Smith, P. S. 1988, *Ap. J. (Letters)*, **329**, L97.

Brainerd, J. J. 1989, *Ap. J.*, **348**, 978.

Chanmugam, G., and Dulk, G. A. 1981, *Ap. J.*, **244**, 569.

Chevalier, R. A., and Imamura, J. N. 1982, *Ap. J.*, **261**, 543.

Cropper, M. 1985, *M. N. R. A. S.*, **212**, 709.

———. 1986, *M. N. R. A. S.*, **222**, 853.

———. 1989, *Space Sci. Rev.*, in press.

Mason, K. O., Rosen, S. R., and Hellier, C. 1988, in *The Physics of Compact Objects, Advances in Space Research* Vol. 8 (2), eds N. E. White and L. Filipov, (Oxford: Pergamon Press) p. 293.

Meggitt, S. M. A., and Wickramasinghe, D. T. 1989, *M. N. R. A. S.*, **236**, 31.

Mukai, K. 1988, *M. N. R. A. S.*, **232**, 175.

Piirola, V., Reiz, A., and Coyne, G. V. 1987, *Astr. Ap.*, **186**, 120.

Robinson, P. A., and Melrose, D. B. 1984, *Australian J. Phys.*, **37**, 675.

Tapia, S. 1977, *Ap. J. (Letters)*, **212**, L125.

Wickramasinghe, D. T., 1989, in *White Dwarfs, I. A. U. Colloquium No. 114*, ed. G. Wegner (Berlin: Springer-Verlag), p. 314.

Wickramasinghe, D. T., and Ferrario, L. 1988, *Ap. J.*, **334**, 412.

Wickramasinghe, D. T., and Meggitt, S. M. A. 1985, *M. N. R. A. S.*, **214**, 605.

Wu, K., and Chanmugam, G. 1989, *Ap. J.*, **344**, 889.

———. 1990, *Ap. J.*, in press.

Polarized Radiation from Extended Polar Caps

Juhan Frank

Max-Planck-Institut für Astrophysik
8046 Garching bei München, F. R. G.

ABSTRACT

The values of magnetic moments in magnetic cataclysmic variables expected from theoretical considerations about spin up/down and binary evolution are reviewed. While observational estimates for magnetic fields exist for many AM Her binaries, BG CMi is the only intermediate polar with direct observational evidence of its magnetic nature. The degree of polarization observed in BG CMi and its wavelength dependence can be understood as a combination of polarized free-free and cyclotron emission from a large polar cap with dilution by the thermal radiation from the underlying heated white dwarf surface. Recent theoretical models suggest a value of 4×10^6 G for the polar field in BG CMi. The evolutionary implications for locking and the relationship between AM Her systems and intermediate polars are discussed.

1 INTRODUCTION

The evolution of compact binaries including magnetic systems has been recently reviewed by King (1988). The evolution of magnetic cataclysmic variables has been the particular focus of a review by Lamb and Melia (1987), and recent developments are discussed by Hameury *et al.* (1989). The similarity between the period distribution of non-magnetic and magnetic cataclysmic variables (MCVs) suggests that the orbital evolution of these systems is similar. This is consistent with the assumption that the evolution of MCVs is driven as in all cataclysmic variables by magnetic braking of the secondary and gravitational radiation because neither of these mechanisms depends on the magnetic field of the white dwarf. Departures from this simple picture adopted here are evaluated critically in the above references.

The MCVs are divided into two subclasses: the synchronous systems or AM Herculis binaries (AM Hers) and the asynchronous systems termed intermediate polars (IPs) or DQ Herculis binaries. If MCVs evolve in the same way as non-magnetic systems, then at least some IPs above the period gap must eventually lock and become AM Hers as the binary shrinks while the magnetospheric radius of the white dwarf increases. This possibility was first pointed out by Chanmugam and Ray (1984) who estimated the minimum polar field required for locking to be around 2 MG. King,

Frank, and Ritter (1985) proposed that the period distribution of IPs and AM Hers could be understood if the white dwarfs in both subclasses had similar magnetic moments ($\mu_{IP} \sim \mu_{AM} \sim 10^{33-34}$ G cm^3). Currently, 11 out of 13 IPs are known with periods longer than 3 hours and 12 out of 17 AM Hers have periods below 2 hours.

Early attempts to estimate the magnetic fields of IPs based on the observed spin-up/down behaviour indicated magnetic fields substantially lower than those measured in AM Hers (Lamb and Patterson 1983). This approach however leads to model-dependent results and was criticized by van Amerongen et al. (1987). See also Patterson (1990). The requirement that the spin period of the white dwarf should not be less than the equilibrium period is somewhat less model dependent and was used by Lamb and Melia (1987) to estimate the magnetic moments of IPs. They obtained $\mu_{IP} \lesssim \frac{1}{10}\mu_{AM}$ and concluded that most IPs would not become AM Hers and that in most cases there would be room for an accretion disk to develop between the magnetospheric boundary and the Roche lobe.

Numerous searches for linear and circular polarization in most IPs have yielded extremely low values consistent with zero polarization (see a summary by Berriman 1989). There is to date only one clear detection of polarization rapidly rising towards near-infrared wavelengths in BG CMi (Penning, Schmidt, and Liebert 1986; West, Berriman, and Schmidt 1987). These observations are of great importance as they constitute the first *direct* evidence of the magnetic nature of IPs and afford the possibility of estimating the magnetic field intensity for this object (Chanmugam et al. 1989, see below).

Our understanding of the relationship between IPs and AM Hers might be summarized as follows: if $\mu_{IP} \sim \mu_{AM}$ then the extremely low degree of polarization observed in IPs requires an explanation (see below); if on the other hand $\mu_{IP} \lesssim \mu_{AM}$ then the observed period distribution of MCVs does not result naturally, leading Lamb and Melia (1987) to appeal to selection effects. Here we take the view that IPs evolve into AM Hers and that the low degree of polarization of IPs is due to dilution effects in a large polar cap.

2 THE KEY OBJECT: BG CANIS MINORIS

The hard X-ray light curves of most IPs display roughly sinusoidal modulation at the spin period. This prompted the suggestion by King and Shaviv (1984) that the polar caps in these systems occupy a large fraction $f \sim 0.25$ of the white dwarf's surface. Chanmugam and Frank (1987) calculated the polarized cyclotron emission from such large polar caps for a range of polar field intensities. These model calculations showed that the polarization in extended polar caps is indeed lower than in a small polar cap with the same polar field because of the spread in field directions and strengths.

The circular polarization detected by Penning et al. (1986) in the optical infrared can be explained by a large polar cap with a high field ($B \gtrsim 50$ MG) radiating in a near optically thick regime (Chanmugam and Frank 1987). However, the detection by West *et al.* (1987) of circular polarization rising rapidly towards the IR, with values of -1.74% and -4.24% in the J and H bands respectively, is contrary to the expectation of decreasing polarization with increasing optical depth and is reminiscent of the behaviour found in AM Hers at optical wavelengths. It suggests a somewhat lower field and *small* optical depth.

In the optically thin regime all relevant diluting fluxes must be taken into account. The observations of BG CMi also showed the presence of 50% photometric modulation (Penning *et al.* 1986) implying that any diluting flux must itself be modulated. The obvious candidate for this flux is the reprocessed radiation from the underlying white dwarf surface which results from the thermalization of about one-half of the accretion luminosity released in the hot post-shock plasma with $T_s \sim 10^8$ K. The characteristic temperature of this blackbody flux is relatively insensitive to assumptions and is typically $T_{bb} \sim 10^5$ K. The circular polarization is due to differences in optical depth for the propagation of the two modes with electric vectors rotating in opposite senses. Quite generally one may write the circular polarization as follows:

$$\frac{V}{I} = \frac{I_O - I_X}{I_O + I_X + I_{bb}} \sim \frac{I_{RJ}(T_s)(\tau_O - \tau_X)}{I_{RJ}(T_s)(\tau_O + \tau_X) + I_{RJ}(T_{bb})},$$

where $I_{RJ}(T)$ is the Rayleigh-Jeans intensity for the temperature T and $\tau_{O,X}$ is the optical depth for the ordinary,extra-ordinary mode. Clearly, dilution by the reprocessed radiation can only be important in reducing the degree of polarization when

$$\tau_O + \tau_X \lesssim T_{bb}/T_s \sim 10^{-3}.$$

Chanmugam *et al.* (1989) have calculated the polarization from large polar caps in the optically thin limit including dilution by the reprocessed radiation, and taking into account bremsstrahlung and cyclotron opacities. They found that the degree of circular polarization observed in BG CMi can be explained by a large polar cap with an aperture of ~ 60 degrees and polar fields ranging from 2 MG to 10 MG if the plasma temperature is adjusted to obtain a fit. If an X-ray temperature $T_s \sim 10$ keV is adopted (McHardy *et al.* 1984) then a value around 4 MG is required to fit the data. The polarization in the optical and optical infrared is dominated by bremsstrahlung, whereas the rise towards near-infrared wavelengths is caused by cyclotron radiation as anticipated by West *et al.* (1987). Work currently in progress indicates that these conclusions also hold for thin rings with a similar aperture of 60 degrees.

3 DISCUSSION AND OUTLOOK

With a polar field of 4 MG, BG CMi would have a dipole magnetic moment in the range $(1-2)\times 10^{33}$ G cm^3. This is near the lower end of the range suggested by King, Frank, and Ritter (1985) but high enough for this system to lock and to become an AM Her below the period gap. With an orbital period of 3.24 hours and the magnetic moment estimated above, no accretion disk should be present in this system (Hameury, King, and Lasota 1986). For the particular case of BG CMi, these two conclusions are also consistent with the theoretical considerations of Lamb and Melia (1987) in spite of their claims to the contrary for IPs *as a class*. In fact, if *all* IPs had magnetic moments near the upper limits allowed by their spin periods then most of them would lock and have no accretion disk.

Clearly, more polarization observations of IPs are desirable, in particular at infrared wavelengths. Modelling of the spectral dependence of the polarization, following for example the method of Chanmugam *et al.* (1989), would then enable better estimates of the magnetic fields and moments in these objects than it is presently possible. Model calculations currently in progress suggest that polarization measurements longward of 2μm may be required to derive reliable estimates or place upper limits on field strengths in the accretion region. A further complication is that higher multipoles may dominate the field near the white dwarf. The polarization observations may determine the strength of this near field whereas the evolution and structure of IPs depend mainly on the dipole moment. These observations and their detailed modelling are necessary for quite a few IPs before one can reach definitive conclusions about the structure and evolution of these objects and their link to AM Hers.

REFERENCES

Berriman, G. 1988, in *Polarized Radiation of Circumstellar Origin*, ed. G. Coyne *et al.* (Vatican City: Vatican Press), p. 281.

Chanmugam, G., and Frank, J. 1987, *Ap. J.*, **320**, 746.

Chanmugam, G., and Ray, A. 1984, *Ap. J.*, **285**, 252.

Chanmugam, G., Frank, J., King, A. R., and Lasota, J.-P., 1989, *Ap. J. (Letters)*, in press.

Hameury, J. M., King, A. R., and Lasota, J. P. 1986, *M. N. R. A. S.*, **218**, 695.

Hameury, J. M., King, A. R., and Lasota, J. P. 1989, *M. N. R. A. S.*, in press.

King, A. R. 1988, *Q. J. R. A. S.*, **29**, 1.

King, A. R., Frank, J., and Ritter, H. 1985, *M. N. R. A. S.*, **213**, 181.

King, A. R., and Shaviv, G. 1984, *M. N. R. A. S.*, **211**, 883.

Lamb, D. Q., and Melia, F. 1987, *Ap. Sp. Sci.*, **131**, 511.

Lamb, D. Q., and Patterson, J. 1983, in *Cataclysmic Variables and Related Objects*, ed. M. Livio and G. Shaviv (Dordrecht: Reidel), p. 229.

McHardy, I. M., Pye, J. P., Fairall, A. P., Warner, B., Cropper, M., and Allen, S. 1984, *M. N. R. A. S.*, **210**, 663.

Patterson, J. 1990, this volume.

Penning, W. R., Schmidt, G. D., and Liebert, J. 1986, *Ap. J.*, **301**, 881.

van Amerongen, S., Augusteijn, T., and van Paradijs, J. 1987, *M. N. R. A. S.*, **228**, 377.

West, S. C., Berriman, G., and Schmidt, G. D. 1987, *Ap. J. (Letters)*, **322**, L35.

A Theory of the Rapid Burster (MXB1730–335)

Tomoyuki Hanawa[1], Kouichi Hirotani[1], and Nobuyuki Kawai[2]

[1]Department of Astrophysics, Nagoya University
[2]Cosmic Radiation Laboratory, The Institute of Physical and Chemical Research

ABSTRACT

The Rapid Burster (MXB1730–335) is a unique X-ray burster from which both type I and II bursts are observed. In the present paper, we construct a model for the Rapid Burster based on the observed properties of type II bursts. In our model, the Rapid Burster is assumed to be a binary system containing a magnetized accreting neutron star. First, we discuss the magnetic field strength of the neutron star. It is estimated to be $B \sim 10^8$ G at the neutron star surface from three independent arguments. Second, we discuss the high-energy power-law component observed in the spectra of type II bursts. In our model, high-energy X-ray photons are produced by Compton scattering with electrons in the accretion flow. Since the accretion velocity is about a half the speed of light, the scattered photons gain energy. It is also shown that our model spectrum fits the observed spectrum quantitatively.

1 INTRODUCTION

The Rapid Burster is an important X-ray source from which a large amount of information has been obtained. Besides normal type I bursts and quasi-periodic oscillations, rapidly-repetitive type II bursts are observed from the Rapid Burster (Lewin et al. 1976). The Rapid Burster has been observed extensively with many X-ray satellites and many interesting properties have been discovered. Compared to the progress in the observational studies, the progress in the theory of the Rapid Burster is slow. Only a few questions imposed by observations are solved theoretically and many interesting characteristics are still open questions. At present, we are sure that the Rapid Burster is a close binary system containing a neutron star and that the energy source of type II bursts is the gravitational energy release of the gas accreting onto the neutron star (Hoffman et al. 1978), that the Rapid Burster has a relatively strong magnetic field among X-ray bursters, and that type II bursts are triggered by some type of magnetohydrodynamical instability (see, e.g., Lamb et al. 1977). It is, however, a pending question which kind of instability is responsible for type II bursts, how the gas is accreted onto the neutron star during a type II burst, how the observed X-rays are radiated, and so on.

In this paper, we discuss the accretion onto the neutron star during a type II burst and the spectrum of type II bursts. In Section 2, we consider the magnetohydrodynamics of the accretion onto a magnetized neutron star and estimate the magnetic field strength of the Rapid Burster by comparing the model with observations. Three independent arguments lead us to a conclusion that the magnetic field is $B \sim 10^8$ G at the neutron star surface. In Section 3, we propose a model for the high-energy power-law component observed in the spectrum of type II bursts (Kawai *et al.* 1990). Our model is a photon analogue of first-order Fermi acceleration. X-ray photons are accelerated by the bulk motion of electrons in the accretion flow through Compton scattering. Section 2 of this paper is based on Hanawa, Hirotani, and Kawai (1989) and Section 3 on Hirotani, Hanawa, and Kawai (1989).

2 MAGNETIC FIELD STRENGTH

Following is the rough scenario for type II bursts. In the Rapid Burster, the region adjacent to the magnetosphere acts as the main reservoir of type II bursts. When the gas in the main reservoir reaches a critical amount, magnetic Rayleigh-Taylor or some other instabilities take place. Then a certain amount of the gas is transferred into a region inside the magnetosphere. We name this region the waiting room. The gas in the waiting room has an angular momentum comparable to the Keplerian one and does not fall onto the neutron star directly. The magnetic fields are sheared in the azimuthal direction by the rotation and the toroidal component is generated from the poloidal component. The magnetic torque due to the toroidal component extracts the angular momentum from the gas in the waiting room and the accretion onto the neutron star proceeds. Then the accretion rate will be regulated by the magnetic fields in our model. This regulation mechanism may account for the observation that the peak luminosity is independent of the total energy release during a type II burst (Kunieda *et al.* 1984a). The former is proportional to the accretion rate and the latter to the total amount of the gas accreted. This magnetic torque-regulation mechanism can also be applied to the flat-topped (long-duration) bursts reported by Inoue *et al.* (1980) and Kunieda *et al.* (1984a). When the gas in the waiting room is exhausted, the type II burst terminates.

In the following subsections we evaluate the magnetic fields of the Rapid Burster by applying the above scenario to observations.

2.1 High Persistent Flux Phase without Type II Burst Activity in 1983

From 1983 August 5 to August 16, no type II bursts were observed while type I bursts and persistent emission were observed (Kunieda *et al.* 1984b; Barr *et al.* 1987). The persistent luminosity then was very high ($L = 1.4 \times 10^{37}$ erg s^{-1}, for an assumed distance of 10 kpc). This fact implies that the magnetohydrodynamical instabilities triggering type II bursts disappear when the average accretion rate is high.

As the accretion rate increases, the magnetosphere shrinks. When the Alfvén radius is smaller than the neutron star radius, the magnetosphere disappears. If the surface magnetic field is as strong as $\sim 10^8$ G, the magnetosphere will disappear when the luminosity is larger than several times of 10^{37} erg s^{-1}. The magnetic field of 10^8 G is much weaker than those of X-ray pulsars. The weak magnetic field hypothesis ($B \sim 10^8$ G) accounts naturally for the period of no type II burst activity in 1983.

2.2 Burst-Emitting Area

The size of the X-ray-emitting area during a burst is as large as the whole neutron star surface (Marshall *et al.* 1979). The burst-emitting area is the region where the gas is accreted and the gravitational energy is released during the burst. Since the gas flows along the magnetic field, it is funneled into the polar regions. If the magnetic field is much stronger than 10^8 G, the burst-emitting area should be much smaller than the whole neutron star surface. Thus, the weak magnetic field hypothesis is favorable also for accounting for the large burst-emitting area.

2.3 Regulation Mechanism of the X-ray Emission during a Type II Burst

The type II burst spectrum is approximately a blackbody in the low-energy range ($E < 10$ keV) (see, e.g., Hoffman, Marshall, and Lewin 1978; Kawai *et al.* 1990). (We discuss the deviation from the blackbody spectrum in the next section.) The color temperature is 1.5–2.0 keV and changes little during the decay, unlike that of a type I burst. The change in the total flux is mainly due to the change in the burst-emitting area. This implies that the X-ray emission per unit area is regulated to be constant. Our scenario explains the regulated X-ray emission by the magnetic torque mechanism. If $B \sim 10^8$ G, the color temperature is regulated to be the observed value, ~ 2 keV.

3 HIGH-ENERGY POWER-LAW COMPONENT OBSERVED IN TYPE II BURST SPECTRUM

While the X-ray spectra of type II bursts are well approximated by a blackbody in the low-energy range ($E < 10$ keV), they have a high-energy power-law component (Kawai *et al.* 1990). We propose the model that the high-energy component is produced by Compton scattering with electrons in the accretion flow.

During a type II burst, the gas is accreted onto the neutron star and thermal X-ray photons are radiated from the neutron star surface. Some of the X-ray photons from the surface are scattered by electrons in the accretion flow. Since the accretion velocity is $v \sim 10^{10}$ cm s^{-1}, the scattered X-ray photons gain energy by the Compton effect. Statistically, a fraction of X-ray photons experience multiple scattering and

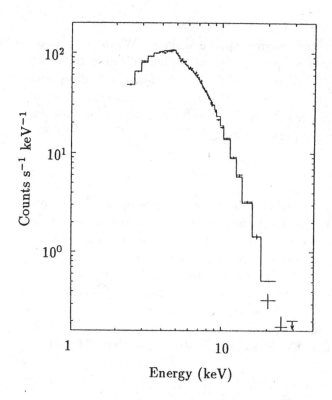

Fig. 1—The spectrum of a type II burst observed with *Tenma* at its peak (Kawai *et al.* 1990). The abscissa and the ordinate are the energy and the count rate per unit energy. The crosses denote the observed flux and the histogram is the best-fit model spectrum.

become high energy photons. We observe both the scattered and the direct components of the emission. The scattered component dominates in the high energy range. Figure 1 shows the comparison of our model with the spectrum observed with *Tenma* (Kawai *et al.* 1990).

REFERENCES

Barr, P., White, N. E., Haberl, F., Stella, L., Pollard, G., Gottwal M., and Parmar, A. N. 1987, *Astr. Ap*, **176**, 69.

Hanawa, T., Hirotani, K., and Kawai, N. 1989, *Ap. J.*, **336**, 920

Hirotani, K., Hanawa, T., and Kawai, N. 1989, *Ap. J.*, submitted.

Hoffman, J. A., Marshall, H. L., and Lewin, W. H. G. 1978, *Nature*, **271**, 630.

Inoue, H., *et al.* 1980, *Nature*, **284**, 358.

Kawai, N., Matsuoka, M., Inoue, H., Ogawara, Y., Tanaka, Y., Kunieda, H., and Tawara, Y. 1990, *Pub. Astr. Soc. Japan*, **42**, in press.

Kunieda, H., *et al.* 1984*a*, *Pub. Astr. Soc. Japan*, **36**, 215.

——. 1984*b*, *Pub. Astr. Soc. Japan*, **36**, 807.

Lamb, F. K. 1977, *Ap. J.*, **217**, 197.

Lewin, W. H. .G., *et al.* 1976, *Ap. J. (Letters)*, **207**, L95.

Marshall, H. L., Ulmer, M. P., Hoffman, A., Doty, J., and Lewin, W. H. G. 1979, *Ap. J.*, **227**, 555.

Recent Advances in the Studies of the Nova Outburst

Sumner Starrfield

IGPP and Theoretical Division, Los Alamos National Laboratory and
Department of Physics and Astronomy, Arizona State University

1 INTRODUCTION

This conference marks 13 years since the first North American Cataclysmic Variable Workshop (NACVW 1) was held in 1976 at the University of Illinois. It, therefore, seems appropriate to briefly review the progress that we have made in understanding the nova outburst. Recent detailed reviews of both the nova binary systems and their outbursts can be found in the proceedings of the Madrid IAU Symposium No. 122 on the "Physics of Classical Novae" (Cassatella and Viotti 1990), Gehrz (1988), Shara (1989), Starrfield (1988, 1989), and Bode and Evans (1989). Because of the existence of these reviews, I will describe only a few of the developments that have had a major impact on studies of the outburst.

2 CHANGES IN OUR UNDERSTANDING OF THE OUTBURST

The list of topics that have strongly influenced our ideas about the cause and evolution of the nova outburst must begin with enhanced abundances in the accreted and ejected material. Some of the other items in the list are: the constant bolometric luminosity, magnetic novae, neon novae, hibernation, shear mixing, diffusion, and boundary layer heating. Observationally, we must consider the infrared studies and the formation of dust and coronal lines, X-ray studies and the turn-off timescale, ultraviolet studies and enhanced abundances, and the relationship between recurrent novae and X-ray novae.

2.1 The Detection of Enhanced Abundances

The optical, infrared, and ultraviolet determinations of abundances in novae ejecta have now convincingly demonstrated both that the CNONeMg nuclei can be strongly enhanced in novae ejecta and, in addition, that there are two abundance classes of novae. However, at NACVW 1, there was, as yet, no reliable evidence for elemental enhancements in novae, since the first reliable abundance measurements did not appear until Williams *et al.* (1978). This emphasizes the predictive nature of the thermonuclear runaway theory (TNR) of the outburst (see Starrfield 1989) since we had already predicted enhanced abundances in novae (Starrfield *et al.* 1972). A major topic at the first workshop, therefore, was a discussion showing how the energetic requirements of a fast nova outburst demanded enhanced abundances.

It was not until 1977 that Williams and his collaborators began to apply standard nebular analysis techniques to the study of the ejected and resolved *shells* of novae (Williams *et al.* 1978). In addition, Ferland and Shields (1978) showed that V1500 Cyg had enhanced CNO in its ejecta.

A significant advance in studies of nova abundances occurred as a result of *IUE* observations. The ability of *IUE* to observe to wavelengths as short as 1100 Å has provided us with a wealth of new data on novae abundances (Starrfield and Snijders 1987). Up to this time, the most detailed papers are on V1668 Cyg, U Sco, V693 CrA, GQ Mus, V1370 Aql, PW Vul, and QU Vul (see Starrfield and Snijders 1987 for specific references). It is also worth pointing out that some of us (Wehrse *et al.* 1990) are working on a spectrum synthesis method that will be used to determine the abundances for species that are present in the spectra of the optically thick expanding pseudo-photosphere seen in the ultraviolet at maximum light. Finally, recent analyses of coronal lines found in the infrared are providing abundances of novae for some highly-ionized species (Greenhouse *et al.* 1990).

2.2 Constant Bolometric Luminosity

At NACVW 1, the prediction of a constant bolometric luminosity phase of the outburst had already appeared in the literature along with confirmation for one nova from ultraviolet observations by Gallagher and Code (1974) of FH Ser. This prediction is based on the fact that only part of the envelope material is ejected during the burst phase of the outburst and the rest remains on the rekindled white dwarf (Starrfield, Sparks, and Truran 1974).

Additional support came from infrared observations of novae both at maximum and in the dust formation stage (Gehrz 1988). The importance of dust is that it acts as a calorimeter of the EUV and soft X-ray photons emitted by the hot white dwarf at a time when the optical brightness has dropped by many magnitudes (Gehrz 1988). Although the dust studies indicated that a nova should be bright in the EUV or soft X-rays for months after the beginning of the optical decline, it was not until *EXOSAT* that any novae were detected at these wavelengths (Ögelman, Krautter, and Beuermann 1987). *EXOSAT* detected 4 novae in outburst and it both provided an estimate of the effective temperature of these novae and, also, showed that they were luminous X-ray emitters for long times.

The infrared studies of novae in outburst can provide distances to novae using the blackbody expansion parallax technique that was applied to V1500 Cygni by Gallagher and Ney (1976). This same method can be used a second time when dust forms later on in the outburst (Gehrz 1988).

2.4 Magnetic Novae

One of the more exciting results in the study of novae has been the discovery that V1500 Cygni contains a strongly-magnetized white dwarf and is, thereby, an AM Her system (Stockman, Schmidt, and Lamb 1988). This nova underwent one of the most unusual outbursts ever studied (Gallagher and Starrfield 1978). I note that the AM Her variables had not even been discovered at the time of NACVW 1. It has recently been proposed that Nova GQ Mus 1983 may also be an AM Her variable and this result may be very important in understanding the characteristics of its outburst.

Livio, Shankar, and Truran (1988) investigated TNRs on magnetic white dwarfs but assumed that the only effect of the strong magnetic field was to reduce the efficiency of convection. This reduction occurs because the strong magnetic fields impact the material flows perpendicular to the magnetic field lines. However, it is also important to consider that the infalling material cannot be spherically symmetric and the distribution of hydrogen-rich material on the surface of the star must be very non-spherical.

2.5 Recurrent Novae

Major changes in our understanding about the nova outburst have come about because of the recent outbursts of recurrent novae such as U Sco, V394 CrA, RS Oph, and V745 Sco. At NACVW 1, it was thought that a recurrent nova was a classical nova that exploded on a short timescale. It now seems likely that the class of recurrent novae contains at least two classes of objects: those with giant secondaries and those with evolved and small-radius secondaries. T CrB, RS Oph, and V745 Sco fall into the giant secondary class and may outburst as a result of either an accretion disk or mass-transfer instability. In contrast, U Sco, V394 CrA, and T Pyx have short orbital periods and, therefore, are in the other class. It seems likely that their outbursts are a direct result of a thermonuclear runaway in the accreted envelope. It also seems likely that this second class is directly related to those LMXRBs that arose out of binary evolution and not capture.

3 SUMMARY

It has not been possible in this short paper to review all of the topics that have affected our understanding of the nova outburst. Nevertheless, it is clear that there have been major changes in this field over the past 13 years and it seems likely that this field will continue to change. I would like to point out that there are a number of ideas "looming on the horizon" that could markedly change our current views about novae. For example, it seems like the application of Smooth Particle Hydrodynamic techniques to the evolution of the thermonuclear runaway will allow us to explore the three-dimensional evolution of nova systems and explosions. The recent advances in calculation of expanding, spherical, stellar atmospheres will allow analyses of nova spectra at very early stages in their outburst. *ROSAT* will discover a large number of cataclysmic variables and provide many new objects for study. Finally, there should soon be major improvements in our understanding of the emission from accretion disks that will significantly improve our understanding of cataclysmic variables.

I would like to express my thanks for many useful discussion to Drs. R. Gehrz, R. Hjellming, J. Krautter, G. Shaviv, S. Shore, E. M. Sion, W. M. Sparks, G. Sonneborn, J. Truran, R. Wehrse, R. Wade, R. M. Wagner, and R. E. Williams. I am also grateful to Drs. S. Colgate, A. N. Cox, C. F. Keller, M. Henderson, and K. Meyer for the hospitality of the Los Alamos National Laboratory and a generous allotment of computer time. This work was supported by NSF Grant AST88-18215 to ASU, by the IGPP at Los Alamos, by NASA grants to ASU, and by the DOE.

REFERENCES

Bode, M. F., and Evans, A. N. 1989, *Classical Novae* (Chichester: Wiley).

Cassatella, A., and Viotti, R. 1990, *Physics of Classical Novae* (Heidelberg: Springer), in press.

Ferland, G. J., and Shields, G. 1978, *Ap. J.*, **226**, 172.

Gallagher, J. S., and Code, A. D. 1974, *Ap. J.*, **189**, 303.

Gallagher, J. S., and Ney, E. P. 1976, *Ap. J. (Letters)*, **204**, L35.

Gallagher, J. S., and Starrfield, S. 1978, *Ann. Rev. Astr. Ap.*, **16**, 171.

Gehrz, R. D. 1988, *Ann. Rev. Astr. Ap.*, **26**, 377.

Greenhouse, M. A., Grasdalen, G. L., Woodward, C. E., Benson, J., Gehrz, R. D., Rosenthal, E., Skrutskie, M. F. 1990, *A. J.*, in press.

Livio, M., Shankar, A., and Truran, J. W. 1988, *Ap. J.*, **330**, 264.

Ögelman, H., Krautter, J., and Beuermann, K. 1987, *Astr. Ap.*, **177**, 110.

Shara, M. M. 1989, *Pub. Astr. Soc. Pac.*, **101**, 5.

Starrfield, S. 1988, in *Mulitwavelength Observations in Astrophysics*, ed. F. A. Córdova, (Cambridge: Cambridge University Press), p. 159.

Starrfield, S. 1989, in *Classical Novae*, ed. M. F. Bode and A. N. Evans (Chichester: Wiley), p. 39.

Starrfield, S., and Snijders, M. A. J. 1987, in *Exploring the Universe with the IUE Satellite*, ed. Y. Kondo (Dordrecht: Reidel), p. 377.

Starrfield, S., Sparks, W. M., and Truran, J. W. 1974, *Ap. J. Suppl.*, **28**, 247.

Starrfield, S., Truran, J. W., Sparks, W. M., and Kutter, G. S. 1972, *Ap. J.*, **176**, 169.

Stockman, H. S., Schmidt, G. D., and Lamb, D. Q. 1988, *Ap. J.*, **332**, 282.

Wehrse, R., Hauschildt, P. H., Shaviv, G., and Starrfield, S. 1990, in *Physics of Classical Novae*, ed. A. Cassatella and R. Viotti (Heidelberg: Springer), in press.

Williams, R. E., Woolf, N. J., Hege, E. K., Moore, R. L., and Kopriva, D. A. 1978, *Ap. J.*, **224**, 171.

Imaging or Resolving the Radio Shells of Novae

Robert M. Hjellming

National Radio Astronomy Observatory, Socorro, NM 87801

1 INTRODUCTION

Several classical novae have been observed as radio sources. HR Del 1967, FH Ser 1970, and V1500 Cyg 1975 have had extensive analysis (Hjellming *et al.* 1979) of their radio light curves which are fit very well by a "Hubble flow" model of a shell with a R^{-2} density distribution and a linear velocity gradient. QU Vul 1984 has been imaged (Taylor *et al.* 1987; Taylor *et al.* 1988) with the 35-km VLA at 14.9 GHz. These images show an evolving, roughly spherically-symmetric, optically thin shell with a mass of $3.6 \times 10^{-4} \, M_\odot$. It has recently been realized (Hjellming 1989) that all classical novae that can be observed with sensitive, high-resolution arrays like the VLA, at good signal-to-noise levels, can be resolved or imaged. This means radio observations of novae have the potential of resolving major questions about the masses and structures of nova shells.

2 RADIO LIGHT CURVES AND MODELS

"Hubble flow" models, with a linear velocity gradient ranging from v_1 at an inner radius (θ_1) to v_2 at an outer radius (θ_2), are used to fit the radio light curves of novae. For a distance of $d_{\rm kpc}$ at a time $t_{\rm yr}$ one has

$$\theta_2 = 0.2'' \, (v_2/1000 \text{ km s}^{-1}) \cdot t_{\rm yr}/d_{\rm kpc} + \theta_{20} \qquad \text{and}$$

$$\theta_1 = 0.2'' \, (v_1/1000 \text{ km s}^{-1}) \cdot (t_{\rm yr} - t_s)/d_{\rm kpc} + \theta_{10} \qquad (\text{for } t > t_s) \ .$$

Ejection occurs between $t_{\rm yr} = 0$, when $\theta_2 = \theta_{20}$, and $t_{\rm yr} = t_s$, when $\theta_1 = \theta_{10}$.

Because of the R^{-2} density gradient in this expanding shell, the radio source evolves through three distinct phases. Phase I is an optically thick expansion phase in which the shell has a disk of radius $\theta_2(t)$ and (electron) brightness temperature T_e. At some point, the radio pseudo-photosphere begins to recede inside the shell with an angular radius corresponding to

$$\theta_{\rm ph}(t) \approx 0.2'' \, \nu_{\rm GHz}^{-0.7} \left(\frac{T_e}{10^4 \text{ K}} \right) \left(\frac{M}{10^{-4} \mu M_\odot} \right) \cdot d_{\rm kpc}^{-5/2} [\theta_2(t) - \theta_1(t)]^{-2/3} \ ,$$

where ν is the observed frequency, M is the shell mass, and μ is the mean molecular weight. During this Phase II, the radio flux is dominated by the emission inside $\theta_{\rm ph}(t)$

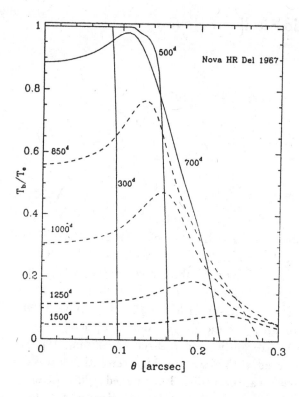

Fig. 1—Normalized 14.9 GHz surface brightness distributions for nova HR Del 1967 at 300, 500, 700, 850, 1000, 1250, and 1500 days after outburst when the radio fluxes were 47, 121, 163, 148, 113, 66, and 40 mJy, respectively.

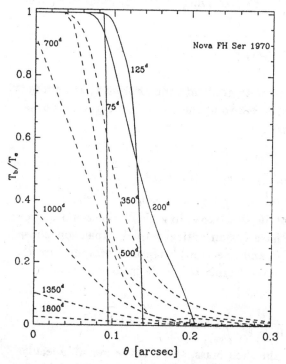

Fig. 2—Normalized 14.9 GHz surface brightness distributions for nova FH Ser 1970 at 75, 125, 200, 350, 500, 700, 1000, 1350, and 1800 days after outburst when the radio fluxes were 45, 83, 101, 79, 57, 40, 22, 9, and 4 mJy, respectively.

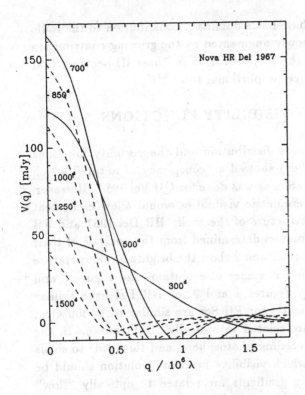

Fig. 3—Visibility functions for nova HR Del 1967 at 14.9 GHz at 300, 500, 700, 850, 1000, 1250, and 1500 days after outburst when the radio fluxes were 47, 121, 163, 148, 113, 66, and 40 mJy, respectively.

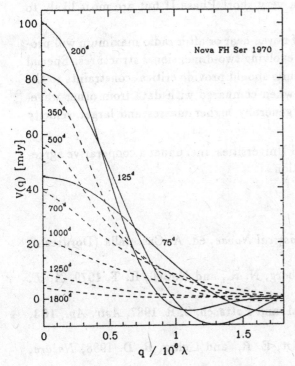

Fig. 4—Visibility functions for nova FH Ser 1970 at 14.9 GHz at 75, 125, 200, 350, 500, 700, 1000, 1350, and 1800 days after outburst when the radio fluxes were 45, 83, 101, 79, 57, 40, 22, 9, and 4 mJy, respectively.

with brightness temperature T_e. This phase has a dominant contribution to the radio flux proportional to $\nu^{0.6}t^{-4/3}$, which is slowly augmented by the growing contribution from optically thin gas. When $\theta_{ph} \leq \theta_1$, the radio emission in Phase III becomes that of a completely optically thin radio source proportional to $\nu^{-0.1}t^3$.

3 EVOLUTION OF IMAGES AND VISIBILITY FUNCTIONS

The evolution of the surface brightness distribution and the visibility functions for nova V1500 Cyg 1975 (Hjellming 1989) showed a "conspiracy of nature" which prevented it from becoming easily imageable, as was done for QU Vul 1984. However, for V1500 Cyg, measurements of interferometric visibilities would allow equivalent determinations of the evolution of the structure of the shell. HR Del 1967 and FH Ser 1970 are two other novae with parameters determined from the analysis of their light curves (Hjellming *et al.* 1979). Figures 1 and 2 show the brightness temperature distributions for these two novae at different stages of evolution, and Figure 3 and 4 show the related visibility functions. Figures 1 and 3 for HR Del have similar properties to QU Vul, while Figures 2 and 4 for FH Ser are similar to V1500 Cyg. Systems with large velocity gradients are related to optically "fast" novae. In the radio they have a long Phase II with a receding photosphere, and this leads to shells that will be difficult to image, but for which visibility function evolution should be observable. Systems with small velocity gradients are related to optically "slow" novae, and in the radio they will have a very short Phase II but are most likely to produced larger, more imageable shells.

High-resolution radio observations of novae near or after radio maximum will provide either images or visibilities to study evolving two-dimensional structures. Special observations will be required, but the results should provide critical constraints to the models fitting radio light curves which, when compared with data from other wavelengths, may decide the validity of the generally higher masses and larger velocity gradients determined from radio data.

The NRAO is operated by Associated Universities, Inc. under a cooperative agreement with the National Science Foundation.

REFERENCES

Hjellming, R. M. 1989, in *Physics of Classical Novae*, ed. A. Cassetalla (Dordrecht: Reidel), in press.

Hjellming, R. M., Wade, C. M, Vandenberg, N. R., and Newell, R. T. 1979, *A. J.*, **84**, 1619.

Taylor, A. R., Seaquist, E. R, Hollis, J. M. and Pottasch, S. R. 1987, *Astr. Ap.*, **183**, 38.

Taylor, A. R., Hjellming, R. M., Seaquist, E. R., and Gehrz, R. D. 1988, *Nature*, **335**, 235.

Spectroscopy of Novae in M31

Austin B. Tomaney and Allen W. Shafter

McDonald Observatory and
Department of Astronomy, University of Texas at Austin

1 INTRODUCTION

Studies of extragalactic novae have received considerable attention recently. Of particular interest are the results from M31 novae observations (Ciardullo *et al.* 1987; Capaccioli *et al.* 1989) which indicate that novae in that galaxy belong primarily to the bulge population. This is a surprising result, since previous Galactic nova observations have constrained novae in our Galaxy to the old disk population, not to the same population as globular clusters (Payne-Gaposchkin 1957; Patterson 1984).

Only a few spectra of novae in M31 have been obtained so far (Ciardullo, Ford, and Jacoby 1983; Cowley and Starrfield 1987). Part of the problem is logistical: typical M31 novae can only be accessed with a large telescope spectroscopically for a few months after discovery. A survey was initiated in 1987 at McDonald Observatory to obtain comprehensive photometric and spectroscopic coverage of newly-discovered novae in M31. We present the preliminary results of this ongoing survey here. The objective of the survey is to provide a comparison of the spectroscopic evolution of M31 (primarily bulge) novae with Galactic (primarily disk) novae, and, where possible, the physical conditions such as temperature, density, and abundances. Such a survey is of interest because the nova outburst properties may be affected by stellar population.

Spectra of four M31 novae are presented here; a more complete sample will be presented in a later publication. Three of the novae were discovered at McDonald Observatory in a narrow-band Hα CCD imaging survey on the 0.7-m telescope. The fourth was discovered in a similar survey at Kitt Peak reported in Ciardullo *et al.* (1987). Long-slit spectra were obtained with the 2.7-m telescope Large Cassegrain Spectrograph using a 300-line grating and a TI CCD. The spectral resolution was about 8 Å and the spectral range is from 4000 to 6700 Å.

2 OBSERVATIONS AND DISCUSSION

The photometric evolution of novae varies greatly from one nova to another. The spectral evolution of Galactic novae was first studied comprehensively by McLaughlin (1960). Various spectral stages were identified and found to be well correlated with the number of magnitudes the nova (continuum) had declined from maximum light. With this outline we can make an initial qualitative comparison of our spectra to Galactic novae observations.

2.1 McD87#1

First detected on 3 October 1987, UT, this nova declined only ~ 1.0 magnitudes in Hα to $m_{H\alpha} \sim 18.2$ at the time the spectra were obtained (25, 26 November 1987). The time of eruption is unknown.

The spectrum showed strong and broad Balmer lines (~ 40 Å, 1800 km s^{-1}) together with lines from Fe II (multiplet 42). Detection of the broad blend of N II, N III, and O II centered at $\lambda 4640$ is indicative that the nova had reached the transition stage of its evolution from a stellar- to a nebular-type spectrum. This is further evidenced by the detection of [O III] $\lambda 4363$, the first of the [O III] lines to be traced at this stage. The latter implies that this nova had probably declined ~ 3.5 mag from maximum. Also consistent with this picture are the weak emissions from the numerous multiplets of N II, He I, and the presence of relatively strong emissions of [O I] $\lambda 6300$, $\lambda 6364$ and [N II] $\lambda 5755$ which reached their peak strengths relative to the continuum at 2.6 and 3.3 magnitudes below maximum, respectively.

2.2 McD87#3

This nova was discovered on 29 October 1987, UT. Prior observations constrain the time of eruption to within three weeks before this date. The Hα maximum was probably seen around this time since observations taken on two consecutive weeks yielded roughly the same $m_{H\alpha} \sim 16.4$. At the time the spectrum was taken (23, 25 November 1987), it had declined to 17.2.

There were some similarities with the previous spectrum, namely strong and broad Balmer lines (~ 33 Å, 1500 km s^{-1}) and the strong Fe II (42) multiplet. However, as we might expect, the spectrum shows evidence of less extensive evolution. The most striking evidence is the sharp blueward absorption features of the Balmer lines at -2100 km s^{-1} and possibly at Na I $\lambda 5890$, 5896. This is most likely to be the latter stage of the diffuse enhanced spectrum (McLaughlin 1960). There is also evidence for a second absorption system, the so-called Orion absorption stage. This consists of diffuse bands of He I, N II, and O II. The strongest features of this system are the groups of N II and O II lines between $\lambda 4600$–$\lambda 4700$ and the N II group at $\lambda 5667$–$\lambda 5686$, all of which can be seen in this spectrum. Thus, the evidence is quite strong that this nova exhibited a typical spectrum seen in Galactic novae between 2 to 2.5 magnitudes below maximum light.

2.3 McD88#1

Originally, this nova was discovered with an $m_{H\alpha} \sim 14.8$ and $m_B \sim 16.6$ on 18 August 1988, UT. The time of eruption is unknown. The continuum was undetectable by 30 September ($m_B > 21$), and the nova decayed precipitously in Hα in the month prior to this spectrum to $m_{H\alpha} \sim 19.4$. The spectrum was taken on the nights of 1 and 9 November 1988, UT. Due to the faintness of the nova, this spectrum had the lowest signal-to-noise ratio.

The narrow Balmer lines (12 Å, 550 km s^{-1}) are probably unresolved and are quite surprising for such a fast nova. With observations of > 4.5 magnitudes in the continuum and 4.5 magnitudes of decline to Hα, we expect this nova to be well into the nebular stage. This is supported by the lack of [O III] emission which decays early in the nebular phase and the detection of [N II] $\lambda6548$ and $\lambda6584$ flanking Hα which are generally seen to strengthen in the later phases. The narrow emission lines resemble more a postnova spectrum.

Most of the spectrum appears featureless; interestingly, however, there appears to be a significant detection of [S II] $\lambda6717$ and $\lambda6731$. Such lines have been seen in the nebular stage of novae (Collin-Souffrin 1977). This pair of lines provides a diagnostic of the electron density of the nebula (e.g., Osterbrock 1974). Gaussian fits to the blend yielded a ratio of $\lambda6717/\lambda6731 \sim 1.5$. For reasonable nebula temperatures (1,000 to 10,000 K), this implies an extremely low electron density of between 10 and 100 cm^{-3}. Such densities are typical for nova shells.

2.4 CFJNS Nova 1986 No. 32

This nova was discovered in the recent Kitt Peak M31 nova survey (Ciardullo et al. 1987) on 4 October 1986, UT; the actual time of eruption is unknown. A spectrum of this object was obtained at the MMT (Cowley and Starrfield 1987). Narrow Balmer lines (~ 700 km s^{-1}) and multiplets of Fe II were the principle emission lines detected. Cowley and Starrfield (1987) concluded that this nova was observed about 1 to 2 magnitudes below maximum light.

Despite being a slow nova, it was somewhat surprising that it was rediscovered the next year at McDonald, having declined only ~ 0.6 magnitudes to $m_{H\alpha} \sim 17.2$. To date, its light curve in Hα has been observed extensively at McDonald and Kitt Peak (Ciardullo et al. 1989). The spectra presented here were taken at two epochs: 22, 26 November 1987 and 2 November 1988, UT, and are shown in Figure 1.

The 1987 spectrum showed significant progress in evolution from that seen by Cowley and Starrfield (1987). We find that the Balmer lines have narrowed slightly to ~ 11 Å (520 km s^{-1}). Fe II multiplets appear to have faded while He I lines appear to have strengthened. Most significantly, however, is the appearance of [O III] $\lambda4363$. This line is first traced at 3.7 magnitudes below maximum in Galactic novae and marks the transition into the nebular phase. A trace detection of the $\lambda\lambda4640$ band

and the absence of the Fe II (42) multiplet constrains the stage of this nova to be fainter than ~ 4.5 magnitudes below maximum light.

By 1988 we see a further progression. The remaining [O III] lines, $\lambda 4959, 5007$ have appeared and become quite strong and trace [N II] $\lambda 6584$ can be seen. Once again we see some consistency with Galactic nova evolution: the equality of $\lambda 5007$ and Hβ seen in this nova occurs at ~ 5.4 magnitudes below maximum. At ~ 5.0 magnitudes below maximum [O III] $\lambda 4363$ starts to exceed Hγ, also seen in this nova. Thus, we conclude from line strengths that this nova's continuum has decayed < 0.9 magnitudes during the year between these two spectra.

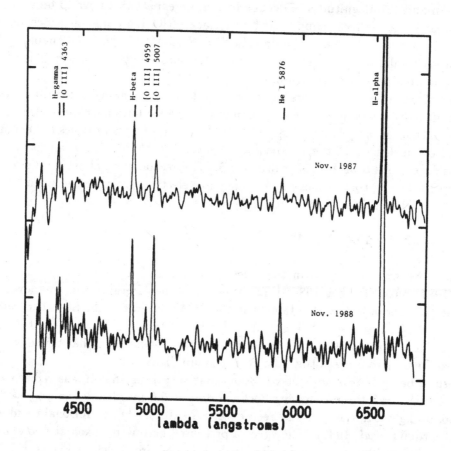

Fig. 1—Spectra of CFJNS Nova 1986 No. 32 obtained in November 1987 and November 1988.

Adopting a foreground differential extinction of $E(B - V) = 0.11$ for M31 (McClure and Racine 1969), the corrected value of the Balmer decrement in 1988 became 5.9, implying high densities for the nova ejecta. We can probe the physical conditions still further in the 1988 spectrum since it provided us access to two important temperature and density diagnostics that were first applied to V1500 Cyg (Ferland and Shields 1978):

$$\frac{I([\text{O III}]\ \lambda\lambda 4959,\ 5007)}{I([\text{O III}]\ \lambda 4363)} \quad \text{and} \quad \frac{I([\text{O III}]\ \lambda\lambda 4959,\ 5007)}{I(\text{He I}\ \lambda 5876)}.$$

The 1988 spectrum yields reddening-corrected values of 2.25 and 3.00 for these ratios, respectively. Using the equations incorporating the revisions to the atomic constants reported in Lance, McCall, and Uomoto (1988) and assuming a typical ejecta temperature of $T \sim 10{,}000$ K, we can solve for the electron density. We obtain $N_e \sim 8 \times 10^7$ cm^{-3}. Such a high density supports the high density implied by the Balmer decrement. A value of 0.013 was found for the abundance ratio, $X(\text{O}^{++})/X(\text{He}^{+})$. This compares with ratios of ~ 0.04 derived for V1500 Cyg (Lance, McCall, and Uomoto 1988) and recently for V1819 Cyg (Whitney and Clayton 1989). If this result indicates a high helium abundance in this nova, this would be consistent with similar measurements found in slow Galactic novae (Truran and Livio 1986).

The observed evolution of this nova into the nebular phase has been particularly important in allowing some estimates of the physical conditions to be made with some assumptions about the temperature of the ejecta. It is hoped that more recent spectra of this object taken at the time of writing will enable the evolving temperature and density conditions to be determined explicitly, thereby providing a unique opportunity to make abundance estimates for a nova outside the Galaxy.

REFERENCES

Capaccioli, M., *et al.* 1989, *A. J.*, **97**, 1622.

Ciardullo, R. B., Ford, H. C., and Jacoby, G. H. 1983, *Ap. J.*, **272**, 92.

Ciardullo, R. B., *et al.* 1987, *Ap. J.*, **318**, 520.

Ciardullo, R. B., *et al.* 1989, *Ap. J.*, submitted.

Collin-Souffrin, S. 1977, in *Novae and Related Objects*, ed. M. Friedjung (Dordrecht: Reidel), p. 123.

Cowley, A. P., and Starrfield, S. G. 1987, *Pub. Astr. Soc. Pac.*, **99**, 854.

Ferland, G. J., and Shields, G. A. 1978, *Ap. J.*, **226**, 172.

Lance, C. M., McCall, M. L., and Uomoto, A. K. 1988, *Ap. J. Suppl.*, **66**, 151.

McClure, R. D., and Racine, R. 1969, *A. J.*, **74**, 1000.

McLaughlin, D. B. 1960, in *Stars and Stellar Systems*, Vol. **6**, *Stellar Atmospheres*, ed. J. L. Greenstein (Chicago: University of Chicago Press), p. 585.

Osterbrock, D. E. 1974, *Astrophysics of Gaseous Nebulae* (San Francisco: Freeman).

Patterson, J. 1984, *Ap. J. Suppl.*, **54**, 443.

Payne-Gaposchkin, C. 1957, *The Galactic Novae* (New York: Dover).

Truran, J. W., and Livio, M. 1986, *Ap. J.*, **308**, 721.

Whitney, B. A., and Clayton, G. C. 1989, *A. J.*, **98**, 297.

IUE Observations of Faint Old Novae

A. Cassatella[1], P. L. Selvelli[2], R. Gilmozzi[3], A. Bianchini[4], and
M. Friedjung[5]

[1]CNR, IUE Observatory, ESA VILSPA, Spain
[2]CNR, Astronomical Observatory of Trieste, Italy
[3]CNR, Space Telescope Science Institute, Baltimore, MD, U. S. A.
[4]Astronomical Observatory of Padova, Italy
[5]Institut d'Astrophysique, Paris, France

1 INTRODUCTION

Optical observations of old novae have generally revealed quite high luminosities indicating mass-accretion rates higher than that required by current models of thermonuclear runaways. The few detailed UV studies of old novae have confirmed this trend, but selection effects might have played an important role since only the brightest members of the old nova class have been observed. Since most accretion luminosity is emitted in the UV, *IUE* observations are crucial for any reliable estimate of L_{disk} and of the mass-accretion rate $\dot{M}$. About 16 old novae are accessible to *IUE* but only a few, i.e., RR Pic, V 603 Aql, HR Del, and GK Per, have been studied in any detail. We here present some preliminary results (including also *IUE* archive data) for 9 old novae. A few objects are still missing here and we plan to observe them in the near future.

2 THE UV LUMINOSITY AND THE MASS-ACCRETION RATE

The reddening-corrected UV luminosities in the 1200–3200 Å spectral range are listed in Table 1. The distances are taken from Duerbeck (1984) and the inclinations from Warner (1987) except for CP Pup, which is not included in Warner's (1987) list. For CP Pup we have used the results of Duerbeck, Seitter, and Duemmler (1987), who, from the observed geometry of the nova ejecta, have reported a lower limit of $\sim 30°$ for the inclination angle. For DK Lac and X Ser the distances have been estimated using the data of Warner (1987). Because of the uncertainties in the reddening correction and in the distances, there might be some degree of error (up to 50%) in the L_{UV} estimates. Figure 1 is a plot of L_{UV} against $\cos i$. Despite the paucity of the data, the trend of Figure 1 suggests a dependence of L_{UV} on $\cos i$. Eclipsing objects (T Aur, BT Mon) have $L_{UV} \sim 1\,L_\odot$, while objects seen at low inclination or nearly pole-on tend to have $L_{UV} \sim 10\,L_\odot$, which can be considered the 'intrinsic' UV luminosity. These findings are in agreement with the conclusions reached by Warner (1987) that the 'observed' M_v of old novae depends on the inclination angle, while

TABLE 1
PARAMETERS OF THE OLD NOVAE IN OUR SAMPLE

Object	d(pc)	E_{B-V}	$\cos i$	$L_{UV}/L_\odot$	α
V 841 Oph	855	0.30	1.0	9.5	2.0
CP Pup	1500	0.27	<0.87	12.0	1.8
DI Lac	895	0.15	<0.87	1.9	1.5
Q Cyg	1485	0.25	0.64	4.6	1.0
V 533 Her	620	0.0	0.47	0.7	1.3
T Aur	600	0.35	0.37	1.7	2.3
BT Mon	1000	0.20	0.10	0.9	1.0
DK Lac	1800	0.35	<1.0	6.0	2.8
X Ser	4300	<0.1	0.20	1.4	0.9

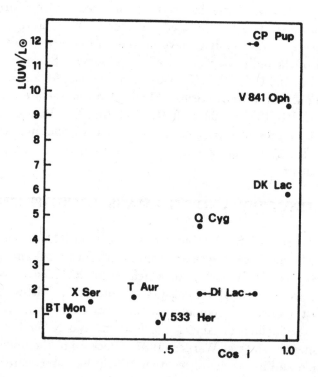

Fig. 1—UV luminosity versus orbital inclination for classical nova remnants.

the 'intrinsic' M_v (at $i \sim 0°$) does not vary greatly from star to star.

A direct estimate of the mass-accretion rate $\dot{M}$ can be obtained if the total accretion luminosity L_{disk} is known. If most of the disk luminosity is emitted in the

UV, as seems to be the case in old novae, then L_{UV} is not much smaller than L_{disk}, and it can be used to provide an estimate of L_{disk}. On the assumption that $L_{disk} \sim 2\,L_{UV}$, as indicated in Wade's (1984) models, our data suggest that the 'intrinsic' disk luminosity of the old novae in our sample is $\sim 20\,L_\odot$. A representative value of the mass-accretion rate in old novae is therefore $\dot{M} \sim 3 \times 10^{17}\ \mathrm{g\ s^{-1}} \sim 4.5 \times 10^{-9}\ \mathrm{M_\odot\ yr^{-1}}$.

The reddening-corrected continuum energy distribution of the old novae in our sample can be represented by a power-law spectrum of the form $F_\lambda \propto \lambda^{-\alpha}$, where α ranges from 0.9 to 2.8. With the exception of T Aur, there is a good correlation between the index α and the inclination i: high values of α are associated with high values of $\cos i$, that is, with systems seen face-on (Fig. 2).

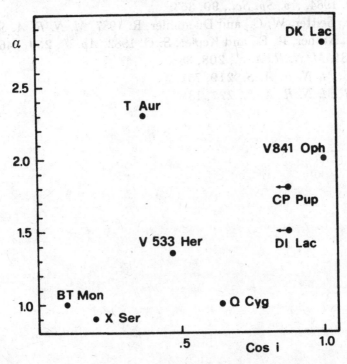

Fig. 2—The UV spectral index α versus orbital inclination for classical nova remnants.

3 REMARKS

1) Apparently there is no correlation between L_{UV} and parameters such as the orbital period P, the rate of decline t_3, and the time elapsed since the last outburst. X-ray data are available for only 5 objects of our sample (Becker 1989). It is remarkable that the two brightest stars in the UV (CP Pup and V 841 Oph) are also the brightest in the X-ray, while the contrary is true for the three objects (T Aur, BT Mon, and V 533 Her) which are the weakest in the UV.

2) BT Mon is optically very bright and, as a consequence, the $\dot{M}$ reported in the

literature from estimates made in the optical is surprisingly high ($\sim 2 \times 10^{-8}\, M_\odot\, yr^{-1}$). There is no way to reconcile this value with our observations even after the different reddening and distance values are taken into account.

3) V 841 Oph, one of the oldest nova remnants, is still very UV bright, 140 years after the outburst.

REFERENCES

Becker, R. H. 1989, in *Classical Novae*, ed. M. F. Bode and A. Evans (New York: Wiley), p. 215

Duerbeck, H. D. 1984, *Ap. Sp. Sci.*, **99**, 363.

Duerbeck, H. D., Seitter, W. C., and Duemmler, R. 1987, *M. N. R. A. S.*, **229**, 653.

Robinson, E. L., Nather, R. E., and Kepler, S. O. 1982, *Ap. J.*, **254**, 646.

Wade, R. A. 1984, *M. N. R. A. S.*, **208**, 381.

Warner, B. 1986, *M. N. R. A. S.*, **219**, 751.

Warner, B. 1987, *M. N. R. A. S.*, **227**, 13.

Observational Selection among Classical Novae in Outburst

H. Ritter[1], M. J. Politano[2], M. Livio[1,3,4], and R. F. Webbink[4]

[1]Max-Planck Institut für Physik und Astrophysik, D-8046 Garching, F. R. G.
[2]Department of Physics, Arizona State University, Tempe, AZ 85287
[3]Department of Physics, Technion, Haifa 32000, Israel
[4]Department of Astronomy, University of Illinois, Urbana, IL 61801

ABSTRACT

We investigate to what extent observational selection can account for the prevalence of massive white dwarfs among classical novae as was originally proposed by Truran and Livio (1986). For this we elaborate on their approach by taking into account a detailed model distribution function for the masses of newly-formed cataclysmic binaries from Politano (1988, 1990), an improved ignition condition for the thermonuclear runaway, as well as effects of the secular evolution of the systems and flux limitation of the observations (including interstellar absorption). Our results agree qualitatively with those obtained by Truran and Livio (1986). However, since Politano's model calculations do not take into account the formation of O-Ne-Mg white dwarfs, we cannot make any quantitative prediction as to their expected abundance among observed novae.

1 INTRODUCTION

Despite the fact that there is not a single white dwarf mass in a postnova binary that is accurately known (with the possible exception of DQ Her, see Horne, Welsh, and Wade 1990), there is strong circumstantial evidence that a substantial fraction of the known classical novae occurred on a massive white dwarf ($M_{WD} \gtrsim 1\,M_\odot$) and a fair fraction of them even on a O-Ne-Mg white dwarf with $M_{WD} \gtrsim 1.3\,M_\odot$ (Truran 1989). Given the fact that intrinsically the majority of the white dwarfs in cataclysmic binaries (CBs) have a much lower mass (see Table 1 and e.g., Politano and Webbink 1989a, b; Politano 1990), we must therefore ask whether observational selection as proposed by Truran and Livio (1986; 1989) and Politano $et\ al.$ (1989) could account for the prevalence of massive white dwarfs among the observed classical novae.

Truran and Livio (1986) were, in fact, the first to recognize that the probability of detecting a classical nova going into outburst increases strongly with the mass of the white dwarf involved in the thermonuclear runaway (TNR). This is because the envelope mass ΔM_{ign} that is required to ignite hydrogen burning decreases rapidly with increasing white dwarf mass. In order to quantify this selection effect, Truran and Livio (1986, 1989) and Politano $et\ al.$ (1989) assume that the TNR is ignited whenever a critical pressure P_{ign} at the base of the accreted envelope is reached. From the equation of hydrostatic equilibrium one then gets

$$\Delta M_{ign} = \frac{4\pi P_{ign}}{G} \frac{R_{WD}^4}{M_{WD}}. \tag{1}$$

As is shown elsewhere (e.g., Politano *et al.* 1989), the resulting selection effect is so strong that about one-third of all observed novae should occur on a white dwarf with $M_{WD} \gtrsim 1.35\,M_{\odot}$ and more than 80% on a white dwarf with $M_{WD} \gtrsim 0.9\,M_{\odot}$, if the intrinsic mass spectrum of the white dwarfs in CBs is similar to that of isolated ones.

2 IMPROVED ANALYSIS OF THE SELECTION EFFECT

Despite the fact that the above results are promising, the approach by Truran and Livio (1986, 1989) and Politano *et al.* (1989) is not entirely satisfactory for a number of reasons. First, as one can infer from results of Nariai and Nomoto (1979), Fujimoto (1982), and MacDonald (1984), ΔM_{ign} depends, in general, not only on M_{WD} but also on the accretion rate $\dot{M}_{WD}$. In fact, these results suggest that $P_{ign} = $ constant is not a good approximation. Second, the sample of novae that we observe is more likely to be magnitude-limited than volume-limited. Therefore, the effects of magnitude limitation, including the influence of interstellar absorption, should be taken into account. Third, the accretion rate $\dot{M}_{WD}$ is not a free parameter but rather subject to constraints from the secular evolution of CBs. In particular, the secular evolution determines the number distribution of systems as a function of M_{WD}. Fourth, one has to take into account the two-dimensional distribution of the birth rate of newly-formed CBs (hereafter zero age CBs = ZACBs) over the initial masses of both components.

Until not very long ago, our knowledge about the intrinsic properties of newly-formed CBs, in particular about the mass spectrum of the white dwarfs, was rudimentary at best. Recently, however, Politano (1988, 1990) and Politano and Webbink (1989*a*, *b*) presented results of extensive model calculations aimed at determining the properties of ZACBs. Specifically they compute the birth rate of ZACBs per unit area of the Galactic plane, $\dot{\Sigma}_{ZACB}(t)$, as a function of time t and the distribution of that birthrate over the initial values of the masses of the binary components $(M_{1,i}, M_{2,i})$. The mass spectrum of the white dwarfs of newly-formed CBs is then given by:

$$\frac{d\dot{\Sigma}_{ZACB}}{dM_{1,i}} = \int_0^{M_{2,crit}} \frac{\partial^2 \dot{\Sigma}_{ZACB}}{\partial M_{1,i} \partial M_{2,i}}\, dM_{2,i}\,, \tag{2}$$

where M_1 and M_2 are, respectively, the mass of the white dwarf primary and of the secondary, and $M_{2,crit}$ is the highest value of M_2 that allows for dynamically- and thermally-stable mass transfer during the subsequent evolution. Information about the mass distribution of the C-O white dwarfs in ZACBs after 10^{10} yr of constant star formation is given in Table 1. These numbers show that the mean white dwarf mass is rather low, i.e., $\langle M_1 \rangle \approx 0.76\,\mathrm{M_\odot}$, that 86% of the white dwarfs have a mass $M_1 < 0.9\,\mathrm{M_\odot}$ and less than 10% a mass $M_1 > 1\,\mathrm{M_\odot}$. For further details, the reader is referred to Politano (1988, 1990) and Politano and Webbink (1989a, b).

Taking all these additional effects into account, the frequency distribution of a visual magnitude-limited sample of classical novae in outburst is approximately (Ritter et al. 1990)

$$\frac{1}{\nu_N}\frac{d\nu_N}{dM_1} = \text{constant} \cdot L_v^n \int_0^{M_{2,\text{crit}}} \frac{1}{\Delta M_{\text{ign}}(M_1, \dot{M}_1)} \int_0^{M_{2,\text{crit}}} \frac{\partial^2 \Sigma_{\text{ZACB}}}{\partial M_1 \partial M_{2,i}}\, dM_{2,i} dM_2. \quad (3)$$

Here, L_v is the visual luminosity of a nova in outburst. As was shown by Ritter (1986), $n = 1$ if the Galactic distribution of novae is disk-like and if interstellar absorption is negligible. Including the influence of interstellar absorption one gets $n \approx 1/2$. For a volume-limited sample, on the other hand, one has $n = 0$.

Assuming now that the mass of the white dwarf does not change as a result of the secular evolution, i.e., $\langle \dot{M}_1 \rangle = 0$, that L_v is approximatively given by the bolometric luminosity derived from the core-mass luminosity relation (e.g., Paczynski 1970; Kippenhahn 1980), i.e., $L_v = L_{\text{bol}}(M_c = M_1)$, and that ΔM_{ign} may be parametrized as

$$\Delta M_{\text{ign}} = \text{constant} \cdot \left(\frac{M_1}{R_1^4}\right)^{-\alpha} \dot{M}_1^{-\beta}, \quad (4)$$

we obtain

$$\frac{1}{\nu_N}\frac{d\nu_N}{dM_1} = \text{constant} \cdot L_{\text{bol}}^n (M_c = M_1) \left(\frac{M_1}{R_1^4}\right)^{\alpha} \langle \dot{M}_1^\beta \rangle F \frac{d\Sigma_{\text{ZACB}}}{dM_1}, \quad (5)$$

where

$$F = \frac{\int_0^{M_{2,\text{crit}}} \int_{M_2}^{M_{2,\text{crit}}} \frac{\partial^2 \Sigma_{\text{ZACB}}}{\partial M_1 \partial M_{2,i}}\, dM_{2,i} dM_2}{\frac{d\Sigma_{\text{ZACB}}}{dM_1}} \approx \text{constant} \cdot q_{\text{crit}}\, M_1 = \text{constant} \cdot M_{2,\text{crit}} \quad (6)$$

and $\langle \dot{M}_1^\beta \rangle$ is an integral mean of $\dot{M}_1^\beta$. Comparing equations (1) and (4), we realize that the case of fixed ignition pressure corresponds to $\alpha = 1$ and $\beta = 0$. While the results of semi-analytical computations by Fujimoto (1982) and MacDonald (1984) agree reasonably well with results of detailed numerical computations (e.g., Nariai and Nomoto 1979; Kovetz and Prialnik 1985), as far as the value of α is concerned (all these computations yield $\alpha \lesssim 0.7$), this is not the case with regard to β. Whereas Mac-Donald's results yield $0.1 \lesssim \langle \beta \rangle \lesssim 0.5$, detailed numerical computations (e.g., Prialnik et al. 1982; Kovetz and Prialnik 1985; Starrfield, Sparks, and Truran 1986; Prialnik, Kovetz, and Shara 1989) yield consistently smaller values, typically $0 \lesssim \beta \lesssim 0.2$. In cases of low $\dot{M}_1$ ($\dot{M}_1 \lesssim 10^{-10}\,\mathrm{M_\odot}\,\mathrm{yr}^{-1}$) where diffusion becomes important (e.g., Kovetz and Prialnik 1985; Prialnik, Kovetz, and Shara 1989), β becomes even as low as -0.2.

TABLE 1
DISTRIBUTION OF WHITE DWARF MASSES

$\dfrac{M_1}{M_\odot}$	intrinsic $\dfrac{\Delta\dot{\Sigma}_{ZACB}}{\dot{\Sigma}_{ZACB}}$	nova frequency $\Delta\nu_N/\nu_N$			
		$n=1/2$ $\alpha=1$	$n=1/2$ $\alpha=0.7$	$n=0$ $\alpha=1$	$n=0$ $\alpha=0.7$
0.5–0.6	$<10^{-3}$	$<10^{-3}$	0.003	0.002	0.007
0.6–0.7	0.416	0.008	0.031	0.017	0.062
0.7–0.8	0.312	0.024	0.084	0.044	0.135
0.8–0.9	0.133	0.025	0.074	0.038	0.097
0.9–1.0	0.062	0.028	0.068	0.036	0.077
1.0–1.1	0.034	0.037	0.075	0.044	0.076
1.1–1.2	0.021	0.061	0.098	0.065	0.090
1.2–1.3	0.013	0.131	0.150	0.128	0.127
1.3–1.4	0.009	0.686	0.418	0.626	0.330
$\dfrac{\langle M_1\rangle}{M_\odot}$	0.76	1.28	1.16	1.25	1.09

In order to illustrate how different values of the parameters n and α influence selection, we show in Table 1 the results for the cases $n=1/2$, 0 and $\alpha=0.7$, 1.0 using Politano's (1990) results for $\dot{\Sigma}_{ZACB}$. Since $\langle\dot{M}_1^\beta\rangle$ depends only weakly on M_1 because β is so small ($-0.2\lesssim\beta\lesssim0.2$), we assume for simplicity $\beta=0$.

As can be seen from Table 1, even in the case where selection is weakest ($n=0$, $\alpha=0.70$), about one-third of all observed novae are expected to occur on a white dwarf with $M_1>1.3\,M_\odot$. On the other hand, the contribution from low-mass white dwarfs, i.e., with masses $M_1<0.9\,M_\odot$, is small but not totally negligible (at most 30% if $n=0$ and $\alpha=0.7$ and still 19% in the most realistic case $n=1/2$, $\alpha=0.7$). In conclusion, the more comprehensive treatment of the selection effect yields qualitatively the same result as previous estimates by Truran and Livio (1986, 1989) and Politano *et al.* (1989). Finally, we should, however, emphasize that because the model calculations by Politano (1988, 1990) do not include a detailed treatment of the formation of O-Ne-Mg white dwarfs in CBs, we cannot make a quantitative prediction of the frequency of occurrence of such white dwarfs to be expected among the observed novae.

H. R. would like to thank the Observatoire de Paris, Section de Meudon, where this paper was written for its hospitality and financial support. This work was also supported by NATO, grant No. 0752/87.

REFERENCES

Fujimoto, M. Y. 1982, *Ap. J.*, **257**, 767.

Horne, K., Welsh, W. F., and Wade, R. A. 1990, this volume.

Kippenhahn, R. 1980, *Astr. Ap.*, **102**, 293.

Kovetz, A., and Prialnik, D. 1985, *Ap. J.*, **291**, 812.

MacDonald, J. 1984, *Ap. J.*, **283**, 241.

Nariai, K., and Nomoto, K. 1979, in *White Dwarfs and Variable Degenerate Stars*, IAU Coll. No. 53, ed. H. M. van Horn and V. Weidemann (Rochester: University of Rochester), p. 525.

Paczynski, B. 1970, *Acta Astr.*, **20**, 47.

Politano, M. J. 1988, Ph.D. thesis, University of Illinois at Urbana-Champaign.

Politano, M. J. 1990, this volume.

Politano, M. J., and Webbink, R. F. 1989*a*, in *White Dwarfs*, IAU Coll. No. 114, ed. G. Wegner, Lecture Notes in Physics 328 (Berlin: Springer), p. 440.

Politano, M. J., and Webbink, R. F. 1989*b*, in *The Physics of Classical Novae*, IAU Coll. No. 122, ed. A. Cassatella (Berlin: Springer), in press.

Politano, M. J., Livio, M., Truran, J. W., and Webbink, R. F. 1989, in *The Physics of Classical Novae*, IAU Coll. No. 122, ed. A. Cassatella (Berlin: Springer), in press.

Prialnik, D., Livio, M., Shaviv, G., and Kovetz, A. 1982, *Ap. J.*, **257**, 312.

Prialnik, D., Kovetz, A., and Shara, M. M. 1989, *Ap. J.*, **339**, 1013.

Ritter, H. 1986, *Astr. Ap.*, **168**, 105.

Ritter, H., Politano, M. J., Livio, M., Webbink, R. F. 1990, in preparation.

Starrfield, S., Sparks, W. M., and Truran, J. W. 1986, *Ap. J. (Letters)*, **303**, L5.

Truran, J. W. 1989, in *The Physics of Classical Novae*, IAU Coll. No. 122, ed. A. Cassetella, (Berlin: Springer), in press.

Truran, J. W., and Livio, M. 1986, *Ap. J.*, **308**, 721.

Truran, J. W., and Livio, M. 1989, in *White Dwarfs*, IAU Coll. No. 114, ed. G. Wegner, Lecture Notes in Physics 328, (Berlin: Springer), p. 498.

On the Mass of Nova DQ Her (1934)

Keith Horne[1], William F. Welsh[2], and Richard A. Wade[3]

[1]Space Telescope Science Institute
[2]Astronomy Department, Ohio State University
[3]Steward Observatory, University of Arizona

Nova eruptions are currently modelled as thermonuclear runaways (TNRs) on accreting white dwarfs. In these models, fast novae occur with (1) a high-mass white dwarf (1.0–1.4 $M_\odot$), whose deep gravitational well compresses the nuclear fuel, and (2) enhanced CNO abundances, so that energy is stored in β-unstable nuclei and released gradually over a period of some hundreds of seconds.

Nova DQ Herculis (1934) displays enhanced He and CNO abundances in its ejecta, and yet it was a moderately slow nova. Thus, the TNR models require a low-mass white dwarf in DQ Her. Published estimates of masses in DQ Her, which range over a factor of 2, are not accurate enough to test this prediction.

We have detected Na I λ8183, 8195 absorption from the secondary star in DQ Her using time-resolved 1.6-Å resolution CCD spectra taken with the Palomar 5-m telescope. Figure 1 shows several of our DQ Her spectra taken near the conjunction and the quadrature phases, the mean spectrum averaged over our entire dataset (8.6-hour exposure), and rotationally-broadened spectra of the dM3 star Gleise 752A.

The DQ Her spectra display a velocity shift of ~460 km s^{-1} between the quadrature phases, implying an orbital velocity of approximately 230 km s^{-1}. Our cross-correlation radial velocity curve, shown in Figure 2, yields $K_2 = 232 \pm 5$ km s^{-1}.

Rotational broadening of the Na I lines from DQ Her's secondary star can be used to constrain the size of the Roche lobe, and hence the mass ratio of the binary, since this star's rotation is tidally locked to the 4.5-hour binary orbit. The expected rotational broadening is evident in Figure 1, where the mean spectrum of DQ Her may be compared with the series of rotationally-broadened spectra of Gleise 752A. (Note that broadening due to orbital motion was avoided by applying appropriate velocity shifts before averaging the individual spectra of DQ Her.) Based on least-squares fits and considering systematic effects due to limb-darkening, spectral type, gravity darkening, and irradiation, we find $V \sin i = 115 \pm 18$ km s^{-1}.

The near-infrared continuum light curve, which is shown in Figure 2, has a deep primary minimum at phase 0 due to the eclipse of the accretion disk by the secondary star. There is also a broad and variable maximum around phase 0.5, which may be attributed to heating on the inward face of the secondary star.

Figure 2 also shows the radial velocity curve and flux deficit curve for the Na I lines. In contrast to the continuum, the Na I lines remain strong during the primary

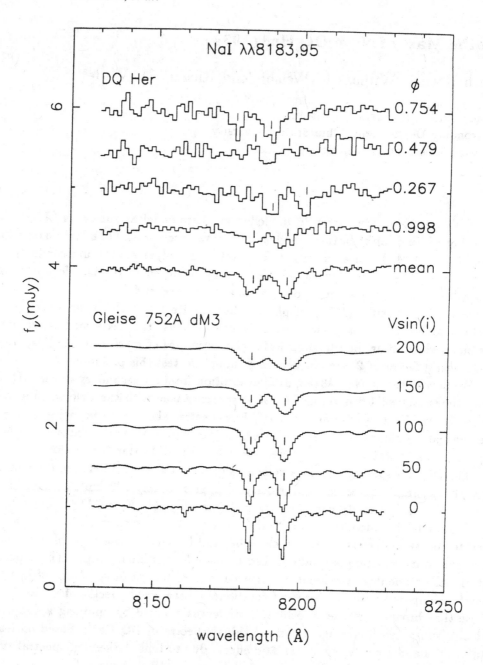

Fig. 1—The Na I λλ8183, 8195 absorption lines in spectra of DQ Her taken near conjunction and quadrature phases, in the mean spectrum averaged over two 4.5-hour binary orbits, and in rotationally-broadened spectra of an M dwarf. The dispersion is 50 or 33 km s⁻¹ pixel⁻¹.

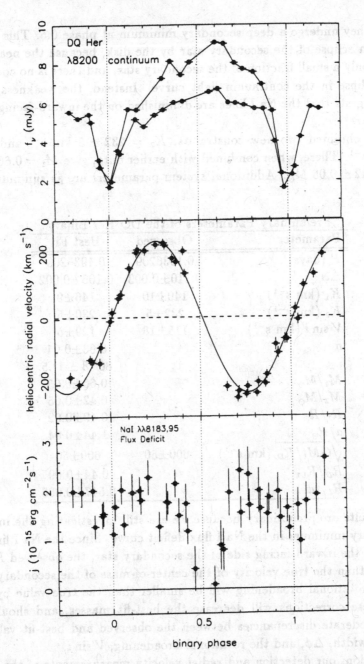

Fig. 2—The continuum light curve and the Na I radial velocity and flux deficit curves of DQ Her. The open and filled symbols distinguish the observations taken on two consecutive nights. The γ velocity is not reliable because the instrumental velocity shift has not yet been applied.

eclipse, but they undergo a deep secondary minimum at phase 0.5. This probably is not due to an eclipse of the secondary star by the disk, because the nearly edge-on disk occults only a small fraction of the secondary star, and there is no corresponding secondary eclipse in the continuum light curve. Instead, the weakness of Na I at phase 0.5 suggests that the Na I lines are diminished on the inward-facing side of the secondary star.

We have obtained two new constraints, $K_2 = 232 \pm 5$ km s^{-1} and $V \sin i = 115 \pm 18$ km s^{-1}. These, when combined with earlier data, give $M_1 = 0.66 \pm 0.05$ M$_\odot$ and $M_2 = 0.42 \pm 0.05$ M$_\odot$. Additional system parameters are as summarized in the table below:

Preliminary Parameters of the DQ Her Binary		
Parameter	Observed	Best Fit
P (days)	0.1936209	0.1936209
$\Delta\phi$	0.110±0.003	0.106±0.002
K_1 (km s^{-1})	140±10	146±9
K_2 (km s^{-1})	232±5	230±5
$V \sin i$ (km s^{-1})	115±18	129±5
q		0.63±0.04
i		86°.8 ± 1°.5
$M_1/M_\odot$		0.66±0.05
$M_2/M_\odot$		0.42±0.05
$R_2/R_\odot$		0.49±0.02
$a/R_\odot$		1.44±0.04
$\sqrt{GM_1/R_D}$ (km s^{-1})	600±50	600±50
R_D/R_{L1}		0.44±0.09
R_{L1}/a		0.547±0.006

These results are preliminary because we are still investigating the implications of the secondary minimum in the Na I flux deficit curve. Since the Na I lines appear to be weak on the inward-facing side of the secondary star, the observed K_2 velocity will be larger than the true velocity of the center-of-mass of the secondary star, and the observed rotational broadening will be smaller than the true value by a similar amount. These corrections will decrease the best-fit masses, and should help to resolve the moderate discrepancies between the observed and best-fit values of the eclipse phase width, $\Delta\phi$, and the rotational broadening, $V \sin i$.

In conclusion, our detection and radial velocity measurements of the secondary star in DQ Her imply a white dwarf mass of ~ 0.6 M$_\odot$. This result is in accord with the prediction of current TNR models, which require a low-mass white dwarf to avoid a fast nova eruption in this system.

Nova V842 Cen 1986 — A Preliminary Report [1]

Joachim Krautter[1] and Marton A. J. Snijders[2]

[1]Landessternwarte, Königstuhl, D-6900 Heidelberg, F. R. G.
[2]Astronomisches Institut, D-7400 Tübingen, F. R. G.

1 INTRODUCTION

Nova V842 Cen 1986 was discovered on November 22, 1986 by McNaught (1986) at a magnitude of $V = 5.6$. It reached its visual maximum ($m_V = 4.6$ mag) two days later on November 24, 1986 being the brightest nova since N V1500 Cyg 1975. The visual and IR (J-band) light curves up to the end of April 1987 was reported by Whitelock (1987). A more complete light curve obtained by the amateur astronomers of the Royal Astronomy Society of New Zealand was presented by F. Bateson at this conference.

2 OBSERVATIONS

Spectroscopy of both high and medium resolution and photometric observations in the visual spectral range were carried out during the first six months after the maximum at the European Southern Observatory, La Silla, Chile and at the South African Astronomical Observatory, Cape Town, South Africa. Ultraviolet observations were carried out on January 4, February 18, and May 21, 1987 with *IUE* using the facilities of the ESA-Villafranca *IUE* satellite tracking station. Spectra were obtained in the low-resolution mode in both the short- and long-wavelength range. A high-resolution spectrum taken on January 4 in the short-wavelength range has reasonable signal-to-noise ratio but 2 high-resolution spectra taken on 21 May have a very low signal-to-noise ratio, and only the strongest emission lines could be used for further evaluation.

3 RESULTS

The visual light curve showed a rather smooth decline for the first 40 days. Afterwards, the V brightness dramatically dropped by about 8 magnitudes within

---

[1]Based on observations collected at the Villafranca Satellite Tracking Station of the European Space Agency by the *International Ultraviolet Explorer* under the Target of Opportunity program, at the European Southern Observatory, La Silla, Chile, and at the South African Astronomical Observatory, Cape Town, South Africa.

20 to 25 days, indicating the formation of circumstellar dust in the expanding shell. The decay time $t_3 \simeq 55$ days could be determined by a linear extrapolation of the early decline light curve. Hence, N V842 Cen can be classified as fast nova, and one obtains an absolute magnitude $M_V = -7.4$ mag.

The interstellar extinction can be derived from the well-known 2200 Å feature in the ultraviolet spectral range. Hereby the continuum energy distribution around 2200 Å is used. Best results for a smooth continuum were obtained for $E_{B-V} = 0.55 \pm 0.05$. With this value, one obtains a distance for V842 Cen of $d = 1200$ pc. This is consistent with Neckel and Klare's (1980) relation of the visual absorption A_V versus distance in the area around V842 Cen. The velocities of the absorption systems appearing during the early phases (principal absorption 900 km s^{-1}, diffuse enhanced absorption 1715 km s^{-1}), as well as the shell expansion velocity (about 1300 km s^{-1}) are in excellent agreement with the classification as fast nova.

4 PHYSICAL PARAMETERS

Physical parameters in the expanding shell like N_e, T_e, and chemical abundances were derived from spectroscopic data in the visual and UV spectral range obtained simultaneously on May 21, 1987, after N V842 Cen had entered the nebular stage. Standard plasma diagnostic methods were applied. However, some uncertainty is created by the unknown amount of circumstellar extinction. As a broad feature in the spectrum, present on 21 May, around 2500 Å suggests, this extinction is mainly caused by amorphous carbon grains whose reddening curve below 2500 Å is highly uncertain (Bussoletti *et al.* 1987). Hence, a constant correction factor, which we obtained from the He II $\lambda1640/\lambda4686$ ratio applying standard recombination theory, was used for the whole UV spectral range.

The electron temperature of the gas in the expanding shell can be determined according to Stickland *et al.* (1981) using fluxes of lines excited by different processes (dielectric recombination versus collisional excitation), and which therefore have different temperature dependences. In the case of V842 Cen, the intensity ratios of the following lines were used: C II $\lambda1335$/C III] $\lambda1909$, C III $\lambda2296$/C IV $\lambda1550$, and N IV $\lambda1718$/N V $\lambda1240$. The electron temperatures T_e (9500 K–13000 K) depend on the ionization stage of the ion used for diagnostic, being higher for ions with higher ionization potentials. In a second step, in order to calculate the electron density N_e, we used the line ratios of the collisionally-excited forbidden [O III] $\lambda\lambda4363$, 4959, and 5007 lines. By doing so, $\log N_e = 6.8$ was determined.

With the derived values of T_e and N_e, chemical abundances were determined. For the helium abundance with respect to hydrogen intensity ratios of He I $\lambda5876$/Hβ and He II $\lambda4686$/Hβ , respectively, have been used. Helium is slightly overabundant, the number ratio He/H = 0.21 is roughly twice the solar value. Strong overabundances of about a factor of 100 or more with respect to solar values were found for C, N, and O. For N emission lines of N II, N III, N IV, and N V, which cover all relevant ionization stages, were available. For C and O, ionization corrections had to be applied. Rough estimates of chemical abundances for the heavier elements Ne, Mg, Al, Si, and Fe yielded about solar abundances, but slight overabundances of up to 5–10 times the solar values cannot be excluded. The abundances found for V842 Cen are consistent with the thermonuclear runaway model of the nova outburst which proposes for fast novae strong overabundances of C, N, and O (cf., e.g., Starrfield and Sparks 1987).

For the shell a mass of $M = 5 \times 10^{-5}\,M_{\odot}$ was found. All the above-mentioned quantities refer to the gas phase of the expanding shell; no attempt has been made yet to derive dust quantities.

A more complete description of these results will be published elsewhere in the near future.

REFERENCES

Bussoletti, E., Colangeli, L., Borghesi, A., Orofino, V. 1987, *Astr. Ap. Suppl.*, **70**, 257.

Neckel, T., Klare, G. 1980, *Astr. Ap. Suppl.*, **42**, 251.

McNaught, R. H. 1986, *IAU Circ.*, No. 4274.

Starrfield, S. G., and Sparks, W. M. 1987, *Ap. Sp. Sci.*, **131**, 379.

Stickland, D. J., Penn, C. J., Seaton, M. J., Snijders, M. A. J., Storey, P. J. 1981, *M. N. R. A. S.*, **197**, 107.

Whitelock, P. A. 1987, *M. N. A. S. S. A*, **46**, 72.

Spectroscopy and Photometry of CP Puppis and T Pyxidis*

N. Vogt[1], L. H. Barrera[1], H. Barwig[2], and K.-H. Mantel[2]

[1]Astrophysics Group, Universidad Católica de Chile, Santiago, Chile
[2]Universitätssternwarte München, F. R. G.

1 OBSERVATIONS

Spectroscopy	Photometry
Du Pont 2.5-m telescopy at Las Campanas Observatory 2D-frutti spectrograph (Shectman et al. 1985) Spectral resolution 3 Å Spectral range 3600–7200 Å Date: Feb. 16–22, 1986.	ESO 1-m telescopy at La Silla Observatory Standard photometer (UBV) and 15-channel Munich photometer with fiber link $UBVRI$ (Barwig et al. 1988) Date: Jan. 31–Feb. 14, 1986.

2 RESULTS

CP Puppis, one of the brightest known classical novae, erupted in 1942 with an extremely rapid decine (Duerbeck 1987). Warner (1985) detected humps in the quiescent light curve and derived a photometric period of about 0.06614 day. Previous spectroscopic observations (Duerbeck et al. 1987, and references therein) revealed periodic radial velocity variations with $P = 0.06142$ day and $K_1 = 92 \pm 18$ km s^{-1}.

Our spectroscopy consists in 131 individual spectrograms obtained in 5 nights. An analysis of the radial velocities from the strongest emission lines (Balmer series, HeII $\lambda 4686$) gives $P_0 = 0.06142$ day, $K_1 = 85 \pm 6$ km s^{-1}, consistent with Duerbeck et al.'s values. It seems that there is a unique spectroscopic period whose accurate value, however, could not yet be determined due to an unfortunate time distribution of the available observations (subsequent runs are either a few days or one year apart). The radial velocity curve of CP Pup is consistent with an orbital inclination of $i \approx 60°$ and masses $M_1 = 0.6$ M$_\odot$ and $M_2 = 0.15$ M$_\odot$ for the stellar components.

Extensive photometric data obtained with ESO standard photometer (UBV in filter sequence, time resolution 100 s, 8.7 observing hours in 3 nights) and with the

---

* Based on observations obtained at the Las Campanas Observatory, Chile, and the European Sourthern Observatory, La Silla, Chile.

Munich 15-channel photometer with fiber links ($UBVRI$ simultaneously, time resolution 2 s, 34.4 observing hours in 5 nights) confirm, in principle, Warner's (1985) hump light curves. However, there are still ambiguities in the cycle counts between subsequent nights in spite of an excellent coverage of 4 to 5 complete cycles in several nights. The reason for this becomes evident when determining periods from individual nights: they range between 0.059 day and 0.070 day, with typical errors of ±0.002 day. Also, Warner's (1985) individual observing runs (4 nights) show significantly different periods on subsequent nights. A preliminary analysis of these night-to-night variations suggests that the photometric period oscillates around a mean value $\bar{P}_p \approx 0.063$ day with a period of approximately 3.8 days, which corresponds to a beat period between P_0 and $\bar{P}_p$. Similar effects have been reported for V 1500 Cyg (Pavlenko and Pelt 1988) and TT Ari (Udalski 1988).

T Pyxidis is one of the prototypes of recurrent novae with a total of five recorded outbursts between 1890 and 1966 (Duerbeck 1987). No orbital period has yet been determined; however, Szkody and Feinswog (1988) obtained an infrared light curve and detected sinusoidal variations with a period of about 100 minutes. Our spectroscopic observations (101 spectrograms in 5 nights) revealed periodic radial velocity variations with $P_0 = 0.1433$ day and $K_1 = 24 \pm 5$ km s^{-1}. P_0 is nearly twice the infrared period of Szkody and Feinswog (1988). If this is taken as evidence for ellipsoidal variations of the secondary star, the orbital inclination must be rather high ($i \gtrsim 50°$). In this case, our relatively small radial velocity amplitude implies that the secondary of T Pyx is slightly evolved rather than a main-sequence star.

A detailed publication is in preparation.

This research was supported by the Chilean "Fondo Nacional de Ciencias y Tecnología" (Grants FONDECYT 369/88 and 481/89). We also acknowledge the support given by Stiftung Volkswagenwerk through a cooperation project between the Astrophysics Group in Santiago and Universitätssternwarte München, as well as to "Fundación Andes," Chile, for a travel grant.

REFERENCES

Barwig, H., Schoembs, R., and Huber, G. 1988, *in Proc. 2nd Workshop on Improvements to Photometry*, NASA Conference Pub. 10015 (Gaithersburg: NBS).

Duerbeck, H. W. 1987, *Space Sci. Rev.*, **45**, No. 1–2, 1.

Duerbeck, H. W., Seitter, W. C., and Duemmler, R. 1987, *M. N. R. A. S.*, **229**, 653.

Pavlenko, E., and Pelt, J. 1988, *IBVS*, No. 3252.

Shectman, S., Price, C., and Thompson, I. 1985, *Annual Report of the Director of the Mount Wilson and Las Campanas Observatories*, Carnegie Institution, p. 52.

Szkody, P., and Feinswog, L. 1988, *Ap. J.*, **334**, 422.

Udalski, A. 1988, *Acta Astr.*, **38**, 315.

Warner, B. 1985, *M. N. R. A. S.*, **127**, 1p.

The Nature of the Recurrent Nova T Coronae Borealis

R. Gilmozzi[1], P. L. Selvelli[2], and A. Cassatella[3]

[1]Space Telescope Science Institute, Baltimore, MD 21218, U. S. A. and
 Istituto di Astrofisica Spaziale, CNR, Frascati, Italy
[2]CNR, Osservatorio Astronomico di Trieste, Italy
[3]IUE Observatory, European Space Agency, Madrid, Spain and
 Istituto di Astrofisica Spaziale, CNR, Frascati, Italy

1 INTRODUCTION

T Cr B is a double-line spectroscopic binary, with period $P = 227.6$ days (Kraft 1958; Kenyon and Garcia 1986), containing an M3 giant and a hotter companion whose nature has been so far elusive. The radial velocity data suggest that $M_{giant} > 2.2\,M_\odot$ and $M_{hot} > 1.6\,M_\odot$, although these values might be affected by the uncertainties in the determination of $K_2 = v_2 \sin i$. T Cr B is a member of the class of recurrent novae, a small group of objects whose precise nature is still controversial. Recently, Webbink *et al.* (1987) have identified two sub-classes of recurrent novae on the basis of their outburst (OB) mechanism: those powered by a thermonuclear runaway (TNR) on a white dwarf and those powered by the transfer of a burst of matter from a red giant to a main-sequence star. One of the conclusions of the Webbink *et al.* study was the interpretation of the behavior of the outburst of T Cr B in terms of accretion onto a main-sequence star of a parcel of matter ejected by the giant companion. We present here some results from eleven years of observations of T Cr B with the *IUE* satellite. The main conclusion of this study is that the overall behavior of T Cr B can find a natural and self-consistent interpretation in terms of accretion onto a white dwarf.

2 THE UV OBSERVATIONS

T Cr B has been monitored at VILSPA from the early phases of *IUE*'s life until very recently. The UV spectra of T Cr B show a broad interstellar dust absorption feature centered around 2175 Å, compatible with $E_{B-V} = 0.15$. The reddening-corrected continuum energy distribution is variable and can in general be represented, at the various epochs, by a single power-law spectrum of the form $F_\lambda \propto \lambda^{-\alpha}$ over the entire *IUE* range. The UV spectral index α ranges from 0.9 to 2.0, with a mean value of 1.3. This value is lower than that generally observed in CVs. The peculiarity of T Cr B is that this low value, similar to that of dwarf novae in quiescence, corresponds to a much higher UV luminosity (see the next section) and a negligible contribution to the optical range.

The changes in the UV continuum show no obvious dependance on the orbital phase. Near phase 0.50 (red giant in front), a possible occultation effect could be present. However, no flux decrease was observed in spectra taken at phase 0.54, 0.48, and even at phase 0.50. A distinctive peculiarity of T Cr B is the fact that the UV and optical variations (as measured with the FES on board *IUE*) are not correlated with each other. In particular, during the UV relative maximum of May 1983, the magnitude was 10.0, while, at the time of a deep UV minimum in 1979, the magnitude was 9.9.

A typical UV spectrum of T Cr B is shown in Figure 1. The emission lines are remarkably intense in comparison with classical novae in quiescence and, in addition to N V, C IV, and He II, include also strong intercombination transitions (e.g., Si III λ1892, C III λ1909) which are usually absent in classical novae. N V, which is always present in T Cr B, is detected only in the hottest old novae and is absent, for example, in V603 Aql.

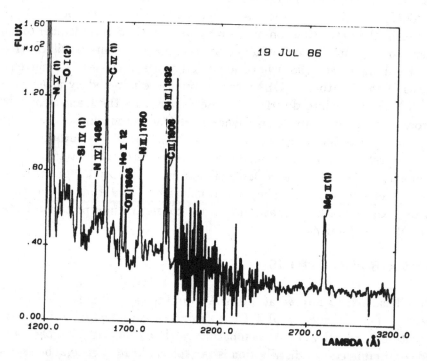

Fig. 1—A typical *IUE* spectrum of T Cr B. The energy distribution is corrected for $E_{B-V} = 0.15$. The flux is in units of 10^{-14} erg cm^{-2} s^{-1} Å^{-1}.

High-resolution spectra obtained with the SWP camera at various epochs indicate that the intercombination transitions have narrow cores and rather broad wings. The high-excitation permitted transitions (C IV λ1550, He II λ1640) show instead only a very shallow and broad profile with HWZI > 900 km s^{-1}.

3 THE UV AND EUV LUMINOSITIES AND THE MASS-ACCRETION RATE

Assuming a distance to T Cr B of 1300 pc, the integrated *IUE* luminosity (from 1980 to 1988) varies from $30 L_\odot$ to $60 L_\odot$ with an average value of about $40 L_\odot$. The mass-accretion rate $\dot{M}$ can be estimated if the disk luminosity and the nature of the accreting object are known. If most of the disk luminosity is emitted in the UV, as is the case for T Cr B, then $L_{\rm UV}$ is not much smaller than $L_{\rm disk}$, and can be used to give a reasonable estimate of $L_{\rm disk}$ and, therefore, to set a (lower) limit on $\dot{M}$. With the conservative assumption that the observed *IUE* flux is about 2/3 of the integrated flux, and after correction for inclination, we obtain an estimate of about $120 L_\odot$ for the intrinsic disk luminosity. In what follows we assume that the accreting object is a white dwarf. This assumption is justified by direct UV observational evidences and by several other considerations (reported in the next section) which are hardly compatible with the presence of a main-sequence companion. With the assumption of a white dwarf accretor and taking indicative values of $M = 1\,{\rm M}_\odot$ and $R = 0.01\,R_\odot$, the derived mass-accretion rate is in the range 3–$5 \times 10^{-8}\,{\rm M}_\odot\,{\rm yr}^{-1}$.

An independent check for the value of $\dot{M}$ can be made through the $\dot{M}$-$\lambda1640$ intensity relation reported by Patterson and Raymond (1985) for CVs. Their Table 2 gives $\dot{M} \sim 10^{19}\,{\rm g\,s}^{-1}$ for a $1\,{\rm M}_\odot$ white dwarf and $L_{1640} \sim 10^{33}\,{\rm erg\,s}^{-1}$, in good agreement with that derived directly from the UV luminosity.

The $\lambda1640$ intensity can also be used to determine the number of photons with energies higher than 55.4 eV to provide an estimate of the EUV/soft X-ray flux in T Cr B. The average $\lambda1640$ luminosity is $\sim 1.3 \times 10^{33}\,{\rm erg\,s}^{-1}$. Assuming case B recombination and that all EUV photons ionize He II, we can estimate the number $Q4$ of photons with energies higher than 55.4 eV: $Q4 \sim 3.3 \times 10^{44}$ photons. If we assign an average typical energy of 75 eV to these photons, the derived EUV luminosity is about $5 \times 10^{34}\,{\rm erg\,s}^{-1}$. This is however a lower limit for $L_{\rm EUV}$ because it is unlikely that all EUV photons will ionize He II and also because the geometry near the boundary layer (where these processes take place) is far from the spherically symmetric case. A value of $L_{\rm EUV}$ in the range 10^{35}–$10^{36}\,{\rm erg\,s}^{-1}$ seems more realistic.

T Cr B has been positively detected with the *Einstein* IPC "hard" X-ray detector (0.2–4 keV) with a luminosity of about $5 \times 10^{31}\,{\rm erg\,s}^{-1}$ (Córdova, Mason, and Nelson 1981). This value is in agreement with the considerations of Patterson and Raymond (1985) who predict that at high accretion rates (such as found in T Cr B) only a small fraction of the boundary layer luminosity is emitted as "hard" X-rays.

Assuming that $L_{\rm UV}$ is produced in the accretion disk, while $L_{\rm EUV}$ and $L_{\rm X}$ are produced in the boundary layer, our results indicate that $L_{\rm disk} \sim L_{\rm BL} \gg L_{\rm X}$. This picture is fully consistent with the theoretical expectations at the $\dot{M}$ found in T Cr B.

4 THE NATURE OF THE ACCRETING COMPONENT AND THE PROBLEM OF RADIAL VELOCITIES

A white dwarf companion has been explicitly or implicitly assumed in our previous considerations. There are indeed several facts which are hardly compatible with the presence of a main-sequence companion while they find a natural and self consistent interpretation in terms of a white dwarf accretor. We recall here briefly:

1) The bulk of the disk luminosity is emitted in the UV with a negligible contribution in the optical range. The UV luminosity is much larger than that of old classical novae. The observed L_{UV} could be also produced with a main-sequence accretor, but this would require an extremely high accretion rate and, as a consequence, the disk would emit mostly in the optical, in contradiction with the observations.

2) A rather strong He II $\lambda1640$ emission is generally present. This emission is indicative of temperatures of the order of 10^5 K and is naturally associated with the boundary layer. The mass-accretion rate associated with the He II $\lambda1640$ luminosity on the basis of the semi-empirical estimates of Patterson and Raymond (1985) is in good agreement with that derived directly from the UV luminosity. Only a white dwarf accretor, at the calculated $\dot{M}$, can explain at the same time both the observed L_{UV} and the high temperature required to produce the He II $\lambda1640$ emission.

3) The X-ray luminosity is of the same order of that found in the X-ray brightest old novae $L_X \sim 5 \times 10^{31}$ erg s^{-1} and is practically impossible to explain in terms of accretion onto a main-sequence star.

4) The profile of the C IV $\lambda1550$ emission line in high-resolution spectra is very wide and shallow. C IV, the strongest emission line in the low-resolution spectra, is hardly evident at high resolution, while weaker lines (e.g., C III $\lambda1909$ and Si III $\lambda1892$) are sharper and clearly present. This indicates that C IV $\lambda1550$ is strongly broadened by rotation because it originates in the innermost disk-boundary layer region. The HWZI of C IV is larger than 800 km s^{-1}, a value not compatible with a main-sequence star.

5) The EUV luminosity emitted below 228 Å is comparable with the observed UV luminosity. If L_{UV} is attributed to the disk and L_{EUV} to the boundary layer, the observations are in agreement with the theoretical predictions for a standard disk around a white dwarf accretor. Also in agreement with the models is the fact that at the large observed $\dot{M}$ only a small fraction of the boundary layer luminosity is emitted as "hard" X-rays.

All these arguments in favor of the white dwarf are "disturbed" by the results of the radial velocities which indicate a mass for the companion higher than that acceptable for a white dwarf. Kraft (1958) used several plates for the determination of K_1 (~ 23 km s^{-1}) from the absorption lines of the giant, but only seven plates for the determination of K_2 (~ 31 km s^{-1}) from the hydrogen emission lines. The results for the masses were $M_1 \sim 2.6\,M_\odot$ and $M_2 \sim 1.9\,M_\odot$, thus placing the hot component above the Chandrasekhar limit. It is important, however, to stress that:

1) The hydrogen emission lines of T Cr B are quite wide (300 km s^{-1}) and severely distorted by the absorptions of the giant and show a composite structure.

2) An entire period of the emission lines was covered by only seven points (plates) at all and only two plates were close to quadratures (velocity maxima).

3) About 30 years have elapsed since the radial velocity measurements. Even nowadays, in spite of the considerable improvements in the measuring techniques, the problem of how to measure K_2 in CVs is serious and difficult.

4) Kraft himself, after the laborious operations for the reconstruction of the emission lines profile, stated explicitly that "a non-negligible degree of error might still exist in the orbit derived from the hydrogen emissions."

5) It is not clear whether the hydrogen emissions can be necessarily associated with the orbital motion of the hot component.

Under these circumstances, an error of about 8 km s^{-1} in excess for the K2 value obtained by Kraft is not unlikely. A reduction of K_2 by this amount would yield $K_2 \sim K_1 \sim 23$ km s^{-1} and a solution $M_1 \sim M_2 \sim 1.4\,M_\odot$ for the masses, thus allowing the accreting object to be a white dwarf very close to the Chandrasekhar limit.

5 REMARKS ON THE OUTBURST OF 1946 AND ON THE SECONDARY MAXIMUM

Webbink *et al.* (1987) have interpreted the 1946 outburst in terms of a burst of mass which was transferred from the giant onto the companion. The reasons for excluding a TNR were both the supposed presence of a main-sequence star and the presence of a secondary maximum in the OB light curve, not usually observed in fast novae. It is our opinion that the OB of T Cr B can be nevertheless interpreted in terms of a TNR on a white dwarf for at least three reasons:

1) The spectral evolution of T Cr B after OB has followed the same pattern observed in other (very fast) novae. The expansion velocity reached -5000 km s^{-1} (Herbig and Neubauer 1946), a value of the order of the escape velocity from a white dwarf.

2) Photometrically, T Cr B has obeyed the same relation $M_v(\max)$-t_3 followed by the classical novae (Warner 1987).

3) The observed $\dot{M}$ (calculated on the assumption of a white dwarf) is almost constant over the entire observing period and is exactly what is required to produce a TNR recurrent nova with the observed recurrence time (Starrfield, Sparks, and Truran 1986; Livio 1988).

This behavior can hardly be a mere coincidence.

The presence of the secondary maximum in the light curve about 100 days after maximum was considered by Webbink *et al.* (1987) as a strong indication against a TNR event and has been interpreted as due to the impact of the orbiting material (after circularization) with the surface of the accreting star. This interpretation faces, however, serious problems when the spectroscopic behavior at the time of the secondary maximum is taken into account. The spectrum was nebular-like until about 100 days after OB (following the normal spectral evolution of a very fast nova), but then the luminosity increased by 1.5 mag, the emission lines (O III, Ne III, He II) became very faint or invisible, and a strong shell absorption spectrum appeared and persisted until the end of the secondary maximum.

This spectral behavior contradicts the OB model proposed by Webbink *et al.* (1987): when the orbiting material strikes the star's surface an increase in excitation, emission line intensity, and continuum temperature is expected, in contrast to what actually observed; note in particular that the He II λ4686 line was very strong BE-FORE the onset of the secondary maximum and then faded to an almost undetectable emission at the epoch of the secondary maximum.

In our opinion, the secondary maximum can be physically explained by the appearance of the optically thick shell surrounding the nova. The newly formed shell has acted like a pseudo-photosphere at rather low (10^4 K) temperature. Thus, the energetic photons emitted by the nova have been converted by the shell into low-energy photons emitted mostly in the optical. This has been the cause of the observed photometric increase and of the secondary maximum.

We think that the OB of T Cr B, with its peculiarities can be explained in terms of a TNR in a low-mass envelope, as expected if the accreting white dwarf is very close to the Chandrasekhar limit. Faster novae occur on more massive white dwarfs in which the envelope mass is quite low. This accounts for the high expansion velocity and the very rapid photometric and spectroscopic changes.

It is interesting that Ne (a signature of TNR on a massive white dwarf; Starrfield, Sparks, and Truran 1986) was a strong component of the emission line spectrum of T Cr B at several epochs. Pre-outburst spectroscopic observations made by Minkowsky (1939) and by Swings and Struve (1943) have always reported the presence of Ne III lines, sometimes even of both Ne III and Ne V. After OB, Sanford (1947) reported the simultaneous presence of Ne III and Ne V as prominent lines. Kraft (1958) described the Ne III emission lines as "fairly strong" while the other nebular lines (i.e., O II λ3727, O III λ4363, O III λ4958, 5007), usually rather strong, were described as very faint or not detectable.

An overabundance of Ne is clearly indicated by these "archive" data. However, the fact that strong Ne lines were observed also before and 10 years after OB is not of easy interpretation. We can tentatively suggest that this was caused by the fact that the giant was re-transferring to the white dwarf some of the material ejected during previous outbursts and intercepted by its outer layers.

6 CONCLUSIONS

UV and other observations have indicated beyond any reasonable doubt that the behavior of T Cr B in quiescence can find a natural interpretation in terms of fast accretion onto a white dwarf. The spectroscopic and photometric observations at the time of the last OB are not in contradiction with this conclusion and actually provide further indications in favor of a TNR event on a massive white dwarf. If new and very accurate radial velocities measurements would confirm beyond any observational error that the companion of the giant is as massive as, say, $1.8\,M_\odot$, we feel that a possible explanation could be sought in a triple system model. The giant's companion could be a normal nova composed of a massive white dwarf and a $0.5\,M_\odot$ low-luminosity main-sequence star, practically undetectable in the light of the giant.

REFERENCES

Córdova, F. A., Mason K. O., and Nelson J. E. 1981, *Ap. J.*, **245**, 609.

Herbig, G. H., and Neubauer, F. J. 1946, *Pub. Astr. Soc. Pac.*, **58**, 196.

Kraft, R. 1958, *Ap. J.*, **127**, 625.

Kenyon, S. J., and Garcia, M. R. 1986, *A. J.*, **91**, 125.

Livio, M. 1988, in *The Symbiotic Phenomenon*, ed. J. Mikolajevska, *et al.* (Dordrecht: Kluwer), p. 323.

Minkowsky, R. 1939, *Pub. Astr. Soc. Pac.*, **51**, 54.

Paczynski, B. 1965, *Acta Astr.*, **15**, 198.

Patterson, J., and Raymond, J. C. 1985, *Ap. J.*, **292**, 550.

Sanford, R. F. 1947, *Pub. Astr. Soc. Pac.*, **59**, 87.

Starrfield, S., Sparks, W. M., and Truran, J. W. 1986, *Ap. J. (Letters)*, **303**, L5.

Swings, P., and Struve, O. 1943, *Ap .J.*, **98**, 91.

Wade, R. A. 1984, *M. N. R. A. S.*, **208**, 381.

Warner, B. 1987, *M. N. R. A. S.*, **227**, 13.

Webbink, R. F. 1976, *Nature*, **262**, 271.

Webbink, R. F., Livio, M., Truran, J. W., and Orio, M. 1987, *Ap. J.*, **314**, 653.

The Common Envelope Phase in Classical Novae

A. Shankar[1], A. Burkert[1], M. Livio[2], and J. W. Truran[1]

[1]Department of Astronomy, University of Illinois at Urbana-Champaign
[2]Department of Physics, Technion, Israel

1 INTRODUCTION

A common envelope (CE) forms in classical nova (CN) systems when the hydrogen envelope accreted by the white dwarf expands dynamically in response to a thermonuclear runaway (TNR) and subsequently to energy deposition by β-decays in the outer layers. The substantial velocities observed in the nova systems imply that the envelope radius reaches the binary dimensions on rapid timescales, engulfing the main-sequence companion and initiating the CE phase. The drag experienced by the secondary as it moves inside the CE dissipates orbital energy and angular momentum. The deposition of this energy and angular momentum in the envelope can have potentially significant consequences for the evolution of the outburst.

The existence of a CE phase in CN has been known for some time. Its relevance to inducing further mass loss from the nova system was investigated first by Mac-Donald (1980). He found that the drag energy deposition was capable of ejecting matter at velocities resembling slow novae at best. In order to continue the investigation of this potentially important phase further, we have carried out a numerical study of the effects of a CE phase on CN outbursts using one- and two-dimensional hydrodynamical techniques.

2 THE OBSERVATIONAL EVIDENCE

For typical binary parameters ($M_{WD} = 1.25 \, M_\odot$, $M_{sec} = 0.5 \, M_\odot$, $P = 3$–5 hours), the orbital separation is roughly of the order of 10^{11} cm. CN at visual maximum have characteristic A–F supergiant spectra, implying photospheric radii of 10^{12}–10^{13} cm. Alternatively, assuming the rise time to be as short a day for all novae, along with the fact that typical observationally-deduced expansion velocities lie in the range 150–3000 km s^{-1}, we can deduce that the envelope radii at visual maximum fall in the same range (10^{12}–10^{13} cm). The secondary must therefore lie deep within the envelope at visual maximum. The conclusion that *all CN undergo a CE phase* thus seems inescapable. Possible exceptions are symbiotic novae, where the large binary separations may preclude a CE phase.

The circumstantial observational evidence for the existence of a CE phase in CN comes from recent optical and X-ray observations: (1) In many novae the spectrum changes to a characteristic nebular type in less than one year (Williams 1989). An example is DQ Her, where the nebular spectrum appeared on a timescale of roughly 100 days. (2) Observations of Nova Muscae 1982, PW Vul, Nova Vul 1984 No. 2 (Ögelmann, Krautter, and Beuermann 1987) show rapid emergence of X-rays on timescales of ~ 100 days. These observations are consistent with the luminosity from a hot ($T_e > 10^5$ K) white dwarf. Both (1) and (2) point toward a very efficient mass-loss mechanism, capable of removing mass at a rate larger than that obtained by the TNR + β-decays or winds. Possible additional evidence comes from the prolate morphologies of the ejected shells, for example in DQ Her (Wade 1989). Such morphologies can be obtained with the aid of an interacting wind model (Soker and Livio 1988), in which a CE phase helps the shaping of the nova shells in a way similar to the planetary nebulae with binary nuclei.

3 THE NUMERICAL MODELS

The tools used to study the CE problem include a one-dimensional, Lagrangian, fully-implicit hydrodynamic code (Kutter and Sparks 1972, 1980; Starrfield, Sparks, and Truran 1985), and a two-dimensional, Eulerian, explicit hydrodynamic code (Burkert and Hensler 1988). The drag energy was calculated assuming the Bondi-Hoyle picture (for details, see Shankar, Livio, and Truran 1989 and Livio *et al.* 1989), in which the drag luminosity is given by

$$L_d = \eta \xi(M) \pi R_A^2 \rho v_{rel}^3 , \tag{1}$$

where $\xi(M)$ is a Mach number (M)-dependent drag coefficient, given by Shima *et al.* (1985); $R_A = 2GM_{sec}/(v_{rel}^2 + c_s^2)$ is the secondary accretion radius; ρ is the density in the interaction region, and $v_{rel} = \sqrt{(v_{out}^2 + v_{orb}^2)}$ is the relative velocity with which the secondary moves through the CE. This energy is deposited self-consistently in a region located at the orbital separation. The size of the "interaction" region is determined by the secondary accretion radius — in the 1D case it is a spherical shell with $\Delta R = R_{acc}$, and a torus with $R = R_{acc}$ in the 2D case. The 1D calculations do not include the deposition of angular momentum in the envelope but the effect of envelope spin-up is simulated by an efficiency factor $\eta < 1$. The 2D calculations are adiabatic and take into account the deposition of both the angular momentum and the drag energy in the CE. The initial 2D envelope has a mass of 5×10^{-6} M$_\odot$ and is taken to be in hydrostatic equilibrium, a condition nearly always satisfied by the inner, slowly-expanding layers for all novae after the initial burst ejection. The primary is taken to be a 1.0 M$_\odot$ white dwarf in both cases. The 2D calculations include the gravitational potentials of the primary and the secondary — the primary is taken to be a point mass at the center and the secondary a ring of total mass M_{sec} located at the orbital separation (in a manner similar to Bodenheimer and Taam 1984). The secondary mass is taken to be 0.5 M$_\odot$ and the binary period $P = 3.5$ and 5 hours, in the 1D and 2D cases respectively.

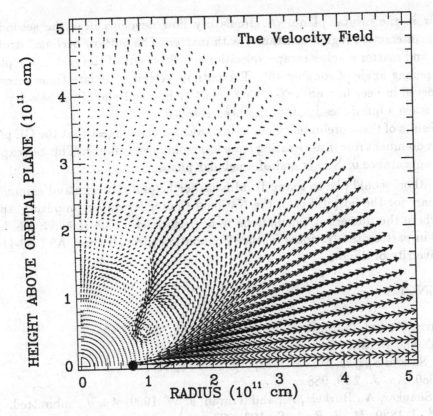

Fig. 1—The velocity field in the common envelope, 4 orbital periods after the energy and angular momentum deposition begins. The black circle represents the location of the secondary.

While at least 2D models are clearly required to study this non spherically symmetric problem, it is not possible to model a full nova outburst using the 2D code. The 1D code on the other hand does follow a full nova development. The 1D results are thus expected to provide the basic hydrodynamics and an input model for the 2D calculations.

4 RESULTS AND CONCLUSIONS

The results of the 1D calculations show that the inclusion of the drag energy indeed accelerates mass loss from the system, increasing the effective mass-loss rate. The ejection velocities are found to resemble slow novae. The results of the 2D calculations show the precise flow geometry (see Fig. 1). It is found that injection of angular momentum in the CE has the effect of rapidly spinning up the envelope in the neighborhood of the secondary, bringing down v_{rel} and thus L_d. A combination of the centrifugal forces and the expansion caused by the energy deposition helps to initiate an outflow in the orbital plane. A circulatory flow develops near the secondary, feeding the interaction region constantly with matter. The outflow becomes stronger with time and matter reaches escape velocities and flows out of the grid in the plane, with an opening angle of roughly 15°. The outflow velocities at late times resemble those observed in very fast novae ($\sim$ 3000 km s^{-1}). Figure 1 shows the velocity field at a time when a quasi-steady state has been reached.

The results of these preliminary numerical calculations suggest that the CE phase may play a dominant role in acclerating mass loss from nova systems. This can explain the rapid appearance of X-rays and of a nebular phase.

The authors would like to thank D. Mihalas for helpful criticism and comments, and L. Smarr for the hospitality of the National Center for Supercomputing Applications, where the numerical calculations were performed. This research has been supported in part by the National Science Foundation through grant AST 86-11500 at the University of Illinois.

REFERENCES

Bodenheimer, P., and Taam, R. E. 1984, *Ap. J.*, **280**, 771.
Burkert, A., and Hensler, G. 1988, *Astr. Ap.*, **199**, 131.
Kutter, G. S., and Sparks, W. M. 1972, *Ap. J.*, **175**, 407.
————. 1980, *Ap. J.*, **239**, 988.
Livio, M., Shankar, A., Burkert, A., and Truran, J. W. 1989, *Ap. J.*, submitted.
MacDonald, J. 1980, *M. N. R. A. S.*, **191**, 933.
Ögelmann, H., Krautter, J., and Beuermann, K. 1987, *Astr. Ap.*, **177**, 110.
Shankar, A., Livio, M., and Truran, J. W. 1989, in preparation.
Shima, E., Matsuda, T., Takeda, H., and Dawada, K. 1985, *M. N. R. A. S.*, **217**, 367.
Soker, N., and Livio, M. 1988, *Ap. J.*, **339**, 268.
Starrfield, S., Sparks, W. M., and Truran, J. W. 1985, *Ap. J.*, **291**, 136.
Wade, R. A. 1989, in I. A. U. Coll. No. 122, *Physics of Classical Novae*, ed. A. Cassatella (Berlin: Springer), in press.
Williams, R. E. 1989, in I. A. U. Coll. No. 122, *Physics of Classical Novae*, ed. A. Cassatella (Berlin: Springer), in press.

The Super-Eddington Phase of Classical Novae

J. Hayes[1], J. W. Truran[1], M. Livio[2], and A. Shankar[1]

[1]University of Illinois at Urbana-Champaign
[2]Department of Physics, Technion, Haifa, Israel

1 INTRODUCTION

The existing observational data on classical novae (cf., Payne-Gaposchkin 1957, hereafter PG) reveal a striking feature of these systems: all novae classified as fast or very fast exhibit peak visual magnitudes substantially above the Eddington limit for a limiting-mass white dwarf. We note at the outset that by "Eddington limit" we mean

$$L_{\mathrm{Edd}} = \frac{4\pi cGM}{\kappa_{\mathrm{es}}} = 3.8 \times 10^4 \, L_\odot \left(\frac{M}{M_\odot}\right). \tag{1}$$

Since this definition assumes hydrostatic equilibrium and purely radiative energy transport, one must realize that L_{Edd} as defined has limited applicability to novae in outburst. However, L_{Edd} is useful as a reference luminosity and serves to define the phase discussed here. We should also note that while many of the peak magnitudes quoted in PG were deduced from an empirical relation between peak magnitude and rate of decline, they are consistent with observations of novae in M 31 (Arp 1956), for which the distance is known, and also of Galactic novae (de Vaucouleurs 1978), for which distances were determined from expansion parallaxes or interstellar medium line intensities.

With the above in mind, the basic theoretical statement of this paper is the following: the existence of the super-Eddington phase in fast classical novae is a direct consequence of abundance enhancements of CNO nuclei in the accreted envelope, the energy released in the positron decays of nuclei formed during CNO-cycle hydrogen burning, and the degree to which convection can mix these positron-unstable isotopes into the surface regions of the envelope during the peak of the runaway.

2 CALCULATIONS AND RESULTS

The dependence upon Z_{CNO} may be illustrated by a simple physical argument (cf., Truran 1982). Once the runaway is initiated, the degeneracy will be lifted when $P_{\mathrm{gas}} = P_{\mathrm{deg}}$, or

$$\frac{\rho kT}{\mu m_{\mathrm{H}}} = 10^{13} \left(\frac{\rho}{\mu_e}\right)^{\frac{5}{3}}. \tag{2}$$

Solving equation (2) for T yields

$$T = 8 \times 10^7 \text{ K} \left(\frac{\rho}{2 \times 10^4 \text{ g cm}^{-3}} \right)^{\frac{2}{3}}.$$ (3)

Thus we see that CNO-cycle hydrogen burning reactions control the outburst behavior. Once T becomes greater than approximately 10^8 K, the rate of energy generation becomes constrained by the positron decay timescales of ^{14}O ($\tau = 100$ s) and ^{15}O ($\tau = 176$ s). Since these timescales are now longer than the proton-capture timescale, two effects result: (1) large accumulations of ^{14}O and ^{15}O will accrue and (2) the energy available on a hydrodynamic timescale of seconds, assuming one capture per CNO nucleus, will be

$$E_{\text{proton capture}} \sim 2 \times 10^{15} \frac{\text{erg}}{\text{g}} \left(\frac{Z_{\text{CNO}}}{Z_{\odot}} \right).$$ (4)

Noting that

$$\frac{GM}{R} \sim 2 \times 10^{17} \frac{\text{erg}}{\text{g}},$$ (5)

we see that high CNO abundances imply a generally more dynamic event, as reflected in higher outflow velocities and more rapid mass ejection. The dependence of nova speed class on CNO abundances has firm observational support. Two limiting examples are the fast nova V1500 Cyg ($Z = 0.30$; Ferland and Shields 1978) and the slow nova HR Del ($Z = 0.08$; Tylenda 1978). The white dwarf mass is also a critical parameter.

Additionally, the development of a strong convective zone throughout the envelope implies that the accumulated ^{14}O and ^{15}O nuclei may be mixed into the envelope surface regions, where they may decay and release their energy. This effectively creates an energy-producing region which extends throughout the envelope, rather than being confined to a thin shell at the envelope base. We find that the release of such large amounts of energy (roughly 10^{13} erg g^{-1} s^{-1}) near the surface has a profound effect on the outburst dynamics.

Numerical simulations of a nova outburst on a 1.25 $M_{\odot}$ white dwarf affirm the critical dependencies on Z_{CNO} and the positron decay energy. Comparison of Figures 1 and 2, which differ only in the presence/absence of the positron decay energy contribution in the region above the convective burning shell, reveals the crucial role played by the positron decays in producing a strong super-Eddington peak. A similar effect results from reducing the CNO abundance by a factor of two, from $Z = 0.50$ to $Z = 0.25$.

The above dependencies on Z_{CNO} and the positron decays hinge upon a degree of convective overturn sufficient to transport the ^{14}O and ^{15}O nuclei to the surface on a short timescale. Calculations involving a range of mixing lengths ($= \alpha \times$ pressure scale height (PSH); α is a pure number) show that the outburst characteristics scale

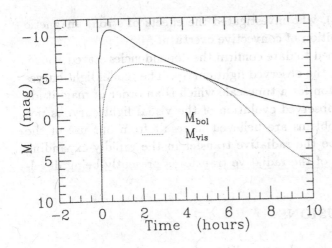

Fig. 1—This model gives the broadest super-Eddington peak presently obtainable numerically; notice that the super-Eddington timescale is an order of magnitude shorter than that observed, and that the visual magnitude is only marginally super-Eddington. Here, $M_{wd} = 1.25\,M_\odot$, and $Z = 0.5$.

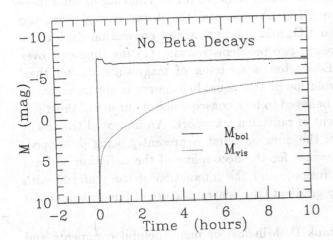

Fig. 2—This model is identical with the one shown in Fig. 1 except that the beta decay energy release in the region above the convective burning shell has been artificially removed; note that the super-Eddington peak has all but vanished.

TABLE 1
EFFECT OF MIXING LENGTH ON OUTBURST
$M_{wd} = 1.25\,M_\odot$, $Z = 0.5$

α (PSH)	Peak M_{bol} (mag)	Max V_{ej} (km s^{-1})
0.0	−7.25	276
0.1	−7.25	403
0.3	−9.47	2204
1.0	−10.65	3692
2.0	−10.85	4019

with α as expected (see Table 1), but that for high Z_{CNO}, one obtains a strong super-Eddington peak unless α is less than about 0.1. This is consistent with the results of

Livio, Shankar, and Truran (1989), who investigated the effects of strong magnetic fields on the outburst via an inhibition of convective overturn.

While the calculations performed to date confirm the dependencies stated above, problems remain with reproducing the observed light curves. The model light curves are (bolometrically) super-Eddington for a timescale which is an order of magnitude less than that observed, and the observed evolution of the visual light curve is not currently reproducible. These problems are believed to result from our use of the diffusion approximation to describe the radiative transfer in the rapidly-expanding envelope. An improved treatment of the radiative transfer is presently being developed.

3 SUMMARY AND CONCLUSIONS

The calculations performed to date clearly indicate the dependence of the super-Eddington phase on the envelope CNO abundances, the decay energy from the positron-unstable isotopes ^{14}O and ^{15}O, and the presence of convection during the early phase of the outburst. However, two problems remain: (1) the timescale over which the models remain super-Eddington is an order of magnitude shorter than that observed and (2) the time evolution of the visual light curve is not numerically reproducible. These problems are believed to be a consequence of our use of the diffusion approximation in our treatment of radiation transport. An improved treatment of the radiation hydrodynamics of the nova outburst is presently being developed. Improvements will include an allowance for the decoupling of the radiation and gas temperatures (non-equilibrium diffusion), and the interaction of the material with spectral lines via the local velocity gradient (the central feature of line-driven stellar winds).

The authors would like to thank D. Mihalas for many helpful comments and discussion, and L. Smarr for the hospitality of the National Center for Supercomputing Applications, where the numerical work was conducted. This research has been supported in part by the National Science Foundation through grant AST 86-11500.

REFERENCES

Arp, H. C. 1956, *A. J.*, **61**, 15.
de Vaucouleurs, G. 1978, *Ap. J.*, **223**, 351.
Ferland, G. J., and Shields, G. A. 1978 *Ap. J.*, **226**, 172.
Livio, M., Shankar, A., and Truran, J..W., 1989, in preparation.
Payne-Gaposchkin, C. 1957, *The Galactic Novae* (Amsterdam: North Holland).
Truran, J. W. 1982, in *Essays in Nuclear Astrophysics*, ed. C. A. Barnes, D. D. Clayton, and D. N. Schramm (Cambridge: Cambridge University Press), p. 467.
Tylenda, R., 1978, *Acta Astr.*, **28**, 333.

Helium Star Cataclysmic Variables, A Theoretical Construct

Icko Iben, Jr.

The Pennsylvania State University
525 Davey Laboratory
University Park, PA 16802

1 HOW ARE HELIUM STAR CATACLYSMIC VARIABLES MADE?

Scenarios for close binary star evolution suggest that there is a finite probability for the formation of systems in which a carbon-oxygen white dwarf is accreting helium from a companion that is burning helium in its core. The most likely way of producing helium star CVs (see, e.g., Tornambè and Matteucci 1986, Iben and Tutukov 1987) begins with systems in which the component masses are both in the range 2–8 $M_\odot$ and the separation A_0 is such that the primary fills its Roche lobe considerably after it has exhausted central helium but before it has ignited helium. Mass transfer will continue until the primary (of initial mass M_1) has lost most of its hydrogen-rich envelope and has evolved into a helium star of mass

$$M_{1R} \sim 0.08\, M_1^{1.4} , \qquad (1)$$

[Unless otherwise noted, masses and distances will here and hereinafter be understood to be in solar units.] If the initial mass ratio $q = M_2/M_1$ is small enough (M_2 is the mass of the secondary), one may expect that the mass-transfer timescale is sufficiently shorter than the thermal timescale of the accretor that a common envelope will be formed and that the mass lost by the primary will also be lost from the system. Frictional interaction between the compact stellar cores (the low-mass secondary and the helium core of the primary) and the common envelope will result in orbital shrinkage to an extent which may be parameterized by

$$A_f \sim \alpha A_0 M_{1R} M_2/M_1^2 , \qquad (2)$$

where A_f is the semimajor axis after the common envelope event and α is of the order of unity (Tutukov and Yungelson 1979).

The helium star will evolve into a carbon-oxygen (CO) white dwarf and, when the secondary exhausts hydrogen at its center and grows to fill its Roche lobe, mass will be transferred to a white dwarf on a timescale short compared to the thermal timescale of the accretor. Again, a common envelope will be formed and frictional interaction between the compact helium core of the secondary and the common envelope will result in orbital shrinkage to an extent which may be parameterized by

$$A_{ff} \sim \alpha A_f M_{1R} M_{2R}/M_2^2 , \qquad (3)$$

where M_{2R} is the mass of the helium star remnant of the secondary. The relationship between initial and final semimajor axis becomes

$$A_{ff} \sim \alpha^2 A_0 (M_{1R}/M_1)^2 (M_{2R}/M_2) \sim 10^{-3.3} M_1^{2.2} M_2^{3.2} A_0 . \qquad (4)$$

The time required for radiation of angular momentum by gravitational waves to bring the helium star and the CO white dwarf, of masses M_{He} and M_{CO}, respectively, close enough together for the helium star to fill its Roche lobe is given by

$$\tau_{GW} \, (\text{yr}) = 10^{8.176} A^4 / (M_{He} M_{CO}[M_{He} + M_{CO}]) . \qquad (5)$$

The core helium-burning lifetime of the primary remnant is given approximately (see, e.g., Iben and Tutukov 1985; Iben 1989) by

$$\tau_{He} \, (\text{yr}) \sim 10^{7.1} M_{He}^{-3.31} . \qquad (6)$$

The helium star will fill its Roche lobe before it exhausts central helium as long as $M_2 \times M_1^{1.4} > 5.1$, a condition which will be satisfied in all cases where the secondary has an initial mass larger or equal to $2 \, M_\odot$.

2 PROPERTIES OF HELIUM STAR CVS DURING QUIESCENCE

The radius of a somewhat evolved helium star is approximately (Savonije, de Kool, and van den Heuvel 1986)

$$R_{He} \sim 0.26 M_{He}^{1.15} . \qquad (7)$$

Assuming that the helium star fills its Roche lobe and that angular momentum loss due to gravitational wave radiation is the only agency that drives mass loss, the mass-transfer rate becomes (Iben and Tutukov 1987)

$$dM/dt = -(M_{He}/8\tau_{GW})(1.355 - q)^{-1} , \qquad (8)$$

where q is the ratio of the mass of the helium star to the mass of the white dwarf. Using equation (7) and an approximate relationship between the Roche-lobe radius of the helium star and orbital parameters, equation (8) becomes (Iben and Tutukov 1987)

$$dM/dt = -10^{-7.88} q^{-0.24} (1 + q)^{-0.76} \times M_{He}^{-0.6} (1.355 - q)^{-1} \, M_\odot \, \text{yr}^{-1} . \qquad (9)$$

For a very wide range of combinations of component masses, this expression gives a mass-transfer rate between $2 \times 10^{-8} \, M_\odot \, \text{yr}^{-1}$ and $4 \times 10^{-8} \, M_\odot \, \text{yr}^{-1}$ (Iben $et \, al.$ 1987), and in further discussion we adopt

$$dM/dt \sim 3 \times 10^{-8} \, M_\odot \, \text{yr}^{-1} . \qquad (10)$$

During the quiescent accretion phase, the rate of energy release from the accretion disk in helium star CVs is approximately

$$L_{disk} \sim (GM_{CO}/R_{CO})dM/dt \sim 100\,L_\odot M_{CO}/r_{CO} \,, \tag{11}$$

where r_{CO} is the radius of the accretor in units of $10^{-2}\,R_\odot$. The "temperature" of the disk is roughly

$$T_{disk} \sim 58{,}000\,\text{K}\; M_{CO}^{1/4}/r_{CO}^{-3/4} \,. \tag{12}$$

The helium star itself has a luminosity (Iben 1990)

$$L_{He} \sim 256\,L_\odot M_{He}^{3.676} \,, \tag{13}$$

and a surface temperature

$$T_e \sim 50{,}000\,\text{K}\; M_{He}^{0.44} \,. \tag{14}$$

The orbital period is given roughly by

$$P_{orb}(\text{hr}) \sim M_{He}(M_{He} + M_{CO})^{0.16} \,. \tag{15}$$

The binary star HZ-Her, with an orbital period near 17 min and an apparently shrinking orbit has been considered as a candidate real analog (Iben and Tutukov 1987), but the uncertainties in the orbital characteristics are such that it remains only as a possible analog.

If we suppose that the birth rate of helium star CVs in the entire Galaxy is approximately 10^{-3} yr^{-1} (Tutukov and Yungelson 1987a, b, 1988a, b), equation (6) suggests that there should be of the order of 10^5 helium star CVs in the quiescent phase at any one time. This corresponds to about one system within 100 pc and to about 300 in a disk of 1 kpc radius about the Sun. These systems must be tightly concentrated to the disk of the Galaxy since the minimum primary progenitor mass is about $2\,M_\odot$ or larger.

It is of interest to compare several characteristics of helium star CVs with those of classical CVs which are in the short period regime where mass transfer is (probably) driven primarily by gravitational wave radiation. Mass-transfer rates are related by

$$(dM/dt)_{He}/(dM/dt)_H \sim 234\,M_H^{-1.24}M_{He}^{-0.6} \,, \tag{16}$$

where the subscripts denote the composition of the transferred matter. Lifetimes of the two classes of cataclysmic variables (CVs) are related by

$$\tau_{He\,CV}/\tau_{H\,CV} \sim (1/442)\,M_{He,0}^{1.84}/M_{H,0}^{1.24} \,, \tag{17}$$

where $M_{He,0}$ and $M_{H,0}$ are the initial masses of the helium star donor and of the low-mass main-sequence donor, respectively.

3 THE THERMAL RUNAWAY AND ITS CONSEQUENCES — SUPER NOVA OR SUPERNOVA?

As matter is accreted by the white dwarf, the temperature and density in the accreted layer continue to increase until helium is ignited. Explicit quasistatic calculations which show that, for $dM/dt = 3 \times 10^{-8}$ $M_\odot$ yr^{-1}, the mass m_e of the accreted layer when ignition occurs is $m_e = 0.152\,M_\odot$ when $M_{CO} = 0.6$; when $M_{CO} = 1.0$, $m_e = 0.136\,M_\odot$ (Iben and Tutukov 1990).

The nature of the ensuing explosion depends on the degree of electron degeneracy in the accreted layer at the onset of the thermal runaway. The larger the degree of degeneracy, the more dramatic are the consequences. Several investigators (Ergma, Rahunen, and Vilhu 1978; Fujimoto 1980; Nomoto 1980; Taam 1980a, b; Fujimoto and Sugimoto 1982; and Nomoto 1980, 1982a, b, 1984) have determined that whether the thermonuclear runaway is mild (as in a classical nova explosion) or whether it develops into a deflagration or detonation (which incinerates the entire star) is a relatively well defined function of the mass-accretion rate and of the CO white dwarf mass, with 2–5×10^{-8} $M_\odot$ yr^{-1} being a critical accretion rate which separates the response to ignition into two types — nova-like or supernova-like (see Iben and Tutukov 1984 for a discussion of this theme).

When the accretion rate is fixed, as in the current case, the outcome of ignition depends only on the mass of the white dwarf core. This becomes apparent from an inspection of Figure 1, where structure curves within models with $M_{CO} = 0.6$ and 1.0 are shown at a time when the thermonuclear runaway is well under way (Iben and Tutukov 1990). Shown also are several curves along which the ratio of ε_F (the Fermi energy of the electrons) to kT is constant.

In the case of accretion onto the white dwarf model of mass $M_{CO} = 1.0$, the situation is very clear cut. The thermal runaway is initiated at a point where $\rho \sim 2 \times 10^6$ g cm^{-3}, $T \sim 8 \times 10^7$ K, and $\varepsilon_F/kT \sim 25$. An increase in pressure sufficient to cause expansion cannot occur until the temperature has increased to $\sim 8 \times 10^8$ K. The timescale for the complete conversion of helium into carbon (neglecting all reactions other than the triple alpha reaction) is

$$\tau_{3\alpha}(\text{sec}) \sim X_4^{-3} T_8^3 \rho_6^{-2} \exp(-14.33 + 43.20/T_8) \,, \tag{18}$$

where T_8 and ρ_6 are the temperature and density in units of 10^8 K and 10^6 g cm^{-3}, respectively, and X_4 is the abundance by mass of ^{4}He. At $\rho_6 = 2$ and $T_8 = 8$, $\tau_{3\alpha} = 0.02$ s, which is significantly smaller than the timescale for dynamical expansion (Taam 1980a, b)

$$\tau_{\text{dyn}} = (24\pi G\rho)^{-1/2} = 0.32 \text{ s } (2 \times 10^6 \text{ g cm}^{-3}/\rho)^{1/2} \,. \tag{19}$$

Hence, nuclear processing continues until iron-peak nuclei have been produced in statistical equilibrium. The amount of energy per gram released in the conversion of ^{4}He into ^{12}C is 6×10^{17} erg and the additional energy released in going all the way to

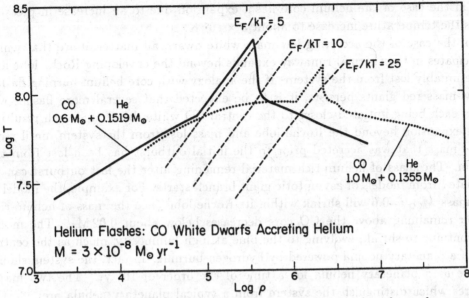

Fig. 1—Structural characteristics of accreting CO white dwarfs at the onset of thermal runaways triggered by helium burning.

statistical equilibrium is 8×10^{17} erg. This is sufficient to increase the temperature to about 2×10^{10} K, if all of the energy released were to go into thermal motions. However, once $kT > \varepsilon_F$, the overpressure is large enough to ensure that dynamical expansion must occur. If all of the nuclear energy released were converted into the kinetic energy of the expanding material, the velocity of this material would be about $v_{max} \sim 1.7 \times 10^4$ km s^{-1}, compared to the local sound velocity of about 10^3 km s^{-1}. However, some fraction of the nuclear energy released must be used up in overcoming the gravitational binding energy. The binding energy per gram of the accreted layer is about 10 percent of the total nuclear energy released.

A compromise is reached between conversion into random thermal energy, directed kinetic energy, and work done in propelling matter outwards against gravity. Too much energy is released too rapidly for the star to respond other than by developing shocks that are driven by the energy released. Flame fronts proceed inwards as well as outwards, imparting greater than escape velocity to all matter left behind. The model explodes as a supernova.

The situation is also fairly clear cut when $M_{CO} = 0.6$. The thermal runaway begins at a point where $\rho \sim 4 \times 10^5$ g cm^{-3} and $T \sim 8 \times 10^7$ K, precisely the density and temperature which prevail at the onset of the thermal runaway in the electron-degenerate helium core of a low-mass red giant (e.g., Despain 1981). In the low-mass red giant, the thermal runaway does not develop into a hydrodynamic event (Fujimoto, Iben, and Hollowell 1990) and it is reasonable to suppose that the development of the helium flash in the case of the accreting low-mass white dwarf will also be of quasistatic character. In both bases, $\varepsilon_F/kT \sim 10$ when ignition occurs and, in the case of the helium core flash, expansion due to an increase in pressure limits the temperature increase to about 3×10^8 K.

In the case of the accreting low-mass white dwarf, all matter above that which participates in the thermal runaway expands beyond the enveloping Roche lobe and is presumably lost from the system. If the analogy with core helium-burning flashes in low-mass red giants persists, it is to be expected that several more flashes will occur, each being initiated closer to the central CO white dwarf, and each resulting in an expansion beyond the Roche lobe and mass loss from the system, until most of the mass that was accreted prior to the initial outburst has been lost from the system. The mass of helium-rich material remaining after the last outburst can be estimated from models of asymptotic giant branch starts. For example, the model of core mass $M_{CO} = 0.6$ will shrink within its Roche lobe once the mass of helium-rich matter remaining above the CO core decreases below about $0.02\,M_\odot$. The model will continue to shrink, evolving to the blue at high luminosity, much as the central star of a planetary nebula powered by hydrogen burning. In fact, the system should survive as a planetary nebula for a time of the order of 10^4 yr. The two major features which distinguish the system from a typical planetary nebula are: (1) the most prominent emission features are due to helium recombination rather than to hydrogen recombination; and (2) the bolometric luminosity of the system is larger than that of a classical nova of comparable mass.

To summarize, helium star cataclysmics in which the accreting white dwarf is massive (say, $M_{CO} > 0.75\,M_\odot$) evolve into supernovae (of type Ib?), whereas those in which the accreting white dwarf is of low mass experience several super nova outbursts.

Following each outburst, once it has contracted within its Roche lobe, the sometimes accretor continues to burn helium at high luminosity for a period of $\sim 10^4$ yr, causing the ejected helium-rich shell to fluoresce as a planetary nebula. However, at an estimated production rate of less than $\sim 3 \times 10^{-3}$ yr^{-1} (assuming ~ 3 outbursts per helium star cataclysmic), compared to a formation rate of ~ 0.5 yr^{-1} for ordinary planetary nebulae, at most only one out 200 or so planetary nebulae will be of this origin and variety.

This work has been supported in part by the National Science Foundation, grant AST 88-07773, and in part by the Eberly Professorship in Astronomy at the Pennsylvania State University.

REFERENCES

Despain, K. H. 1981, *Ap. J.*, **251**, 639.

Ergma, E., Rahunen, T., and Vilhu, O. 1978, *Nauch. Inf.*, **45**, 159.

Fujimoto, M. Y. 1980, in *Type I Supernovae*, ed. J. C. Wheeler (Austin: University of Texas), p. 115.

Fujimoto, M. Y., Iben, I. Jr., and Hollowell, D. 1990, *Ap. J.*, **349**, in press.

Fujimoto, M. Y., and Sugimoto, D. 1982, *Ap. J.*, **257**, 291.

Iben, I. Jr., Nomoto, K., Tornambé, A., and Tutukov, A. V. 1987, *Ap. J.*, **317**, 717.

Iben, I. Jr., and Tutukov, A. V. 1984, in *High Energy Transients in Astrophysics*, ed. S. E. Woosely (New York: AIP), p. 11.

Iben, I. Jr., and Tutukov, A. V. 1985, *Ap. J. Suppl.*, **58**, 661.

Iben, I. Jr., and Tutukov, A. V. 1987, *Ap. J.*, **313**, 727.

Iben, I. Jr., and Tutukov, A. V. 1990, in preparation.

Nomoto, K. 1980, in *Type I Supernovae*, ed. J. C. Wheeler (Austin: University of Texas), p. 164.

Nomoto, K. 1982*a*, *Ap. J.*, **253**, 798.

Nomoto, K. 1982*b*, *Ap. J.*, **257**, 780.

Nomoto, K. 1984, *Ap. J.*, **277**, 791.

Savonije, G. J., de Kool, M., and van den Heuvel, E. P. J. 1986, *Astr. Ap.*, **155**, 51.

Taam, R. E. 1980*a*, *Ap. J.*, **237**, 142.

Taam, R. E. 1980*b*, *Ap. J.*, **242**, 749.

Tornambè, A., and Matteucci, F. 1986, *M. N. R. A. S.*, **233**, 69.

Tutukov, A. V., and Yungelson, L. 1979, *Acta Astr.*, **29**, 666.

Tutukov, A. V., and Yungelson, L. 1987*a*, *Comments on Astrophysics*, **XII**, 57.

Tutukov, A. V., and Yungelson, L. 1987*b*, in *The Second Conference on Faint Blue Stars*, ed. A. G. Davis Philip, D. S. Hayes, and J. Liebert (Schenectady: L. Davis), p. 435.

Tutukov, A. V., and Yungelson, L. 1988*a*, *Nauch. Inf.*, **65**, 30.

Tutukov, A. V., and Yungelson, L. 1988*b*, *Sov. Astr. Lett.*, **14**, 265.

Accreting Main-Sequence Stars Within a Common Envelope

Michael S. Hjellming and Ronald E. Taam

Northwestern University and Lick Observatory, UC Santa Cruz

1 INTRODUCTION

In the last decade, a resolution of the paradox of high white dwarf masses and short orbital periods in cataclysmic variables (CVs) has been suggested involving a common envelope phase of evolution. First applied to CVs by Paczynski (1976), this phase involves the expansion of an intermediate-mass giant beyond the orbit of its lower-mass companion, hereafter called the secondary. The asynchronous "stirring" by the secondary within the giant's envelope is believed to convert orbital energy into kinetic energy of the giant's envelope. In the process, the secondary spirals towards the giant's compact core and the envelope can be ejected. The dynamics of this interaction have been studied by several groups (Livio and Soker 1988; Taam and Bodenheimer 1989). However, very little attention has been given to the consequences of common envelope evolution on the secondary. Observations of CVs indicate that the secondaries are low-mass main-sequence stars with a normal Population I composition (Patterson 1984). Theoretical arguments have been made by Webbink (1988) to explain why this should be so.

The dynamical calculations of common envelope evolution have indicated that the accretion rates can be as high as $10^{-1} M_\odot \, yr^{-1}$. Given the estimated duration of this phase, 10^3–10^4 years, it would be possible for the secondary to accumulate a significant amount of mass. However, these accretion rates were based on a Bondi-Hoyle description of the accretion in which the secondary moves through a "wind" at the orbital velocity. An important factor not considered was the high opacity of the envelope gas and the resulting Eddington limit on the accretion rate, which is closer to $10^{-4} M_\odot \, yr^{-1}$. Consequently, the mass accumulated by the secondary would be on the order of $0.1 \, M_\odot$. This paper presents preliminary results from calculations of accretion onto zero-age main-sequence stars under conditions appropriate to common envelope evolution.

2 METHOD

The effects of accretion during common envelope evolution are calculated with a spherically symmetric, hydrostatic stellar evolution code. A model of a $5\,M_\odot$ AGB star, with a $0.94\,M_\odot$ carbon core, is used to determine the outer boundary conditions of a $1.25\,M_\odot$ ZAMS model. The secondary begins the calculation at an envelope depth of $0.05\,M_\odot$, where the local envelope temperature equals the secondary's initial surface temperature. The usual boundary conditions, the blackbody relation and optical depth of unity, are replaced by the density and temperature of the giant at the current separation, A. When the secondary's radius, R_*, is less than its Roche lobe radius, R_L, mass can be accreted. When $R_* = R_L$, mass is removed as R_L decreases with orbital shrinkage.

Since the secondary moves through the giant's envelope at slightly supersonic speeds, a bow shock forms with a rather wide opening angle. Accretion of the post-shock gas is roughly spherical and subsonic. The orbital evolution is calculated as follows. The separation of the binary decreases by frictional dissipation. The rate of dissipation produced by the motion of the secondary is $\pi R_{\mathrm{acc}}^2 \rho v_{\mathrm{orb}}^3$. The energy is deposited into the giant's envelope at the expense of the binding energy, $GM_g M_*/A$. In these two expressions, R_{acc} is the secondary's accretion radius, ρ is the local density in the giant's envelope, v_{orb} is the orbital velocity, M_g is the mass of the giant within a radius equal to the current separation, and M_* is the current mass of the secondary. The timescale for decay of the orbit is then the ratio of the two expressions. The accretion radius may vary between the Bondi-Hoyle radius, $R_{\mathrm{BH}} = 2GM_*/v_{\mathrm{orb}}^2$, and the stellar radius, R_*. Neither reflects the physical circumstances very well. To maintain consistency between accretion and overflow, $R_{\mathrm{acc}} = R_*$ was chosen. The actual mass-accretion rate is less than the rate implied by this accretion radius, $\pi R_*^2 \rho v_{\mathrm{orb}}$. The high radiative opacity, κ, within the giant's envelope limits the accretion rate to the Eddington value, $4\pi c R_*/\kappa$. The giant's local density and temperature at A are used to determine the opacity of the gas. In this preliminary calculation, the effect of accretion on the state of the gas is not considered.

3 RESULTS

The time between the immersion of the secondary and the end of the calculation is approximately 920 years, of which 900 years are spent at $\log A(\mathrm{cm}) \gtrsim 13.25$. Figure 1 shows the radial response of the secondary as the orbit shrinks. The onset of Roche lobe overflow is indicated by the dotted line. The mass-accretion and -loss rates are shown in Figure 2. Most of the material accretes from the outer layers of the giant's envelope where the majority of the time is spent. The temporary decrease in the accretion rate corresponds to the opacity maximum within the giant's envelope. During accretion, the original surface layers become nearly isothermal while the new layers have normal temperature gradients. Very high luminosities are

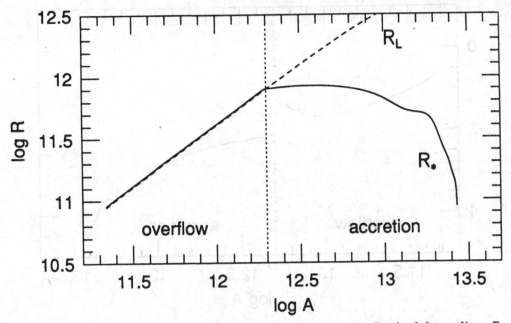

Fig. 1—The response of the stellar radius, R_*, and the Roche lobe radius, R_L, as the secondary spirals towards the giant's compact core. Note the secondary's expansion during the accretion phase.

produced in the new mass layers forcing the surface to expand by more than a factor of 10. Only $0.05\,M_\odot$ is gained before Roche lobe overflow. The mass-loss rate is initially rather large due to the removal of a thin convective region of $0.005\,M_\odot$ which forms during the latter part of the accretion phase. The calculation ends when the radius decreases to its initial value, and the secondary has nearly penetrated to the hydrogen-burning shell source. By this time the giant's envelope may have been expelled if energy deposition in the envelope had been considered. The mass lost during overflow amounts to roughly $0.04\,M_\odot$. Hence, the $1.25\,M_\odot$ model finishes with a net addition of $0.01\,M_\odot$.

4 CONCLUSIONS

In this first calculation of main-sequence secondaries during common envelope evolution, a radiative $1.25\,M_\odot$ model accumulates only $0.01\,M_\odot$ after phases of accretion and Roche lobe overflow. This small addition of mass would not noticeably affect the appearance of such a secondary in a CV. However, several important accretion effects were neglected: compression and heating of the accreted gas and the energy feedback into the giant's envelope. The first effect will cause the accreted gas to arrive at the surface with higher entropy. The radius of the secondary should expand more rapidly and accrete less mass before Roche lobe overflow. Second, the giant's

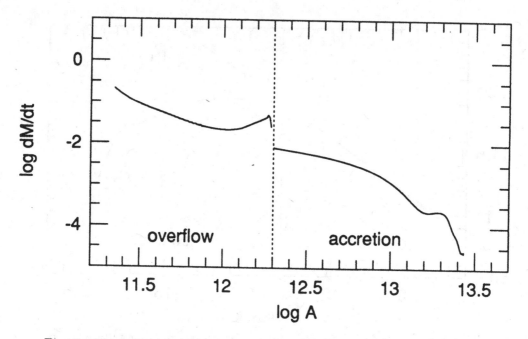

Fig. 2—The behavior of the mass-accretion or -loss rate during common envelope evolution. Roche lobe overflow occurs to the left of the dotted line.

convective envelope responds to energy deposition by expanding outwards, thereby
lowering the density to an approximately constant value through the envelope (Taam,
Bodenheimer, and Ostriker 1978), which may lengthen the decay timescale. Further
work is planned with lower-mass main-sequence models of 0.75, 0.50, and 0.30 $M_\odot$
which are more commonly found in CVs.

M. S. H. would like to thank the organizers for financial assistance in attending
the Workshop. This work has been funded in part by NSF grant AST 86-08291.

REFERENCES

Livio, M., and Soker, N. 1988, *Ap. J.*, **329**, 764.
Paczynski, B. 1976, in *IAU Symposium 73, The Structure and Evolution of Close
    Binary Systems*, ed. P. Eggleton, S. Mitton, and J. Whelan (Dordrecht: Reidel),
    p. 75.
Patterson, J. 1984, *Ap. J. Suppl.*, **54**, 443.
Taam, R. E., and Bodenheimer, P. 1989, *Ap. J.*, **337**, 849.
Taam, R. E., Bodenheimer, P., and Ostriker, J. P. 1978, *Ap. J.*, **222**, 269.
Webbink, R. F. 1988, in *Critical Observations vs. Physical Models for Close Binary
    Systems*, ed. K. C. Leung (New York: Gordon and Beach), p. 403.

Theoretical Statistics of Zero-Age Cataclysmic Variables

Michael Politano

Department of Physics and Astronomy, Arizona State University and
Theoretical and Applied Theoretical Physics Divisions, Los Alamos National
Laboratory

1 INTRODUCTION

A zero-age cataclysmic variable (ZACV) I define as a binary system at the onset of interaction as a cataclysmic variable. I present here the results of calculations of the distribution of white dwarf masses and of orbital periods in ZACVs due to binaries present in a stellar population which has undergone continuous, constant star formation for 10^{10} years. The white dwarf mass distribution calculated for ZACVs is compared to a distribution of masses in single white dwarfs. I also present estimates of the present formation rate, total number, and space density of CVs in the Galaxy based on these calculations, as well as an estimate of the average recurrence time for classical novae. Lastly, I identify the orbital parameter ranges of CV progenitors on the zero-age main sequence (ZAMS) that are indicated by these calculations.

2 METHOD

Consider binaries which formed t years ago. Some fraction of these binaries will become CVs. Of these, a certain subset will have secondaries which are just filling their Roche lobes for the first time at the present time; these I have denoted as ZACVs. In fact, if a Cartesian coordinate system is defined with $\log M_{wd}$ (mass of the white dwarf), $\log q_f$ (mass ratio, defined as secondary mass/white dwarf mass at the present time) and $\log P_f$ (orbital period at the present time) as the x, y, and z axes, respectively, then the orbital period at which the secondary just fills its Roche lobe defines a surface in this 3-dimensional orbital parameter space for a given t. By assuming some initial distribution of the orbital parameters in ZAMS binaries and by relating the orbital parameters in the ZAMS progenitor binaries to their ZACV offspring, the density of systems (i.e., the number of systems per unit volume) along the surface described may be calculated. Combining this density with the rate at which the secondary expands due to nuclear evolution and the rate at which the radius of the Roche lobe shrinks as a result of orbital angular momentum losses (i.e., gravitational radiation or magnetic braking) allows the flux of systems (the number of binaries passing through the surface per unit time per unit area) at any point on the surface to be calculated. From the definition of the surface, this flux is just the contribution to the present birthrate of CVs per unit area of the surface from those binaries which were formed t years ago. Projecting this flux onto the $\log M_{wd}$-$\log q_f$ plane then gives the 2-dimensional distribution of the birthrate over $\log M_{wd}$ and $\log q_f$ (i.e., the number of CVs forming per unit time per unit per $\log M_{wd}$ per $\log q_f$). Projecting the surface onto this plane and integrating the two-dimensional distribution over $\log q_f$ between the boundaries on this surface which actually enclose stable CVs yields the distribution of this birthrate over $\log M_{wd}$ for those binaries which were formed t years ago. The distribution of the birthrate over $\log P_f$ is calculated in a similar fashion except that the flux and surface are projected onto the $\log M_{wd}$-$\log P_f$ plane and the integration is over $\log M_{wd}$. These distributions are computed for values of $\log t$ in intervals of 0.1 from $\log t = 7.4$ (the youngest age at which a viable CV can form) to $\log t = 10.0$ (the assumed age of the Galactic disk). The distribution of white dwarf masses and orbital periods in ZACVs at the present time due to all binaries formed within the last 10^{10} years is then calculated by summing up the contributions from the respective distributions for individual values of $\log t$.

3 KEY ASSUMPTIONS

The following main assumptions were made in these calculations:

(1) The rate of star formation in the Galaxy has remained constant.

(2) The distribution of the orbital parameters in ZAMS binaries may be written as the product of the distribution of the masses in ZAMS stars (Miller and Scalo 1979), the distribution of the mass ratios in ZAMS binaries (cf. Popova et $al.$ 1982), and the distribution of the orbital periods in ZAMS binaries (Abt 1983).

(3) All of the orbital energy deposited into the envelope during the common envelope phase goes into unbinding the envelope.

(4) The magnetic braking law in Rappaport, Verbunt, and Joss (1983) with $\gamma = 2$ applies.

(5) Magnetic braking is turned off if the secondary is completely convective.

4 RESULTS

The distribution of white dwarf masses in ZACVs (the solid curve in Fig. 1) contains two components: a low-mass component composed of systems with He white dwarfs, and a high-mass mass component composed of systems with C-O white dwarfs. Systems with He white dwarfs comprise a significant fraction, one-third to one-half, of all ZACVs. The helium white dwarfs have masses which range from 0.27 to $0.46 \, M_\odot$. The C-O white dwarfs have masses which range from a minimum mass of $0.54 \, M_\odot$ up to the Chandrasekhar limit. (Note: systems with O-Ne-Mg white dwarfs are not distinguished from systems with C-O white dwarfs in this calculation. Presumably, these O-Ne-Mg systems comprise the upper end of the white dwarf mass

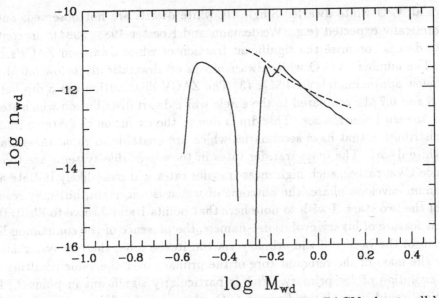

Fig. 1—The distribution of the white dwarf masses in ZACVs due to all binaries formed within the last 10^{10} years (*solid curve*), and a theoretical distribution of masses in single white dwarfs (*dashed curve*; generated as described in the text). M_{wd} is the mass of the white dwarf in solar masses and n_{wd} is the number of systems forming per year per log white dwarf mass per unit area of the Galactic plane (in pc^2). The single-star distribution has been normalized such that the total formation rate of single white dwarfs is equal to the total formation rate of white dwarfs in ZACVs.

distribution.) The distribution of white dwarf masses in ZACVs is not sufficient to account for the observed high white dwarf masses in CVs. The mean white dwarf mass for all ZACVs is $0.49\,M_\odot$, and even the mean white dwarf mass just for those ZACVs with C-O white dwarfs is only $0.74\,M_\odot$. This is compared to a mean white dwarf mass of about $1\,M_\odot$ in observed CVs, emphasizing the possible importance of selection effects in CVs (Ritter and Burkert 1986; Politano, Ritter, and Webbink 1989; Ritter *et al.* 1990 and references therein).

Figure 1 also shows a theoretical single-star white dwarf mass distribution (dashed curve) which I constructed using the same initial mass function and the same mass-loss rate due to stellar winds on the asymptotic giant branch as was used for the primaries in the ZACV calculation. Comparison of this distribution to the distribution of white dwarf masses in ZACVs then illustrates most directly the effects of binary evolution (although comparison to any other single white dwarf distribution in the literature [theoretical or observational] would not essentially alter the conclusions reached below.) This comparison shows the following: (1) The distribution of white dwarf masses in ZACVS is similar to, although steeper than, the distribution of white dwarf masses in single white dwarfs for masses above approximately $0.7\,M_\odot$. (2) A significant fraction of single white dwarfs have masses below $0.6\,M_\odot$, although these are all C-O white dwarfs. Single He white dwarfs are not observed, and are not theoretically expected (e.g., Weidemann and Koester 1983, 1984). In contrast, He white dwarfs comprise the significant fraction of white dwarfs in ZACVs below $0.6\,M_\odot$. The number of C-O white dwarfs drops off dramatically below $0.6\,M_\odot$ and is zero below approximately $0.54\,M_\odot$. (3) The ZACV distribution has a dip between about 0.6 and $0.7\,M_\odot$ compared to the single white dwarf distribution which steadily increases toward lower masses. This dip is due to the exclusion of systems from the ZACV distribution that have secondaries which are unstable to rapid mass transfer to the white dwarf. The mass-transfer rates in these unstable systems are too high to produce CVs; rather, such high mass-transfer rates will most likely initiate a second common envelope phase, the outcome of which is uncertain, but may result in merger of the two stars. I wish to note here that points 1 and 2 serve to illustrate an important feature of binary evolution—namely, the presence of the companion limits the growth of the radius of the primary via the Roche lobe, thereby systematically reducing the mass of the remnant core of the primary over the value resulting from isolated evolution of the primary. This is particularly significant in point 2. Most stars which would normally end up as C-O white dwarfs of mass less than about $0.6\,M_\odot$ end up as He white dwarfs because the primary is forced to fill its Roche lobe on the giant branch, before the ignition of helium.

The distribution of orbital periods in ZACVs, shown in Figure 2, identifies five main subsets of CVs: (1) ultrashort-period systems containing degenerate (hydrogen) secondaries, (2) short-period systems formed below the period gap containing helium white dwarfs, (3) systems containing carbon-oxygen white dwarfs whose secondaries are convectively stable against rapid mass transfer to the white dwarf, (4) systems containing carbon-oxygen white dwarfs whose secondaries are radiatively stable against rapid mass transfer to the white dwarf, and (5) long-period systems containing evolved secondaries. Minima occur in the orbital period distribution between approximately 2 and 3 hours, arising from the discontinuity in core masses between systems with helium white dwarfs and systems with carbon-oxygen white dwarfs (cf. Webbink 1979), and between approximately 3 and 5 hours, arising from the exclusion of systems with secondaries that are unstable to rapid mass transfer to the white dwarf. I find that CVs form in period gap (between 2 and 3 hours), although at a rate reduced by about an order of magnitude over the peak rates of formation in Figure 2. This may be relevant to the recent discovery of CVs in the period gap (Shafter and Abbott 1989; Shafter et al. 1990); it is not clear whether these systems were formed in the gap or somehow entered the gap as a result of secular evolution.

The contributions to the total distributions from binaries of specific ages (from $\log t = 7.4$ to 10.0, i.e, the distributions that were summed over to produce the distributions shown in Figs. 1 and 2) provide information about the dependence of the mass and period distributions of ZACVs in Figures 1 and 2 on the age of the constituent binaries. (The distributions due to binaries for individual values of $\log t$

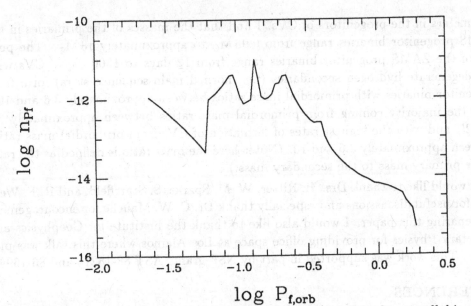

Fig. 2—The distribution of the orbital periods in ZACVs due to all binaries formed within the last 10^{10} years. $P_{f,\mathrm{orb}}$ is the orbital period in days and n_{Pf} is the number of CVs forming per year per log orbital period per unit area of the Galactic plane (in pc^2).

are not shown here because of space considerations, but see Politano 1988, 1989; Politano and Webbink 1989.) These individual distributions indicate the following: (1) CVs first appear in a stellar population at an approximate age of $\log t = 7.4$. This is the smallest age at which binaries leave white dwarf remnants. (2) CVs forming from binaries less than 10^9 years old do not contain He white dwarfs. (3) The formation rate of CVs is maximized when the nuclear evolutionary timescale of the initial primary is comparable to the timescale for angular momentum loss during the pre-CV state. In this case, pre-CV systems are brought into contact roughly as fast as they are formed (emerge from the common envelope state).

Integration of the mass or orbital period distribution in Figures 1 and 2 gives the present formation rate of CVs per unit area of the Galactic plane. Multiplying this by the area of the Galactic plane then gives an estimate of the present formation rate of CVs, which I find to be 0.001 per year. Taking an age for the Galaxy of 10^{10} years, this formation rate implies a total number of CVs in the Galaxy of order 10^7. It further implies a space density of CVs in the Galaxy of approximately 10^{-4} per cubic parsec. I also estimate the average recurrence time for classical novae to be 3×10^4 years, based on the above formation rate, and assuming a mean secondary mass of $1\,M_\odot$ and a nova rate of about 30 per year (Arp 1956). All of these numbers are in good agreement with other estimates (Ford 1978; Bath and Shaviv 1978; Patterson 1984; Iben and Tutukov 1984; Ritter and Burkert 1986).

In addition to providing information about the distribution of the orbital parameters in ZACVs, these calculations provide better identification of the orbital parameters in the progenitors of CVs. I find that the masses of the primaries in the ZAMS progenitor binaries range from $0.95 \, M_\odot$ to approximately $13 \, M_\odot$. The periods of the ZAMS progenitor binaries range from 12 days to 1300 days. CVs with non-degenerate hydrogen secondaries (i.e., normal main-sequence stars) form from progenitor binaries with primordial mass ratios between approximately 3.6 and 100, with the majority coming from primordial mass ratios between approximately 3.6 and 25, and with the highest rates of formation of CVs from primordial mass ratios between approximately 3.6 and 10. (Note: here the mass ratio is defined as the ratio of the primary mass to the secondary mass.)

I would like to thank Drs. H. Ritter, W. M. Sparks, S. Starrfield, and R. F. Webbink for useful discussions and especially thank Dr. C. W. Mauche for encouragement in preparing this paper. I would also like to thank the Institute for Geophysics and Planetary Physics for providing office space at Los Alamos where this talk was prepared. This work was supported in part by NSF grants AST 88-18215 and 86-16992.

REFERENCES

Abt, H. 1983, *Ann. Rev. Astr. Ap.*, **21**, 343.

Arp, H. 1956, *A. J.*, **61**, 15.

Bath, G. T., and Shaviv, G. 1978, *M. N. R. A. S.*, **183**, 515.

Ford, H. C. 1978, *Ap. J.*, **219**, 595.

Iben, I., Jr., and Tutukov, A. V. 1984, *Ap. J. Suppl.*, **54**, 335.

Miller, G. E., and Scalo, J. M. 1979, *Ap. J. Suppl.*, **41**, 513.

Patterson, J. 1984, *Ap. J. Suppl.*, **54**, 443.

Politano, M. 1988, Ph.D. thesis, University of Illinois at Urbana-Champaign.

Politano, M. 1989, in preparation.

Politano, M., Ritter, H., and Webbink, R. F. 1989, in IAU Coll. No. 114, *White Dwarfs*, ed. G. Wegner (Berlin: Springer), p. 465.

Politano, M., and Webbink, R. F. 1989, in IAU Coll. No. 114, *White Dwarfs*, ed. G. Wegner (Berlin: Springer), p. 440.

Popova, E. I., Tutukov, A. V., and Yungelson, L. R. 1982, *Ap. Sp. Sci.*, **88**, 55.

Rappaport, S., Verbunt, F., and Joss, P. C. 1983, *Ap. J.*, **275**, 713.

Ritter, H., and Burkert, A. 1986, *Astr. Ap.*, **158**, 161.

Ritter, H., Politano, M., Livio, M., and Webbink, R. F. 1990, this volume.

Shafter, A. W., and Abbott, T. M. C. 1989, *Ap. J. (Letters)*, **339**, L75.

Shafter, A. W., Robinson, E. L., Crampton, D., Warner, B. and Prestage, R. M. 1990, this volume.

Webbink, R. F. 1979, in IAU Coll. No. 53, *White Dwarfs and Variable Degenerate Stars*, eds. H. M. Van Horn and V. Weidemann (Rochester: University of Rochester Press), p. 426.

Weidemann, V., and Koester, D. 1983, *Astr. Ap.*, **121**, 77.

Weidemann, V., and Koester, D. 1984, *Astr. Ap.*, **132**, 195.

The Turn-On of Mass Transfer in Cataclysmic Binaries

F. D'Antona[1,2], I. Mazzitelli[1], and H. Ritter[3]

[1]Istituto di Astrofisica Spaziale C.N.R., C.P. 67, I-00044 Frascati, Italy
[2]Osservatorio Astronomico di Roma, I-00040 Monte Porzio, Italy
[3]Max-Planck Institut für Physik und Astrophysik, D-8046 Garching, F. R. G.

ABSTRACT

It is shown how the results of detailed numerical computations by D'Antona, Mazzitelli, and Ritter (1989) of the turn-on of mass transfer in cataclysmic binaries and structurally similar systems can be understood qualitatively on the basis of the same simple model of mass transfer as was used to perform the numerical computations.

1 INTRODUCTION

In a recent paper, D'Antona, Mazzitelli, and Ritter (1989, hereafter DMR) presented results of detailed numerical computations, involving full stellar models, of the turn-on of mass transfer in cataclysmic binaries (CBs) and structurally similar systems. Thereby, the mass-transfer rate was computed by modelling the mass transfer as an isothermal, optically thin, subsonic flow of gas through the inner Lagrangian point L_1 that reaches sound velocity near L_1 (e.g., Ritter 1988). Accordingly, the mass-transfer rate can be written as

$$-\dot{M}_2 = \dot{M}_o \, e^{-\frac{R_{2,R}-R_2}{H_p}} \, , \quad R_2 \lesssim R_{2,R} \, , \tag{1}$$

where M_2, $R_{2,R}$, R_2, and H_p are, respectively, the mass, the critical Roche radius, the photospheric radius, and the effective pressure scale height of the secondary (the donor). For the purpose of our discussions below we shall assume that $\dot{M}_o$ is a constant, as was in fact shown to be the case to a very good approximation in the computations of DMR. If the donor is a low-mass star ($M_2 \lesssim 1\,M_\odot$) near the main sequence, $\dot{M}_o \approx 10^{-8}\,M_\odot\,\mathrm{yr}^{-1}$ (see Table 1 of DMR). The numerical computations by DMR, which started from a detached system with a nuclearly unevolved low-mass secondary ($M_2 = 0.2, 0.35, 0.6$, and $1\,M_\odot$) in thermal equilibrium, showed that initially $-\dot{M}_2$ increases exponentially in all cases studied. The subsequent evolution depends, however, on M_2 and on the timescale of angular momentum loss. In systems 1, 2 ($M_2 = 0.2\,M_\odot$), and 4 ($M_2 = 0.6\,M_\odot$) of DMR, the mass-transfer rate peaks after the initial exponential growth and thereafter decays on a longer timescale and becomes eventually stationary at about a factor of 2 below the peak value. In system

3 ($M_2 = 0.35\,\mathrm{M_\odot}$) of DMR, on the other hand, $-\dot{M}_2$ levels off at a rather high value after the initial exponential growth, without showing an intermediate peak. Yet another behaviour is seen in system 6 ($M_2 = 1\,\mathrm{M_\odot}$), where the initial exponential growth of $-\dot{M}_2$ is followed by a further increase on an ever-increasing timescale. (For further details the reader is referred to DMR). We shall now show below how these different types of behaviour can be understood in terms of equation (1) and the internal structure of the secondary.

2 THE FOUR PHASES OF THE TURN-ON

The turn-on of mass transfer may be subdivided conveniently into four different phases. How these come about is easily seen when we take the time derivative of equation (1), assuming that $\dot{M}_o = $ constant. This yields

$$\ddot{M}_2 = \dot{M}_2 \frac{R_2}{H_p} \left\{ (\zeta_{2,s} - \zeta_{R,2}) \frac{\dot{M}_2}{M_2} + \left(\frac{\partial \ln R_2}{\partial t} \right)_{\mathrm{nuc}} + \left(\frac{\partial \ln R_2}{\partial t} \right)_{\mathrm{th}} - 2 \frac{\partial \ln J}{\partial t} \right\}. \quad (2)$$

Here

$$\zeta_{2,s} = \left(\frac{\partial \ln R_2}{\partial \ln M_2} \right)_S \quad (3)$$

is the adiabatic mass-radius exponent of the secondary,

$$\zeta_{R,2} = \left(\frac{\partial \ln R_{2,R}}{\partial \ln M_2} \right)_J \quad (4)$$

is the mass-radius exponent of its Roche radius (at constant orbital angular momentum J), $(\partial \ln R_2/\partial t)_{\mathrm{nuc}}$ and $(\partial \ln R_2/\partial t)_{\mathrm{th}}$ are, respectively, the relative temporal change of the secondary's radius due to nuclear evolution and thermal relaxation, and $\partial \ln J/\partial t < 0$ is the relative systemic angular momentum-loss rate. For the following discussion, we shall assume that the system is dynamically stable against mass transfer, i.e., that $(\zeta_{2,s} - \zeta_{R,2}) > 0$, and that nuclear evolution is unimportant, i.e., that $(\partial \ln R_2/\partial t)_{\mathrm{nuc}} = 0$, which is true to a very good approximation in all the cases studied by DMR. Furthermore, we note that our discussion is only valid in the limit that the relative mass lost from the secondary remains small, i.e., that $\Delta M_2/M_2 \ll 1$. The four phases of the turn-on are now characterized as follows:

1) <u>The Early Turn-On Phase</u>

At the beginning of the evolution, the secondary is in thermal equilibrium, hence $(\partial \ln R_2/\partial t)_{\mathrm{th}} = 0$, and the system sufficiently detached, i.e., $-\dot{M}_2$ so small, that $|(\zeta_{2,s} - \zeta_{R,2})\dot{M}_2/M_2| \ll |\partial \ln J/\partial t|$. Therefore, it follows from equation (2) that the mass-transfer rate increases exponentially on the timescale (Ritter 1988)

$$\tau_{\dot{M}_2} = \frac{\dot{M}_2}{\ddot{M}_2} = \frac{1}{2} \frac{H_p}{R_2} \left(\frac{\partial t}{\partial \ln J} \right). \quad (5)$$

2) Stationary Adiabatic Mass Loss Phase

With the mass-transfer rate increasing, the term $(\zeta_{2,S} - \zeta_{R,2})\dot{M}_2/M_2$ in equation (2) becomes more and more important. If the increase is fast enough such that thermal relaxation of the secondary remains negligible, i.e., if $|\partial t/\partial \ln J|$ is much shorter than $|\partial t/\partial \ln R_2|_{th}$, then the mass-transfer rate levels off and becomes stationary ($\ddot{M}_2 = 0$) at the adiabatic mass-loss rate

$$\left(\dot{M}_2(\ddot{M}_2 = 0)\right)_{ad} = \frac{M_2}{\zeta_{2,S} - \zeta_{R,2}} \cdot 2\frac{\partial \ln J}{\partial t}. \tag{6}$$

3) Non-Stationary Thermal Relaxation Phase

It is clear that the phase of stationary adiabatic mass loss must end when the thermal relaxation term in equation (2) becomes non-negligible. In fact, this phase may not exist at all, if the timescale given by equation (5) is long enough for the secondary to react thermally, i.e., if $|\partial \ln R_2/\partial t|_{th} \lesssim |\partial \ln J/\partial t|$. Once thermal relaxation becomes important, the subsequent change of $\dot{M}_2$ depends on details of the internal structure of the donor. If the star is either fully convective or has a deep outer convective envelope, $\zeta_{2,S} < \zeta_{2,e}$, where $\zeta_{2,e} = (\partial \ln R_2/\partial \ln M_2)_{\dot{S}=0}$ is the thermal equilibrium mass-radius exponent, and thus $(\partial \ln R_2/\partial t)_{th} < 0$. If, on the other hand, the donor is mainly radiative, $\zeta_{2,S} > \zeta_{2,e}$ and therefore $(\partial \ln R_2/\partial t) > 0$. Consequently, after the initial exponential growth, $-\dot{M}_2$ will peak and thereafter decay to reach eventually a lower, quasi-stationary value (see below), if the star is mainly convective and if $\tau_{\dot{M}_2}$ as given by equation (5) is short enough. If, on the other hand, it is mainly radiative, the early turn-on phase is followed by a decelerated increase of $-\dot{M}_2$ until it reaches eventually the quasi-stationary phase.

4) Quasi-Stationary Mass Loss with Thermal Relaxation

With fully-developed thermal relaxation, the mass-transfer rate reaches another quasi-stationary value ($\ddot{M}_2 = 0$ again) of

$$(\dot{M}_2\,(\ddot{M}_2 = 0))_{th} = \frac{M_2}{\zeta_{2,S} - \zeta_{R,2}} \left[2\left(\frac{\partial \ln J}{\partial t}\right) - \left(\frac{\partial \ln R_2}{\partial t}\right)_{th}\right]. \tag{7}$$

By comparing equations (6) and (7) it is seen that $(-\dot{M}_2)_{ad} > (-\dot{M}_2)_{th}$ if $(\partial \ln R_2/\partial t)_{th} < 0$, i.e., if the donor is mainly convective, and $(-\dot{M}_2)_{ad} < (-\dot{M}_2)_{th}$ if the donor is mainly radiative. If $(-\dot{M}_2)_{th}$ as given by equation (7) is not too high, such that the donor does not deviate too much from thermal equilibrium, the change of its radius with mass is approximatively given by the thermal equilibrium mass-radius exponent $\zeta_{2,e}$ and, therefore,

$$(\dot{M}_2)_{th} \approx \frac{M_2}{\zeta_{2,e} - \zeta_{R,2}} 2\frac{\partial \ln J}{\partial t}. \tag{8}$$

Thus, $(-\dot{M}_2)_{ad} > (-\dot{M}_2)_{th}$ if $\zeta_{2,S} < \zeta_{2,e}$ and vice versa. For a donor near the main sequence, $\zeta_{2,S} \approx \zeta_{2,e}$ for a mass of $\sim 0.8\,M_\odot$ (Webbink 1985; Hjellming 1988). Therefore, an intermediate mass-transfer peak, which requires that $(-\dot{M}_2)_{ad} > (-\dot{M}_2)_{th}$, can only be expected if the mass of the donor is $M_2 \lesssim 0.8\,M_\odot$.

3 DISCUSSION

The results obtained by DMR can now be understood in the framework of the above discussion as follows:

Systems 1, 2 ($M_2 = 0.2\,M_\odot$), and 4 ($M_2 = 0.6\,M_\odot$) go through the phases 1, 3, and 4 enumerated in Section 2. Since thermal relaxation becomes non-negligible towards the end of phase 1, phase 2 is skipped. Instead, because $M_2 < 0.8\,M_\odot$, $(-\dot{M}_2)$ peaks, but below $(-\dot{M}_2)_{ad}$. In system 3 ($M_2 = 0.35\,M_\odot$), the early turn-on phase is so short that the secondary has virtually no time to react thermally. Thus, phase 1 is followed by phase 2, during which the computations were stopped. Had they been continued for a much longer time, system 3 would also have gone through phases 3 and 4. Finally, system 6 ($M_2 = 1\,M_\odot$) is distinguished from the others in that the donor star is essentially fully radiative and, for the same reason, the angular momentum-loss rate is relatively small. Therefore, $\tau_{\dot{M}_2}$ as given by equation (5) is so long that thermal relaxation becomes important very early. Consequently, phase 1 is immediately followed by phase 3, whereas phase 4 was not reached during that computation. Since the donor star has $M_2 > 0.8\,M_\odot$, the mass-transfer rate does not show an intermediate peak.

H. R. would like to thank the Observatoire de Paris, section de Meudon, where this paper was written, for its hospitality and financial support.

REFERENCES

D'Antona, F., Mazzitelli, I., and Ritter, H. 1989, *Astr. Ap.*, **225**, 391, DMR.
Hjellming, M. 1988, Ph.D. thesis, University of Illinois at Urbana-Champaign.
Ritter, H. 1988, *Astr. Ap.* **202**, 93.
Webbink, R. F. 1985, in *Interacting Binary Stars*, ed. J. E. Pringle and R. A. Wade (Cambridge: Cambridge University Press), p. 39.

The Disrupted Magnetic Braking Model for the Period Gap of Cataclysmic Variables

Ronald E. Taam

Northwestern University and Lick Observatory, UC Santa Cruz

1 INTRODUCTION

Upon examining the period distribution of cataclysmic variables, it is apparent that there is a marked deficiency of systems in the period interval ranging from 2.3 hrs to 2.8 hrs (see for example, Ritter 1986). The fact that only two systems out of ~ 100 systems with known orbital periods lies within the period gap (the exceptions are V Per at 2.57 hrs and V795 Her at 2.6 hrs; see Shafter and Abbott 1989 and Shafter et al. 1990, respectively) suggests that the long-term evolution of these binary systems has somehow been interrupted at the upper edge. A number of proposals have been advanced as possible solutions to this problem, and the most popular involve the sudden switch-off of angular momentum losses associated with magnetic braking (see Rappaport, Verbunt, and Joss 1983; Spruit and Ritter 1983).

The basis for this paradigm stems from the observation made by Robinson et al. (1981) in which it was pointed out that the location of the upper edge of the period gap lies near the orbital period for which a main sequence star would become fully convective. If magnetic braking is disrupted at this point then a gap can be produced. The general idea is described as follows. For sufficiently high mass-transfer rates promoted by the angular momentum losses, the main sequence-like star exceeds its thermal equilibrium size for orbital periods longward of the gap. Once magnetic braking is switched off, the star will shrink more rapidly than its Roche lobe in its attempt to regain thermal equilibrium. Hence, the system will enter into a detached state where mass transfer is temporarily interrupted. Further angular momentum losses associated with gravitational radiation eventually force the Roche lobe to shrink and the system reemerges as a mass-transferring cataclysmic variable at a shorter orbital period.

2 SECULAR EVOLUTION WITH MAGNETIC BRAKING

The detailed evolutionary history of such systems has been studied by a number of investigators. Although the results are somewhat sensitive to the stellar input physics (see Hameury 1990), the results of McDermott and Taam (1989) are sufficient to outline the general features common to all evolutionary calculations. From a systematic study of such binary evolutionary calculations, it has been found that for an initially unevolved main sequence star, the donor always becomes fully convective for binary orbital periods greater than ~ 2.65 hrs. This is found to be the case independent of the average mass-transfer rates above the period gap and, hence, is relatively insensitive to the functional form and magnitude of the angular momentum loss prescription. The result provides striking verification of the suggestion that the long-period edge of the gap is to be identified with the transition of a main sequence-like star from a radiative core-convective envelope structure to one which is fully convective. At an average mass-transfer rate of $\dot{M} \sim 1.9 \times 10^{-9}\,\mathrm{M_\odot yr^{-1}}$, the observed gap location and width are approximately reproduced. For average mass-transfer rates less than this critical value, the resumption of mass transfer occurs within the period gap. On the other hand, for high mass-transfer rates ($\dot{M} \sim 10^{-8}\,\mathrm{M_\odot yr^{-1}}$) the upper edge of the gap shifts to longer orbital periods ($P \sim 4$ hrs). This implies that the average mass-transfer rates of the majority of cataclysmic variables cannot be so high for, otherwise, the upper edge of the gap would be located at periods greater than 2.8 hrs. However, the results do not exclude some systems from having such large mass-transfer rates since the period at which mass transfer resumes does not depend sensitively upon the mass-transfer rates in this regime. Thus, provided that the average mass-transfer rate exceeds $\sim 2 \times 10^{-9}\,\mathrm{M_\odot yr^{-1}}$, the systems are expected to become semi-detached at similar periods (~ 2.1–2.2 hrs).

3 POSSIBLE MECHANISM FOR DISRUPTIVE MAGNETIC BRAKING

The sudden decline in angular momentum loss associated with magnetic braking has been argued to be a consequence of a vanishing radiative core in the main sequence-like star as it evolves to lower mass. This belief stems from our current theoretical ideas of the solar dynamo process. From a number of reasons based upon the observations of the solar photosphere, it is thought that the dynamo operates from the base of the convective envelope. Stars which lack a radiative core may be less susceptible to dynamo action of the solar type, and it is conceivable that the magnetic activity of a fully convective star may be significantly reduced. However, there is no strong observational evidence to support the hypothesis that magnetic activity discontinuously declines at the spectral type corresponding to fully convective stars. In fact, the existence of magnetically-active flare stars among the late M dwarfs is a counterexample to this suggestion. For a review of the observational situation see the paper by Taam and Spruit (1989).

A less drastic hypothesis and one that is, perhaps, more likely is that the dynamo changes character when the mass-losing star undergoes a transition to a fully convective state. That is, the dynamo would operate throughout the convective zone rather than at the radiative-convective interface. Such a dynamo could conceivably produce a quite different magnetic field topology. The resultant fields would probably exhibit large-scale fluctuations about the mean leading to the possibility that the surface fields in these stars would consist of high-order multipole components. Given the lack of observational evidence to substantiate a sudden decline in magnetic activity, Taam and Spruit (1989) investigated the consequences of such a rearrangement of the large-scale surface magnetic field on the angular momentum loss process.

Based upon an analysis similar to Mestel and Spruit (1987), it was found that the change in the angular momentum-loss rate associated with a rearrangement of the magnetic field configuration is significant at short periods ($P \leq 5$ hrs) for high multipole order ($n \geq 7$, where $B \propto r^{-n}$). As an illustration, at an orbital period of 3 hrs, the angular momentum-loss rate declines by a factor of 2000 between a dipole field and an asymptotically high-order field. This decline is attributable to the decrease in the fraction of open field lines (or equivalently a decrease in the fraction of surface magnetic flux that extends out to the Alfvén radius). The substantial differences at short orbital periods are due to the effect of the centrifugal acceleration on the determination of the size of the closed field region (known as the dead zone). In particular, the radius of the dead zone shifts closer to the stellar surface with increasing multipole order as a result of the more rapid decline in magnetic field strength with distance. At short orbital periods the pressure in the dead zone is strongly influenced by the centrifugal acceleration, which increases rapidly with distance from the stellar surface. The effect of the centrifugal acceleration on the fraction of open field lines for higher multipole components is smaller than at low order and, hence, fewer field lines are open. Since the angular momentum-loss rate (as well as the mass-loss rate) is proportional to the number of open field lines, the loss rate decreases with increasing multipole order.

4 CONCLUSIONS

From the detailed computations of the secular evolution of cataclysmic variables, it has been shown that the disrupted magnetic braking hypothesis can reproduce the observed location and width of the period gap of cataclysmic variable binary systems, lending support to the basic picture advanced in earlier works. The results, furthermore, make a firm prediction; namely, for those cataclysmic variable systems which evolve from periods longward of the gap, the mass of the main sequence-like component of the detached binary within the gap is expected to lie in a very narrow mass interval. For example, from the work of McDermott and Taam (1989), the mass may be in the range 0.25 to 0.28 $M_\odot$. The fact that two cataclysmic variable systems are transferring mass at high rates within the gap is not necessarily inconsistent with the above results since the mass-losing components in these systems may have evolved cores. In this case the transition to a fully convective state occurs at lower masses and, hence, shorter orbital periods (Taam 1983).

A new mechanism for disrupting the magnetic braking has been studied involving the rearrangement of the stellar magnetic field from a regular low-order structure to a highly tangled, disorganized one at the point where the main sequence-like star enters a fully convective state. It has been shown that the magnetic braking can become significantly less effective due to a reduction in the stellar wind associated with the decrease in the number of open field lines. This particular hypothesis is of considerable interest since the reduction in the rate of angular momentum loss is associated with the decline of the stellar wind-loss rate without a sudden decline in magnetic activity. We note that Hameury *et al.* (1987), in studying the locking of magnetic cataclysmic variables, also suggest that the wind loses its effectiveness in the gap.

In order to obtain a better understanding of the secular evolution of cataclysmic binaries, constraints on the mass-transfer rates should also be provided by other theories. In particular, the recent work on the nova outburst phenomenon by Kutter and Sparks (1989) suggests that the mass-transfer rates required to reproduce the gap may not be inconsistent with nova theory when account is taken of shear induced mixing between the nuclear burning regions and the white dwarf carbon oxygen core. In addition, the assistance provided by the common envelope phase in nova outbursts in ejecting the envelope (Livio *et al.* 1989) also suggests that higher mass-transfer rates may be compatible with those required for the secular evolution. Future work directed toward resolving the desired mass-transfer rates required for explaining the subclasses of cataclysmic variable systems and the existence of the period gap holds promise for producing a consistent picture for these systems.

This work was supported in part by the National Science Foundation under grant AST-8608291.

REFERENCES

Hameury, J. M. 1990, this volume.
Hameury, J. M., King, A. R., Lasota, J. P., and Ritter, H. 1987, *Ap. J.*, **316**, 275.
Kutter, G. S., and Sparks, W. M. 1989, *Ap. J.*, **340**, 985.
Livio, M., Shankar, A., Burkert, A., and Truran, J. W. 1989, preprint.
McDermott, P. N., and Taam, R. E. 1989, *Ap. J.*, **342**, 1019.
Mestel, L., and Spruit, H. C. 1987, *M. N. R. A. S.*, **226**, 57.
Rappaport, S., Verbunt, F., and Joss, P. C. 1983, *Ap. J.*, **275**, 713.
Ritter, H. 1986, in *The Evolution of Galactic X-Ray Binaries*, ed. J. Trümper, W. H. G. Lewin, and W. Brinkmann (Dordrecht: Reidel), p. 271.
Robinson, E. L., Barker, E. S., Cochran, A. L., Cochran, W. D., and Nather, R. E. 1981, *Ap. J.*, **251**, 611.
Shafter, A. W., and Abbott, T. M. C. 1989, *Ap. J. (Letters)*, **339**, L75.
Shafter, A. W., *et al.* 1990, this volume.
Spruit, H. C., and Ritter, H. 1983, *Astr. Ap.*, **124**, 267.
Taam, R. E. 1983, *Ap. J.*, **268**, 361.
Taam, R. E., and Spruit, H. C. 1989, *Ap. J.*, **345**, 972.

CV Evolution: Importance of Detailed Stellar Models

J. M. Hameury

DAEC, Observatoire de Paris, Section de Meudon,
F-92195 Meudon Principal cédex, France

1 INTRODUCTION

The evolution of short period systems with main-sequence companions is driven by orbital angular momentum losses due to braking by a magnetically-coupled stellar wind. The effect of mass transfer is to drive the secondary slightly out of thermal equilibrium, as the mass-transfer timescale is comparable to the Kelvin-Helmholtz time t_{KH}, and the star becomes slightly oversized. It is generally assumed that when the radiative core disappears, the secondary magnetic field vanishes, or, more probably, the field rearranges from a low-order to a higher-order multipole (Taam and Spruit 1989), leading to a sudden decrease of angular momentum losses. The binary evolution time is now longer than t_{KH}, and the secondary contracts towards the main sequence, detaching from the Roche lobe. Accretion ceases until the contraction of the system under the effect of gravitational radiation and any residual magnetic braking brings the system back into contact.

The theoretical basis for this explanation of the period gap is, however, extremely weak because there is a complete theory for neither the stellar dynamo nor the strength of the secondary wind; there are nevertheless several indirect arguments in favor of this model. In order to check it, Rappaport, Verbunt, and Joss (1982) and McDermott and Taam (1989) have computed the binary evolution and found that the period gap can be explained in terms of the disrupted magnetic braking model, although the limits of the period gap are somewhat different from the observed values.

Rappaport *et al.* (1982) and McDermott and Taam (1989) have considered the effect of the magnetic braking law on the location and width of the period gap; I will focus here on the effect of the influence of the assumed stellar physics (equation of state, opacities, etc.) in the period gap, and show that its location is relatively sensitive to it, so that if one could have enough confidence in the model, one would get another observational handle on the structure of low mass stars.

2 THE MODEL

The internal structure of the secondary is calculated using a code derived from Eggleton (1971, 1972). The chemical composition is that of Population I stars ($X =$

0.70, $Z = 0.02$), with an initial mass fraction of ^{3}He of 8.5×10^{-5}. The equation of state (EOS) was interpolated from the tables of Fontaine, Graboske, and Van Horn (1977). The opacities were taken from Cox and Tabor (1976) for temperatures above 10^4 K, and from Alexander (1975) at lower temperatures. Nuclear energy is provided by the PP I chain, and the rates are taken from Harris *et al.* (1983) and references therein, with screening corrections from Graboske *et al.* (1973); the ^{3}He(^{3}He,2p)^{4}He reaction was considered separately, so that the departure of ^{3}He from nuclear equilibrium could be treated. The Schwarzschild criterion was used for testing the stability against convection; the mixing length theory of Henyey, Vardya, and Bodenheimer (1965) was used in both optically thin and thick regions, with $\ell/H_p = 1.5$. The photospheric and subphotospheric layers were treated as indicated in Dorman, Nelson, and Chau (1989): the Krishna-Swamy (1966) $T(\tau)$ relation was used to take into account the departure from the diffusion approximation. With these ingredients, ZAMS models obtained were in excellent agreement with those of Dorman *et al.* (1989) who use essentially the same inputs.

Evolution is driven by angular momentum losses due to gravitational radiation and magnetic braking. We use the Mestel and Spruit (1986) law for the latter, with $n = 1.45$ (see Hameury *et al.* 1988), n being the index relating the coronal X-ray luminosity to the magnetic field: $L_x \propto B^n$. When the radiative core disappears, magnetic braking is not suppressed, but strongly reduced (Taam and Spruit 1989); we then take $n = 0.5$.

When the secondary fills its Roche lobe, mass is transferred at a rate proportional to $\exp[(R_2 - R_L)/H_p]$, where R_L is the Roche lobe radius and M_2 and R_2 are the mass and radius of the secondary. For numerical convenience, I have taken $R_2/H_P = 200$ instead of 10^4 (D'Antona, Mazzitelli, and Ritter 1989). The turn-on and turn-off phases are therefore not treated accurately and last 50–100 times longer than normally; this has no effect on the secular evolution.

3 RESULTS

Figure 1 shows the evolution of a system with $M_1 = 0.7\ M_\odot$ and $M_2 = 0.6\ M_\odot$ initially. Mass transfer starts at $P = 5.3$ hr. The secondary becomes fully convective when its mass is reduced to $0.20\ M_\odot$; the corresponding period is 3.01 hr. Mass transfer resumes at a period of 2.00 hr. Note that the edges of the gap are not perfectly steep; this is due to the assumed dependence of mass-transfer rate versus $R_2 - R_L$; the values quoted here correspond to the point where $R_2 = R_L$ exactly.

These results are slightly different from those of McDermott and Taam (1989) who find that the secondary becomes fully convective for a mass of 0.25–0.28 $M_\odot$. As a consequence, the lower edge of the gap is always located at $P \geq 2.1$ hr, even for very high mass-transfer rates ($\langle \dot{M} \rangle = 9 \times 10^{-9}\ M_\odot\ \mathrm{yr}^{-1}$); for more typical values of $\dot{M}$, their period gap starts at ~ 2.7 hr and ends at ~ 2.2 hr. This difference is not due to evolutionary effects: their secondary becomes fully convective for a mass of 0.37 $M_\odot$

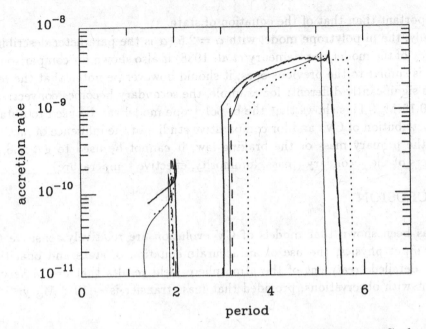

Fig. 1—Accretion rate versus period for several models: Fontaine, Graboske, and Van Horn (1977) EOS with accurate photospheric condition (*solid curve*); Paczyński (1969) EOS (*dashed curve*) and $\kappa P = 2/3g$ at the surface (*dotted curve*). Also shown is the bi-polytropic model (*thin solid curve*).

when angular momentum losses are assumed to be due only to gravitational radiation, i.e., when the star is very close to the main sequence, while I find that main-sequence stars loose their radiative core for a mass of 0.30 M$_\odot$. This is probably due to the different opacity tables used in both codes (McDermott and Taam (1989) use Cox and Stewart (1970) who did not include molecular contributions), and also to the different boundary condition used. They take $\kappa P = 2/3g$ at the surface, and therefore neglect the effect of convection in the atmosphere, as well as deviations from the radiative diffusion approximation in regions where the atmosphere is convectively stable.

In order to investigate the importance of the input stellar physics on the position of the period gap, evolution has been calculated with a different boundary condition and with a different equation of state. The dashed curve of Figure 1 has been computed using the Paczyński (1969) EOS, in which pressure ionization and Coulomb corrections are neglected. The diagram $\dot{M}(P)$ is almost the same for periods above 4 hr; however, the period gap starts at 3.2 hr and ends at 1.92 hr. The dotted curve illustrates the importance of the photospheric boundary condition: condition $\kappa P = 2/3g$ has been used. The evolution is now different even at long periods: the secondary is bigger, and mass transfer is initiated at $P = 5.7$ hr; the period gap starts at 3.52 hr and ends at 1.91 hr. This shows that surface opacities and boundary conditions have a significant effect on the location of the period gap; it is actually more important than that of the equation of state.

Finally, the bi-polytrope model with $\alpha = 2.5$ (α is the parameter describing the inaccuracy of the model, see Hameury *et al.* 1988) is also shown for comparison. This diagram is similar to the previous ones; it should however be noted that the relation $M_2(P)$ is significantly different: for example, the secondary becomes convective for a mass of 0.15 $M_\odot$. This shows that the bi-polytrope model can be used to obtain the period distribution of CVs and for comparative studies of the influence of parameters such as the primary mass or the braking law; it cannot be used to get the stellar parameters of the secondary (mass, luminosity, effective temperature).

4 CONCLUSION

It has been shown that models of CV evolution are relatively sensitive to the assumed input physics; the use of an accurate equation of state and opacities, as well as a detailed treatment of the atmosphere yield results that are in very good agreement with observations, provided that mass transfer is $\sim 10^{-9}$ $M_\odot$ yr^{-1} for P = 3 hr.

REFERENCES

Alexander, D. R. 1975, *Ap. J. Suppl.*, **29**, 363.

Cox, A. N., and Stewart, J. N. 1970, *Ap. J. Suppl.*, **19**, 243.

Cox, A. N., and Tabor, J. E. 1976, *Ap. J. Suppl.*, **31**, 271.

D'Antona, F., Mazzitelli, I., and Ritter, H. 1989, *Astr. Ap.*, in press.

Dorman, B., Nelson, L. A., and Chau, W. Y. 1989, *Ap. J.*, **342**, 1003.

Eggleton, P. P. 1971, *M. N. R. A. S.*, **151**, 351.

Eggleton, P. P. 1972, *M. N. R. A. S.*, **156**, 361.

Graboske, H. C., Jr., DeWitt, H. E., Grossman, A. S., and Cooper, M. S. 1973, *Ap. J.*, **181**, 457.

Fontaine, G., Graboske, H. C., Jr., and Van Horn, H. M. 1977, *Ap. J. Suppl.*, **35**, 293.

Hameury, J. M., King, A. R., Lasota, J. P., and Ritter, H. 1988, *M. N. R. A. S.*, **231**, 535.

Harris, M. J., Fowler, W. A., Caughlan, G. R., and Zimmerman, B. A. 1983, *Ann. Rev. Astr. Ap.*, **21**, 165.

Henyey, L., Vardya, M. S., and Bodenheimer, P. 1965, *Ap. J.*, **142**, 841.

Krishna-Swamy, K. S. 1966, *Ap. J.*, **145**, 174.

McDermott, P. N., and Taam, R. E. 1989, *Ap. J.*, **342**, 1019.

Mestel, L., and Spruit, H. C. 1987, *M. N. R. A. S.*, **226**, 57.

Paczyński, B. 1969, *Acta Astr.*, **19**, 1.

Rappaport, S., Verbunt, F., and Joss, P. C. 1982, *Ap. J.*, **275**, 713.

Taam, R. E., and Spruit, H. C. 1989, *Ap. J.*, **345**, 972.

ROSAT and CV Evolution

A. R. King[1], J. M. Hameury[2], and J. P. Lasota[3]

[1]Astronomy Group, University of Leicester, Leicester LE1 7RH, U. K.
[2]DAEC, Observatoire de Paris, Section de Meudon,
 F-92195 Meudon Cedex, France
[3]GAR, CNRS, DARC, Observatoire de Paris, Section de Meudon,
 F-92195 Meudon Cedex, France

ABSTRACT

In the context of currently-popular ideas of CV evolution, the existence of the AM Her period spike implies very tight mass limits on these systems. The comparison with present observational estimates is encouraging, especially the high mass of UZ For. ROSAT is likely to discover $\gtrsim$ 60 new AM Her systems. This should allow decisions on this question, on whether CVs cross the period gap at all, and on whether AM Her systems are usually born as intermediate polars. If magnetic CVs are able to synchronize with lower field strengths than those estimated for the known AM Her systems, a new class of CVs may be discovered, having all the properties of AM Her systems other than optical polarization.

1 ROSAT AND CVS

In its survey mode, the ROSAT Wide Field Camera should be at least as sensitive as the low-energy detectors on EXOSAT. With 5% sky coverage, the latter instrument serendipitously discovered 3 new AM Her systems. A simple extrapolation then suggests that ROSAT will turn up at least 60 new AM Her systems; more sophisticated calculations give rather larger estimates. It should be feasible to identify these sources in a reasonable time, as the Wide Field Camera is expected to detect only a few thousand sources in its very soft bandpass. The orbital period of an AM Her system is always found once the system is identified. This will make the period histogram of AM Her systems roughly as well-populated as that of all CVs is at present. Obviously, this must transform our understanding of every aspect of these systems: here we concentrate on the impact on the study of CV evolution.

2 THE PERIOD SPIKE

Hameury et al. (1988) (hereafter HKLR) explained the well-known AM Her period spike at $P_{\text{orb}} = 114$ min as marking the resumption of mass transfer after passage through the period gap, the discovery probability being particularly high at this point. This is confirmed by more detailed modelling (D'Antona, Mazzitelli, and Ritter 1989; McDermott and Taam 1989). In the currently-popular picture, the period $P_{orb} = P_l$ at the lower edge of the gap is fixed by the degree of thermal disequilibrium suffered by the secondary at the upper edge of the gap, $P_{orb} = P_u$. This in turn depends on the ratio of the mass-transfer timescale,

$$t_{\dot{M}} = -M_2/\dot{M}_2 \approx (4/3 - M_2/M_1)(-J/\dot{J}) \tag{1}$$

(M_1, M_2 = white dwarf and secondary masses, J =orbital angular momentum), to the secondary's Kelvin-Helmholtz time t_{KH}: the smaller $t_{\dot{M}}/t_{KH}$, the lower P_l. Since the ratio in general depends on M_1, it is clear that this type of explanation of the spike demands a white dwarf mass distribution with very little dispersion for most of the systems at this period. Hameury, King, and Lasota (1988) estimate $M_1 \approx 0.6$–$0.7\,M_\odot$ for the spike. The exceptions are any systems at this period which were born at periods $\lesssim 3.5$ hrs, i.e., very close to or in the gap. Such systems cease mass transfer, if at all, only for a rather short period range. For them, $P_{orb} = 114$ min has no particular significance, being just one of the periods they evolve through, and there is no reason to expect their white dwarf masses to be constrained to the same value as the other spike members.

The mass estimates by Webbink (1990) are of interest in this connection. These result from the adoption of an empirical main-sequence relation for the secondary stars in CVs. Webbink assigns values 0.28, 0.43, 0.6, 0.7, and $1.14\,M_\odot$ to five of the six systems in the spike. If we accept the first of these five masses at face value, it is too low for stable mass transfer at periods $\gtrsim 3$ hrs, so we would have to conclude that this system (BL Hyi) is probably an "interloper" in the spike as described above. However, the mass $M_1 = 0.43\,M_\odot$ assigned to ST LMi is rather lower than quoted in papers reporting measurements of the secondary star's radial velocity: both Schmidt, Stockman, and Grandi (1983) and Bailey *et al.* (1985) give $M_1 > 0.5\,M_\odot$, with upper limits $0.8\,M_\odot$ and $1.1\,M_\odot$, respectively. The errors on these results certainly do not exclude $M_1 = 0.6$–$0.7\,M_\odot$. There is then only one discrepant mass ($1.14\,M_\odot$, WW Hor) in Webbink's list. However, this mass is in conflict with the radial velocity amplitude of 388 km s^{-1} quoted by Beuermann *et al.* (1987); adopting the inclination $i = 87°$ given by Webbink (which is very reasonable, as the system eclipses), this yields $M_1 = 0.7\,M_\odot$. Hence, *all four mass estimates for genuine spike members are consistent with expectation.* Further, Webbink's quoted *average* M_1 for 12 magnetic CVs below the period gap is $0.70 \pm 0.10\,M_\odot$, and the dispersion will be even smaller with the adjustments suggested above. HKLR's hypothesis is thus not yet contradicted by the available data. However, in view of the possibility of random or systematic error, one should suspend judgement until the sample size has increased substantially. A simple scaling suggests that *ROSAT* will raise the spike membership to ~ 25. (Note that if we abandon HKLR's explanation of the spike, we would have to conclude either that $t_{\dot{M}}$ is independent of M_1, or that many AM Her systems are simply born with just the right secondary mass to come into contact for the first time at $P_{orb} = 114$ min.)

3 DO CVS CROSS THE PERIOD GAP?

In the remainder of this article we will adopt the explanation for the period spike proposed by HKLR. The dependence of $t_{\dot{M}}$ on M_1 leads to a simple test, not only of this suggestion, but also of the basic idea that CVs cross the period gap. From equation (1), for any likely angular momentum-loss mechanism giving $\dot{J}$, the main dependence of $t_{\dot{M}}$ on M_1 is through the denominator. Thus, $t_{\dot{M}}$ is larger for large M_1, and *one expects the period gap to be narrower for systems with high white dwarf masses.* For this reason, Hameury, King, and Lasota (1988) predicted that the newly-discovered AM Her system EXO 033319-2554.2 (now called UZ For) must have a significantly larger white dwarf mass, $M_1 \gtrsim 1.2 \, M_\odot$. Observations by Beuermann, Thomas, and Schwope (1988) and Ferrario *et al.* (1989) confirm this. It is worth noting that the two eclipsing systems (thus probably giving the most reliable mass estimates) in and near the spike, namely WW Hor and UZ For, do have very different radial velocity amplitudes for the secondary. This implies a much heavier white dwarf in UZ For (factors ~ 2), as expected. For the larger sample expected from *ROSAT*, Hameury, King, and Lasota (1989) show that the M_1-distributions on each side of the gap must be mirror images; those systems closest to the edges must have the largest values of M_1. This effect is potentially observable and offers a clearcut test of the idea that CVs cross the period gap. This paper also points out three other evolutionary tests which will become possible. First, there should be a spike at the minimum period, simply because $\dot{P}_{orb}$ vanishes there. The spike position will again depend on $t_{\dot{M}}$ and hence on M_1. If the minimum period is greater for high-mass systems than for low, this is consistent with angular momentum losses driven by gravitational radiation alone. The opposite result will indicate that some other mechanism, such as some residual magnetic braking, is acting. Second, the relative importance of the various spikes will show whether the AM Her systems form a flux-limited sample or not. Finally, if the enlarged AM Her distribution still cuts off above $P_{orb} \sim 4$ hrs, this will force us to conclude that AM Her systems are born at longer periods in some other form: the only likely possibility would be the intermediate polars.

4 A NEW CLASS OF CVS?

The relation between synchronous and non-synchronous magnetic CVs touched on above has been the subject of intense debate over recent years. What seems generally agreed is that systems having somewhat lower magnetic moments $\mu \lesssim 10^{34}$ G cm^3 than those of the AM Her systems should become synchronous (i.e., lock the spin and orbital rotations) at periods below the gap. BG CMi (Chanmugam *et al.* 1989) may be an example: it has a field of ~ 4 MG and should lock for $P_{orb} < 2$ hr. Clearly, such systems would not show the optical polarization characteristic of AM Hers. However, there is little reason to conclude that a small (factors $\lesssim 2$) decrease in μ should have any effect on the other AM Her characteristic: strong soft

X-ray emission. Simple estimates of the polecap fraction suggest that the soft X-ray temperature should vary only as $T_{\text{eff}} \propto \mu^{1/7}$. Accordingly, we would expect that if systems with $\mu \lesssim 10^{34}$ G cm^3 are at all common, *ROSAT* should turn them up in large numbers. They should appear identical to the AM Her systems except for the absence of optical polarization, but might well show such polarization in the near infrared. There are two ways of escaping from this conclusion. First, Warner and O'Donoghue (1987) have suggested that V2051 Oph ($P_{\text{orb}} = 90$ min) is a synchronous low-field system. It does not show soft X-ray emission, and might therefore constitute a counterexample to the argument above, suggesting that the presence of such emission might be somehow extremely sensitive to the strength of the magnetic moment. However, there is no obvious reason for this, or any way of deciding whether all low-field systems will be similarly deficient. The lack of soft X-rays might alternatively constitute a reason for doubting the identification as a synchronous system. The second reason why low-field systems may not turn up is simply that such fields may be intrinsically uncommon in CVs, i.e., that most magnetic CVs have $\mu \gtrsim 10^{34}$ G cm^3.

ARK thanks the Royal Society and the organizers for support in attending the Workshop, and Lilia Ferrario for valuable discussions.

REFERENCES

Bailey, J. A., *et al.* 1985, *M. N. R. A. S.*, **215**, 179.

Beuermann, K., Thomas, H. C., Giommi, P., and Tagliaferri, G. 1987, *Astr. Ap.*, **175**, L9.

Beuermann, K., Thomas, H. C., and Schwope, A. 1988, *Astr. Ap.*, **195**, L15.

Chanmugam, G., Frank, J., King, A. R., and Lasota, J. P. 1989, *Ap. J. (Letters)*, in press.

D'Antona, F., Mazzitelli, I., and Ritter, H. 1989, *Astr. Ap.*, in press.

Ferrario, L., Wickramasinghe, D. T., Bailey, J., Tuohy, I. R., and Hough, J. H. 1989, *Ap. J.*, **337**, 832.

Hameury, J. M., King, A. R., Lasota, J. P., and Ritter, H. 1988, *M. N. R. A. S.*, **231**, 535. (HKLR)

Hameury, J. M., King, A. R., and Lasota, J. P. 1988, *Astr. Ap.*, **195**, L12.

Hameury, J. M., King, A. R., and Lasota, J. P. 1989, *M. N. R. A. S.*, in press.

McDermott, P. N., and Taam, R. E. 1989, *Ap. J.*, **342**, 1019.

Schmidt, G. D., Stockman, H. S., and Grandi, S. A. 1983, *Ap. J.*, **271**, 735.

Warner, B., and O'Donoghue, D. 1987, *M. N. R. A. S.*, **224**, 733.

Webbink, R. F. 1990, this volume.

Angular Momentum of Accreting White Dwarfs: Implications for Formation of LMXRBs and Recycled Pulsars

Robert Popham and Ramesh Narayan

Steward Observatory, University of Arizona

1 INTRODUCTION

Low mass X-ray binaries (LMXRBs) with $\leq 1\,M_\odot$ stellar companions, and recycled pulsars (RPs) with low mass ($\leq 0.4\,M_\odot$) white dwarf companions, are found both in the disk of the Galaxy and in globular clusters. The RPs are dominated by low-field millisecond pulsars. A scenario where the neutron star is formed in a Type II supernova explosion runs into difficulties for the following reasons. First, because the companion has a low mass, the explosion is expected to disrupt the binary system. Second, even if the system remains bound, it will still receive a substantial recoil velocity which will propel it to high z in the disk or remove it from the globular cluster, in conflict with observations.

An alternative method for forming LMXRBs and RPs involves the accretion-induced collapse (AIC) of an O-Ne-Mg white dwarf (or possibly a C-O white dwarf — see Nomoto 1987 and Isern *et al.* 1990). In such a scenario, an accretion disk is formed, and the accreted matter, apart from adding mass M to the white dwarf, also imparts substantial angular momentum J (Grindlay and Bailyn 1988). We have calculated the angular momentum transferred to the white dwarf for a range of accretion rates $\dot{M}$, initial white dwarf masses $M_{\rm init}$, and white dwarf magnetic fields B in order to investigate the periods and fields expected for neutron stars produced by AIC.

We use the accretion torque theory developed by Ghosh and Lamb (1979 a, b) and modified by Wang (1987). We assume that the white dwarf rotates rigidly and include the effects of rotation on the white dwarf structure, using the calculations of Hachisu (1986). We also assume that the magnetic flux BR^2 is conserved throughout the accretion and collapse; we use this quantity to specify the magnetic field strength in our models. A more detailed account of our methods and choice of input parameters is given in Narayan and Popham (1989).

2 RESULTS

Accreting white dwarfs will evolve in the M-J plane until they reach either the critical rotation or the collapse line. We further subdivide the systems which reach the collapse line according to their angular momenta at the time of collapse. Assuming a standard moment of inertia for neutron stars of 10^{45} g cm^2, white dwarfs with $J > 6 \times 10^{48}$ g cm^2 s^{-1} would form neutron stars with periods of less than 1 ms. Such rapid rotation is not allowed by most equations of state, so the collapse will be halted midway and the star will form a "fizzler" (Tohline 1984). Such an object is expected to contract more slowly to a neutron star as it loses angular momentum by magnetic dipole radiation and/or gravitational radiation from nonaxisymmetric distortions (Michel 1987). We label the three possible outcomes for an accreting white dwarf as follows:

"**Case I**" refers to direct collapse to a neutron star,

"**Case II**" to fizzler collapse, and

"**Case III**" to systems which reach the critical rotation.

Usually, the initial mass of the white dwarf will be substantially below the Chandrasekhar limit, and the evolution will be as in Figure 1. Weak magnetic fields have little effect upon the rotation, and the white dwarf will reach the critical rotation after accreting 0.1–0.15 $M_\odot$ of material. If the field is sufficiently strong, the white dwarf soon reaches an equilibrium spin at which the torque is essentially zero. From then on, it continues to accrete mass until it reaches the collapse line. An interesting result is that even after the star achieves spin equilibrium with the accretion disk, it continues to spin up slowly; at no stage does the white dwarf spin down.

3 DISCUSSION

Our calculations differ from those of Lamb and Patterson (1983) in that we follow the time evolution of accreting white dwarf systems and include the effects of the accreted mass and angular momentum on the structure of the white dwarf. In general, our calculations predict that for constant accretion rates, accreting white dwarfs should always be spinning up. Two DQ Her systems, FO Aqr and V1223 Sgr, have been observed to be spinning down. This behavior must be due to changes in the accretion rate; for instance, a system at the equilibrium period would respond to a drop in the accretion rate by spinning down to a new equilibrium.

Of the possible outcomes for accreting O-Ne-Mg white dwarfs, Case I, involving accretion-induced collapse to form a neutron star directly, is probably the best understood. Unless the white dwarf starts very close to the limiting mass, this should only occur for highly-magnetized white dwarfs; the required field varies from about $BR^2 \sim 10^{24}$ G cm^2 for $\dot{M} = 10^{-7}$ $M_\odot$ yr^{-1} to a few $\times 10^{22}$ G cm^2 for $\dot{M} = 10^{-10}$ $M_\odot$ yr^{-1}. Thus, the neutron stars produced would have fields of a few $\times 10^{10}$ G to a few $\times 10^{12}$ G. The combination of strong field and millisecond period in these neutron stars would produce a luminosity of 10^{39}–10^{43} erg s^{-1} in magnetic dipole radiation for 10^2–10^5 yr. This rapid loss of rotational energy would mean that these stars would only have millisecond periods for a few million years or less.

Case II, where the collapse is halted to form a "fizzler," is less well understood. According to Tohline (1984), the actual nature of the collapse depends largely on

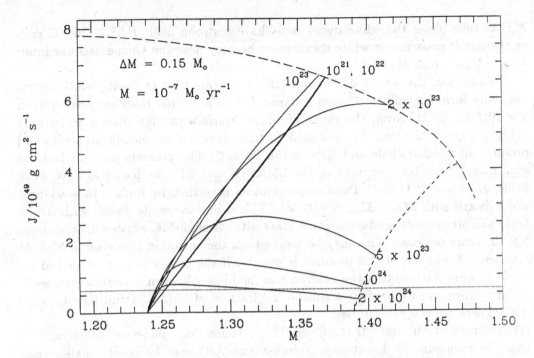

Fig. 1—The mass and angular momentum of white dwarfs which start out $0.15M_\odot$ below the limiting mass, accreting at 10^{-7} $M_\odot$ yr^{-1}. Each evolutionary track is labeled with the magnetic flux BR_{WD}^2. White dwarfs which reach the long-dashed line at top are critically rotating, while those which reach the short-dashed line at right collapse.

the value of the adiabatic exponent Γ as a function of density. In any case, once the collapse is initiated, a neutron star should eventually be formed with a period of around 1 ms.

Case III, where the white dwarf reaches the critical rotation, is probably the least understood outcome. At this point, the white dwarf is on the verge of losing mass due to rotation, but more mass is being added by the continuing accretion. Thus, the star must either eject mass or lose angular momentum and accrete more mass. The latter possibility may occur if the star is able to lose angular momentum back through the disk or through a magnetically-coupled wind; in this case the star may stay on the critical rotation line, losing angular momentum and accreting mass until it reaches the critical central density and collapses as a fizzler.

Case III should be the most common situation (Papaloizou and Pringle 1978), since most white dwarfs have weak fields, and it is important to investigate the fate of these systems. If we assume that Case III white dwarfs are unable to accrete because of their large angular velocities (but see Pringle 1989), then they will be unable to collapse to neutron stars. Our results would then place severe constraints on the properties of any potential white dwarf progenitors for LMXRBs and RPs, viz., for

AIC to take place, the white dwarf must have a strong field, $BR^2 \geq 10^{23}$ G cm², or the initial mass of the white dwarf must be very near the Chandrasekhar limit, $M_{Ch} - M_{init} \leq 0.03$ M$_\odot$.

In addition, the results of Nomoto (1987) indicate that O-Ne-Mg white dwarfs may only form neutron stars upon collapse if their accretion rates are greater than 4×10^{-8} M$_\odot$ yr^{-1}. Given the rarity of white dwarfs with high masses and strong fields in high-$\dot{M}$ binaries, it seems unlikely that there will be enough progenitors to produce all the LMXRBs and RPs in the Galaxy. The problem may, in fact, be even more severe because most of the LMXRBs and RPs are found to have weak fields, $BR^2 \sim 10^{21}$ G cm². Possible progenitors may then be limited to weak-field white dwarfs with $M_{Ch} - M_{init} \leq 0.03$ M$_\odot$. This is because white dwarfs with strong fields will presumably produce neutron stars with strong fields, which will take about 5×10^7 years to decay. This may be longer than the accretion timescale in high-$\dot{M}$ binaries. (A way out of this problem is possible if the binary remains detached for $\sim 10^8$ yr after AIC, so that the neutron star field is weak when accretion resumes.)

How then can we produce a substantial fraction of LMXRBs and RPs through AIC? There are two possibilities:

(*i*) It may turn out that Case III systems do undergo collapse.

(*ii*) The minimum $\dot{M}$ for steady accretion and AIC may be much smaller than 4×10^{-8} M$_\odot$ yr^{-1} as we have assumed.

We are not aware of any compelling evidence in favor of either of these possibilities, but more work in these directions is clearly worthwhile.

We benefitted from helpful discussions with Richard Wade, Don Lamb, and Hans-Walter Rix. This work was supported in part by NSF grant AST-88-14725.

REFERENCES

Ghosh, P., and Lamb, F. K. 1979*a*, *Ap. J.*, **232**, 259.

Ghosh, P., and Lamb, F. K. 1979*b*, *Ap. J.*, **234**, 296.

Grindlay, J. E., and Bailyn, C. D. 1988, *Nature*, **336**, 48.

Hachisu, I. 1986, *Ap. J. Suppl.*, **61**, 479.

Isern, J., Labay, J., Canal, R., and García, D. 1990, this volume.

Lamb, D. Q., and Patterson, J. 1983, in *Cataclysmic Variables and Related Objects*, ed. M. Livio and G. Shaviv (Dordrecht: Reidel), p. 229.

Michel, F. C. 1987, *Nature*, **329**, 310.

Narayan, R., and Popham, R. 1989, *Ap. J. (Letters)*, **346**, L25.

Nomoto, K. 1987, in *The Origin and Evolution of Neutron Stars*, ed. D. J. Helfand and J.-H. Huang (Dordrecht: Reidel), p. 281.

Papaloizou, J., and Pringle, J. E. 1978, *M. N. R. A. S.*, **182**, 423.

Pringle, J. E. 1989, *M. N. R. A. S.*, **236**, 107.

Tohline, J. E. 1984, *Ap. J.*, **285**, 721.

Wang, Y.-M. 1987, *Astr. Ap.*, **183**, 257.

On the Origin of Low Mass X-ray Binaries

Jordi Isern[1,4], Javier Labay[2,4], Ramón Canal[2,4], and Domingo García[3,4]

[1]Centre d'Estudis Avancats Blanes (CSIC)
[2]Departament de Física de l'Atmosfera, Astronomia i Astrofísica
[3]Departament de Física i Enginyeria Nuclear (UPC)
[4]Grup d'Astrofísica (IEC)

1 INTRODUCTION

It has been suggested that the neutron star in low mass X-ray binary sources is due to the accretion-induced collapse (AIC) of a white dwarf (Taam and Van den Heuvel 1986). Two types of candidates have been proposed thus far: CO white dwarfs (Canal and Schatzman 1976; Canal and Isern 1979; Canal, Isern, and Labay 1980), and ONeMg white dwarfs (Miyaji et al. 1980; Miyaji and Nomoto 1987). In both cases, the feasibility cannot be regarded as secure.

There is ample consensus that CO white dwarfs are the progenitors of Type Ia supernovae. Current models (Nomoto, Thieleman, and Yokoi 1984; Sutherland and Wheeler 1984; Woosley and Weaver 1986) do not predict the formation of any compact remnant. A possible way out of this dilemma comes from the consideration that white dwarfs are, at least, partially solid objects. Solidification has two main consequences: (1) the onset of thermonuclear runaway is delayed up to high densities (Hernanz et al. 1988) and (2) the propagation velocity of a burning front inside a solid likely corresponds to the conduction velocity only, implying lower velocities ($v_{bf} \simeq 10^{-2} c_s$, where c_s is the local sound speed) than in a fluid, since Rayleigh-Taylor instabilities cannot develop, which allows the energy loss by electron captures behind the front to win over the energy release by advance of the burning front (Canal and Isern 1979).

2 RESULTS AND DISCUSSION

It has been shown by Isern et al. (1983) and by Hernanz et al. (1988) that central thermonuclear runaway can indeed be delayed up to densities $\rho \geq 10^{10}$ g cm^{-3} in mass-accreting CO white dwarfs provided that they are initially cold and massive ($M_{wd} > 1.15 \, M_\odot$) and the accretion rates are either $\dot{M} \geq 10^{-7} \, M_\odot$ yr^{-1} or $\dot{M} \leq 10^{-9} \, M_\odot$ yr^{-1}. Central densities at which runaway occurs inside a central solid core span the range 9.5×10^9 g cm$^{-3} \leq \rho_c \leq 1.5 \times 10^{10}$ g cm^{-3} or 6×10^9 g cm$^{-3} \leq \rho_c \leq 1.3 \times 10^{10}$ g cm^{-3}, depending on the adopted approximation ("static" or "relaxed") to compute the nuclear reaction rates in the pycnonuclear regime (Salpeter and Van Horn 1969).

The density at which the nuclear runaway starts and the velocity of the burning front are critical in determining the final fate of the white dwarf. In order to elucidate this point, we have performed a study of the dynamical evolution of a CO white dwarf for two different central ignition densities: $\rho_c = 1.13 \times 10^{10}$ g cm^{-3} and $\rho_c = 9.5 \times 10^9$ g cm^{-3}. We have adopted a parameterized burning front velocity in terms of the local sound speed. For each ignition density we have evolved four models corresponding to burning front velocities of 0.1, 0.03, 0.01, and 0.005 times c_s. The last velocity roughly corresponds to the estimate of conductive burning front velocities by Woosley and Weaver (1986). In the models with the highest central ignition densities, the transition from explosion to collapse takes place between 0.03 and 0.01 times c_s. Calculations were stopped when the collapse to nuclear densities was the only possible outcome. In the second set of models, corresponding to the lowest central ignition densities, the white dwarf exploded even in the case of a velocity equal to the conductive velocity.

A further relevant process to be considered here is heating of the solid layers induced by the interaction of the relativistic electrons with the neutrinos emitted by electron captures in the growing central incinerated regions, since this heating can melt the crystal before the start of dynamical contraction. Taking into account the latent heat of melting, this typically gives 1 s for the lifetime of the solid layers after ignition.

In a second series of calculations, we have taken into account the change in burning propagation velocity that must result from the emergence of the burning front out of the central solid core into the fluid layers. In the solid core, we adopted Woosley and Weaver's (1986) expression for the conductive velocities. In the fluid layers, we switched to a turbulent front velocity given by Sutherland and Wheeler's (1984) prescription for the propagation of Rayleigh-Taylor instabilities. For a model with $\rho_c = 1.13 \times 10^{10}$ g cm^{-3} and initial solid core mass of $M_{core} \simeq 0.6 M_{star}$, with the burning front velocity prescriptions given above, collapse also ensues, but this is no longer the case for central ignitions starting at lower densities or burning fronts propagating at velocities higher than the quoted conductive velocity. We thus see that AIC on CO white dwarfs is possible but only for a limited range of ignition densities and solid core sizes (depending on the still-uncertain values of conductive velocities). That, in turn, sets narrow limits on the allowed initial masses, temperatures, and mass-accretion rates.

The case of the ONeMg white dwarfs is not necessarily better. Miyaji *et al.* (1980) and Miyaji and Nomoto (1987) calculated that ignition would happen at a density of the order of $\rho_c \simeq 2 \times 10^{10}$ g cm^{-3}. This prediction was based on strict adoption of Schwarzschild's criterion for the onset of the convective instability induced by electron captures on Mg and Ne. However, electron captures also create a positive Y_e gradient that has a stabilizing effect (Mochkovitch 1984). If Ledoux's criterion is applied, runaway starts at $\rho \simeq 9.5 \times 10^9$ g cm^{-3}. As no solid layers can remain, hydrodynamic instability should thus grow and a turbulent flame front would propagate. In a series of calculations, we have evolved models with a parameterized flame velocity given by: $v_{burn} \simeq F c_s (1 - e^{-r/R_0})$, where c_s is the local sound speed, r is the distance to the center, and F and R_0 are parameters. For values of $F = 0.5$ and $R_0 = 2 \times 10^7$ cm (suggested for CO dwarfs; see Woosley and Weaver 1986), the runaway leads to complete disruption of the star. However, as the specific energy released when incinerating a ONeMg mixture is smaller than a CO mixture, the buoyancy should also be lower and the hydrodynamic instability might grow more slowly. For that reason, we evolved two models with $F = 0.3$ and $R_0 = 5 \times 10^7$ cm and 10^8 cm, respectively. In the first case, we obtained an explosion, and in the second a collapse. If a semiconvective instability develops and produces mixing, the ignition would happen for a central density somewhere in the interval 9.5×10^9 g cm$^{-3} < \rho_c < 2 \times 10^{10}$ g cm^{-3} . Depending on this value and on the velocity of the burning front, either collapse to a neutron star or supernova outburst can be obtained.

Therefore, we conclude that AIC might either result from a small fraction of mass-accreting CO white dwarfs or from most of the ONeMg white dwarfs or from both. It might also turn out to be physically impossible in all cases. Astronomical evidence points to the contrary and further work is necessary to clarify the problem.

This work has been supported in part by CICYT grants PB0304, PB87-0147, and PB87-0150.

REFERENCES

Canal, R., and Isern, J. 1979, in *White Dwarfs and Variable Degenerate Stars*, ed. H. M. van Horn and V. Weidemann (Rochester: University of Rochester Press), p. 53.

Canal, R., Isern J., and Labay, J. 1980, *Ap. J. (Letters)*, **241**, L33.

Canal, R., and Schatzman, E. 1976, *Astr. Ap.*, **46**, 229.

Hernanz, M., Isern, J., Canal, R., Labay, J., and Mochkovitch, R. 1988, *Ap. J.*, **324**, 331.

Isern, J., Labay, J., Canal, R., and Hernanz, M. 1983, *Ap. J.*, **273**, 320.

Miyaji, S., and Nomoto, K. 1987, *Ap. J.*, **318**, 307.

Miyaji, S., Nomoto, K., Yokoi, K., and Sugimoto, D. 1980, *Pub. Astr. Soc. Japan*, **32**, 303.

Mochkovitch, R. 1984, in *Problems of Collapse and Numerical Relativity*, ed. D. Banzel and M. Signore (Dordrecht: Reidel), p. 125.

Nomoto, K., Thielemann, F., and Yokoi, K. 1984, *Ap. J.*, **286**, 644.

Salpeter, E. E., and Van Horn, H. M. 1969, *Ap. J.*, **155**, 183.

Sutherland, P. G., and Wheeler, J. C. 1984, *Ap. J.*, **280**, 282.

Taam, R. E., and Van den Heuvel, E. P. J. 1986, *Ap. J.*, **305**, 235.

Woosley, S. E., and Weaver, T. A. 1986, *Ann. Rev. Astr. Ap.*, **24**, 205.

Are LMXRBs the Progenitors of Low Mass Binary Pulsars?

Ramesh Narayan[1], A. S. Fruchter[2], S. R. Kulkarni[3], and R. W. Romani[4]

[1]Steward Observatory, University of Arizona
[2]Department of Terrestrial Magnetism, Carnegie Institution of Washington
[3]Department of Astronomy, California Institute of Technology
[4]School of Natural Sciences, Institute for Advanced Study

ABSTRACT

We have estimated the number and birthrate of low mass binary pulsars in the disk of the Galaxy and in the globular clusters. In both regions of the Galaxy, the birthrate of these systems exceeds that estimated for their supposed progenitors, the low mass X-ray binaries, by a factor of order 100. The discrepancy could be mitigated by reducing the assumed lifetime of the X-ray binaries; for instance, one could invoke X-ray irradiation of the secondaries, and hence accelerated evolution, during the X-ray phase. Alternatively, one could invoke a non-X-ray-producing progenitor for the low mass binary pulsars, e.g. accretion-induced collapse of a binary white dwarf directly to a rapidly rotating neutron star whose luminosity inhibits further accretion.

1 INTRODUCTION

Low mass binary pulsars (LMBPs) are radio pulsars with low mass ($\lesssim 0.4 \, M_\odot$) white dwarf companions. These are long-lived systems, some older than 10^9 yr, and in many cases the pulsar is found to be spinning very rapidly with millisecond periods. It is widely believed that the systems observed as LMBPs today acquired their rapid spin during an earlier episode of mass transfer from their companion stars. Since during the spin-up the system would have had the characteristics of a low mass X-ray binary (LMXRB), it has been suggested that all LMBPs were typical LMXRBs at some time in the past (e.g., van den Heuvel 1987; Verbunt 1988 and references therein).

We present results that throw some doubt on this hypothesis. We have calculated the birthrates of LMBPs in the disk of the Galaxy and in the globular clusters and find that LMBPs are born about 100 times more frequently than LMXRBs. We discuss some possible ways of resolving this discrepancy.

2 BIRTHRATES IN THE GALACTIC DISK

Table 1 lists the four pulsars with $P < 10$ ms that have been discovered so far

TABLE 1

MILLISECOND PULSARS IN THE GALACTIC DISK

Pulsar	P (ms)	$\dot{P}/10^{-20}$	P_{orb} (days)	$\log L_{400}$ (mJy·kpc²)	τ (10^9 yr)	$N(P, L)$	Birthrate (per 10^6 yr)
PSR 1855+09	5.4	2	12	0.4	4	75000	9.4
PSR 1937+21	1.6	10	–	3.4	0.1	300	1.5
PSR 1953+29	6.1	3	190	2.0	3	300	0.05
PSR 1957+20	1.6	1.5	0.3	1.2	2	40000	12

in the Galactic disk. One of these, PSR 1937+21, is single, but we nevertheless consider it to be an LMBP because most scenarios for its formation involve a low-mass companion some time in the past. The companion has since disappeared, possibly through pulsar-driven ablation.

In Table 1 we give the period P, period derivative $\dot{P}$, the orbital period P_{orb}, the radio luminosity at 400 MHz, L_{400}(mJy · kpc²), and the spin-down age, $\tau \equiv P/2\dot{P}$, of the disk millisecond pulsars. Column 7 gives $N(P, L)$, which represents for each pulsar our estimate of the number of such systems that are present in the whole Galaxy. To calculate this, we have carefully modeled all the selection effects in pulsar searches (Narayan 1987; Kulkarni and Narayan 1988) and estimated the (weighted) fractional volume of the Galactic disk that has been effectively searched for pulsars of the given P and L. $N(P, L)$ is then the reciprocal of this fraction. The estimates of $N(P, L)$ listed here include the results of a recent very deep millisecond pulsar search (Fruchter 1989) and are more up-to-date than the estimates given by Kulkarni and Narayan (1988). Also, the present values, derived from our modeling of search selection effects, agree well with those obtained more directly through beam-area by beam-area estimates of sensitivity (Fruchter 1989; Fruchter and Taylor 1990).

The last column in Table 1 gives the birthrate of each species of LMBP for the whole disk, estimated by the quantity $N(P, L)/2\tau$ (see Kulkarni and Narayan 1988 for details). We obtain a net birthrate of ~23 LMBPs per 10^6 yr.

To obtain the birthrate of LMXRBs, we note that there are ~100 LMXRBs known in the disk and bulge of the Galaxy. This is likely to represent the total population because these bright X-ray sources are visible anywhere in the Galaxy and the catalogs are essentially complete (Lewin and Joss 1983). Of the 100 sources, only a fraction ($< 1/2$) have orbital periods in the range required to be progenitors of the LMBPs listed in Table 1, particularly the high-birthrate systems PSR 1855+09, PSR 1937+21, and PSR 1957+20. This subset is believed to have X-ray lifetimes ~10^9 yr. Thus, the birthrate of the progenitor LMXRBs is estimated to be $< 100/10^9$ yr^{-1} = 0.1 per 10^6 yr.

We see that the discrepancy between the birthrates of the LMBPs and LMXRBs is extremely large, a factor of over 100. It is possible that the smallness of the

LMBP sample as well as systematic effects contribute to some part of this discrepancy. However, we have been quite conservative in our choice of parameters. For instance, the number $N(P, L)$ quoted in Table 1 for PSR 1855+09 has been revised downward by a factor ~ 2.5 because we have used a recent distance estimate of 500 pc (J. H. Taylor, private communication), obtained through an uncertain parallax measurement, instead of the earlier dispersion-derived estimate of 350 pc (Segelstein *et al.* 1986). Also, we have neglected any beaming factor in estimating $N(P, L)$ of the LMBPs. The inclusion of this would increase the discrepancy by a factor of 5 if we used the standard factor. In any case, we note that 3 of the 4 pulsars in Table 1 have estimated birthrates that are individually much larger than the birthrate of all the LMXRBs combined. This is much too large an effect to be a statistical fluctuation.

3 BIRTHRATES IN THE GLOBULAR CLUSTERS

A number of radio pulsars have been discovered in the globular clusters (GCs). Some of these are in binaries and others are single, and their rotation periods range from 3 ms to 110 ms. All of the pulsars are believed to be old and may be classified as LMBPs. This includes the single ones, which probably lost their companions through three-body scattering in the cluster core (Romani, Kulkarni, and Blandford 1987; Rappaport, Putney, and Verbunt 1989) or pulsar-induced ablation (*e.g.*, Kluzniak *et al.* 1988).

We have recently estimated the birthrate of LMBPs in the GCs using detailed models of pulsar evolution and radio luminosity, and taking account of the selection effects in the various pulsar searches (Kulkarni, Narayan, and Romani 1989). We included the results of several searches that had been completed and whose data had been fully analyzed at the time of our calculation. However, we only had preliminary results from the on-going 400 MHz search with the Arecibo radio telescope. Our final conclusion is that there are between 2000 and 17,000 LMBPs in the GCs, depending on assumptions on the beaming and other factors. The corresponding birthrate is in the range 0.2–1.7 LMBPs per 10^6 yr. We take a birthrate of 1 LMBP per 10^6 yr as a representative estimate.

In the case of the LMXRBs, there are $\lesssim 10$ such sources in all the GCs, each with an estimated lifetime $\sim 10^9$ yr. The birthrate is thus ~ 0.01 LMXRBs per 10^6 yr.

Once again, as in the disk, we see that the birthrate of the LMBPs exceeds that of the LMXRBs by a factor ~ 100.

4 DISCUSSION

We take the view that the discrepancies discussed above in the disk and the GCs are far too large to be explained away by poor statistics in the LMBPs. How then do we resolve this problem? We see two possibilities.

The first possibility is that the X-ray lifetime of $\sim 10^9$ yr that we used is in error. This estimate comes partly from theoretical estimates based on Roche lobe overflow of a sub-giant companion and gravitational radiation-driven evolution with a main-sequence companion. These estimates do not allow for any feedback from the accretion luminosity. If one includes the effect of the X-rays from the primary on the secondary then clearly evolution will be speeded up (cf., Kluzniak *et al.* 1988; Ruderman, Shaham, and Tavani 1989). The effect will be strongest in short orbital-period ($P_{orb} \lesssim 1$ day) LMXRBs, precisely the proposed progenitors of the LMBPs such as PSR 1937+21 and PSR 1957+20. This therefore appears to be a promising solution. The question however is whether the effect is strong enough to reduce the X-ray lifetime by a factor of ~ 100. This might be difficult, especially since Eddington-limited acceptance of mass over the resulting 10^7yr lifetime would barely allow accrual of the $0.1\,M_\odot$ needed to spin pulsars up to the $\sim$ millisecond periods seen.

The other possible resolution is to give up the hypothesis that all LMBPs are formed from LMXRBs and instead to consider non-X-ray producing progenitors for the bulk of the LMBPs. (A weak objection to this is that since the birthrate discrepancy is of the same order of magnitude in both the disk and the GCs there must be a connection between LMBPs and LMXRBs; however, the statistical uncertainties are currently too large to give much weight to this argument.) One possibility is that the LMBPs are formed directly from white dwarf binaries such as the cataclysmic variables through accretion-induced collapse (*e.g.*, Chanmugam and Brecher 1987; Grindlay and Bailyn 1989; Bailyn and Grindlay 1989). The resulting pulsars will be rapidly spinning and may have low magnetic fields, exactly the characteristics of millisecond pulsars (however, see Narayan and Popham 1989 and Popham and Narayan 1990 for a discussion of difficulties in producing such collapse). X-ray emission could be avoided after the pulsar is formed if the pulsar luminosity is so high that it prevents accretion, instead blowing away the envelope of the companion star. Note that this second solution contains elements of the first, since it is precisely those systems in which the mass-transfer lifetime is shortened and $\dot{M}$ is large, *i.e.*, close, accelerated-evolution binaries and very wide giant-fed systems, that should undergo successful accretion-induced collapse (Romani 1989a). However, it is again unclear if the number of such systems is large enough to explain the formation of the bulk of the LMBPs.

Since the discrepancy in birthrates is present in both the disk and the GCs, and is, if anything, stronger in the disk, it is preferable to look for explanations that operate in both regions of the Galaxy. This implicates a general modification to mass-transfer binary evolution as discussed above, and rules out solutions specific to tidal capture scenarios which work only in the GCs. Nevertheless, the globular cluster sample may provide a useful laboratory for untangling some of the evolutionary uncertainties, since, as we have shown (Romani 1989b), particular evolutionary scenarios should have characteristic signatures in the population of pulsar products;

other authors (*e.g.*, Kluzniak and Ray 1990) have made related observations. Such studies should cast light on the corresponding problem for the Galactic disk.

This work was supported in part by NSF grants PHY86-20226 (RWR), AST88-14725 (RN), and a grant from the Corning Glass works (RWR). RN and SRK are also supported by Presidential Young Investigator awards from the NSF, and ASF is supported by a Teagle Fellowship at the Carnegie Institution of Washington.

REFERENCES

Bailyn, C. D., and Grindlay, J. E. 1989, *Ap. J.*, submitted.

Chanmugam, G., and Brecher, K. 1987, *Nature*, **329**, 696.

Fruchter, A. S. 1989, Ph.D. thesis, Princeton University.

Fruchter, A. S., and Taylor, J. H. 1990, in preparation.

Grindlay, J. E., and Bailyn, C. D. 1989, *Nature*, **336**, 48.

Kluzniak, W., Ruderman, M., Shaham, J., and Tavani, M. 1988, *Nature*, **334**, 225.

Kluzniak, W., and Ray A. 1990, this volume.

Kulkarni, S. R., and Narayan, R. 1988, *Ap. J.*, **335**, 755.

Kulkarni, S. R., Narayan, R., and Romani, R. W. 1989, *Ap. J.*, in press.

Lewin, W. G. H., and Joss, P. C. 1983, in *Accretion-Driven Stellar X-ray Sources*, ed. W. G. H. Lewin and E. J. P. van den Heuvel (Cambridge: Cambridge University Press), p. 189.

Narayan, R. 1987, *Ap. J.*, **319**, 162.

Narayan, R., and Popham, R. 1989, *Ap. J. (Letters)*, **346**, L25.

Popham, R., and Narayan, R. 1990, this volume.

Rappaport, S., Putney, A., and Verbunt, F. 1989, *Ap. J.*, in press.

Romani, R., Kulkarni, S. R., and Blandford, R. D. 1987, *Nature*, **329**, 309.

Romani, R. W. 1989*a*, in *Proceedings of Goa Workshop on SN and Stellar Evolution*, ed. A. Ray, in press.

Romani, R. W. 1989*b*, *Ap. J.*, submitted.

Ruderman, M., Shaham, J., and Tavani, M. 1989, *Ap. J.*, **336**, 507.

Segelstein, D. J., Stinebring, D. R., Fruchter, A. S., and Taylor, J. H. 1986, *Nature*, **322**, 714.

van den Heuvel, E. P. J. 1987, in *Proceedings of IAU Symp. No. 125: The Origin and Evolution of Neutron Stars*, ed. D. J. Helfand and J. H. Huang (Dordrecht: Reidel), p. 393.

Verbunt, F. 1988, in *The Physics of Compact Objects: Theory vs. Observation*, ed. N. E. White and L. Fillipov (Oxford: Pergamon Press), p. 529.

Evolutionary Origin of Globular Cluster Pulsars

A. Ray[1] and W. Kluźniak[2]

[1]Tata Institute of Fundamental Research
[2]Physics Department, Columbia University

1 ABUNDANCES AND BIRTHRATES

The observed statistics of radio pulsars and low mass X-ray binaries (LMXRBs) in Globular Clusters (GCs) have three distinct aspects: 1) the abundance of LMXRBs, 2) the birthrate of binary/recycled pulsars vis à vis that of the LMXRBs, and 3) the orbital period distribution of the radio pulsars. Although there are ten LMXRBs in GCs, the inferred number of radio pulsars present is much greater, leading to a discrepancy (Kulkarni, Narayan, and Romani 1989; Romani 1990) of a factor of 10 to 100 between the birthrate of radio pulsars and that of their presumed progenitors: the canonical LMXRBs. We point out that the orbital period distribution of radio pulsars (Table 1) is incompatible with progenitors being either the late stages of LMXRBs having rapid mass transfer in self-excited winds (Ruderman et $al.$ 1989) or the accretion induced collapse (AIC) of white dwarfs in wide binaries alone (Grindlay and Bailyn 1988). Instead, the GC pulsars may have two distinct classes of progenitors: the short orbital period pulsars and some of the solitary pulsars being descendants of rapidly-evolving LMXRBs, while the long-period pulsars having been formed from binaries undergoing (mainly) AIC of white dwarfs and nuclear evolution of the secondary.

Only 10^{-4} of the Galactic mass is contained in GCs, yet about ten percent of all Galactic LMXRBs are located there. This overabundance of LMXRBs in GCs (Clark et $al.$ 1975) has led workers to suggest (Fabian, Pringle, and Rees 1975) that these are formed by the tidal capture of a cluster star. The number N_X of currently-observable LMXRBs in GCs is related to the number of neutron star binaries n_b^* which have formed in each GC core over its lifetime τ_{GC} ($\sim 10^{10}$ yr) and undergone a phase of steady X-ray emission for τ_X year by:

$$N_X = N_{GC} n_b^* (\tau_X / \tau_{GC}) . \qquad (1)$$

Here, N_{GC} is the number of GCs. The computed rate (Ray, Kembhavi, and Antia 1987) of tidal captures over a GC lifetime indicates that roughly three percent of the neutron stars in the core will undergo a tidal capture if the number density of the target main-sequence stars has the canonical value of the order of 10^4 pc^{-3} in the GC core. However, many GCs may have already collapsed cores (Spitzer 1987) where stellar densities are higher; as many as 20% of the GCs which have unresolved cores (Djorgovski and King 1986) could have undergone collapse. In these, possibly up to 50% of the primordial neutron stars could have captured a companion (Bailyn and Grindlay 1990). Hence, although the actual number of GCs is nearly 150, N_x is dominated by about a dozen GCs which produce a large n_b^* on the average. Thus, since the currently-observable $N_x \sim 10$, $(1000/\langle N_{GC} n_b^* \rangle) \times \tau_x \sim 10^8$ yr. If the number in parenthesis is close to unity (as a few high central density and low velocity clusters may give rise to a large number of binaries) then the X-ray lifetime would be of the order of 10^8 yr. In fact, the models of stellar evolution in close binaries (Taam 1983; Webbink,Rappaport, and Savonije 1983) for the GC burst sources (of which there are 9), give mass-transfer timescales of the order of a few times 10^8 yr, where the mass-transfer rate on neutron stars from stripped giants can be as high as $10^{-8} M_\odot$ yr^{-1} but more often the average $\langle \dot{M} \rangle$ is a few times $10^{-9} M_\odot$ yr^{-1}. Such timescales of mass transfer already pose a problem for the birthrate of LMXRBs in relation to that of the pulsars as referred to above; if the timescale is any longer, the discrepancy in the birthrates is further aggravated. A suggestion (Kulkarni and Narayan 1988) to ameliorate the problem in the case of systems appearing in the disk (where the discrepancy of the birthrates are even larger) is to decrease the X-ray lifetimes to nearly 10^7 yrs, i.e., by a factor of more than 10 times relative to the standard model. A similarly decreased lifetime in the GC LMXRBs would make the birthrates of the progenitor-progeny systems nearly equal (Kulkarni, Narayan, and Romani 1989; Romani 1990). In particular, models (Ruderman et al. 1989; Kluźniak et al. 1988; Tavani, Ruderman, and Shaham 1989) involving a self-sustained wind from a companion star by a neutron star primary for the eclipsing binary PSR 1957+20 and Cygnus X-3 do indeed involve such a strong mass-transfer phase.

However, while a simple shortening of the X-ray active lifetime of LMXRBs in a GC may solve the birthrate problem, it will pose a problem towards the observed number of GC LMXRBs unless the number of neutron star binaries which formed by tidal capture is also found to be larger (see eq. [1]). The latter would require a larger number of neutron stars to be retained in the GC core over its lifetime, which in turn requires a less steep initial mass function (IMF) of the GCs. Alternately, the higher capture efficiency of the neutron stars in the core collapsed clusters having large central density and low velocity dispersion in a few clusters could contribute a large number of neutron star binaries. Indeed, it has been argued (Verbunt and Hut 1987) that about a dozen such clusters contribute a high fraction of the total number of neutron star binaries (including LMXRBs) to the globular cluster population. In addition, the total number of binaries formed in a globular cluster is affected by mass segregation effects (Verbunt and Meylan 1988) in core collapsed clusters as well as velocity weighting of the number of neutron stars retained after formation in the globular cluster core. Both these effects can possibly increase the estimates of the number of LMXRBs ever formed in GCs. These would be amenable to better theoretical estimates when further observational data of a quality comparable to that of Omega-Cen and 47 Tuc become available for a large number of clusters.

2 ORBITAL PERIOD DISTRIBUTION

The currently observed distribution of binary/recycled pulsar orbital periods also pose certain problems for the pulsar evolutionary scenarios involving purely the LMXRBs. If for the sake of argument, one assumed that all low mass binary pulsars (including the solitary ones) in GCs were the product of LMXRB evolution, where for the short orbital period ($P_{orb} < 1$ day) systems the high mass transfer induced by the self-excited wind was operative, while the long P_{orb} systems were products of LMXRBs having the mass transfer induced by the nuclear evolution of the secondary, where the mass-transfer timescale τ_m, given by (Webbink, Rappaport, and Savonije 1983), $\tau_m = 1.1 \times 10^9 \ P_i^{-1}$ yr, then the birthrate of the long P_{orb} radio pulsars would be 10 to 100 times smaller than that of the short orbital period systems which would have a mass-transfer timescale of 10^7 yr. This is clearly not the case as the number of observed long and short orbital period radio pulsars are roughly similar and there are no obvious selection effects which strongly favor one over the other. While it is true that high acceleration in a short period binary can be a selection effect against these pulsar surveys, Kulkarni, Narayan, and Romani (1989) estimate from the data on the M15 and M3, that the factor β by which the short P_{orb} pulsars could have been missed in the estimate of N_{PSR} is $\beta \sim 1.5$.

In addition, the most frequent collision and tidal capture of a neutron star in a GC core is with a main-sequence star. The maximum orbital period P_{orb}^i following capture (after orbital circularization with constant angular momentum) when the target star has approximately the cluster turnoff mass, is less than a day. To evolve to longer orbital period, the binary has to evolve according to the scenario of Webbink, Rappaport, and Savonije (1983), under the nuclear evolution of a secondary more massive than the cluster turn-off mass, in order to evolve substantially within a GC lifetime. However, when the binary starts off with an initial period $P_{orb} \approx 1$ day, at the end of the transfer of roughly 0.6 $M_\odot$, it expands only to a final orbital period of the order of 10 days. Therefore, it is hard to explain the two long period systems PSR 1620−26 and 0021−72B (both of which have periods in excess of 10 days) on the basis of the capture and subsequent nuclear evolution of a main-sequence dwarf by a neutron star.

Although the capture of a red giant by a neutron star can in principle give the right post-capture (initial) orbital period for the binary to evolve to required long periods in the case of some of the GC PSRs, the number of such systems formed would again be small, both because of the paucity of red giants as well as their would-be partners, the neutron stars. Instead, the capture of a red giant by a massive white dwarf and the subsequent accretion induced collapse to a neutron star in a wide orbit seems statistically more probable since the number of white dwarfs retained in the GC core is larger than the number of neutron stars retained. In this AIC scenario, the amount of mass lost to the white dwarf in the neutron star formation process is believed to be small, such that the binary is not disrupted, yet it undergoes a small orbit expansion so that the companion no longer fills the Roche lobe. At that stage this long P_{orb} system would not become an LMXRB and could be detected only through the pulsar's radio emission.

All the known pulsars in GCs belong to the class of millisecond pulsars (we assume the pulsars in M15 have been spun down to the current values of P_{spin}). Any pulsars formed in supernova explosions of primordial massive cluster stars would have crossed the death-line (Ruderman and Sutherland 1975) (i.e., turned off in the radio) a long time ago. The GC pulsars must then either be "recycled" neutron stars whose magnetic field has decayed to low values and spun up by accretion from a binary companion (Alpar *et al.* 1982) or they must have been formed in accretion induced collapse of massive white dwarfs (Bailyn and Grindlay 1990), where the magnetic field of the newly-born pulsar may be low (Chanmugam and Brecher 1987). In the case of solitary pulsars, the companion may have been subsequently evaporated (Ruderman, Shaham, and Tavani 1989; Kluźniak *et al.* 1988) or disrupted (Ray and Kembhavi 1988). "Ionization" rates of binaries computed by Rappaport, Putney, and Verbunt (1989) indicate that in M15 (but not in M28) the single pulsar could instead have been liberated by collisional disruption of the binary. Therefore, all these pulsars need to be considered in a single framework. The data in Table 1 therefore suggest that the number of long- and short-period pulsars are comparable.

Accretion induced collapse is envisaged to happen under rather special circumstances. In particular, the white dwarf has to be sufficiently massive to begin with such that it can grow to the Chandrasekhar limit and collapse to form the neutron star. Furthermore, the mass-transfer rate needs to be high enough that the white dwarf mass actually increases and does not decrease in nova flash ejections. A mass-transfer rate higher than $4 \times 10^{-8}\,M_{\odot}\,\mathrm{yr}^{-1}$ is required for AIC (Nomoto 1987). Such a transfer rate is possible in red giant-fed systems (van den Heuvel 1984) and requires long orbital periods $P_{orb} \geq 15$ days. AIC can also take place in a common envelope phase where the mass-transfer rates can be large.

As far as the shorter orbital period binaries containing white dwarfs and main-sequence stars are concerned, only a small number of the former can undergo accretion induced collapse since otherwise the ratio of the number of long P_{orb} pulsars to the short P_{orb} pulsars would be quite small. Indeed, the fraction of all tidal captures which involve a giant is about 7% according to Verbunt and Hut (1987). Thus, AIC of white dwarfs in short orbital period systems containing main-sequence companions must be very inefficient. Bailyn and Grindlay (1990) argue that the number of massive white dwarfs which manage to remain in GC cores and form binaries through tidal captures is 25 to 50 times the number of binaries formed by neutron stars. The number of binaries where a white dwarf has captured a giant in a GC core is then roughly twice ($\sim 0.07 \times 25$) the number of those having neutron stars with main-sequence companions. Since the self-excited wind mass-transfer rate is comparable

TABLE 1
GLOBULAR CLUSTER PULSARS

Pulsar	Cluster	P_{spin} (ms)	P_{orb} (day)	e	Likely $M_c/M_\odot$
1821−24*	M28	3.05	–	–	–
1516+02A*	M5	5.5	–	–	–
1639+36*	M13	10.	–	–	–
0021−72A	47Tuc	4.48	0.022	0.32	4×10^{-3} or 0.8
0021−72C†	47Tuc	5.76	†		
2127+11C	M15	30.	~ 0.3		0.2–0.4
1310+18†	M53	33.	†		
2127+11B*	M15	56.	–	–	–
1746−20*	NGC6440	288.6	–	–	–
2127+11A*	M15	110.66	–	–	–
0021−72B	47Tuc	6.13	7–95		0.2–0.4
1620−26	M4	11.08	195	0.025	0.3–0.4
1516+02B‡	M5	7.9	‡		

Notes:

* Solitary pulsar.

† Solitary or very wide binary.

‡ $v_{orb} > 20$ km s^{-1}.

to the minimum rate required for the AIC, the birthrate of the long P_{orb} systems via AIC should be comparable to the birthrate of short P_{orb} systems formed through self-excited winds in neutron star binaries. Thus, in this scenario, there should be roughly the same number of long P_{orb} binary pulsars as there are the short P_{orb} (or single) systems, which is in fact the case.

To conclude, the pulsar orbital period distribution suggests the binary/recycled pulsars have two simultaneous sets of progenitors: one through the LMXRBs undergoing mass transfer through self-excited winds; the other through the AIC of white dwarfs undergoing mass transfer from nuclearly-evolved secondaries in wider binaries and generally invisible in the X-ray band except in deep surveys. The combination of a larger number of white dwarfs compared to the neutron stars and a smaller number of nuclearly-evolved subgiant or giant stars relative to main-sequence stars would make the number of the progeny pulsars from AIC roughly equal to those arising from the rapidly-evolving LMXRBs via the self-excited wind route.

This work was started at the Aspen Workshop on the Evolution of Compact Stellar Remnants in June 1989. We thank the Aspen Center for Physics for its hospitality and the Workshop participants, especially Charles Bailyn, Josh Grindlay, and

Ramesh Narayan for several discussions. A. R. thanks Roger Romani for discussions at the Goa meeting on Supernova and Stellar Evolution, March 1989. He also thanks Stan Woosley for hospitality at University of California at Santa Cruz where this work was completed. This work was supported in part by the NSF grants INT 87-15411-A01-TIFR (A. R.) and AST 86-02831 (W. K.).

REFERENCES

Alpar, M. A., Cheng, A. F., Ruderman, M. A., and Shaham, J. 1982, *Nature*, **300**, 728.

Bailyn, C. D., and Grindlay, J. E. 1990, *Ap. J.*, submitted.

Chanmugam, G., and Brecher, K. 1987, *Nature*, **329**, 696.

Clark, G. W. 1975, *Ap. J. (Letters)*, **199**, L143. Djorgovski, S. G., and King, I. R. 1986, *Ap. J. (Letters)*, **305**, L61.

Grindlay, J. E., and Bailyn, C. D. 1988, *Nature*, **336**, 48.

Fabian, A. C., Pringle, J. E., and Rees, M. J. 1975, *M. N. R. A. S.*, **172**, 15p.

Kluźniak, W., Ruderman, M., Shaham, J., and Tavani, M. 1988, *Nature*, **334**, 225.

Kulkarni, S. R., and Narayan, R. 1988, *Ap. J.*, **335**, 755.

Kulkarni, S. R., Narayan, R., and Romani, R. 1989, preprint.

Nomoto, K. 1987, in *13th Texas Symposium on Relativistic Astrophysics*, ed. M. P. Ulmer (Singapore: World Scientific), p. 519.

Romani, R. 1990, in *Supernovae and Stellar Evolution, School and Workshop at Goa, March 1989*, ed. A. Ray and T. Velusamy (Singapore: World Scientific), in press.

Rappaport, S., Putney, A., and Verbunt, F. 1989, *Ap. J.*, in press.

Ray, A., and Kembhavi, A. K. 1988, *Mod. Phys. Lett.*, **A3**, 229.

Ray, A., Kembhavi, A. K., and Antia, H. M. 1987, *Astr. Ap.*, **184**, 164.

Ruderman, M., Shaham, J., and Tavani, M. 1989, *Ap. J.*, **336**, 507.

Ruderman, M., Shaham, J., Tavani, M., and Eichler, D. 1989, *Ap. J.*, **343**, 292.

Ruderman, M. A., and Sutherland, P. G. 1975, *Ap. J.*, **196**, 51.

Spitzer, L. 1987, *Dynamical Evolution of Globular Clusters*, (Princeton: Princeton University Press).

Taam, R. 1983, *Ap. J.*, **270**, 694.

Tavani, M., Ruderman, M. A., and Shaham, J. 1989, *Ap. J. (Letters)*, **342**, L31.

van den Heuvel, E. P. J. 1984, *Ap. Astr.*, **5**, 209.

Verbunt, F., and Hut, P. 1987, in IAU Symp. No. 125, *Origin and Evolution of Neutron Stars*, ed. D. Helfand and J. Huang (Dordrecht: Reidel), p. 187.

Verbunt, F., and Meylan, G. 1988, *Astr. Ap.*, **203**, 297.

Webbink, R., Rappaport, S., and Savonije, G. 1983, *Ap. J.*, **270**, 678.

Evolution of Very Short Orbital Period Binaries in Globular Clusters

Shigeki Miyaji

Department of Natural History, Chiba University, Japan

1 INTRODUCTION

A recent survey of millisecond (msec) pulsars revealed that globular clusters host 8 msec pulsars. Compared to the number of other field msec pulsars, the enhancement factor in globular clusters is about 100. Among these, the existence of PSR 0021−72A (47 Tuc; Ables *et al.* 1988) is unique because it is in a binary system with a very short orbital period (VSOP; $P_{orb} = 1924$ s). This system is the fourth-shortest VSOP binary known.

The shortest VSOP binary 4U1820−30 ($P_{orb} = 685$ s) is a low mass X-ray binary (LMXRB) burster and is also in a globular cluster (NGC 6626; Stella, Priedhorsky, and White 1987). The enhancement factor relative to the field for LMXRBs (8 bright bulge sources and 7 bursters) in globular clusters is about 4.

These high enhancement factors of both msec pulsars and LMXRBs support the existence of a special mechanism for the creation of such systems in globular clusters, i.e., tidal captures and/or three-body encounters. These mechanisms result because globular clusters have a high star density compared to other field stars.

A high eccentricity of $e = 0.33$ is observed in PSR 0021−72A (Ables *et al.* 1988). It is very hard to imagine that this pulsar has experienced two tidal captures (making a msec pulsar and the present binary system) since, if this were true, the possibility of tidal captures in globular clusters would be very high ($\sim 1/8$) and, therefore, a large number of both LMXRBs and msec pulsars would be observed. If a third body encountered a 4U1820−30-like system, it is possible to distort the orbit of the secondary, enlarge its separation (in order to terminate mass transfer), and leave a PSR 0021−72A-like system. However, such a possibility can only be evaluated by numerical simulations. I will discuss the effect of tidal captures in Section 2 and present the results of three-body numerical simulations in Section 3.

2 EVOLUTION TOWARD THE SHORTEST ORBITAL PERIOD

The possibility of tidal captures in the globular clusters was estimated by Verbunt (1987) and Bailyn (1988). Since the resultant binary from a tidal capture has a very small separation (< 3 times the radius of the captured companion), about half of the captured systems start dynamical mass transfer from the companion because it already fills its Roche lobe at the time of the formation of the binary. The other half will start cases A and/or AB mass transfer onto a degenerate star. Since the degenerate primary allows only a small accumulation rate (determined by either the core-luminosity relation for a white dwarf or by the Eddington luminosity of a neutron star), a common envelope will be formed. Then, we need to address three questions concerning the tidal capture scenario for VSOP binaries 4U1820−30 and PSR 0021−72A: (1) Whether it is possible to remain a binary system after the common envelope stage? (2) Whether its orbital period becomes as short as 10 minutes? and (3) Whether such a parameter range is only possible in the case of tidal captures?

When a common envelope is formed and fills the outer Lagrangian (L_2) point of the system, a large amount of angular momentum ($l \approx 1.7a^2\Omega$ per mass; Nariai and Sugimoto 1976; Sawada, Hachisu, and Matsuda 1984) is removed along with the mass lost from the L_2 point. Miyaji (1983) tried to discuss the origin of LMXRBs in accordance with this mass and angular momentum loss and needed to introduce a supernova explosion as a mechanism to prevent a merging of the system. However, for the cases of 4U1820−30 and PSR 0021−72A, the probability of a supernova explosion during the common envelope phase is too small.

Recently, Hachisu and Saio (1989) studied the dynamical mass stripping mechanism from the L_2 point, including the internal structure of the companion star. They showed that, when dynamical mass loss takes place, the system shrinks so rapidly that the gas leaving the L_2 point would not be sufficiently accelerated by the tidal field of the binary system, i.e., $l < 1.7$. Even though, if the companion has a homogeneous structure, it should be totally disrupted and a ring would be formed around a merged star because there is no mechanism to terminate mass transfer and the resultant mass loss from the system. On the other hand, if the companion has a core-halo structure, the shrinkage is stopped when the Roche lobe of the companion approaches its core surface; in other words, when its hydrogen shell burning is extinguished because of a decrease in pressure so that the energy source to support its rapid expansion is diminished. As a result, a binary consisting of two degenerate stars with a ring would be formed. In this case, a small amount of hydrogen ($M \approx 0.03\,M_\odot$) is left on the surface of the companion star (see Tutukov *et al.* 1987).

The definite existence of core-halo structure gives a lower bound of a helium core mass of $M_{\mathrm{He}} \approx 0.17\,M_\odot$. After the extinction of hydrogen burning, the helium white dwarf companion fills its Roche lobe again according to the angular momentum lost through gravitational wave radiation. This timescale should be smaller than the age of the universe and gives an upper limit of $M_{\mathrm{He}} \approx 0.23\,M_\odot$ (Hachisu and Saio 1989). Therefore, an accretion induced collapse of a cataclysmic variable could not leave the above-mentioned two VSOP binaries. Since the present neutron star should initially be a massive star (larger than 4 $M_\odot$ even as the result of an accretion induced collapse), so that the initial separation for the dynamical mass transfer, $\sim 30\,R_\odot$, should be well inside of the common envelope during its red giant stage. Therefore, a similar discussion given above concludes that in such an extended common envelope the companion should be totally disrupted by the dynamical mass loss.

3 THREE-BODY INTERACTION

Here, I present the results of numerical simulations of three-body interactions. For a simplicity, I took a $P_{orb} = 3$ hr binary system consisting of 1 $M_\odot$ and 0.1 $M_\odot$ degenerate stars (Star 1 and 2, respectively) and a 1 $M_\odot$ third body (Star 3, also assumed degenerate). The mass center of the binary was placed at the origin and the third body starts to collide along the x-coordinate from a distance of $x = 10^{14}$ cm with a velocity of $v_x = 10^7$ cm s^{-1} (about the virial velocity of the components of the globular cluster). The initial orbit of the binary was assumed to be circular and 10 phases of it were examined. Collision angles of 0, 45, and 90 degrees were examined and the collision radius is a parameter of $y = (1\text{--}10) \times 10^{11}$ cm. When the separation between any pair of the three stars became larger than 2×10^{14} cm, I stopped further numerical integration. When the Roche lobe radius of a star with the nearest component becomes smaller than the stellar radius, a merger will be formed and I stopped further computing. The results are shown in Table 1. The total angular momentum and the total energy of the system were conserved within 1 percent accuracy.

TABLE 1
RESULTS OF NUMERICAL SIMULATION

Result $y = 10^{11}$	0 degrees 1 2 3 4 5 6 7 8 9 10	45 degrees 1 2 3 4 5 6 7 8 9 10	90 degrees 1 2 3 4 5 6 7 8 9 10
S	- 8 9 9 9 8 6 8 9 8	0 1 0 0 1 1 4 6 8 6	1 0 2 2 1 2 4 5 4 5
L	- 2 0 0 1 2 4 2 1 2	0 0 0 0 0 2 4 3 2 4	0 0 0 0 1 0 0 2 6 5
E1S	- 0 0 0 0 0 0 0 0 0	0 2 0 0 0 0 0 0 0 0	0 2 0 1 1 0 0 0 0 0
E1L	- 0 0 0 0 0 0 0 0 0	0 1 1 0 0 0 0 0 0 0	0 0 1 0 0 0 0 0 0 0
E2S	- 0 0 0 0 0 0 0 0 0	0 0 0 0 0 0 0 0 0 0	0 0 1 1 0 0 0 1 0 0
E2L	- 0 1 0 0 0 0 0 0 0	1 0 0 0 3 0 0 0 0 0	1 3 2 2 4 1 0 0 0 0
M12	- 0 0 0 0 0 0 0 0 0	0 2 1 5 4 6 2 1 0 0	0 0 0 0 1 4 2 1 0 0
M23	- 0 0 0 0 0 0 0 0 0	6 3 6 4 1 1 0 0 0 0	0 4 2 1 0 2 0 0 0 0
M13	- 0 0 0 0 0 0 0 0 0	2 0 1 0 0 0 0 0 0 0	7 0 1 0 0 0 0 1 0 0
D	- 0 0 0 0 0 0 0 0 0	1 1 1 1 1 0 0 0 0 0	1 1 1 3 2 1 4 0 0 0

The meaning of these results are: S; shorten, L; enlarged, Ei; i-th star is exchanged, Mij; i and j stars are merged, and D; the binary system is destroyed.

It is apparent from Table 1 that the most distinct effect of the three-body interaction is to distort the binary, especially shorten its separation. High eccentricity of $e \approx 0.4$ is very common in both cases of shorten and enlarged binary orbits.

The merging cross section is much larger than that of tidal capture and about 10 times of the radius of a main-sequence star. If the collision velocity is smaller, the possibility of merging is much higher. After the formation of a merger, further evolution of the system is ambiguous because a large amount of kinetic energy will be dissipated in the envelope of the merger.

Observed periastron advance of PSR 0021−72A allows two possibilities for its companion mass. In the case of large mass, its mass should be the mass of an ordinary white dwarf and/or a neutron star, and in the case of small mass it should be 0.2–0.3 $M_\odot$ and similar to the companion mass of 4U1820−30. Further observations of PSR 0021−72A could determine which is the case.

4 DISCUSSION

The three-body encounter is also be a good mechanism to create a merger of a VSOP binary, i.e., msec pulsars in the globular clusters. This effect may be more realistic than the possibility of evaporating companions suggested by van den Heuvel and van Paradijs (1988).

Certainly, further observations in both X-ray and radio bands are needed to improve the statistics of VSOP binaries and to reveal their evolutional paths. A systematic survey of msec pulsars of other globular clusters is sorely needed. Finding a periodicity of X-ray bursters in globular clusters is also a key to investigate VSOP binaries because the X-ray burst could give information on the companion.

The numerical simulation was performed on the FACOM M780 of the National Astronomical Observatory. This work has been supported in part by the Grant-in-Aid for Scientific Research (63302015).

REFERENCES

Ables, J. G., et al. 1988, IAU Circ., No. 4602.

Bailyn, C. 1988, Nature, **332**, 330.

Hachisu, I, and Saio, H. 1989, private communication.

Miyaji, S. 1983, in Cataclysmic Variables and Related Objects, ed. M. Livio and G. Shaviv (Dordrecht: Reidel), p. 263.

Nariai, K., and Sugimoto, D. 1976, Pub. Astr. Soc. Japan, **28**, 593.

Stella, L., Priedhorsky, W., and White, N. E. 1987, Ap. J. (Letters), **312**, L17.

Sawada, K., Hachisu, I., and Matsuda, T. 1984, M. N. R. A. S., **206**, 673.

Tutukov, A. V., et al. 1987, Soviet Astr. Lett., **13**, 780.

van den Heuvel, E. P. J., and van Paradijs, J. 1988, Nature, **334**, 227.

Verbunt, F. 1987, Ap. J. (Letters), **312**, L23.

A New Class of Strongly-Magnetic CVs in Globular Clusters

G. Chanmugam[1] and A. Ray[2]

[1]Department of Physics and Astronomy, Louisiana State University
 Baton Rouge, LA 70803
[2]Tata Institute of Fundamental Research, Bombay 400005, India

1 INTRODUCTION

Magnetic cataclysmic variables (MCVs) are a subclass of cataclysmic variables (CVs) in which the magnetic field strength of the white dwarf is sufficiently strong that the accreting material is channeled onto the magnetic poles of the white dwarf (Berriman 1988; Cropper 1989). All the known MCVs lie in the solar neighborhood and little is known about their existence in the more-distant globular clusters.

The MCVs may be divided into two subclasses: the AM Herculis binaries which contain synchronously-rotating white dwarfs and the DQ Herculis binaries which contain asynchronously-rotating ones. The magnetic fields of the AM Her binaries have been determined from the detection of cyclotron lines (Cropper *et al.* 1989), Zeeman lines (Schmidt *et al.* 1986), and models for the strong optical polarization seen in these systems arising as a result of cyclotron emission (Chanmugam and Dulk 1981); these show that the magnetic field strengths of these systems lie between about 20 and 50 MG (Cropper 1989). On the other hand, the DQ Her binaries emit weak or no optical polarization and do not show evidence of Zeeman or cyclotron lines. In one case only (BG CMi) weak optical and infrared circular polarization been detected (West, Berriman, and Schmidt 1987). Theoretical fits to the polarization suggest a magnetic field of about 4 MG (Chanmugam *et al.* 1990). Since these binaries generally show X-ray pulsations corresponding to their rotation periods, their magnetic fields must be greater than about 5×10^4 G (0.05 MG) in order for the accreting matter to be channeled onto the magnetic poles. Chanmugam and Ray (1984) showed that the DQ Her binaries would evolve into AM Hers if their magnetic fields are larger than about 3 MG. They also argued that some of them could have fields of order 10 MG while others could have weaker fields (this is misquoted by Wickramasinghe 1988). Lamb and Patterson (1983) estimate that the average fields of DQ Her binaries are about 1 MG from their spin-up properties, while King, Frank, and Ritter (1985) suggest on evolutionary grounds that they have magnetic moments comparable to those of the AM Hers. Spectroscopic searches have also failed to reveal magnetic fields in other known CVs (Stockman, unpublished). If the magnetic field is very large ($B \gtrsim 100$ MG) then the source should produce significant UV radiation as a result of optically thick cyclotron emission. Searches for such strong UV emission with *IUE* among various CVs in the solar neighborhood have failed to reveal any such sources (Bond and Chanmugam 1982). Thus, the known MCVs in the Galactic disk have magnetic fields $B \lesssim 50$ MG.

About 20% of the known CVs are magnetic (Morris *et al.* 1987). By contrast, only 3–5% of isolated white dwarfs have been found to be magnetic (Angel, Borra, and Landstreet 1981) with field strengths 1 MG $\lesssim B \lesssim 500$ MG (Schmidt 1989). In globular clusters a majority of MCVs are expected to be formed by tidal capture of white dwarfs by red dwarfs and hence their field distribution will resemble that of isolated magnetic white dwarfs rather than that of the primordially-formed CVs. We therefore predict that globular clusters will contain a significant new class of MCVs with high fields (50 MG $\lesssim B \lesssim 500$ MG) which is unknown at present. We suggest that X-ray measurements with the next generation of X-ray satellites combined with optical/UV measurements with *HST* should reveal the brightest of these systems.

2 FORMATION OF WHITE DWARF BINARIES IN GLOBULAR CLUSTERS

In order to estimate the number of MCVs in globular clusters we make several assumptions: (1) They are mainly formed through tidal capture. (2) The distribution of magnetic fields of the white dwarfs in MCVs resemble single white dwarfs in the cluster. (3) The distribution of magnetic fields of single white dwarfs in globular clusters is similar to those in the Galactic disk and hence different from known MCVs which are formed, in general, without having undergone tidal capture.

The number of white dwarfs in the cluster depends on the initial mass function, $dN(M)/dM \propto M^{-\alpha}$, and the core density, which can be large if there is core collapse and mass segregation (Verbunt and Meylan 1987). The number of white dwarf binaries in the core of the cluster is equal to $\Gamma \tau_{GC} r_c^3 n_{WD}$, where τ_{GC} ($\sim 10^{10}$ yr) is the globular cluster lifetime, n_{WD} is the number density of white dwarfs in the core of radius r_c, and the rate of tidal capture in the core per compact star is given by:

$$\Gamma = n_T \sigma v = 2.9 \times 10^{-12} \, \text{yr}^{-1} \left(\frac{M_T}{2 \, M_\odot} \right) \left(\frac{R_m}{3.84 \, R_*} \right)$$
$$\times \left(\frac{R_*}{0.65 \, R_\odot} \right) \left(\frac{v_{\text{rms}}}{10 \, \text{kms}^{-1}} \right)^{-1} \left(\frac{n_T}{10^4 \, \text{pc}^{-3}} \right) \qquad (1)$$

(Ray, Kembhavi, and Antia 1987). Here, σ is the tidal capture cross-section, v is the relative velocity of the target star and the white dwarf set equal to the velocity dispersion of stars in the globular cluster core v_{rms}, n_T is the number density of target stars, R_m is the periastron distance, R_* is the radius of the main-sequence star, and M_T is the total mass of the system.

The tidal capture rate predicts that the relative contribution of a cluster towards binary formation normalized to the contribution from all clusters would scale as $W_1 \propto n_*^2 r_c^3 / v_{\text{rms}}$. Here, n_* is the number density of stars in the cluster core of radius

r_c and v_{rms} is the root-mean-square velocity. We hence deduce that the globular cluster system would contain $\sim$ 100,000 white dwarf binaries, most of which will be in the high-weight clusters (see Ray and Chanmugam 1990 for details). If the magnetic white dwarf binary system X-ray lifetime $\tau_X = 4 \times 10^8$ yr $\sim 0.04\tau_{GC}$, the total number of MCVs currently observable in globular clusters by their X-ray emission is:

$$N_{MCV} = 100 \left(\frac{N_{WD}}{10^5}\right) \left(\frac{f_{CV}}{0.6}\right) \left(\frac{f_{MWD}}{0.03}\right) \left(\frac{\tau_X/\tau_{GC}}{0.04}\right). \qquad (2)$$

Here, f_{CV} is the fraction of the binary white dwarfs that enter the CV phase and f_{MWD} the fraction of white dwarfs which are magnetic. These fractions are likely to be greater than 0.6 (Ray, Kembhavi, and Antia 1987) and 0.03, respectively.

3 HARD X-RAY LUMINOSITY OF MCVS

In order to estimate the number of systems which can be detected by an X-ray instrument of a given sensitivity, we need to know the luminosity distribution of the MCVs. The only unbiased survey that can be used for the probable number of objects detectable is the hard X-ray band in which both the AM Her and DQ Her binaries emit. The *EXOSAT* survey in the 2–10 keV hard X-ray band detected 13 MCVs in the solar neighborhood and found upper limits in 9 other cases (Watson 1986; Norton and Watson 1989). The brightest 3 of these 22 systems have $\log L_X \geq 32.5$, where the hard X-ray luminosity L_X is given in erg s^{-1}. If the luminosity distribution of globular cluster MCVs is the same as that of the known MCVs and the *EXOSAT* survey is relatively complete for the portion of the sky observed down to the luminosity level indicated, then we can estimate the number of MCVs likely to be observable in a similar survey of the globular clusters. As the globular clusters are a few kiloparsecs away, only the fraction ($\sim$ 3/22) of sources in the highest luminosity bin may be detectable. Since there are approximately 100 currently active MCVs in globular clusters, less than approximately 15 hard X-ray sources may be detectable by surveys like that of *EXOSAT*.

4 SUMMARY

In this paper we have proposed that most of the MCVs present in globular clusters are formed by tidal captures of isolated magnetic white dwarfs by main-sequence stars. We estimate (Ray and Chanmugam 1990) that globular clusters should contain $\sim$ 100 MCVs, with magnetic field strengths 1 MG $\lesssim B \lesssim$ 500 MG. About 1/4 of them should have strong magnetic fields (50 MG $\lesssim B \lesssim$ 500 MG, see Schmidt 1989). Note that this assumes that MCVs with such strong fields would not evolve more quickly than normal CVs and have similar X-ray lifetimes (c.f., Schmidt, Stockman, and Grandi 1986; Hameury, King, and Lasota 1989). About 7 clusters are likely to contain almost half of the MCV population, and when the distance-weighted detection probability at a given flux level is estimated, it is found that Ter 5, Lil 1, and NGC 6440 are the best targets.

Hard X-ray observations in the future with the next generation of X-ray satellites (e.g., *ASTRO-D*, *AXAF*) should reveal about 10 of the brightest of these MCVs with luminosities $\gtrsim 10^{33}$ erg s^{-1}. Those with $B \gtrsim 100$ MG should produce significant UV radiation as a result of optically thick cyclotron emission and soft X-ray emission (c.f., Hertz and Grindlay 1983). Soft X-ray detectors on *ROSAT* should be able to detect the strongly-magnetized MCVs among the brightest AM Her-type systems.

This research was supported by NSF grants AST-8822954 to LSU and INT87-15411-A01-TIFR to the Tata Institute. We thank H. Ritter for a useful suggestion.

REFERENCES

Angel, J. R., Borra, E. F ., and Landstreet, J. D. 1981, *Ap. J. Suppl.*, **45**, 457.

Berriman, G. 1988, in *Polarized Radiation of Circumstellar Origin*, ed. G. V. Coyne, *et al.* (Vatican City State: Vatican Observatory), p. 281.

Bond, H. E., and Chanmugam, G. 1982, in *Advances in Ultraviolet Astronomy : Four Years of IUE Research*, NASA CP 2238, ed. Y. Kondo, J. M. Mead, and R. D. Chapman (NASA), p. 530.

Chanmugam, G., Frank, J., King, A. R., and Lasota, J. P. 1990, *Ap. J. (Letters)*, in press.

Chanmugam, G., and Ray, A. 1984, *Ap. J.*, **285**, 252.

Cropper, M. 1989, *Space Sci. Rev.*, in press.

Cropper, M., *et al.* 1989, *M. N. R. A. S.*, **236**, 69p.

Hameury, J. M., King, A. R., and Lasota J. P. 1989, *M. N. R. A. S.*, **237**, 845.

Hertz, P., and Grindlay, J. E. 1983, *Ap. J.*, **275**, 105.

King, A. R., Frank, J., and Ritter, H. 1985, *M. N. R. A. S.*, **219**, 597.

Lamb, D. Q., and Patterson, J. 1984, in *Cataclysmic Variables and Related Objects, I. A. U. Colloq. No. 72*, ed. M. Livio and G. Shaviv (Dordrecht: Reidel), p. 299.

Morris, S. L., *et al.* 1987, *Ap. J.*, **314**, 641.

Norton, A. J., and Watson, M. G. 1989, *M. N. R. A. S.*, **237**, 715.

Ray, A., and Chanmugam, G. 1990, *Ap. J. (Letters)*, in press.

Ray, A., Kembhavi, A. K., and Antia, H. M. 1987, *Astr. Ap.*, **184**, 164.

Schmidt, G. D. 1989, in *White Dwarfs, I. A. U. Colloq. No. 114*, ed. G. Wegner (Berlin: Springer-Verlag), p. 305.

Schmidt, G. D., Stockman, H. S., and Grandi, S. A. 1986, *Ap. J.*, **300**, 804.

Watson, M. G. 1988, in *The Physics of Accretion onto Compact Objects*, ed. K. O. Mason, M. G. Watson, and N. E. White (Berlin: Springer-Verlag), p. 97.

Verbunt, F., and Meylan, G. 1988, *Astr. Ap.*, **203**, 297.

West, S. C., Berriman, G., and Schmidt, G. D. 1987, *Ap. J. (Letters)*, **322**, L35.

Wickramasinghe, D. T. 1988, in *Polarized Radiation of Circumstellar Origin*, ed. G. V. Coyne, *et al.* (Vatican City State: Vatican Observatory), p 200.

SUBJECT INDEX

STAR INDEX

Printed in the United States
By Bookmasters